VAR MODELS IN MACROECONOMICS – NEW DEVELOPMENTS AND APPLICATIONS: ESSAYS IN HONOR OF CHRISTOPHER A. SIMS

ADVANCES IN ECONOMETRICS

Series Editors: Juan Carlos Escanciano, Thomas B. Fomby, R. Carter Hill, Eric Hillebrand, David Jacho-Chávez, Ivan Jeliazkov, Daniel L. Millimet and Rodney Strachan

Recent Volumes:

Volume 22: Econometrics and Risk Management – Edited by Jean-Pierre Fouque, Thomas B. Fomby and Knut Solna

Volume 23: Bayesian Econometrics – Edited by Siddhartha Chib, Gary Koop, Bill Griffiths and Dek Terrell

Volume 24: Measurement Error: Consequences, Applications and Solutions – Edited by Jane Binner, David Edgerton and Thomas Elger

Volume 25: Nonparametric Econometric Methods – Edited by Qi Li and Jeffrey S. Racine

Volume 26: Maximum Simulated Likelihood Methods and Applications – Edited by R. Carter Hill and William Greene

Volume 27A: Missing Data Methods: Cross-Sectional Methods and Applications – Edited by David M. Drukker

Volume 27B: Missing Data Methods: Time-Series Methods and Applications – Edited by David M. Drukker

Volume 28: DSGE Models in Macroeconomics: Estimation, Evaluation and New Developments – Edited by Nathan Balke, Fabio Canova, Fabio Milani and Mark Wynne

Volume 29: Essays in Honor of Jerry Hausman – Edited by Badi H. Baltagi, Whitney Newey, Hal White and R. Carter Hill

Volume 30: 30th Anniversary Edition – Edited by Dek Terrell and Daniel Millimet

Volume 31: Structural Econometric Models – Edited by Matthew Shum and Eugene Choo

ADVANCES IN ECONOMETRICS VOLUME 32

VAR MODELS IN MACROECONOMICS – NEW DEVELOPMENTS AND APPLICATIONS: ESSAYS IN HONOR OF CHRISTOPHER A. SIMS

EDITED BY

THOMAS B. FOMBY

Department of Economics, Southern Methodist University, Dallas, TX, USA

LUTZ KILIAN

Department of Economics, University of Michigan, Ann Arbor, MI, USA

ANTHONY MURPHY

Research Department, Federal Reserve Bank of Dallas, Dallas, TX, USA

United Kingdom – North America – Japan
India – Malaysia – China

Emerald Group Publishing Limited
Howard House, Wagon Lane, Bingley BD16 1WA, UK

First edition 2013

British Library Cataloguing in Publication Data
A catalogue record for this book is available from the British Library

ISBN: 978-1-78190-752-8
ISSN: 0731-9053 (Series)

CONTENTS

DEDICATION

Volume 32 of *Advances in Econometrics* is dedicated to Professor Christopher A. Sims, co-winner of the 2011 Nobel Prize in Economics, with Thomas J. Sargent. The Royal Swedish Academy of Sciences cited Sims and Sargent "for their empirical research on cause and effect in the macroeconomy." In his seminal 1980 paper "Macroeconomics and Reality" Professor Sims used VAR models along with his creation of impulse response analysis and forecast error variance decompositions to demonstrate how to identify unexpected economic and policy shocks and their impacts on the macroeconomy. Following his introduction of these methods of innovation accounting, many other economists adopted and extended his methods to address a myriad of issues not the least of which are the impacts of monetary and fiscal policy on the macroeconomy. As this volume concentrates on applications and theoretical developments in the area of VARs, the editors of this volume are pleased to dedicate this volume to Professor Sims in honor of the seminal and continuing impacts of his research on our profession.

The majority of the research papers contained in this volume were presented at the 12th *Advances in Econometrics* Conference held at Southern Methodist University in Dallas, Texas, November 2–4, 2012. On Saturday, November 3, the Conference honored Professor Sims with a special recognition dinner in the beautiful Jones Great Hall in the Meadows Museum of Art on the SMU campus. The first picture below is of Professor Sims showing the audience the *Advances in Econometrics* Lifetime Achievement award given to him by the Editorial Board of *Advances in Econometrics* and Emerald Publishing, Ltd, the publisher of the *Advances in Econometrics* series. The second picture is of Professor Sims that same night lecturing a very large audience on the nature of his research and other topics including current economic policy. A facsimile of the inscription on Professor Sims' *Advances in Econometrics* Lifetime Achievement Award is in the third frame below. For videos of Professor Sims' Dinner Address and an informal interview of Professor Sims during the conference one can go to the

Richard B. Johnson Center for Economics Studies website at http://www.smu.edu/Dedman/Academics/InstitutesCenters/RBJCenter and look under "Recent Multimedia."

Thomas B. Fomby, Lutz Kilian, and Anthony Murphy
Editors

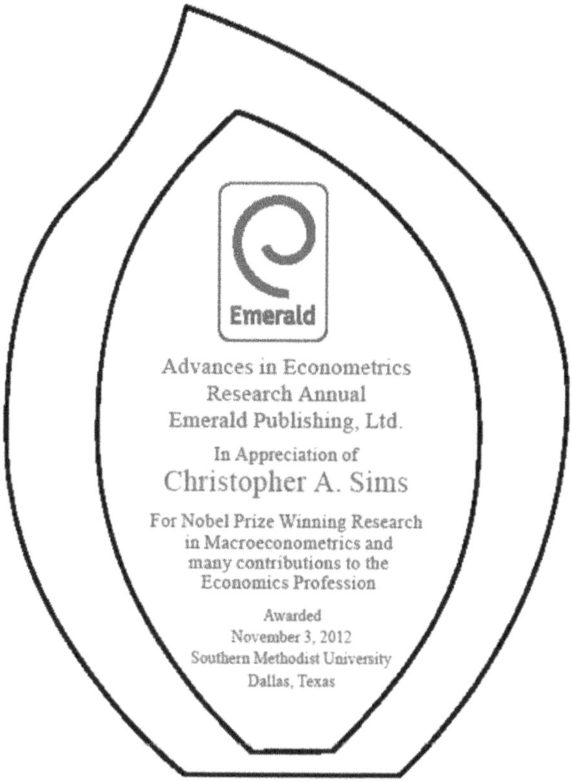
Emerald
Advances in Econometrics
Research Annual
Emerald Publishing, Ltd.
In Appreciation of
Christopher A. Sims
For Nobel Prize Winning Research
in Macroeconometrics and
many contributions to the
Economics Profession
Awarded
November 3, 2012
Southern Methodist University
Dallas, Texas

LIST OF CONTRIBUTORS

Bertrand Candelon	Department of Economics, Maastricht University, Maastricht, Netherlands
Fabio Canova	European University Institute and CEPR, Firenze, Italy
Matteo Ciccarelli	European Central Bank, Frankfurt, Germany
Todd E. Clark	Economic Research Department, Federal Reserve Bank of Cleveland, Cleveland, OH, USA
Elena-Ivona Dumitrescu	Department of Economics, European University Institute, Firenze, Italy
Thomas B. Fomby	Department of Economics, Southern Methodist University, Dallas, TX, USA
Claudia Foroni	Norges Bank, Research Department, Oslo, Norway
Eric Ghysels	Department of Economics, University of North Carolina, Chapel Hill, NC, USA, Department of Finance, Kenan-Flagler School of Business and CEPR
Raffaella Giacomini	Department of Economics, University College London, London, UK
Nikolay Gospodinov	Research Department, Federal Reserve Bank of Atlanta, Atlanta, GA, USA
Thomas B. Götz	Department of Quantitative Economics, Maastricht University, Maastricht, Netherlands

Refet S. Gürkaynak Department of Economics, Bilkent University, Ankara, Turkey

Alain Hecq Department of Quantitative Economics, Maastricht University, Maastricht, Netherlands

Ana María Herrera Department of Economics, University of Kentucky, Lexington, KY, USA

Kirstin Hubrich Research Department, European Central Bank, Frankfurt, Germany

Christophe Hurlin Department of Economics, University of Orleans, Orleans, France

Ivan Jeliazkov Department of Economics, University of California, Irvine, CA, USA

Lutz Kilian Department of Economics, University of Michigan, Ann Arbor, MI, USA

Burçin Kısacıkoğlu Department of Economics, Johns Hopkins University, Baltimore, MD, USA

Helmut Lütkepohl Free University Berlin and DIW Berlin, Berlin, Germany

Massimiliano Marcellino Department of Economics, Bocconi University, Milan, Italy, IGIER and CEPR

Michael W. McCracken Federal Reserve Bank of St. Louis, Research Division, St. Louis, MO, USA

Anthony Murphy Research Department, Federal Reserve Bank of Dallas, Dallas, TX, USA

Franz C. Palm Department of Quantitative Economics, Maastricht University, Maastricht, Netherlands

Elena Pesavento Department of Economics, Emory University, Atlanta, GA, USA

Barbara Rossi	ICREA-Univ. Pompeu Fabra, Barcelona GSE and CREI, Universitat Pompeu Fabra, Barcelona, Spain
Timo Teräsvirta	Aarhus University, Center for Research in Econometric Analysis of Time Series (CREATES), Aarhus, Denmark
Jean-Pierre Urbain	Department of Quantitative Economics, Maastricht University, Maastricht, Netherlands

INTRODUCTION

Vector autoregressive (VAR) models are among the most widely used econometric tools in the fields of macroeconomics and financial economics. Much of what we know about the response of the economy to macroeconomic shocks and about how various shocks have contributed to the evolution of macroeconomic and financial aggregates is based on VAR models. VAR models also have been used successfully for economic and business forecasting, for modeling risk and volatility, and for the construction of forecast scenarios.

Since the introduction of VAR models by C. A. Sims in 1980, the VAR methodology has continuously evolved. Even today important extensions and reinterpretations of the VAR framework are being developed. Examples include VAR models for mixed-frequency data, VAR models as approximations to DSGE models, factor-augmented VAR models, new tools for the identification of structural shocks in VAR models, panel VAR approaches, and time-varying parameter VAR models. This volume collects contributions from some of the leading VAR experts in the world on VAR methods and applications. Each article highlights and synthesizes a new development in this literature in a way that is accessible to practitioners, to graduate students, and to readers in other fields. Below we present a brief overview of the articles in this volume.[1]

One of recurrent questions in the literature has been the relationship between the VAR model and dynamic stochastic general equilibrium (DSGE) models. The first article "The Relationship between DSGE and VAR Models" by Raffaella Giacomini reviews the conditions under which there is a mapping between these two approaches. Starting with a given DSGE model, there are three distinct steps involved in establishing the mapping to a reduced-form VAR model: (1) from the DSGE model to a state-space model, (2) from the state space representation to the VAR(∞) model, and (3) from the VAR(∞) model to a finite-order reduced-form VAR approximation. In addition, care must be taken that the identifying assumptions used in the structural VAR model representation are consistent with the DSGE model. It is not clear, however, why a DSGE model should necessarily be the starting point in comparisons with VAR models because the theoretical rigor of the DSGE models often comes at the expense of the model's ability to fit the data. If a given DSGE model is empirically implausible, there is little interest in the question of whether it can be approximated by a VAR model. The article stresses that indeed a good case can be made for more flexible approaches to combining theory and data.

Proponents of DSGE models sometimes suggest that structural VAR models are not structural, implying that DSGE models are the only valid methodology in macroeconomics. This argument is misleading. Both estimated DSGE models and structural VAR models are structural in the sense defined by the Cowles Commission. The difference is that estimated DSGE models impose additional microeconomic structure in estimation. To the extent that this structure is correct, the cross-equation restrictions on the VAR representation implied by DSGE models will compared to unrestricted VAR models. DSGE models also allow us to explore a wider range of structural shocks than standard VAR models. They are not without limitations, however. One problem is that many of the assumptions about functional form and market structure required in specifying explicit microeconomic foundations are ad hoc and that in general equilibrium the misspecification of any part of the DSGE model will contaminate the estimates of all structural model parameters. Another problem is that the structural parameters of DSGE models often are at best weakly identified (see, e.g., Guerron-Quintana et al., 2013). A third problem is that DSGE models in practice require the specification of latent exogenous state variables. This practice is difficult to justify from an economic point of view. Thus, the DSGE approach and the structural VAR approach are best viewed as complementary. Neither approach dominates the other. They each have their advantages and shortcomings.

Under suitable conditions, however, the reduced-form representation of many linearized DSGE models can be approximated by reduced-form VAR models, as the first article shows, making the VAR a natural benchmark when fitting DSGE models. This interpretation has recently been challenged by Del Negro and Schorfheide (2013) who suggest that DSGE models forecast more accurately out of sample than VAR representations of the same DSGE model estimated with Bayesian shrinkage methods. They make the case that DSGE models should simply replace reduced-form VAR models as a benchmark for DSGE model evaluation exercises. The article "Do DSGE Models Forecast More Accurately Out-of-Sample than VAR Models?" by Refet S. Gürkaynak, Burçin Kısacıkoğlu, and Barbara Rossi examines this claim using a real-time data set that reflects the data actually available to forecasters at the time the forecast is generated. Taking the priors of the Smets–Wouters model as originally estimated as given, they compare the out-of-sample accuracy of this model to VAR and other reduced-form forecasting models. Their conclusion is that the Bayesian DSGE model forecast often is less accurate than that of simple autoregressive time series models. In fact, the Bayesian VAR (BVAR) benchmark against which Del Negro and Schorfheide evaluated their DSGE model forecasts is by no means a particularly accurate forecasting model. Not only are other VAR model forecasts often more accurate than DSGE model forecasts, but there is no single most accurate model, as the relative forecast accuracy of these models tends to evolve over time. Moreover, forecast rankings are also sensitive to the variable to be forecast and to the forecast horizon.

One benefit of the debate about the relationship between explicitly micro-founded DSGE models and VAR models is that users of VAR models have become more aware of the role of assumptions about units roots and cointegration in specifying structural VAR models. VAR models may be specified in levels, in growth rates, or as vector error correction (VEC) models, for example. A common practice is to decide between these specifications based on pre-tests for exact unit roots and cointegration. Such pre-tests were severely criticized by Elliott (1998) who showed that the two-step approach favored by many practitioners may result in highly inaccurate inference in the empirically relevant case of models with roots close to unity. Nevertheless, this practice has persisted, perhaps owing to the perception that VAR models specified in levels, while often asymptotically robust even in the presence of unit roots and cointegration, as shown by Sims, Stock, and Watson (1990), are not reliable in small samples. The article "Unit Roots, Cointegration and Pre-Testing in VAR Models" by Nikolay Gospodinov, Ana María Herrera, and Elena Pesavento investigates this tradeoff in great detail. The article shows that when the order of integration or the nature of the cointegration, if any, is not known, VAR models specified in levels tend to produce more accurate impulse response estimates than VAR models whose specification is based on pre-tests for unit roots and cointegration. This result holds regardless of whether the model is identified by short-run or long-run restrictions, but pre-tests are shown to be particularly misleading for structural VAR models based on long-run restrictions.

Traditionally, VAR models have played an important role in constructing structural impulse responses, historical decompositions, and forecast error variance decompositions. Their use as out-of-sample forecasting models has been more controversial. Indeed, early on advocates of VAR forecasting models including Chris Sims emphasized the need for Bayesian estimation methods in constructing VAR forecasts (see Doan, Litterman, & Sims, 1984). The concern was that unrestricted VAR models would be too imprecisely estimated for accurate out-of-sample forecasts. While the development of Bayesian priors for VAR forecasts continues, recent research has shown that especially at monthly frequency unrestricted VAR models may generate even more accurate out-of-sample forecasts than standard BVAR models (see, e.g., Baumeister & Kilian, 2012). At the same time, there has been increasing recognition of the importance of providing formal tests of the statistical significance of reductions in mean-squared prediction errors (MSPEs). Conventional tests of the null of equal predictive accuracy, as it turns out, are actually tests of the null of no predictability in population, which makes them inappropriate for testing the hypothesis of equal MSPEs in finite samples. As a result, such tests reject the null of equal predictive ability far too often. The article "Evaluating the Accuracy of Forecasts from Vector Autoregressions" by Todd E. Clark and Michael W. McCracken reviews what is known about this problem, contrasting tests for single-equation forecasts and tests for iterated VAR forecasts and outlining directions for future research.

The central question in all structural VAR models is one of identification. This is also an area that has rapidly evolved since 1980 (for a review see Kilian, 2013). One intriguing new idea in recent years has been to exploit information from high-frequency financial data for identification. Another new approach has been the use of sign restrictions on structural impulse response coefficients. A third new idea has been to exploit patterns of conditional or unconditional heteroskedasticity in the data for identification. The article "Identifying Structural Vector Autoregressions via Changes in Volatility" by Helmut Lütkepohl explains how this latter approach works, when it works, and what its limitations are. The basic assumptions are that changes in volatility occur during the sample period, that these changes only affect the volatility of the VAR errors, but not the other model parameters, and that they are sufficiently heterogeneous across variables. The article provides two concrete macroeconomic illustrations of these methods.

The next group of articles considers important extensions of the standard VAR framework. The article "Panel Vector Autoregressive Models: A Survey" by Fabio Canova and Matteo Ciccarelli reviews the growing literature on panel VAR models. Standard macroeconomic VAR models tend to focus on the economy of a given country. For many applications this approach is too restrictive because it leaves unmodelled the fact that economic outcomes in one country also depend on economic conditions in other countries. This is a particular concern when modeling global interdependencies (or interdependencies across sectors or firms) in international economics, macroeconomics, banking, and finance. Panel VAR models are designed to capture both static and dynamic interdependencies across firms, industries, and countries, while recognizing that each member of the panel is potentially different. This survey article not only provides a comprehensive overview of different types of panel VAR models, estimation methods for panel VAR models, and empirical applications, but it also discusses how these models are related to economic theory and it compares the panel VAR approach with related methodologies including large scale BVAR models, factor models, spatial econometric models, global VAR models, and bilateral VAR models, highlighting the pros and cons of each approach.

The article "Mixed-Frequency Vector Autoregressive Models" by Claudia Foroni, Eric Ghysels, and Massimiliano Marcellino summarizes the literature on VAR models for data of mixed frequency. Often researchers have at their disposal data of different frequency, say quarterly and monthly data. Ghysels in a series of papers showed how these data can be combined in a parsimonious manner. This mixed-data sampling (MIDAS) approach has become a standard forecasting tool. More recently, interest has grown in estimating VAR models that incorporate higher-frequency data. One example would be a VAR model of quarterly real GDP augmented by monthly data on industrial production. Another situation would be a VAR model that combines monthly macroeconomic aggregates with weekly or daily financial data. The article reviews what is known about such mixed frequency VAR models, including both Bayesian and frequentist approaches to estimation. The focus is on how mixed frequency data may help in constructing forecasts as well as impulse responses.

The standard VAR model is linear in the model parameters. There is a growing literature on nonlinear generalizations of the VAR approach including, for example, Markov switching VAR models, time-varying parameter (TVP) VAR models, VAR models with fixed thresholds, and VAR models with smooth thresholds. The article "Thresholds and Smooth Transitions in Vector Autoregressive Models" by Kirstin Hubrich and Timo Teräsvirta provides a taxonomy of the latter class of nonlinear VAR models. Thresholds arise naturally in many economic models. Examples include transaction costs limiting arbitrage, exchange rate target zones, or state-contingent pricing. Smooth thresholds typically arise after aggregating across economic agents with different thresholds. Much of the early literature has worked with single-equation models. This survey reviews VAR models of threshold behavior with special emphasis on model selection, estimation, and inference.

The article "Nonparametric Vector Autoregressions: Specification, Estimation, and Inference" by Ivan Jeliazkov provides a set of statistical tools for evaluating the adequacy of the linear approximation underlying standard reduced-form VAR models. His objective is to develop a Bayesian multivariate nonparametric estimator of the VAR parameters. Under the null hypothesis of linearity, these nonparametric estimators should form approximately a line. Allowing for estimation uncertainty, linearity would be violated unless we can fit a linear line within the confidence band surrounding the nonparametric estimates. The proposed methodology is still at an experimental stage. An unresolved problem to date is that the confidence band based on which linearity is judged is pointwise rather than joint, which may cause users to overstate the evidence against linearity. Nor is it clear at this point how to judge small departures from linearity or how to respond to evidence of violations of linearity.

Finally, the volume includes two applied articles. The article "Testing for Common Cycles in Nonstationary VARs with Varied Frequency Data" by Thomas Götz, Alain Hecq, and Jean-Pierre Urbain proposes a new approach to detecting the presence of common cyclical features when several time series are sampled at different frequencies. The article proposes a VEC representation of this model with the objective of improving forecast accuracy. The approach is illustrated in the context of a model including quarterly real GDP and monthly industrial production data.

Discrete choice variables are commonly used for forecasting economic events such as recessions, crises, or devaluations. Standard applications of this model are univariate. The article "Multivariate Dynamic Probit Models: An Application to Financial Crisis Mutation" by Bertrand Candelon, Elena-Ivona Dumitrescu, Christophe Hurlin, and Franz C. Palm extends this idea to the multivariate setting. The article proposes an exact maximum likelihood estimator for this class of models and illustrates the benefits of their approach by predicting economic crises in a range of countries.

Thomas B. Fomby
Lutz Kilian
Anthony Murphy
Editors

NOTE

1. The articles in this volume of Advances in Econometrics were presented initially at the 12th Advances in Econometrics Conference held in Dallas, TX, on the Southern Methodist University campus during November 2–4, 2012. The theme of the conference was *VAR Models in Macroeconomics: New Developments and Applications*. The editors would like to acknowledge the generous financial support provided by the Research Department of the Federal Reserve Bank of Dallas, the Richard B. Johnson Center for Economic Studies and the Department of Economics at Southern Methodist University.

REFERENCES

Baumeister, C., & Kilian, L. (2012). Real-time forecasts of the real price of oil. *Journal of Business and Economic Statistics*, *30*, 326–336.

Del Negro, M., & Schorfheide, F. (2013). Forecasting with DSGE models: Theory and practice. In G. Elliott & A. Timmermann (Eds.), *Handbook of economic forecasting* (Vol. 2). Amsterdam: Elsevier.

Doan, T., Litterman, R., & Sims, C. A. (1984). Forecasting and conditional projection using realistic prior distributions. *Econometric Reviews*, *3*, 1–100.

Elliott, G. (1998). On the robustness of cointegration methods when regressors almost have unit roots. *Econometrica*, *66*, 149–158.

Guerron-Quintana, P., Inoue, A., & Kilian, L. (2013). *Frequentist inference in weakly identified DSGE models*. Quantitative Economics, *4*, 197–229.

Kilian. (2013). Structural vector autoregressions. In N. Hashimzade & M. Thornton (Eds.), *Handbook of research methods and applications in empirical macroeconomics*. Camberley, UK: Edward Elgar.

Sims, C. A., Stock, J. H., & Watson, M. W. (1990). Inference in linear time series with some unit roots. *Econometrica*, *58*, 113–144.

THE RELATIONSHIP BETWEEN DSGE AND VAR MODELS

Raffaella Giacomini

ABSTRACT

This article reviews the literature on the econometric relationship between DSGE and VAR models from the point of view of estimation and model validation. The mapping between DSGE and VAR models is broken down into three stages: (1) from DSGE to state-space model; (2) from state-space model to VAR(∞); (3) from VAR(∞) to finite-order VAR. The focus is on discussing what can go wrong at each step of this mapping and on critically highlighting the hidden assumptions. I also point out some open research questions and interesting new research directions in the literature on the econometrics of DSGE models. These include, in no particular order: understanding the effects of log-linearization on estimation and identification; dealing with multiplicity of equilibria; estimating non-linear DSGE models; incorporating into DSGE models information from atheoretical models and from survey data; adopting flexible modeling

VAR Models in Macroeconomics – New Developments and Applications: Essays in Honor of Christopher A. Sims
Advances in Econometrics, Volume 32, 1–25

ISSN: 0731-9053/doi:10.1108/S0731-9053(2013)0000031001

approaches that combine the theoretical rigor of DSGE models and the econometric model's ability to fit the data.

Keywords: Invertibility; fundamentalness; linearization; impulse-response function; model evaluation

JEL classifications: C18; C52

INTRODUCTION

This article reviews the literature on the econometric relationship between DSGE models and VAR models. The main focus is on formal statistical methods for estimation and validation of DSGE and VAR models, in particular via the use of impulse-response analysis. This focus brings up different issues from those involved in using DSGE and VAR models for forecasting, which are discussed in Giacomini (2013) and in the article by Gürkaynak, Kısaçıkoğlu, and Rossi.

Understanding if a DSGE model can be represented as a reduced-form VAR is important for both estimation and model validation. The topic is relevant for empirical practice as it answers the question of whether the structural shocks ε_t can be recovered from a VAR analysis, that is, whether the following equivalence holds:

$$\varepsilon_t = \Omega(Y_t - E(Y_t|Y^{t-1})) \tag{1}$$

where Y_t is a vector of observable macroeconomic variables, Y^{t-1} is the history of Y_t up to time $t-1$, $Y_t - E(Y_t|Y^{t-1})$ the forecast error from a VAR, and Ω a rotation matrix. Uncovering such a relationship by imposing identifying restrictions on Ω is the objective of structural VAR (SVAR) analysis, but it also has important implications for both estimation of DSGE models (when it is carried out by matching impulse-response functions from the VAR) and for their validation, which often involves assessing whether the impulse responses from the DSGE model can replicate those from a SVAR. If the model shocks cannot be recovered from the SVAR shocks, model estimation and validation become meaningless. This issue has been hotly debated in the literature, with Chari, Kehoe, and McGrattan (2005) in one camp arguing that SVAR models are not suitable for model validation and estimation and Christiano, Eichenbaum, and Vigfusson (2006) in the opposite camp defending SVAR models as a useful tool but cautioning against their incorrect use.

The main goal of this article is to present a selective review of the literature in order to clarify how and when a mapping between DSGE and VAR models can be obtained. I will show that the mapping consists of three stages. In the first stage, the equilibrium conditions from a DSGE model are mapped into a linear state-space model; in the second, the state-space model is represented as a VAR with an infinite number of lags; in the last stage, the VAR(∞) is either shown to be a VAR with a finite number of lags if the model satisfies some testable conditions or the VAR(∞) is approximated by a finite-order VAR.

I will show in detail how the mapping is obtained in the context of the prototypical DSGE model of An and Schorfheide (2007), whose log-linearized version turns out to have an exact VAR(1) representation.

My focus throughout the article will be on highlighting what can go wrong at each stage and on drawing attention to the many assumptions that underlie the analysis. This will bring up some interesting open questions that are discussed at the end of the article, and inevitably will end up pointing to my ongoing work on the econometrics of DSGE models.

A notable omission in this article is a discussion of the literature on identification of structural parameters in DSGE models, which is fast growing and will soon require a separate survey article. A partial, must-read list includes Canova and Sala (2009), Komunjer and Ng (2011), Iskrev (2010), and Guerron-Quintana, Inoue and Kilian (forthcoming).

STAGE 1: FROM DSGE TO STATE-SPACE MODEL

As a first step toward understanding the relationship between DSGE and VAR models, I will begin by discussing the mapping between the equilibrium conditions of a DSGE model and a linear state-space model written in the ABCD form, which is the starting point for most of the analyses in the literature:

$$\begin{aligned} X_t &= AX_{t-1} + B\varepsilon_t \\ Y_t &= CX_{t-1} + D\varepsilon_t \end{aligned} \tag{2}$$

Here X_t are the state variables of the model, Y_t the observable variables and ε_t the structural shocks.

The discussion will assume that the parameters are either known (e.g., if the purpose of the exercise is to simulate a calibrated DSGE model to see if it matches moments implied by a VAR) or are fixed at a particular value, for cases in which stage 1 is part of an optimization routine whose ultimate goal is to estimate the model's parameters.

The article uses as an example the prototypical DSGE model of An and Schorfheide (2007, henceforth AS), which is a representative of the class of models currently used in the analysis of monetary policy at most central banks and policy institutions. The model is a simplified version of Smets and Wouters (2003) and Christiano, Eichenbaum, and Evans (2005), and is characterized by six equilibrium conditions:

$$1 = \beta E_t[e^{-\tau\hat{c}_{t+1}+\tau\hat{c}_t+\hat{R}_t-\hat{z}_{t+1}-\hat{\pi}_{t+1}}] \tag{3}$$

$$\frac{1-\nu}{\nu\phi\pi^2}(e^{\tau\hat{c}_t}-1)=(e^{\hat{\pi}_t}-1)\left\{\left[1-\frac{1}{2\nu}\right]e^{\hat{\pi}_t}+\frac{1}{2\nu}\right\} - \beta E_t\left[(e^{\hat{\pi}_{t+1}}-1)e^{-\tau\hat{c}_{t+1}+\tau\hat{c}_t+\hat{y}_{t+1}-\hat{y}_t+\hat{\pi}_{t+1}}\right] \tag{4}$$

$$e^{\hat{c}_t-\hat{y}_t}=e^{-\hat{g}_t}-\frac{\phi\pi^2 g}{2}(e^{\hat{\pi}_t}-1)^2 \tag{5}$$

$$\hat{r}_t=\rho_R\hat{r}_{t-1}+(1-\rho_r)\psi_1\hat{\pi}_t+(1-\rho_r)\psi_2(\hat{y}_t-\hat{g}_t)+\sigma_r\varepsilon_{r,t} \tag{6}$$

$$\hat{z}_t=\rho_z\hat{z}_{t-1}+\sigma_z\varepsilon_{z,t} \tag{7}$$

$$\hat{g}_t=\rho_g\hat{g}_{t-1}+\sigma_g\varepsilon_{g,t} \tag{8}$$

The model has six state variables $X_t=(\hat{c}_t,\hat{z}_t,\hat{g}_t,\hat{\pi}_t,\hat{r}_t,\hat{y}_t)$ where

- $\hat{c}_t=\ln(c_t/c)$, $c_t=C_t/A_t$ is detrended consumption relative to aggregate productivity A_t, assumed to evolve as $\ln A_t=\ln\gamma+\ln A_{t-1}+\ln z_t$, and $c=(1-\nu)^{1/\tau}$ is the steady state of detrended consumption
- $\hat{z}_t=\ln z_t$ is the innovation to the process governing aggregate productivity
- $\hat{g}_t=\ln(g_t/g)$, where g_t is government spending, assumed to evolve as $\ln g_t=(1-\rho_g)\ln g+\rho_g\ln g_{t-1}+\sigma_g\varepsilon_{g,t}$
- $\hat{\pi}_t=\ln(\pi_t/\pi)$, where π_t is inflation and π the steady-state inflation rate

- $\hat{r}_t = \ln(r_t/r)$, where r_t is gross interest rate and $r = \frac{\gamma}{\beta}\pi^*$, with $\frac{\gamma}{\beta}$ the steady-state real interest rate and π^* the target inflation rate, which in equilibrium equals π
- $\hat{y}_t = \ln(y_t/y)$, where $y_t = Y_t/A_t$ is detrended output and $y = g(1-\nu)^{1/\tau}$ is the steady-state of detrended output.

The model is driven by three independent exogenous shocks: $\varepsilon_t = (\varepsilon_{r,t}, \varepsilon_{z,t}, \varepsilon_{g,t})$, assumed to be i.i.d. N(0,1).

Eqs. (3)–(8) represent a system of six nonlinear expectational equations for six state variables X_t and three shocks ε_t whose solution is an equation of the form:

$$\text{State transition equation: } X_t = f(X_{t-1}, \varepsilon_t, \theta) \tag{9}$$

where $\theta = (\beta, \nu, \phi, \tau, \pi, \psi_1, \psi_2, \rho_r, \rho_z, \rho_g, \sigma_r, \sigma_z, \sigma_g)$ are the structural parameters of the model.

Once the state transition Eq. (9) is pinned down either by finding a closed-form expression for $f(\cdot)$ or by numerical approximation, the parameters of the model can be calibrated and one can simulate paths of the state variables. This would be sufficient if one's sole objective were to understand the model's internal propagation mechanism or to derive qualitative implications about the dynamics and the moments of some of the variables in the system.

Increasingly in recent years, the focus in the literature has shifted toward the adoption of formal econometric methods in the analysis of DSGE models. An early example is the literature that tries to formalize the calibration approach, such as Watson (1993), Canova (1994), Diebold, Ohanian, and Berkowitz (1998), DeJong, Ingram, and Whiteman (2000), and Dridi, Guay, and Renault (2007). More recently, the development of models that are rich enough (and embed enough frictions) to generate realistic time series behavior has generated a demand for formal econometric methods for the estimation and evaluation of DSGE models.

The literature has considered several approaches to estimation, ranging from limited information methods such as GMM and minimum-distance estimation based on matching impulse-response functions (see the survey by Ruge-Murcia, 2007) to full-information likelihood-based methods. Perhaps because Smets and Wouters (2003) made the case that a DSGE model estimated by Bayesian methods could fit the data as well as a VAR model, the literature on the Bayesian estimation of DSGE models has grown tremendously in recent years and these models have been widely adopted by central banks around the world (see, respectively, the reviews by An and Schorfheide, 2007 and Tovar, 2009). In this article, I will accordingly focus on likelihood-based estimation of DSGE models.

In order to derive a likelihood for the DSGE model (3)–(8), one needs to first choose a vector of observable variables Y_t and then specify a measurement equation which links the observables to the state variables, which typically takes the following form:

$$\text{Measurement equation:}\quad Y_t = HX_t \tag{10}$$

Eqs. (9) and (10) form a state-space model which, under suitable conditions, gives a likelihood that can be computed using filtering methods. There are filtering methods for nonlinear state-space models such as the particle filter (e.g., Pitt & Shephard, 1999) and they have been adopted by some authors (e.g., Aruoba, Fernandez-Villaverde, & Rubio-Ramirez, 2006; Fernandez-Villaverde & Rubio-Ramirez, 2007) to perform Bayesian estimation of nonlinear DSGE models. The vast majority of the literature, however, resorts to log-linearization of the model around the steady state in order to obtain a linear version of the state transition equation:

$$\text{Linearized state transition equation:}\quad X_t = AX_{t-1} + B\varepsilon_t \tag{11}$$

where the coefficient matrices A and B implicitly depend on the structural parameters θ. Assuming normality of ε_t, the linear state-space model (10) and (11) can be estimated using the Kalman filter.

In the case of the AS model, log-linearization of Eqs. (1)–(6) around the steady state gives the following system of linear expectational equations:

$$\hat{y}_t = E_t\hat{y}_{t+1} + \hat{g}_t - E_t\hat{g}_{t+1} - \frac{1}{\tau}(\hat{r}_t - E_t\hat{\pi}_{t+1} - E_t\hat{z}_{t+1}) \tag{12}$$

$$\hat{\pi}_t = \beta E_t\hat{\pi}_{t+1} + \frac{\tau(1-v)}{v\pi^2\phi}(\hat{y}_t - \hat{g}_t) \tag{13}$$

$$\hat{c}_t = \hat{y}_t - \hat{g}_t \tag{14}$$

$$\hat{r}_t = \rho_r\hat{r}_{t-1} + (1-\rho_r)\psi_1\hat{\pi}_t + (1-\rho_r)\psi_2(\hat{y}_t - \hat{g}_t) + \sigma_r\varepsilon_{r,t} \tag{15}$$

$$\hat{z}_t = \rho_z \hat{z}_{t-1} + \sigma_z \varepsilon_{z,t} \tag{16}$$

$$\hat{g}_t = \rho_g \hat{g}_{t-1} + \sigma_g \varepsilon_{g,t} \tag{17}$$

which can be solved using, for example, the algorithms of Blanchard and Kahn (1980), Uhlig (1999), or Sims (2002) to obtain an equation of the form (11).

The next step involves choosing the set of observables, which in the AS model are

$$Y_t = (\hat{r}_t, \hat{y}_t, \hat{\pi}_t) \tag{18}$$

Putting together Eqs. (11) and (10) one can obtain the ABCD representation (2) of the log-linearized DSGE model, which is a useful starting point for discussing the relationship between DSGE and VAR models.

In the case of the AS model, for example, as shown by Komunjer and Ng (2011) and Morris (2012), the minimal state representation of the model (i.e., the representation of the model with the smallest number of state variables) for parameters calibrated as

$$\begin{aligned} \theta &= (\beta, \nu, \phi, \tau, \pi, \psi_1, \psi_2, \rho_r, \rho_z, \rho_g, \sigma_r, \sigma_z, \sigma_g) \\ &= (0.995, 0.1, 53.68, 2, 1.01, 1.5, 0.5, 0.75, 0.9, 0.95, 0.002, 0.003, 0.006) \end{aligned} \tag{19}$$

is given by

$$\begin{aligned} \underbrace{\begin{bmatrix} \hat{z}_t \\ \hat{g}_t \\ \hat{r}_t \end{bmatrix}}_{X_t} &= \underbrace{\begin{bmatrix} 0.90 & 0 & 0 \\ 0 & 0.95 & 0 \\ 0.55 & 0 & 0.51 \end{bmatrix}}_{A} \begin{bmatrix} \hat{z}_{t-1} \\ \hat{g}_{t-1} \\ \hat{r}_{t-1} \end{bmatrix} + \underbrace{\begin{bmatrix} 1 & 0 & 0 \\ 0 & 1 & 0 \\ 0.61 & 0 & 0.69 \end{bmatrix}}_{B} \underbrace{\begin{bmatrix} \varepsilon_{zt} \\ \varepsilon_{gt} \\ \varepsilon_{rt} \end{bmatrix}}_{\varepsilon_t} \\ \underbrace{\begin{bmatrix} \hat{r}_t \\ \hat{y}_t \\ \hat{\pi}_t \end{bmatrix}}_{Y_t} &= \underbrace{\begin{bmatrix} 0.90 & 0 & 0 \\ 0 & 0.95 & 0 \\ 0.55 & 0 & 0.51 \end{bmatrix}}_{C} \begin{bmatrix} \hat{z}_{t-1} \\ \hat{g}_{t-1} \\ \hat{r}_{t-1} \end{bmatrix} + \underbrace{\begin{bmatrix} 1 & 0 & 0 \\ 0 & 1 & 0 \\ 0.61 & 0 & 0.69 \end{bmatrix}}_{D} \begin{bmatrix} \varepsilon_{zt} \\ \varepsilon_{gt} \\ \varepsilon_{rt} \end{bmatrix} \end{aligned} \tag{20}$$

with error covariance matrix

$$\Sigma = \begin{bmatrix} 0.003^2 & 0 & 0 \\ 0 & 0.006^2 & 0 \\ 0 & 0 & 0.002^2 \end{bmatrix} \tag{21}$$

Summary of Stage 1 and Discussion

To summarize, stage 1 consists of the following steps: (a) start with the equilibrium conditions of a nonlinear DSGE model; (b) linearize the equations around the non-stochastic steady state; (c) use a solution algorithm to find the log-linearized state transition equation (11); (d) choose a vector Y_t of observable variables and write down the measurement equation (10); (e) put (c) and (d) together and find matrices A, B, C, D in the representation (2).

There are several issues that can arise at each of the above steps, as well as some implicit assumptions that one needs to make when considering the ABCD representation as the starting point of the analysis.

The first issue is that a meaningful discussion of the relationship between DSGE and VAR models can only be carried out in the context of log-linearized DSGE models, taking for granted the adequacy of the linear approximation and ignoring possible nonlinear dynamics that cannot be replicated by linear VAR models.

Perhaps due to the high computational costs of using nonlinear filters in the likelihood evaluation, estimation of nonlinear DSGE models is still a small proportion of the literature, but it is nonetheless growing fast (e.g., Amisano & Tristani, 2011; Chen, Petralia, & Lopes, 2010; Fernandez-Villaverde & Rubio-Ramirez, 2007). This makes it possible that in a few years nonlinear models will be the standard and thus much of the discussion contained in this article will become obsolete. At the time of writing, however, it is fair to say that the profession at large is not yet fully convinced of the need for nonlinear solution methods for DSGE models and routinely continues to rely on log-linearized models that are estimated by the Kalman filter. Nevertheless, the development of nonlinear methods continues at a rapid pace. For example, in ongoing research with Ron Gallant and Giuseppe Ragusa (Gallant, Giacomini, & Ragusa, 2013), we develop new estimation methods for nonlinear DSGE models that do not require approximating the model in order to obtain a likelihood. Our limited-information state-space methods will allow one to estimate nonlinear DSGE models directly from the equilibrium conditions.

In the context of linear vs. nonlinear DSGE models, a further important issue that has been scarcely investigated is the effect of log-linearization on the identification and estimation of the structural parameters of the model. Much of the discussion in the literature that I report in the rest of the article will be about whether and when the state-space model (2) can be written as or approximated by a finite-order VAR, and about trying to understand the economic significance of the possible approximation error. Perhaps the same effort should be dedicated to understanding the importance of the log-linearization error – as it is possible that the latter might dominate the former in applications. See Fernández-Villaverde, Rubio-Ramrez, and Santos (2006) and den Haan and de Wind (2009) for some notable contributions on this topic.

The second issue that arises when finding a solution to the log-linearized DSGE model (12)–(17) is the possible existence of multiple solutions, which is commonly referred to in the literature as indeterminacy (see McCallum, 2003 for a differing view on why it might be helpful to keep the two concepts distinct). The issue of multiplicity of solutions in dynamic rational expectation models is well known in the literature, and several different criteria have been suggested for selecting one model among alternative model solutions, for example, the expectational stability criterion of Evans (1986), the minimum state variable criterion of McCallum (1983), and the stability criterion advocated e.g., by Sargent (1987) and Blanchard and Kahn (1980). As discussed by McCallum (1999), even though it is often unclear which of the selection criteria is adopted in a particular study, the most popular approach in the literature is to utilize the stability criterion and rule out unstable solutions by restricting the parameter space. For notable exceptions, see, for example, Benhabib and Farmer (2000) and Lubik and Schorfheide (2004), who show that indeterminacy might be empirically relevant.

The issue of the multiplicity of solutions is relevant for our discussion because it means that in many cases which variables enter X_t and what the matrices A and B are is not uniquely determined by the model, and the choices made by the researcher about whether and how to deal with multiple solutions will result in different state-space representations. This in turn may have an impact on the analysis in the following sections since a great part of the discussion will be about the relative sizes of X_t and Y_t and the values of the matrices A, B, C and D. To give a concrete example, the local identification analysis of Komunjer and Ng (2011) requires one to start from the minimal state-variable representation of the state-space system (2), and in their application to the AS model the authors rule out indeterminacy. See Qu and Tkachenko (2012) for an analysis of the interplay between indeterminacy and identification in DSGE models.

The third, related, issue is the choice of observable variables Y_t, which is even more arbitrary than the choice of how to deal with multiple solutions discussed above. Again, the interaction between the agent's information set (i.e., what is contained in X_t) and the econometrician's information set (what is in Y_t) will be a key factor in discussing the relationship between the state-space model (2) and a VAR, so it is important to pay attention to issues of data selection. Fortunately, this is an area where the econometrician has both more control (as she can decide which data to use) and a better understanding of the consequences of her choices (as there is some literature on the topic, which I discuss in the next section).

STAGE 2: FROM STATE-SPACE MODEL TO VAR(∞)

This section discusses the mapping between the state-space model written in the ABCD form (2) and a VAR(∞), and is the starting point of most of the literature on the topic.

Let us suppose that the model has been mapped into the state-space representation (2) where X_t is $n_x \times 1$, Y_t is $n_y \times 1$ and ε_t is $n_\varepsilon \times 1$, so that A is $n_x \times n_x$, B is $n_x \times n_\varepsilon$, C is $n_y \times n_x$ and D is $n_y \times n_\varepsilon$. The general representation of the log-linearized DSGE model is thus a VARMA (see, e.g., Aoki, 1990) so the question of whether the DSGE can be written as a VAR is equivalent to asking whether the VARMA model can be inverted and written as a VAR(∞). For the case in which there are as many observables as shocks (and thus D is non-singular), the invertibility condition can be formulated in terms of the matrices A, B, C and D (Fernández-Villaverde, Rubio-Ramirez, Sargent, & Watson, 2007; Hannan & Deistler, 1988) and is easy to check. We have the following result:

Proposition 1. If the eigenvalues of $A - BD^{-1}C$ are strictly less than one in absolute value the state-space model in Eq. (2) can be written as a VAR(∞):

$$Y_t = C\sum_{j=1}^{\infty}(A - BD^{-1}C)^j BD^{-1}Y_{t-j} + D\varepsilon_t \tag{22}$$

Fernandez-Villaverde et al. (2007) show that the condition in Proposition 1 does not hold in important classes of models, such as the permanent income consumption model (Sargent 1987, Chapter XII). The literature on nonfundamentalness – which we will see in the next section is related to the same condition – has also pointed out several cases of empirical interest in which the mapping between model shocks and VAR shocks breaks down. We discuss this literature below.

Relationship with Fundamentalness

Another way to state the condition in Proposition 1 is to relate the invertibility of VARMA models to the issue of fundamentalness of MA representations (see the reviews by Alessi, Barigozzi, & Capasso, 2011; Sims, forthcoming). As in the previous section, a discussion of fundamentalness is carried out in the context of square systems in which there are as many observable variables as shocks. We have the following definition from Alessi, Barigozzi, and Capasso (2011):

Definition (Fundamentalness)

A covariance stationary process Y_t has a fundamental MA representation $Y_t = \theta(L)\varepsilon_t$ if $\theta(L)$ has no poles inside the unit circle and $\det(\theta(L))$ has no roots inside the unit circle.

As discussed by Alessi et al. (2011), in practice detecting nonfundamentalness in the case of DSGE models is the same as checking the invertibility condition in Proposition 1. To understand the implications of nonfundamentalness, note that if an MA representation is fundamental the inverse of the MA lag polynomial $\theta(L)$ depends only on nonnegative powers of L, and thus the model can be equivalently written as a VAR(∞). It is important to keep the issue of fundamentalness separate from invertibility as it can happen (when $\det(\theta(L))$ has roots inside the unit circle) that the MA polynomial can still be inverted, but $\theta(L)^{-1}$ depends on negative powers of L. In this case, we say that there is invertibility in the future, which means that a VAR(∞) representation does not exist and thus one cannot recover nonfundamental shocks from SVAR analysis.

The problem of nonfundamentalness has been known and discussed in the macro literature for some time, at least since Hansen and Sargent (1980, 1991) and Lippi and Reichlin (1993, 1994). These authors showed that nonfundamental representations matter empirically, and can arise either as a feature of the model – for example in the context of permanent income models (Blanchard & Quah, 1993) or rational expectations (Hansens & Sargent, 1980) models – or because of the way in which the exogenous variables are modeled (Lippi & Reichlin, 1993). In the former case nonfundamentalness is often caused by a situation in which the information set of the agents is larger than the information set of the econometrician. An example of the latter case is considered by Lippi and Reichlin (1993), who show that a simple and plausible extension of the model considered by Blanchard and Quah (1993) with dynamics in productivity gives rise to nonfundamental representations and to radically different conclusions from impulse-response analysis and forecast error variance decompositions.

A second example for how the problem of nonfundamentalness may be addressed within a VAR model is Kilian and Murphy (forthcoming), who observe that nonfundamentalness is a problem for oil market VAR models when traders act on information about the future not available to the econometrician. They show that this problem may be overcome by the inclusion of data on oil inventories in the VAR model combined with suitable identifying restrictions derived from economic theory.

Finally, a field in which the issue of nonfundamentalness has come recently to the forefront of the academic discussion is that of VAR analysis aimed at understanding the effect of fiscal shocks, where nonfundamentalness can arise because of the failure to account for the effects of fiscal news (Leeper, Walker, & Yang, 2008; Mertens & Ravn, 2010).

When a model does not satisfy the fundamentalness condition, a natural question to ask is: what should one do? One possibility is to estimate nonfundamental representations of the econometric model and see whether they yield plausible impulse-response functions that can be used for estimation or model validation. This is the route taken by Lippi and Reichlin (1994), who show how nonfundamental representations can be obtained applying Blaschke matrices to MA processes. See also Lanne and Saikkonen (2011) for an approach to estimating noncausal VAR models that are implied by nonfundamentalness. A second way to deal (at least in part) with nonfundamentalness that is caused by a model in which the agents have access to future information that the econometrician does not observe is to try to enlarge the econometrician's information set, for example, by adding variables that capture information used by market participants (Kilian & Murphy, forthcoming) or by exploiting the cross-sectional dimension via factor models (Boivin & Giannoni, 2006; Giannone, Reichlin, & Sala, 2006).

Summary of Stage 2 and Discussion

Once the DSGE has been log-linearized and written in the state-space form (2), which is equivalent to a VARMA, the question of whether the impulse-response functions implied by the DSGE can be meaningfully compared to the impulse-response functions obtained by a VAR(∞) crucially depends on whether a VAR(∞) representation of the model exists. The condition to check is that the eigenvalues $A - BD^{-1}C$ are strictly less than one in absolute value. If the condition is satisfied, the MA representation of the system is fundamental, and it is thus possible to recover the structural shocks and their associated impulse responses from the innovations and the impulse responses computed from a VAR, by imposing suitable identifying restrictions. This in turn means that it is meaningful to use SVAR impulse responses for DSGE model estimation and validation.

An issue that arises in stage 2 is that it cannot be performed when the model implies stochastic singularity, which occurs when there are fewer shocks than observables, i.e., $n_\varepsilon < n_y$. This is a common occurrence in DSGE models, where a small number of structural shocks typically drive a larger set of variables, linked by identities that might not hold in the data. The literature has dealt with stochastic singularity in three ways: (1) by only using a subset of the observable variables Y_t in the estimation of the VAR model; (2) by increasing the number of structural shocks in the DSGE model; and (3) by using all available observables but adding measurement errors in the DSGE model (e.g., Ireland, 2004). Each approach has drawbacks and involves arbitrary decisions on the part of the econometrician which may have important effects on the results of the analysis. Dropping variables from the estimation results in a loss of information and opens up the possibility that the results will differ depending on which subset of variables is used. Shocks that are added to the DSGE model may lack a structural interpretation and, from an econometric perspective, increase the number of parameters and the chance of misspecification. For example, it is common to add preference shocks which are merely violations of first-order conditions. Like measurement error, they lack microfoundations and undermine the perceived advantage of DSGE models over VAR models, namely the advantage of being rigorously derived from micro foundations. That in turns raises the question of why we would take such semi-micro founded models as our starting point for the analysis. Measurement errors can similarly be viewed as an ad hoc econometric device that lacks economic interpretation. In general, any arbitrary modification of the original model whose sole purpose is to make the model match empirical features can make the parameters lose structural interpretability and at the very least cast doubts on the model's predictions.

Canova, Ferroni, and Matthes (2013) propose to deal with the problem of non-singularity by selecting observables in a way that either optimizes parameter identification or that minimizes the distance between singular and non-singular models. King, Plosser, Stock, and Watson (1991) advocate using reduced-rank models as a remedy to non-singularity, but the issue of identification should be better investigated in this context.

STAGE 3: FROM VAR(∞) TO VAR(p)

In this final stage of the analysis, let us suppose that all the previous conditions for the mapping between a DSGE and a VAR(∞) have been met. Even though a VAR(∞) representation exists, in practice one typically estimates a finite-order VAR(p) and it is thus important to understand the impact of this approximation error on the resulting analysis. This problem is investigated in Ravenna (2007) and Morris (2012). Morris (2012) analyzes the question in the context of the representation (2) of the log-linearized DSGE and first of all points out that one trivial case in which an exact (restricted) VAR(1) representation exists is when X_t and Y_t are observable and thus the state-space system can be equivalently written as

$$\begin{bmatrix} X_t \\ Y_t \end{bmatrix} = \begin{bmatrix} A & 0 \\ C & 0 \end{bmatrix} \begin{bmatrix} X_{t-1} \\ Y_{t-1} \end{bmatrix} + \begin{bmatrix} B \\ D \end{bmatrix} \varepsilon_t \tag{23}$$

In the more general and realistic case in which some state variables are unobservable, Morris (2012) further gives the following two conditions that can be used to assess whether a DSGE written in the ABCD form (2) can be expressed respectively as a VAR(1) or as a VAR(p), with $p > 1$ but finite.

Proposition 2 (VAR(1)). If Y_t is a linear function of X_t, $n_y \geq n_x$ and A and C are full column rank, then the DSGE model (2) has a VAR(1) representation:

$$Y_t = \Phi Y_{t-1} + D\varepsilon_t \tag{24}$$

with

$$\Phi = CA(C'C)^{-1}C' \tag{25}$$

Proposition 3 (VAR(p)). The DSGE model (2) has a VAR(p) representation if and only if:

$$[\gamma_2 \cdots \gamma_{p+1}] = [\gamma_1 \cdots \gamma_p] \begin{bmatrix} \gamma_0 & \gamma_1 & \cdots & \gamma_{p-1} \\ \gamma_1' & \gamma_0 & \cdots & \gamma_{p-2} \\ \vdots & \vdots & \ddots & \vdots \\ \gamma_{p-1}' & \gamma_{p-2}' & \cdots & \gamma_0 \end{bmatrix}^{-1} \begin{bmatrix} \gamma_1 & \gamma_2 & \cdots & \gamma_p \\ \gamma_0' & \gamma_1 & \cdots & \gamma_{p-1} \\ \vdots & \vdots & \ddots & \vdots \\ \gamma_{p-2}' & \gamma_{p-3}' & \cdots & \gamma_1 \end{bmatrix} \tag{26}$$

where $\gamma_i = E(Y_t Y'_{t-i})$. In this case, the VAR(p) representation is

$$Y_t = \Phi[Y_{t-1}' \cdots Y_{t-p}']' + u_t \tag{27}$$

with

$$\Phi = [\gamma_1 \cdots \gamma_p] \begin{bmatrix} \gamma_0 & \gamma_1 & \cdots & \gamma_{p-1} \\ \gamma_1' & \gamma_0 & \cdots & \gamma_{p-2} \\ \vdots & \vdots & \ddots & \vdots \\ \gamma_{p-1}' & \gamma_{p-2}' & \cdots & \gamma_0 \end{bmatrix}^{-1} \tag{28}$$

and

$$E(u_t u_t') = \gamma_0 - [\gamma_1 \cdots \gamma_p] \begin{bmatrix} \gamma_0 & \gamma_1 & \cdots & \gamma_{p-1} \\ \gamma_1' & \gamma_0 & \cdots & \gamma_{p-2} \\ \vdots & \vdots & \ddots & \vdots \\ \gamma_{p-1}' & \gamma_{p-2}' & \cdots & \gamma_0 \end{bmatrix}^{-1} \begin{bmatrix} \gamma_1' \\ \gamma_2' \\ \vdots \\ \gamma_p' \end{bmatrix} \tag{29}$$

Ravenna (2007) similarly derives conditions for the existence of a finite-order VAR representation of a DSGE model written in a different state-space form, which makes the distinction between endogenous state variables (W_t) and exogenous state variables (Z_t) explicit:

$$\begin{aligned} W_t &= RW_{t-1} + SZ_t \\ Y_t &= PW_{t-1} + QZ_t \\ Z_t &= \Phi Z_{t-1} + \varepsilon_t \end{aligned} \tag{30}$$

.

For the AS model, for example, Z_t contains the technology and government spending processes $Z_t = (\hat{z}_t, \hat{g}_t)$ and W_t includes all remaining variables in the model.

Let W_t be $n \times 1$, Z_t be $m \times 1$ and Y_t be $r \times 1$. Under the assumptions that all elements of W_t and Y_t are observable and that $m = n + r$, the DSGE model (30) has a restricted VAR(2) representation:

$$\tilde{Y}_t = (A + B\Phi B^{-1})\tilde{Y}_{t-1} - (B\Phi B^{-1}A)\tilde{Y}_{t-2} + B\varepsilon_t$$

$$\tilde{Y}_t = \begin{bmatrix} W_t \\ Y_t \end{bmatrix}, \quad A = \begin{bmatrix} R & 0 \\ P & 0 \end{bmatrix}, \quad B = \begin{bmatrix} S \\ Q \end{bmatrix}$$

In the most realistic case in which some of the the state variables W_t are unobservable, we have the following result.

Proposition 4 (VAR(p)). A necessary and sufficient condition for a DSGE model written in the form (30) to have a finite-order VAR representation is that the determinant of $[I - (R - SQ^{-1}P)L]$ is of degree zero in L.

When some endogenous state variables are unobservable and the condition in Proposition 4 does not hold but the condition in Proposition 1 is valid, it is important to understand the impact of approximation error when one estimates (as one does in practice) a finite-order VAR when the true data-generating process is a VAR(∞). Starting from the DSGE representation (30), when the condition of fundamentalness is satisfied, the VAR(∞) has the form:

$$Y_t = Q\Phi Q^{-1}Y_{t-1} - [Q\Phi Q^{-1}PL - P]\sum_{j=0}^{\infty}(R - SQ^{-1}P)^j L^{j+1} SQ^{-1}Y_t + Q\varepsilon_t$$

and thus a finite-order VAR may still be a good approximation if the second term is small, which occurs when $Q\Phi Q^{-1}PL - P$ is small and/or $(R - SQ^{-1}P)^j$ converges to zero fast enough.

How important the approximation error is in practice can only be assessed on a case by case basis, but some general conclusions can still be reached about whether truncation affects the approximating VAR performance only via a *pure truncation bias* channel or if in addition it induces *identification bias*. The former refers to the bias in estimated coefficients and impulse responses due to the omission of higher order lags; the latter can occur when these biased VAR coefficients are also used to identify structural shocks from the reduced form innovations. Note that the identification bias in this case is not a result of incorrect identifying assumptions, nor is it only a matter of small sample bias. Using a calibrated RBC model with a VAR(∞) representation, Ravenna (2007) shows that truncation can have sizable effects on the impulse-response analysis based on the model.

A further consideration is that the choice of asymptotic or bootstrap inference for impulse-response functions matters when it is made in the context of VAR models that are an approximation of a true infinite-order VAR. This issue is considered by Inoue and Kilian (2002) who show that the residual-based bootstrap is still valid in this situation.

Summary of Stage 3 and Discussion

In this final stage of the mapping between a DSGE and a VAR, we saw that a finite-order VAR(p) representation of a DSGE model exists when either all endogenous state variables are observable or when the coefficients in the state-space representations (2) or (30) satisfy the conditions in Propositions 2 or 3 or 4, from Ravenna (2007) and Morris (2012).

Given a particular calibration or estimation of a DSGE model expressed in form (2) or (30) one can check the conditions for the existence of a finite-order VAR representation of the model. For example, for the minimal state representation of the AS model calibrated as in (20), we have that $n_y = n_x$ and the coefficient matrices A and C are full rank, which means that Proposition 2 holds and the AS model has the following VAR(1) representation:

$$\begin{bmatrix} \hat{r}_t \\ \hat{y}_t \\ \hat{\pi}_t \end{bmatrix} = \underbrace{\begin{bmatrix} 0.79 & 0 & 0.25 \\ 0.19 & 0.95 & -0.46 \\ 0.12 & 0 & .62 \end{bmatrix}}_{\Phi = CAC^{-1}} \begin{bmatrix} \hat{r}_{t-1} \\ \hat{y}_{t-1} \\ \hat{\pi}_{t-1} \end{bmatrix} + \underbrace{\begin{bmatrix} 0.61 & 0 & 0.69 \\ 1.49 & 1 & -1.10 \\ 1.49 & 0 & -0.75 \end{bmatrix}}_{D} \begin{bmatrix} \varepsilon_{zt} \\ \varepsilon_{gt} \\ \varepsilon_{rt} \end{bmatrix}$$

with variance $D\Sigma D'$.

As discussed by Morris (2012), one can similarly show that the model of Christiano et al. (2005) has a VAR(1) representation, whereas the popular model of Smets and Wouters (2007) does not, and caution should therefore be exercised when using impulse responses from estimated VAR models to validate Smets and Wouters' (2007) model.

Even if a finite-order VAR representation of a DSGE model exists, there is still the possibility that the VAR model one estimates in practice does not provide a representation of the data that is consistent with the DSGE model. I am referring in particular to the issue of data transformations and their effect on model estimation and validation. Gorodnichenko and Ng (2010) point out the fact that DSGE models implicitly embed assumption about whether the model variables have deterministic or stochastic trends and that even mild violations of these assumptions can severely affect estimation. They then propose estimation methods that do not require a researcher to take a stand on the nature of the persistence found in the data.

One crucial feature of the robust procedure proposed by Gorodnichenko and Ng (2010) is the fact that they apply the same transformations to both the data and the model variables. This is an important consideration to take into account when comparing DSGE and VAR models, as pointed out by Sims (2003) in a comment to Smets and Wouters (2003). Sims (2003) expresses the suspicion that the pre-processing of the data utilized by Smets and Wouters (2003) implicitly favors the DSGE over the VAR and Giacomini and Rossi (2010) confirm the correctness of this conjecture by applying their test for time-varying comparison of model fit to test the relative performance of Smets and Wouters' (2003) model and a Bayesian VAR (BVAR). They find that the way the data is detrended severely affects the conclusion of which model fits the data best at different time periods, and show that Smets and Wouters' (2003) model is outperformed by the BVAR once the detrending is done in a way that correctly takes into account the information that was available at the time of interest.

Canova (2012) similarly cautions against applying preliminary data transformations when estimating DSGE models, as they may introduce bias and distort a model's policy implications. He proposes an alternative approach to estimation which does not use some cross-equations restrictions from the model and instead builds a flexible link between model and data that captures some features of the data, such as trends, which no longer need to be removed before estimation.

One further issue is that the identifying restrictions in the SVAR model whose responses we compare to the DSGE model's responses have to be consistent with the DSGE model. That can be difficult especially when using delay restrictions for identification, as exemplified by Rotemberg and Woodford (1998), who tweaked the timing in their DSGE model to make it conform with their SVAR model.

Consolo, Favero, and Paccagnini (2009) add to the discussion on the suitability of reduced form VAR as benchmarks for evaluating DSGE models by resorting to the notion of statistical identification

(Spanos, 1990). Spanos (1990) advocates validating structural models using reduced form models that are themselves valid, in the sense that they pass a battery of tests. Consolo et al. (2009) point out that in the literature on DSGE validation, instead, the choice of the reduced form model is solely driven by that of the structural model, without considerations about how accurately the reduced form model describes the data. The critique is applied in particular to the model evaluation approach of Del Negro and Schorfheide (2004, 2006), who obtain the reduced form model by relaxing the cross-equation restrictions imposed by the DSGE, but the point Consolo et al. (2009) make is relevant in our discussion as well. If the mapping between DSGE and VAR is done in the rigorous way mapped out in this article, the benchmark VAR will be low-dimensional, it will omit potentially relevant variables, it will not have time-varying parameters, and as a result it may not be a good representation of the data. Consolo et al. (2009) propose to overcome this limitation by comparing DSGE models to factor-agumented VAR, but the mapping from the theoretical to the reduced form model will in this case be lost and it will be difficult to understand whether the model is rejected because of misspecified restrictions or because of omitted variables.

WHAT NEXT?

The work of Canova (2012), Consolo et al. (2009), and Del Negro and Schorfheide (2004, 2006) share the concern that DSGE models are too stylized and that the restrictions they impose on their reduced form counterparts are too restrictive for the outcomes of their estimation and validation to be taken too seriously. A more radical view is expressed by Howitt et al. (2008) and Pesaran and Smith (2011), who call for the profession to move beyond the narrow and ad hoc modeling framework of DSGE models, echoing the widely expressed concern that these models ignored important financial, housing and foreign channels that were crucial during the 2007 crisis. Howitt et al. (2008) advocate a return to atheoretical econometric methods, and Pesaran and Smith (2011) suggest a flexible approach where theory guides the choice of long-run relationships among variables while the short run dynamics are less restricted, where trade and financial variables are incorporated via the use of global VAR models and where theoretical restrictions are used to discipline estimation in these necessarily high-dimensional models. As it is not clear how to identify structural shocks in this framework, Pesaran and Smith's (2011) approach has a "non-structural" flavor.

Caldara, Harrison, and Lipinska (2012) focus on the missing channels in DSGE models and provide a similarly "semi-structural" approach to formalize what they argue is the current tendency at central banks to expand DSGE models to include missing channels and mechanisms of interest. They advocate augmenting the DSGE by including proxies for the missing channels (such as house prices) which are typically derived from auxiliary econometric models, such as VAR.

An almost opposite, "back to basics" approach is adopted by Giacomini and Ragusa (in press), who advocate starting from an econometric model which is known to provide a good description of the data and forcing it to satisfy some of the (nonlinear) equilibrium conditions implied by theory using exponential tilting. The approach is flexible and general, and can be used to incorporate only a subset of the equations from a DSGE model (e.g., the Euler equation), thus addressing the concern that not all the equations in the DSGE model may have the same "theoretical content" and that the researcher does not have equal faith in all aspects of the DSGE model.

CONCLUSION

Now that the dust has settled on the debate about the usefulness of reduced form time series methods for DSGE model estimation and validation, the emerging picture from the literature is that VAR analysis is a useful tool for estimation and validation of DSGE models, but of course as any tool it should be used with caution. As this article made clear, a rigorous mapping between the structural model and the reduced form model can only be established under very stringent assumptions. The literature offers some insight into whether such a mapping is possible for a given model and parameterization, but there are many situations in which the formal methods discussed here are not applicable.

The running theme throughout the article has been to advocate moving beyond the narrow question of whether a DSGE model can be written as a VAR, and to focus on understanding the impact of the many assumptions and arbitrary decisions that underlie the current practice of estimating and validating DSGE models.

First, the DSGE model has to be log-linearized, but the effects of log-linearization have been scarcely investigated, and it might well be that this approximation error will end up dominating the approximation error from using a finite-order VAR on which the literature solely concentrates. More research is needed on this topic.

Second, the log-linearized model is solved and multiple solutions are typically ruled out, again without clear motivation or an understanding of the effects of this choice on the model estimates. This is another area that would benefit from deeper investigation.

Third, the conclusions depend on which arbitrary choices the researcher makes when bringing the log-linearized model to the data: the choice of observable variables, whether to drop observables, add shocks or measurement errors to avoid stochastic singularity, whether to rule out nonfundamental representations.

Fourth, even if a finite-order VAR representation exists, the outcome of estimation and validation could be severely affected by preliminary data transformations such as demeaning and detrending and by the assumptions made about whether the trends are deterministic or stochastic.

Finally, a choice of benchmark VAR model that is purely driven by the initial DSGE model and by the selected observables, and not by the VAR model's ability to fit the data, casts doubts on the validity of the VAR's predictions. It is also not clear why a specific DSGE model should be the starting point of the analysis and why VAR models should be adapted to fit a given DSGE specification. DSGE models rely on many arbitrary choices about functional form or market structure, about the exogeneity of driving processes (such as government spending, productivity, or monetary aggregates) and about the dynamic specification of the latent driving shocks. DSGE models might be rigorous, but they are not necessarily realistic. For example, changing the dynamic specification of the technology process in a real business cycle model from an AR(1) process to an MA(2) process changes the appropriate lag order specification of the implied VAR.

As a partial response to these concerns, the article concluded by listing recent contributions to the literature that advocate moving beyond a dogmatic belief in the DSGE model specification and restrictions, and drawing instead on the vast and better understood experience of fitting the data with atheoretical models. Examples are the idea of combining the DSGE model and its reduced form VAR model as a way of relaxing all theoretical restrictions in the DSGE model (Del Negro & Schorfheide, 2004, 2006); the idea of relaxing only some theoretical restrictions and letting the data "speak" about some features of the model such as trends (Canova, 2012) or short-run dynamics (Pesaran & Smith, 2011); the idea of starting from a DSGE model and adding information from econometric models to capture missing channels (Caldara et al., 2012), or the idea of starting from an atheoretical model that fits the data well and forcing it to satisfy some of the equilibrium conditions implied by theory (Giacomini & Ragusa, in press).

ACKNOWLEDGMENT

I thank Lutz Kilian for useful comments and suggestions and gratefully acknowledge financial support from the British Academy Mid-Career Fellowship and from the Economic and Social Research Council through the ESRC Centre for Microdata Methods and Practice grant RES-589-28-0001.

REFERENCES

Alessi, L., Barigozzi, M., & Capasso, M. (2011). Non-fundamentalness in structural econometric models: A review. *International Statistical Review*, *79*(1), 16–47.

Amisano, G., & Tristani, O. (2011). Exact likelihood computation for nonlinear DSGE models with heteroskedastic innovations. *Journal of Economic Dynamics and Control*, *35*(12), 2167–2185.

An, S., & Schorfheide, F. (2007). Bayesian analysis of DSGE models. *Econometric Reviews*, *26*(2–4), 113–172.

Aoki, M. (1990). *State space modeling of time series*. Heidelberg: Springer Berlin.

Aruoba, S. B., Fernandez-Villaverde, J., & Rubio-Ramirez, J. (2006). Comparing solution methods for dynamic equilibrium economies. *Journal of Economic Dynamics and Control*, *30*(12), 2477–2508.

Benhabib, J., & Farmer, R. E. A. (2000). The monetary transmission mechanism. *Review of Economic Dynamics*, *3*(3), 523–550.

Blanchard, O. J., & Kahn, C. M. (1980). The solution of linear difference models under rational expectations. *Econometrica*, *48*(5), 1305–1311.

Blanchard, O. J., & Quah, D. (1993). The dynamic effects of aggregate demand and supply disturbances: Reply. *American Economic Review*, *83*(3), 653–658.

Boivin, J., & M. Giannoni. (2006). *DSGE models in a data-rich environment*. NBER Working Paper No. 12772.

Caldara, D., Harrison, R., & Lipinska, A. (2012). *Practical tools for policy analysis in DSGE models with missing channels, finance and economics discussion series 2012-72*. Board of Governors of the Federal Reserve System.

Canova, F. (1994). Statistical inference in calibrated models. *Journal of Applied Econometrics*, *9*(1), 123–144.

Canova, F. (2012). *Bridging cyclical DSGE models and the raw data*. Working Paper.

Canova, F., Ferroni, F., & Matthes, C. (2013). *Choosing the variables to estimate singular DSGE models*. CEPR Discussion Paper 9381.

Canova, F., & Sala, L. (2009). Back to square one: Identification issues in DSGE models. *Journal of Monetary Economics*, *56*(4), 431–449.

Chari, V. V., Kehoe, P. J., & McGrattan, E. R. 2005. *A critique of structural VARs using real business cycle theory*. Working Papers 631. Federal Reserve Bank of Minneapolis.

Chen, H., Petralia, F., & Lopes, H. F. (2010). *Sequential monte carlo estimation of DSGE models*. University of Chicago Working Paper.

Christiano, L. J., Eichenbaum, M., & Evans, C. L. (2005). Nominal rigidities and the dynamic effects of a shock to monetary policy. *Journal of Political Economy*, *113*(1), 1–45.

Christiano, L. J., Eichenbaum, M., & Vigfusson, R. (2006). Assessing structural VARs. *NBER Macroeconomics Annual*, *21*, 1–106.

Consolo, A., Favero, C., & Paccagnini, A. (2009). On the statistical identification of DSGE models. *Journal of Econometrics*, *150*(1), 99–115.

DeJong, D. N., Ingram, B. F., & Whiteman, C. H. (2000). A Bayesian approach to dynamic macroeconomics. *Journal of Econometrics*, *98*(2), 203–223.

Del Negro, M., & Schorfheide, F. (2004). Priors from general equilibrium models for VARs. *International Economic Review*, *45*, 643–673.

Del Negro, M., & Schorfheide, F. (2006). *How good is what you've got? DSGE-VAR as a toolkit for evaluating DSGE models*. Federal Reserve Bank of Atlanta Economic Review.

den Haan, W., & de Wind, J. (2009). *How well-behaved are higher-order perturbation solutions?* DNB Working Paper No. 240.

Diebold, F. X., Ohanian, L. E., & Berkowitz, J. (1998). Dynamic equilibrium economies: A framework for comparing models and data. *Review of Economic Studies*, *65*(3), 433–451.

Dridi, R., Guay, A., & Renault, E. (2007). Indirect inference and calibration of dynamic stochastic general equilibrium models. *Journal of Econometrics*, *136*, 397–430.

Evans, G. W. (1986). Selection criteria for models with non-uniqueness. *Journal of Monetary Economics*, *18*, 147–157.

Fernández-Villaverde, J., & Rubio-Ramírez, J. (2007). Estimating macroeconomic models: A likelihood approach. *Review of Economic Studies*, *74*, 1059–1087.

Fernández-Villaverde, J., Rubio-Ramírez, J., & Santos, M. S. (2006). Convergence properties of the likelihood of computed dynamic models. *Econometrica*, *74*(1), 93–119.

Fernández-Villaverde, J., Rubio-Ramírez, J., Sargent, T. J., & Watson, M. W. (2007). ABCs (and Ds) of understanding VARs. *American Economic Review*, *97*(3), 1021–1026.

Gallant, R., Giacomini, R., & Ragusa, G. (2013). *Generalized method of moments with latent variables*. Working Paper.

Giacomini, R. (2013). *Economic theory and forecasting: Lessons from the literature*. Working Paper.

Giacomini, R., & Ragusa, G. (in press). Theory-coherent forecasting. *Journal of Econometrics*.

Giacomini, R., & Rossi, B. (2010). *Model selection in unstable environments*. Working Paper.

Giannone, D., Reichlin, L., & Sala, L. (2006). VARs, common factors, and the empirical validation of equilibrium business cycle models. *Journal of Econometrics*, *132*(1), 257–279.

Gorodnichenko, Y., & Ng, S. (2010). Estimation of DSGE models when the data are persistent. *Journal of Monetary Economics*, *57*(3), 325–340.

Guerron-Quintana, P., Inoue, A., & Kilian, L. (2013). Frequentist inference in weakly identified DSGE models. *Quantitative Economics*, *4*(2), 197–229.

Hannan, E. J., & Deistler, M. (1988). *The statistical theory of linear systems*: New York: Wiley.

Hansen, L. P., & Sargent, T. J. (1980). Formulating and estimating dynamic linear rational expectations models. *Journal of Economic Dynamics and Control*, *2*(2), 7–46.
Hansen, L. P., & Sargent, T. J. (1991). Two difficulties in interpreting vector autoregressions. In L. P. Hansen & T. J. Sargent (Eds.), *Rational expectations econometrics* (pp. 77–120). Boulder, CO: Westview Press.
Howitt, P., Kirman, A., Leijonhufvud, A., Mehrling, P., & Colander, D. (2008). *Beyond DSGE models: Toward an empirically based macroeconomics*. Middlebury College Economics Discussion Paper No. 08-08.
Ireland, P. (2004). A method for taking models to the data. *Journal of Economic Dynamics and Control*, *28*, 1205–1226.
Inoue, A., & Kilian, L. (2002). Bootstrapping smooth functions of slope parameters and innovation variances in VAR(∞) models. *International Economic Review*, *43*(2), 309–332.
Iskrev, N. (2010). Local identification in DSGE models. *Journal of Monetary Economics*, *57*(2), 189–202.
Kilian, L., & Murphy, D. P. (in press). The role of inventories and speculative trading in the global market for crude oil. *Journal of Applied Econometrics*.
King, R. G., Plosser, C. I., Stock, J. H., & Watson, M. W. (1991). Stochastic trends and economic fluctuations. *American Economic Review*, *81*(4), 819–840.
Komunjer, I., & Ng, S. (2011). Dynamic identification of dynamic stochastic general equilibrium models. *Econometrica*, *79*(6), 1995–2032.
Lanne, M., & Saikkonen, P. (2011). Noncausal autoregressions for economic time series. *Journal of Time Series Econometrics*, *3*(3), 1–32. Article 2.
Leeper, E. M., Walker, T. B., & Yang, S. S. (2008). *Fiscal foresight: Analytics and econometrics*. CAEPR Working Paper 2008-13. Indiana University.
Lippi, M., & Reichlin, L. (1993). The dynamic effects of aggregate demand and supply disturbances: Comment. *American Economic Review*, *83*(3), 644–652.
Lippi, M., & Reichlin, L. (1994). VAR analysis, non-fundamental representations, Blaschke matrices. *Journal of Econometrics*, *63*, 307–325.
Lubik, T. A., & Schorfheide, F. (2004). Testing for indeterminacy: An application to U.S. monetary policy. *American Economic Review*, *94*(1), 190–217.
McCallum, B. T. (1983). On non-uniqueness in rational expectations models: An attempt at perspective. *Journal of Monetary Economics*, *11*, 139–168.
McCallum, B. T. (1999). Role of the minimal state variable criterion in rational expectations models. In P. Isard, A. Razin, & A. Rose (Eds.), *International finance in turmoil: Essays in honor of Robert P. Flood*. Boston, MA: Kluwer Academic.
McCallum, B. T. (2003). Multiple-solution indeterminacies in monetary policy analysis. *Journal of Monetary Economics*, *50*(5), 1153–1175.
Mertens, K., & Ravn, M. O. (2010). Measuring the impact of fiscal policy shocks in the face of anticipation. *Economic Journal*, *120*, 393–413.
Morris, S. D. (2012). *VAR(1) representation of DSGE models*. University of California, San Diego Working Paper.
Pesaran, M. H., & Smith, R. P. (2011). *Beyond the DSGE straightjacket*. Cambridge Working Papers in Economics 1138, Faculty of Economics, University of Cambridge.
Pitt, M., & Shephard, N. (1999). Filtering via simulation: Auxiliary particle filter. *Journal of the American Statistical Association*, *94*, 590–599.
Qu, Z., & Tkachenko, D. (2012). Identification and frequency domain quasi-maximum likelihood estimation of linearized dynamic stochastic general equilibrium models. *Quantitative Economics*, *3*(1), 95–132.

Ravenna, F. (2007). Vector autoregressions and reduced form representations of DSGE models. *Journal of Monetary Economics*, *54*(7), 2048–2064.

Rotemberg, J. J., & Woodford, M. (1998). *An optimization-based econometric framework for the evaluation of monetary policy: Expanded version*. NBER Working Paper No. 233.

Ruge-Murcia, F. J. (2007). Methods to estimate dynamic stochastic general equilibrium models. *Journal of Economic Dynamics and Control*, *31*(8), 2599–2636.

Sargent, T. J. (1987). *Macroeconomic theory* (2nd ed.). New York, NY: Academic Press.

Sims, C. A. (2002). Solving linear rational expectations models. *Computational Economics*, *20*(1), 1–20.

Sims, C. A. (2003). *Comment on Smets and Wouters*. mimeo. Retrieved from https://sims.princeton.edu/yftp/Ottawa/SWcommentSlides.pdf

Sims, E. (forthcoming). News, non-invertibility, and structural VARs. *Advances in Econometrics*.

Smets, F., & Wouters, R. (2003). An estimated dynamic stochastic general equilibrium model of the euro area. *Journal of the European Economic Association*, *1*(5), 1123–1175.

Smets, F., & Wouters, R. (2007). Shocks and frictions in US business cycles: A Bayesian DSGE approach. *American Economic Review*, *97*(3), 586–606.

Spanos, A. (1990). The simultaneous-equations model revisited: Statistical adequacy and identification. *Journal of Econometrics*, *44*, 87–105.

Tovar, C. E. (2009). DSGE models and central banks. *Economics – The Open-Access, Open-Assessment E-Journal (Kiel Institute for the World Economy)*, *3*(16), 1–31.

Uhlig, H. (1999). A toolkit for analyzing nonlinear dynamic stochastic models easily. In R. Marimon & A. Scott (Eds.), *Computational methods for the study of dynamic economies* (pp. 30–61). Oxford University Press.

Watson, M. (1993). Measures of fit for calibrated models. *Journal of Political Economy*, *101*(6), 1011–1041.

DO DSGE MODELS FORECAST MORE ACCURATELY OUT-OF-SAMPLE THAN VAR MODELS?☆

Refet S. Gürkaynak, Burçin Kısacıkoğlu, and Barbara Rossi

ABSTRACT

Recently, it has been suggested that macroeconomic forecasts from estimated dynamic stochastic general equilibrium (DSGE) models tend to be more accurate out-of-sample than random walk forecasts or Bayesian vector autoregression (VAR) forecasts. Del Negro and Schorfheide (2013) in particular suggest that the DSGE model forecast should become the benchmark for forecasting horse-races. We compare the real-time forecasting accuracy of the Smets and Wouters (2007) DSGE model with that of several reduced-form time series models. We first demonstrate that none of the forecasting models is efficient. Our second finding is that there is no single best forecasting method. For example, typically simple AR models are most accurate at short horizons and

☆The views expressed in this article are those of the authors.

VAR Models in Macroeconomics – New Developments and Applications: Essays in Honor of Christopher A. Sims
Advances in Econometrics, Volume 32, 27–79

ISSN: 0731-9053/doi:10.1108/S0731-9053(2013)0000031002

DSGE models are most accurate at long horizons when forecasting output growth, while for inflation forecasts the results are reversed. Moreover, the relative accuracy of all models tends to evolve over time. Third, we show that there is no support to the common practice of using large-scale Bayesian VAR models as the forecast benchmark when evaluating DSGE models. Indeed, low-dimensional unrestricted AR and VAR forecasts may forecast more accurately.

Keywords: Forecasting; DSGE; Bayesian VAR; real-time data; forecast optimality

JEL classifications: C22; C52; C53

INTRODUCTION

The goal of this article is to study what the benchmark should be used in assessing macroeconomic forecasts from dynamic stochastic general equilibrium (DSGE) models. This question has gained importance as DSGE models are increasingly used in forecasting and indeed are judged by their forecasting performance – see the reviews by Del Negro and Schorfheide (2013) and Christoffel, Coenen, and Warne (2011).[1] In particular, Del Negro and Schorfheide (2013) argued that DSGE models should be the new benchmark in forecasting. We focus on a benchmark DSGE model and evaluate both its forecast efficiency (or optimality) as well as its forecasting ability relative to reduced-form models using real-time vintages of data. Our main objective is to determine whether one forecasting method systematically outperforms the others.

We find that, on average over the sample, the DSGE model forecasts output better than autoregressive methods, especially at longer horizons. When forecasting inflation, in contrast, the DSGE model is less accurate on average and, in several instances, the DSGE model's performance worsens toward the end of the sample relative to its competitors. In general, we find that no single model is most accurate at all times. We conclude that the forecasting method with the lowest root mean squared forecast error (RMSFE) depends on the sample and the forecast horizon, implying there is no obvious forecast benchmark. Our analysis does not support the common practice of using the random walk forecasts or Bayesian vector autoregression (VAR) model forecasts as benchmarks for judging the accuracy of DSGE model forecasts. For example, we find that standard large Bayesian VARs appear over-parameterized, making them poor benchmarks against which to judge other forecasting models. In fact, unrestricted small-scale reduced-form models typically outperform the large-scale Bayesian VAR model. This is the case for interest rate forecasts as well as forecasts of inflation and output growth. The random walk model forecasts well both inflation (at horizons greater or equal to two) and the interest rate (up to four quarter-ahead) but not output growth.

Our analysis is based on the DSGE model of Smets and Wouters (2007). We focus on this particular model due to its prominence in the literature and its wide use at central banks for forecasting. We evaluate the out-of-sample accuracy of forecasts generated from this model for the period 1992–2007. As competitors, we consider a series of reduced-form models: univariate autoregressions (ARs), vector autoregressions (VARs), Bayesian VARs (BVARs), and forecast combinations. VAR, AR and random walk models are considered because they are a natural benchmark and are often used in practice when forecasting, although not in conjunction with DSGE model-based forecasts. We report results for AR and VAR forecasts produced via a direct method, although the main results are qualitatively robust to using an iterated forecast method.[2] We also explore forecast combinations, which often, but not always, perform well (e.g., Stock & Watson (2004) and Timmermann (2006) show that forecast combinations work well in forecasting several macroeconomic and financial data; Rossi (2013) shows that forecast combinations perform well in forecasting inflation and output growth in the United States; on the other hand, Inoue, and Kilian (2008) show that bagging predictors provides more accurate inflation forecasts than equally weighted forecasts). BVARs are the benchmark used in Smets and Wouters (2007). Importantly, all models' forecasts are obtained using real-time vintages of data to mimic the information set that was actually available to forecasters in real-time.

The most important macroeconomic variables to be forecast are inflation, output growth, and the interest rate, and we focus on these variables in this article. Medium-scale DSGE models have many more model-defined variables (some of which are unobservable) and also use a relatively large number of observables, making it difficult to include the same information in a VAR, due to overfitting concerns. For example, the Smets and Wouters (SW) model uses seven observables in estimation, which would mean 105 parameters to estimate in a second order, seven variable reduced-form VAR. Hence, the literature has employed BVARs as the forecast benchmark to improve the out-of-sample forecasting accuracy of the VAR model. We show in this article that moving to a smaller, three variable VAR or to an AR model actually reduces the mean squared forecast error (MSFE) of the macroeconomic aggregates we consider compared with larger BVAR models.

The accuracy of the forecasts is evaluated using standard tests of forecast efficiency as well predictive accuracy. In particular, we use Mincer–Zarnowitz's (1969) regressions to evaluate the forecast efficiency of the models. We also compare the models' relative forecasting ability by comparing models' relative MSFEs. Importantly, we focus on studying how the forecasting ability of the models has evolved over time, both in terms of their efficiency and their relative predictive ability.

Our main question is whether there is a forecasting method based on reduced-form time series models or DSGE models which has better properties than others in all samples and for all forecast horizons. Such a model would be the "model to beat" and a natural benchmark. We find, however, that no such model exists among the standard candidates. The models' forecast accuracy in our study depends on the sample and the forecast horizon.

Regarding the models' relative forecasting performance, we find that, with some exceptions, a simple autoregression performs best at very short horizons (i.e., horizons of one quarter), whereas DSGE models perform the best at long horizons (i.e., horizons of two years) when forecasting output growth, and vice versa for inflation. When looking at how the forecasting ability evolved over time, however, several interesting aspects are worth noting. First, when forecasting output growth, the better forecasting ability of the AR relative to the DSGE model is evident in the 1990s but disappears toward the end of the sample (in mid-2000), when the difference between the two models first becomes insignificant and then the DSGE model begins to dominate. Results are similar for forecast combinations. Regarding inflation, results are different: there is a clear tendency for the DSGE model's forecasts to worsen over time relative to that of reduced-form models at most horizons. Similar results hold for interest rates at short and medium horizons.

Regarding the models' forecast efficiency, we find that, with few exceptions, the models' forecasts of output growth and inflation are not efficient – i.e., either the forecasts are biased (i.e., on average the forecast is different from the realized value) or the forecasts are not correlated one-to-one with the actual realizations, or both. Only in the case of output growth, the DSGE model forecasts are efficient at long horizons at some point over the sample. In the case of inflation, the DSGE model forecasts' lack of efficiency shows up consistently over time, whereas that of the ARs and forecast combinations shows up mainly in the early 2000s at some forecast horizons, while they are more efficient in the most recent period. Similarly, interest rate forecasts suffer from lack of efficiency, although VAR-based interest rate forecasts suffer less from lack of efficiency than the DSGE model at short horizons and more at long horizons.

An important finding of this article is the comparative advantage that a simple three variable VAR has in forecasting inflation and output compared with the seven variable BVAR. The BVAR-based forecast, both because it employs the same observables as the DSGE model and because it is estimated using Bayesian methods similar to those used in the estimation of the DSGE model, is the standard reduced-form benchmark of choice against the DSGE model forecast.[3] We show that while this benchmark makes intuitive sense for those reasons, the large-scale BVAR appears over-parameterized and constitutes a weak benchmark for the evaluation of other forecasts. If the variables of interest are inflation and GDP growth, a three variable VAR may be a more suitable benchmark for forecasting than the BVAR. Similarly, the AR model also forecasts better than the BVAR.

Another interesting finding we present is the sample dependency of the most accurate forecast (in terms of root mean squared forecast error, RMSFE). In particular, in several cases the performance of the DSGE model deteriorates toward the end of the sample relative to its competitors, possibly because it corresponds to a truly out-of-sample period, after the publication of Smets and Wouters (2007). Note that our goal is to evaluate the true out-of-sample forecasting ability of the DSGE model. Even when the data are real-time, as the priors are chosen at the time the model was constructed, some ex-post information affects the model forecasts. More subtly, even if one had a way of using real-time priors, as the model is built to fit the data in a certain period (the data about up to 2004 in the Smets–Wouters model's case), modeling choices are made to maximize this fit. Hence, the real out-of-sample period begins at the time the model is built, regardless of the real-time nature of the data used to estimate the model. In the present case, this leaves a short true out-of-sample period as the financial crisis began shortly after the inception of this model and limited the data available for the forecasting exercise. But the deterioration in the model forecasting ability in the true out-of-sample period is nonetheless striking.

Our article is related to the recent contributions by Adolfson, Linde, and Villani (2007), Edge and Gürkaynak (2010), Edge, Kiley, and Laforte (2010), Edge, Gürkaynak, and Kısacıkoğlu (2013), Giacomini and Rossi

(2012), and Del Negro and Schorfheide (2013). Adolfson et al. (2007) compare the forecasting ability of an open economy DSGE model for the Euro area with that of several reduced-form models; however, they focus on the Euro area and only study the MSFE of the competing forecasts, so it is unclear whether the differences in forecast performance of competing models are statistically significant. In contrast, we perform statistical tests of relative forecast comparisons as well as forecast efficiency. Moreover, unlike Adolfson et al. (2007), we evaluate the forecasting ability of the DSGE model using a real-time database, rather than ex-post revised data.

Edge and Gürkaynak (2010) have been the first to perform a thorough analysis of the actual forecasting ability of DSGE models using real-time data based on vintages. They show that the performance of DSGE models is comparable or slightly superior to that of a constant mean model, but both DSGE models' and Blue Chip forecasts are biased: the reason why they perform similarly is because, during the Great Moderation period, volatility was low and most variables were therefore unpredictable. Edge et al. (2013), whose data we utilize, extend the sample of Edge and Gürkaynak and study the model's forecasting ability against the model's own implication of how well it should forecast by generating data from the model and examining the model's ability to forecast model-generated data and its ability to forecast real data. Edge et al. (2010) find that the Federal Reserve Board DSGE model produces competitive forecasts relative to those of the Federal Reserve Board staff; however, they do not consider the Smets and Wouters (2007) model, nor the same sample period we consider.

Giacomini and Rossi (2012) compare the performance of DSGE models and BVARs over time from an in-sample fit perspective. They show that the DSGE model's performance seems to deteriorate over time, once the trend in the model is estimated in real time.[4] Unlike Giacomini and Rossi (2012), we focus on the models' out-of-sample forecasting performance and on U.S. rather than Euro area data. Del Negro and Schorfheide (2013) consider DSGE models and Blue Chip forecasts in their forecasting tests, whereas we compare the forecasts of the DSGE model to those of several reduced-form models. In addition, Del Negro and Schorfheide (2013) only forecast over the full out-of-sample period; we instead analyze the forecast performance of the models as it evolves over time. This allows us to identify periods in time where each of the models were the best in forecasting specific macroeconomic variables.

The remainder of the article is organized as follows. The second section describes the data, the methods, and the models we use for forecasting, as well as the test statistics we rely upon to evaluate the models' forecasting ability. The third section reports empirical results for three of the most important macroeconomic variables in the Smets and Wouters (2007) DSGE model: output growth, inflation, and the interest rate. We discuss both relative forecasting performance of the models as well as their forecast efficiency. The fourth section concludes.

DATA, FORECASTING METHODS AND MODELS, AND TEST STATISTICS

The Data

The data used in this article come from Edge et al. (2013), who extend and improve the quarterly data set of Edge and Gürkaynak (2010), which itself closely follows the data choices of Smets and Wouters (2007). We refer the readers to the detailed data appendix of Edge et al. (2013) for details of the data and provide only a brief outline here.

Edge and Gürkaynak (2010) collected real-time vintages of U.S. data on the same series used in Smets and Wouters (2007). The series are the following: the per capita, real GDP growth rate (quarter on quarter, non-annualized); the GDP deflator inflation rate (quarter on quarter, non-annualized); the interest rate level; real consumption; real investment; the real wage; hours worked (in log); the growth rate of the GDP deflator; and the federal funds rate. They use real GDP, the GDP price deflator, the nominal personal consumption expenditure, and the nominal fixed private investment from the national income and product accounts (NIPA); compensation per hour in the non-farm business sector from the Bureau of Labor Statistics' quarterly Labor Productivity and Costs (LPC) release; average weekly hours of production, civil employment, and population from the Bureau of Labor Statistics Employment Situation Summary (ESS). The realized data, or the "true" outcome that is to be forecast is the first final release corresponding to the last observation in the third release of NIPA data and the second release of LPC data (which are the final release of the data before they are revised in either an annual or a comprehensive revision). For ESS releases, they use the last available observation in the second revision of the data. Edge et al. (2013) extend this data set in sample coverage and also provide a better treatment of the population, which makes an important difference in the output growth forecasts.

The data we use is in the form of 20-year rolling windows. Rolling window estimates help capture changes in parameters (regime shifts), and forecasting tests we use below have desirable properties when the underlying forecasts are made using rolling window estimates. In the extended data set we use, the out-of-sample period used for forecast evaluation starts in 1992:I and ends in 2005:IV for one-step-ahead forecasts, 2006:I for two-step-ahead forecasts, and 2007:III for eight-step-ahead forecasts. The end of the sample is dictated by the financial crisis, which a model without a housing sector or a financial sector cannot be expected to forecast. The sample includes the Great Moderation period as well as the recession of 2001. The Great Moderation is characterized by lower permanent (predictable) and higher temporary (unpredictable) fluctuations in macroeconomic aggregates, which makes inference about model validity based on forecasting ability a tricky affair, as discussed by Edge and Gürkaynak (2010). In this article, our aim is to evaluate the forecast accuracy of DSGE models in a comparative setting without drawing conclusions about the models themselves.

In short, our data are real time versions of the Smets and Wouters (2007) database, employing not only their data series choices but also their priors for the Bayesian estimation of the DSGE model and the BVAR. The only material difference is in the population series, the proper treatment of which helps the DSGE model output forecasts as the model produces per capita output forecasts which are then multiplied by population growth to obtain aggregate output growth forecasts.

Forecasting Models

The Smets and Wouters (2007) DSGE model is a real business cycle model with nominal rigidities (sticky prices and wages) as well as real rigidities (such as habit formation in consumption and cost of adjustment in investment). A detailed description of the model is provided in the first section in Smets and Wouters (2007), to which we refer the interested reader. The model includes seven observable variables and seven shocks.[5] In the Bayesian estimation of the model parameters we use the same priors as Smets and Wouters (2007).

In addition to the Smets and Wouters (2007) model, we consider several reduced-form time series models. The first reduced-form model we consider is a three-variable VAR with output growth (Δy_t), inflation (π_t), and the interest rate (i_t). Let $Y_t = (\Delta y_t, \pi_t, i_t)'$, then the VAR is:

$$A(L)Y_t = A_0 + u_{VAR,t} \tag{1}$$

This represents the basic VAR specification, which we use to forecast output growth, inflation, and the interest rate. This basic specification follows Stock and Watson (2001).[6]

Let the k-th variable in Y_t be denoted by $Y_t^{(k)}$, for $k = 1, 2, 3$. We also consider an AR model for the variables of interest; for example, for $Y_t^{(k)}$, we estimate:

$$Y_t^{(k)} = a_0 + a_1 Y_{t-1}^{(k)} + \ldots + a_p Y_{t-p}^{(k)} + u_{AR,t} \tag{2}$$

Rather than relying on iterated AR and VAR forecasts from Eqs. (1) and (2), we construct direct forecasts instead, as described in the section "Forecast Evaluation Test Statistics." We will label these direct forecasts as AR and VAR forecasts (even though they are technically not obtained from an AR or VAR model) and we demonstrate the robustness of the qualitative results to iterated AR and VAR forecasts in Appendix B.[7]

The Random Walk (labeled "RW") model for $Y_t^{(k)}$ is the following:

$$Y_t^{(k)} = Y_{t-1}^{(k)} + u_{RW,t} \tag{3}$$

We also consider equal-weighted forecast averaging, which has been shown to provide competitive forecasts for output growth and inflation (see Rossi, forthcoming). The equal-weighted forecast combinations ("FC") are obtained by averaging (using equal weights) the forecasts of the autoregressive model with those of additional univariate models that contain an additional predictor, $x_{i,t}$ (the procedure follows Stock & Watson, 2003). For example, when forecasting $Y_t^{(k)}$, we estimate:

$$Y_t^{(k)} = a_0 + a_1 Y_{t-1}^{(k)} + \ldots + a_p Y_{t-p}^{(k)} + b_1 x_{i,t-1} + \ldots + b_k x_{i,t-k} + u_{ARX_i,t} \tag{4}$$

for $i = 1, \ldots, 6$, and then combine the forecasts of models (4) for the set of variables in Smets and Wouters (2007), $x_{1t}, \ldots, x_{6,t}$, and that of the AR model, Eq. (2), giving each of them a weight equal to $1/7$.

Forecasting Methods

The forecasts are based on a model that is characterized by the $(k \times 1)$ parameter vector γ. The forecasts are obtained by dividing the sample of size

$T+h$ observations into an in-sample portion of size R and an out-of-sample portion of size P. The size of the estimation window, R, equals 20 years, corresponding to 80 quarterly observations that provide information for the model estimation at each forecast date.[8] The sequence of P out-of-sample forecast errors depends on the realizations of the forecasted variable and on the in-sample parameter estimates, $\hat{\gamma}_{t,R}$. These parameters are re-estimated at each $t=R,\ldots,T$ over a window of R observations including data indexed $t-R+1,\ldots,t$ (also known as the "rolling scheme").[9]

The forecasts of the VAR and AR models are obtained using a direct forecast method. The number of lags in the VAR and AR specifications is recursively chosen by the Bayesian information criterion (BIC).[10] The number of lags in Eq. (4) is estimated conditional on selecting p by BIC in the univariate AR specification, Eq. (2), and then applying the BIC to Eq. (4) to select the number of lags of the extra predictor, k. To guarantee parsimony, the maximum number of lags imposed in the BIC is two.

It is important to note that it has been shown that the BIC does not necessarily select the best forecasting model (see Inoue & Kilian, 2006). An alternative criterion is the Akaike information criterion (AIC). In Appendix A, we discuss results based on the AIC and show that our results are robust to the choice of the criterion.[11]

Forecast Evaluation Test Statistics

Let the estimated forecast error associated with the h-step-ahead direct forecast made at time t be denoted by ε_{t+h}. For example, in the case of a simple linear AR(1) regression model with h-period lagged regressors, we estimate the regression: $y_t=\gamma_0+\gamma_1 y_{t-h}$ using the most recent R data, and obtain the parameter estimates at time t: $\hat{\gamma}_{0,t,R}, \hat{\gamma}_{1,t,R}$. The forecast made at time t is equal to $y^f_{t+h}=\hat{\gamma}_{0,t,R}+\hat{\gamma}_{1,t,R}y_t$ and the forecast error associated with the direct forecast is $\hat{\varepsilon}_{t+h}=y_{t+h}-\hat{\gamma}_{0,t,R}-\hat{\gamma}_{1,t,R}y_t$, and $\hat{\gamma}_{t,R}\equiv(\hat{\gamma}_{0,t,R},\hat{\gamma}_{1,t,R})'$. We perform a series of tests on the sequence of the forecast errors, $\{\hat{\varepsilon}_{t+h}\}_{t=R}^{T}$.

First, we focus on testing forecast efficiency (under quadratic loss), based on the following regression:

$$\hat{\varepsilon}_{t+h}=\alpha+\beta y^f_{t+h}+\eta_{t+h},\ t=R,\ldots,T \tag{5}$$

where $\hat{\varepsilon}_{t+h}$ is the estimated forecast error. We test whether $\alpha=0$ (forecast unbiasedness), $\beta=0$, and $\alpha=\beta=0$. We refer to the latter test (the joint test for α, β equal to zero) as the Mincer and Zarnowitz (1969) forecast efficiency test.

Second, we note that, as discussed in Rossi and Sekhposyan (2012), traditional forecast efficiency test results suffer from the drawback that they focus on whether forecasts are efficient on average over the whole out-of-sample period. Thus, we follow Rossi and Sekhposyan (2012) and also report a sequence of forecast efficiency tests obtained by estimating regression (5) using data from $t-m+1$ up to t, and construct a *sequence* of Mincer and Zarnowitz (1969) tests at time t, for $t=m, \ldots, P$.[12]

Third, we consider tests of relative forecasting performance. We report results for comparing models' forecasts based on the difference of the MSFEs of competing models using the Diebold and Mariano (1995) test statistic and using the critical values proposed by Giacomini and White (2006).

To conclude, we also report results for comparing models' forecasting ability over time based on the fluctuation test introduced by Giacomini and Rossi (2010). The latter propose to calculate the Diebold and Mariano (1995) test statistic over rolling windows in an effort to keep track of the models' relative forecasting ability over time. In other words, we calculate a sequence of Diebold and Mariano's (1995) test statistics over time by estimating regression (5) using data from $t-m+1$ up to t, and use the results to perform the test at time t, for $t=m, \ldots, P$. To evaluate whether the models are statistically significantly different, we rely on the critical values in Giacomini and Rossi (2010).

EMPIRICAL RESULTS ON FORECASTING OUTPUT GROWTH, INFLATION, AND THE INTEREST RATE

In this section we focus on forecasting output growth, inflation, and the short-term interest rate, showing the results for output growth and inflation (which policymakers are most interested in) first, and showing the results for interest rates separately later. The forecasts are reported in Fig. 1. The figure reports the realized value of the target variable (output growth in panel (a) and inflation in panel (b)), labeled "Actual Data," together with forecasts of the following models: the VAR (labeled "VAR"), the AR (labeled "AR"), Smets and Wouters (2007, labeled "DSGE"), equal-weighted forecast combinations (labeled "Combin. Forecast"), the BVAR (labeled "BVAR"), and the random walk (labeled "RW"). Each graph

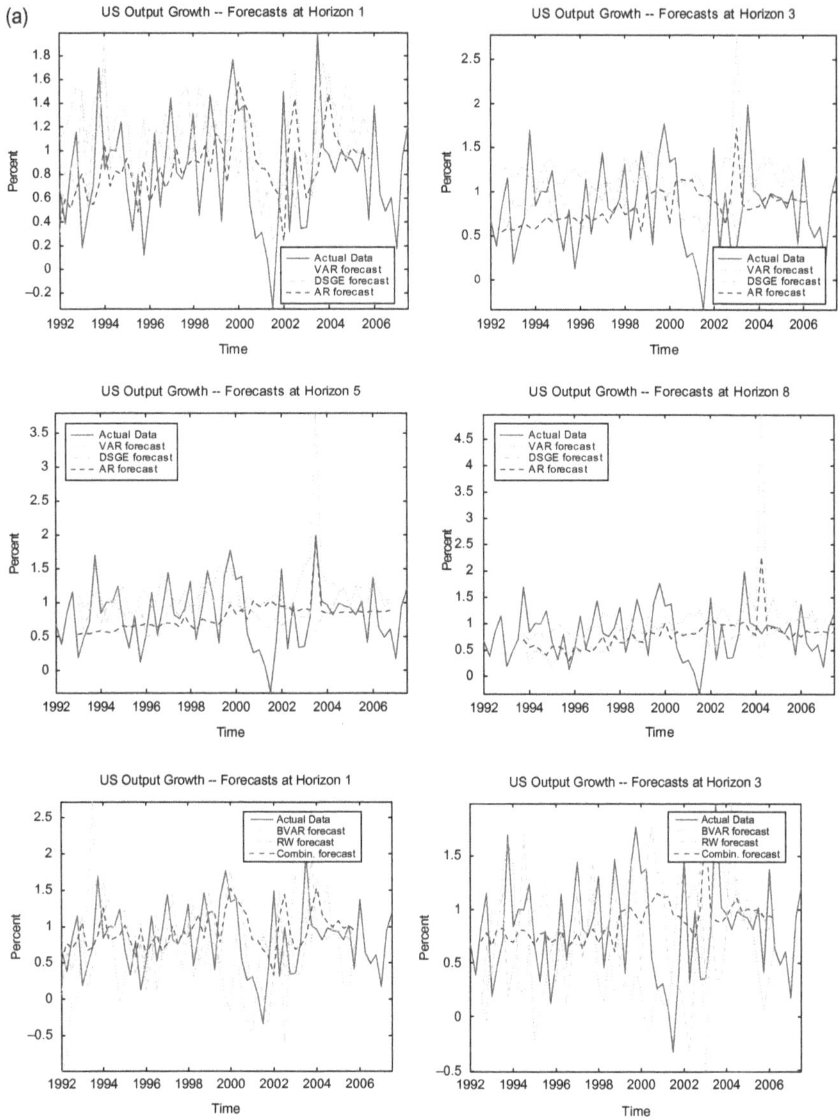

Fig. 1. (a) Output growth forecasts. (b) Inflation forecasts. *Notes*: The figure reports the realized value of the target variable (labeled in the title – e.g., output growth in Fig. 1(a)), labeled "Actual Data," together with forecasts of the VAR (labeled "VAR"), AR (labeled "AR"), Smets and Wouters (2007, labeled "DSGE"), Bayesian VAR (labeled "BVAR"), forecast combinations (labeled "Combin. Forecast") and the random walk (labeled "RW") models. Each graph reports results for different horizons, reported in each graph's title.

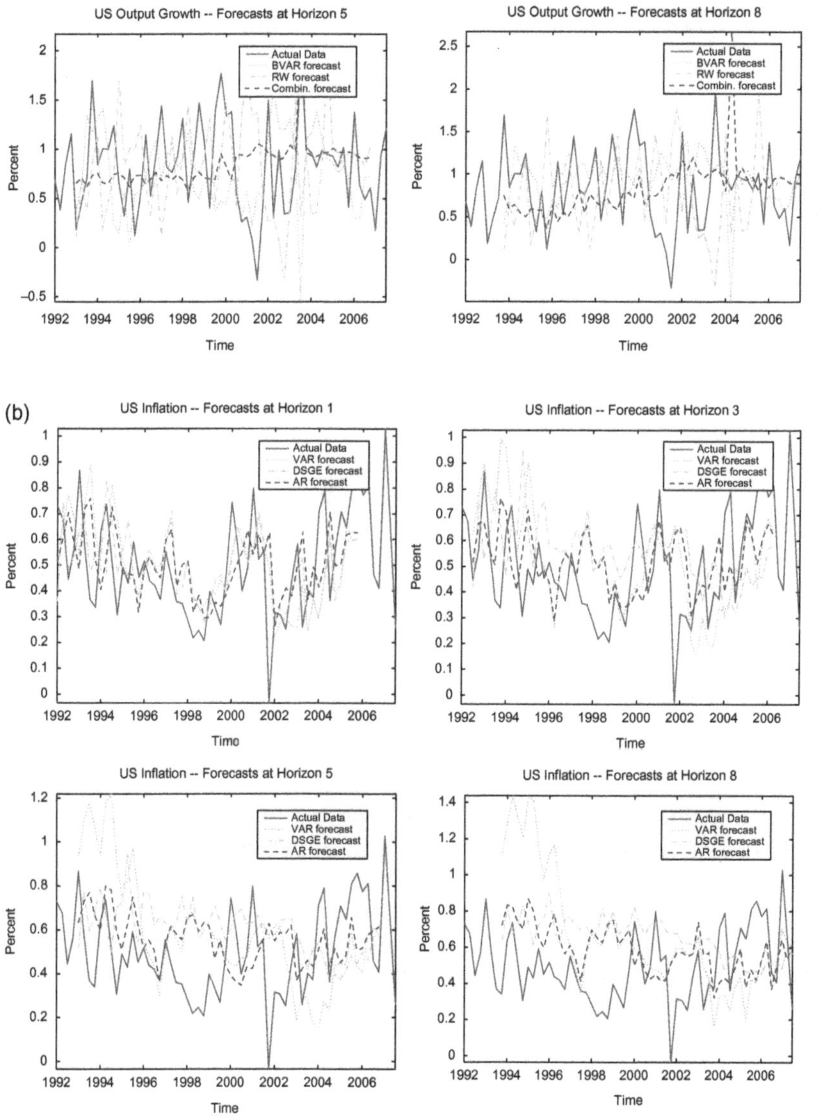

Fig. 1. Continued.

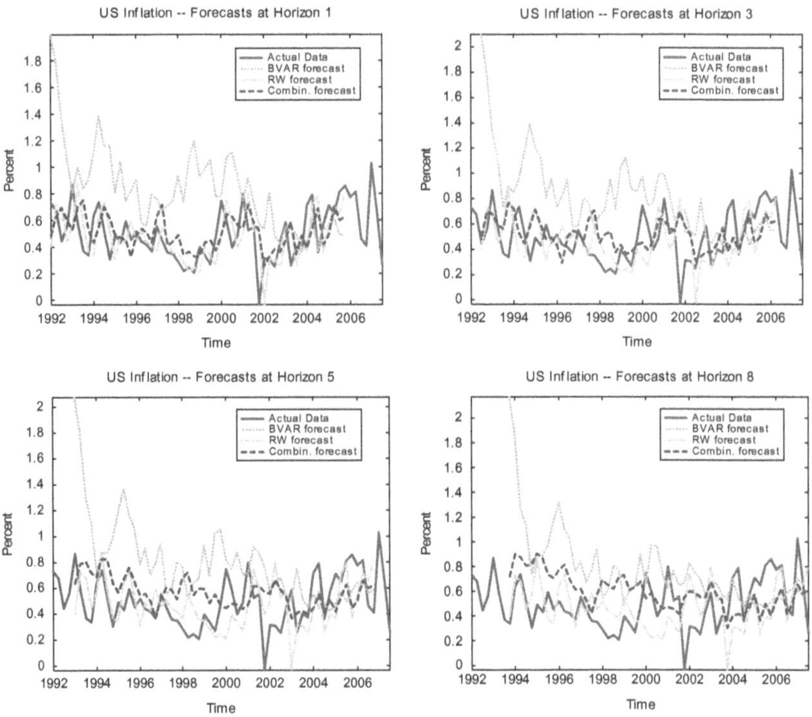

Fig. 1. *Continued.*

reports results for different horizons, reported in each graph's title. From Fig. 1(a), it appears that AR forecasts of output growth track the realized values more closely than the DSGE model at the one-quarter-ahead forecast horizon, but they worsen as the forecast horizon becomes larger. It also appears that the longer the horizon, the more the VAR underestimates output growth. In this section, we will use formal statistical tools to investigate whether this visual impression is correct.

EMPIRICAL RESULTS BASED ON TESTS OF FORECAST EFFICIENCY

In this subsection we investigate whether the forecasts of the models that we study are efficient. It is worthwhile to reflect briefly on what it means for the forecast of an atheoretical reduced-form model or a DSGE model to be efficient. Could a practitioner who is trying to forecast, say, output improve over the forecast of any one of these models by using information available in real time? If so, the practitioner would not use the forecast of the model as is, but modify the model. This is the sense in which we use the term forecast efficiency for the models under study.

Table 1 reports results of forecast efficiency tests for several models and target variables (output growth on the left and inflation on the right). Each panel reports, for several horizons (reported in the first column), the p-values of the Mincer and Zarnowitz (1969) forecast efficiency test (i.e., the joint test that $\alpha=\beta=0$ in Eq. (5), labeled "MZ p-value"), $\hat{\alpha}$ in regression (5) together with the p-value of the test that the constant equals zero (in parenthesis), $\hat{\beta}$ in regression (5) together with the p-value of the test that the slope equals zero (in parenthesis). Table 1 shows, in general, that the forecasts are not efficient, no matter which model we consider. There are some exceptions, however. The tests do not reject joint forecast efficiency for output growth at some intermediate horizons (four or five quarter-ahead) in some models. Note that most reduced-form forecasts underpredict the target variables.

To investigate whether the efficiency of the forecasts has changed over time, Fig. 2 reports the p-value of the sequence of Mincer and Zarnowitz (1969) tests over time, using a rolling window equal to half of the number of out-of-sample forecasts (i.e., approximately 7 years) for each test. This window size is used in all of the sequence of tests of forecast comparisons and efficiency *over time* that we use below.[13] The figure displays results for forecasts at various horizons (reported in the legend) and several models (reported in the title). The target variable is output growth in panel (a) and inflation in panel (b). Fig. 2(a) shows that the large rejections of forecast efficiency for output growth are mainly driven by the end of the sample for the statistical models, whereas those of the DSGE model are mainly driven by both the beginning and the end of the sample, although the bias is more pronounced at the end. The intermediate horizon forecasts of the Smets and Wouters (2007) model are consistently efficient over time, although the p-values of the forecast efficiency test are very small at the end of the sample, raising concerns of lack of efficiency in the late 2000s. Efficiency is uniformly rejected over time for the random walk at all horizons and for the BVAR forecasts at most horizons.

Results are similar for inflation, except that, for inflation, the DSGE model forecasts are not efficient, no matter which sample period or forecast horizon we consider (except for one-step-ahead forecasts in the late 1990s).

Table 1. Forecast Optimality Tests: Forecasting Output Growth and Inflation.

Panel A: VAR Model

	Output Growth Forecasts					Inflation Forecasts				
h	MZ p-value	$\hat{\alpha}$		$\hat{\beta}$		MZ p-value	$\hat{\alpha}$		$\hat{\beta}$	
1	0.00	0.55	(0.10)	0.24	(0.00)	0.00	0.33	(0.00)	0.30	(0.00)
2	0.00	0.94	(0.01)	−0.09	(0.00)	0.00	0.41	(0.00)	0.15	(0.00)
3	0.00	1.07	(0.00)	−0.19	(0.00)	0.00	0.37	(0.00)	0.23	(0.00)
4	0.00	0.78	(0.00)	0.04	(0.00)	0.00	0.41	(0.00)	0.14	(0.00)
5	0.00	0.32	(0.02)	0.43	(0.00)	0.00	0.48	(0.00)	0.00	(0.00)
6	0.00	0.67	(0.00)	0.14	(0.00)	0.00	0.54	(0.00)	−0.08	(0.00)
7	0.00	0.74	(0.00)	0.09	(0.00)	0.00	0.52	(0.00)	−0.05	(0.00)
8	0.00	0.87	(0.00)	−0.03	(0.00)	0.00	0.50	(0.00)	−0.02	(0.00)

Panel B: AR Model

	Output Growth Forecasts					Inflation Forecasts				
h	MZ p-value	$\hat{\alpha}$		$\hat{\beta}$		MZ p-value	$\hat{\alpha}$		$\hat{\beta}$	
1	0.00	0.63	(0.00)	0.22	(0.00)	0.01	0.30	(0.03)	0.36	(0.01)
2	0.00	0.83	(0.00)	0.01	(0.00)	0.01	0.40	(0.01)	0.17	(0.01)
3	0.00	1.22	(0.00)	−0.46	(0.00)	0.03	0.30	(0.04)	0.37	(0.03)
4	0.00	1.02	(0.00)	−0.22	(0.00)	0.00	0.38	(0.00)	0.21	(0.00)
5	0.23	0.50	(0.29)	0.41	(0.38)	0.00	0.56	(0.00)	−0.13	(0.00)
6	0.00	0.93	(0.00)	−0.13	(0.00)	0.00	0.65	(0.00)	−0.30	(0.00)
7	0.00	0.83	(0.00)	0.00	(0.00)	0.00	0.69	(0.00)	−0.35	(0.00)
8	0.00	0.85	(0.00)	−0.01	(0.00)	0.00	0.60	(0.00)	−0.20	(0.00)

Panel C: DSGE Model

	Output Growth Forecasts					Inflation Forecasts				
h	MZ p-value	$\hat{\alpha}$		$\hat{\beta}$		MZ p-value	$\hat{\alpha}$		$\hat{\beta}$	
1	0.00	0.46	(0.12)	0.36	(0.00)	0.01	0.24	(0.06)	0.50	(0.02)
2	0.00	0.23	(0.53)	0.62	(0.10)	0.00	0.47	(0.00)	0.02	(0.00)
3	0.00	0.30	(0.51)	0.57	(0.20)	0.00	0.68	(0.00)	−0.34	(0.00)
4	0.05	0.27	(0.60)	0.62	(0.35)	0.00	0.71	(0.00)	−0.39	(0.00)
5	0.01	0.50	(0.27)	0.36	(0.10)	0.00	0.78	(0.00)	−0.50	(0.00)
6	0.05	0.60	(0.17)	0.26	(0.10)	0.00	0.78	(0.00)	−0.47	(0.00)
7	0.00	0.96	(0.00)	−0.15	(0.00)	0.00	0.98	(0.00)	−0.76	(0.00)
8	0.00	1.03	(0.00)	−0.23	(0.00)	0.00	0.93	(0.00)	−0.68	(0.00)

Table 1. (*Continued*)

Panel D: Random Walk Model

	Output Growth Forecasts					Inflation Forecasts				
h	MZ p-value	$\hat{\alpha}$		$\hat{\beta}$		MZ p-value	$\hat{\alpha}$		$\hat{\beta}$	
1	0.00	0.46	(0.12)	0.36	(0.00)	0.01	0.24	(0.06)	0.50	(0.02)
2	0.00	0.23	(0.53)	0.62	(0.10)	0.00	0.47	(0.00)	0.02	(0.00)
3	0.00	0.30	(0.51)	0.57	(0.20)	0.00	0.68	(0.00)	−0.34	(0.00)
4	0.05	0.27	(0.60)	0.62	(0.35)	0.00	0.71	(0.00)	−0.39	(0.00)
5	0.01	0.50	(0.27)	0.36	(0.10)	0.00	0.78	(0.00)	−0.50	(0.00)
6	0.05	0.60	(0.17)	0.26	(0.10)	0.00	0.78	(0.00)	−0.47	(0.00)
7	0.00	0.96	(0.00)	−0.15	(0.00)	0.00	0.98	(0.00)	−0.76	(0.00)
8	0.00	1.03	(0.00)	−0.23	(0.00)	0.00	0.93	(0.00)	−0.68	(0.00)

Panel E: Forecast Combinations Model

	Output Growth Forecasts					Inflation Forecasts				
h	MZ p-value	$\hat{\alpha}$		$\hat{\beta}$		MZ p-value	$\hat{\alpha}$		$\hat{\beta}$	
1	0.00	0.62	(0.01)	0.22	(0.00)	0.00	0.25	(0.05)	0.45	(0.01)
2	0.00	0.87	(0.00)	−0.04	(0.00)	0.00	0.42	(0.01)	0.13	(0.00)
3	0.00	0.26	(0.00)	−0.47	(0.00)	0.01	0.32	(0.02)	0.32	(0.02)
4	0.00	0.04	(0.00)	−0.24	(0.00)	0.00	0.33	(0.00)	0.28	(0.00)
5	0.21	0.53	(0.22)	0.34	(0.22)	0.00	0.52	(0.00)	−0.06	(0.00)
6	0.00	0.90	(0.00)	−0.08	(0.00)	0.00	0.68	(0.00)	−0.33	(0.00)
7	0.00	0.81	(0.00)	0.03	(0.00)	0.00	0.70	(0.00)	−0.35	(0.00)
8	0.00	0.89	(0.00)	−0.06	(0.00)	0.00	0.59	(0.00)	−0.17	(0.00)

Panel F: Bayesian VAR Model

	Output Growth Forecasts					Inflation Forecasts				
h	MZ p-value	$\hat{\alpha}$		$\hat{\beta}$		MZ p-value	$\hat{\alpha}$		$\hat{\beta}$	
1	0.00	0.81	(0.00)	0.02	(0.00)	0.00	0.44	(0.00)	0.00	(0.00)
2	0.00	0.80	(0.00)	0.07	(0.00)	0.00	0.48	(0.00)	0.01	(0.00)
3	0.00	0.74	(0.00)	0.16	(0.00)	0.00	0.49	(0.00)	−0.00	(0.00)
4	0.00	0.70	(0.00)	0.20	(0.00)	0.00	0.44	(0.00)	0.06	(0.00)
5	0.00	0.91	(0.00)	−0.11	(0.00)	0.00	0.45	(0.00)	0.04	(0.00)
6	0.00	0.62	(0.01)	0.24	(0.00)	0.00	0.52	(0.00)	−0.05	(0.00)
7	0.00	0.66	(0.00)	0.20	(0.00)	0.00	0.57	(0.00)	−0.11	(0.00)
8	0.00	0.93	(0.00)	−0.10	(0.00)	0.00	0.53	(0.00)	−0.05	(0.00)

Note: The table reports results for forecasting output (on the left) and inflation (on the right). Each panel reports, for several horizons (h, reported in the first column), the values of the Mincer and Zarnowitz (1969) forecast efficiency test (labeled "MZ p-value"), the estimate of $\hat{\alpha}$ in regression (5) together with the p-value of the test that the constant equals zero (in parentheses), and the estimate of $\hat{\beta}$ in regression (5) together with the p-value of the test that the slope equals zero (in parentheses).

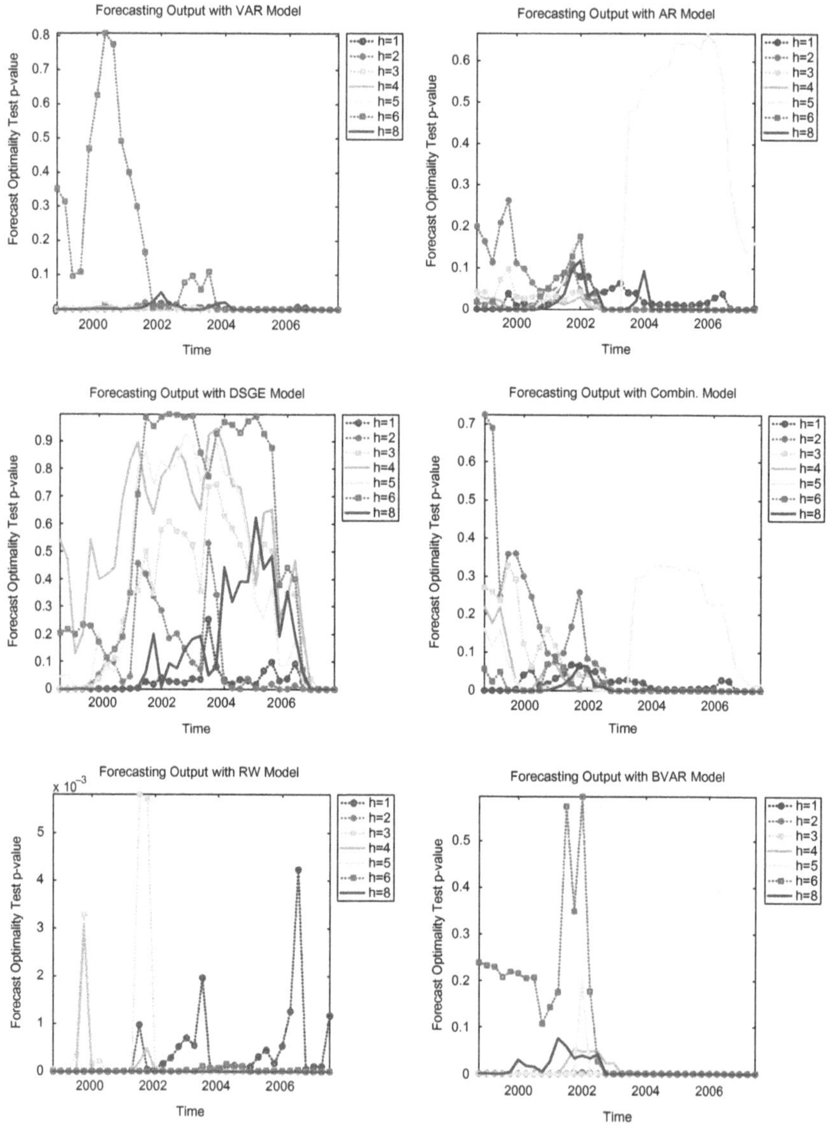

Fig. 2. (a) Forecast optimality test over time: Output growth. (b) Forecast optimality test over time: Inflation. *Notes*: The figure reports the p-value of the Mincer and Zarnowitz (1969) test in Eq. (5) over time. The figure reports results for forecasts at various horizons (see the legend) and several models (see the title). The target variable is output growth in panel (a) and inflation in panel (b).

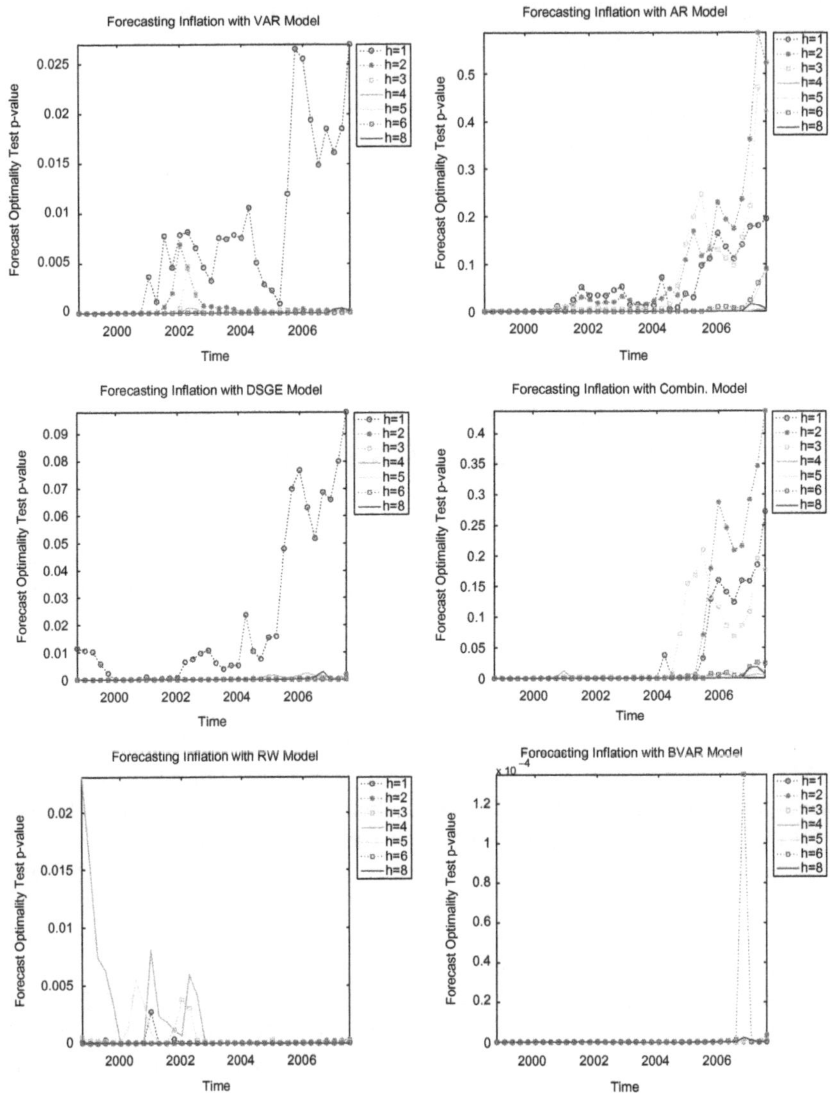

Fig. 2. Continued.

Forecasts of the VAR, AR, and forecast combination models are efficient toward the end of the sample, instead.

Our findings are related to Edge and Gürkaynak (2010), who find that the DSGE model forecasts poorly over the sample period they consider. They attribute this finding to their focus on the Great Moderation sample, a period of stability and low macroeconomic volatility, where most macroeconomic variables have become largely unforecastable (see Stock & Watson, 2007). Our results show that not only the DSGE model, but also the majority of reduced-form models had difficulties forecasting output and inflation over the sample that we consider.

EMPIRICAL RESULTS BASED ON TESTS OF RELATIVE FORECASTING PERFORMANCE

Table 2 reports results for comparing the models' forecasting ability. The columns labeled "RMSFE" in the table report the RMSFE of Smets and Wouters (2007) DSGE model's forecasts (labeled "DSGE"). The remaining columns labeled "RMSFE" in the table report the RMSFE of the following models *relative to* that of the DSGE model: autoregression ("AR"), forecast combinations ("FC"), vector autoregression ("VAR"), Bayesian VAR ("BVAR"), and the random walk ("RW"). For example, the column labeled "VAR" reports the RMSFE of the VAR model divided by the RMSFE of the DSGE model. Values less than unity denote forecasts that have lower RMSFE than the DSGE benchmark. The columns labeled "DM" in the table report the *p*-value of the Diebold and Mariano's (1995) test for comparing the specified forecasts (e.g., the column labeled "DM, FC vs. DSGE" reports the *p*-value of the Diebold and Mariano (1995) test for comparing the forecast of the forecast combination with those of the Smets and Wouters (2007) model).

Table 2 shows that, when considering output growth, the DSGE model forecasts are less accurate than the AR model at the shortest horizon, and more accurate at longer horizons (although differences are not significant in either case). The VAR forecasts are always worse than the DSGE ones and also worse than the forecasts of the AR model. Forecast combinations produce forecasts that are better than the DSGE model at the one-quarter-ahead horizon, but worse at longer horizons, again insignificantly so. The BVAR forecasts are worse than the DSGE model at all horizons. When forecasting inflation, the AR model has lower RMSFE than the DSGE

Table 2. The Diebold and Mariano Test: Forecasting Output Growth and Inflation.

	Output					Inflation				
	RMSFE			DM: Model-DSGE		RMSFE			DM: Model-DSGE	
h	DSGE	VAR	AR	VAR	AR	DSGE	VAR	AR	VAR	AR
1	0.54	1.09	0.94	0.45	0.61	0.19	1.12	1.06	0.31	0.46
2	0.46	1.31	1.11	0.04	0.20	0.23	1.05	0.93	0.64	0.22
3	0.46	1.41	1.16	0.05	0.20	0.26	0.95	0.80	0.34	0.00
4	0.46	1.46	1.15	0.06	0.16	0.26	1.09	0.83	0.42	0.00
5	0.48	1.24	0.98	0.06	0.87	0.28	1.20	0.85	0.30	0.01
6	0.48	1.39	1.09	0.22	0.29	0.29	1.38	0.90	0.20	0.11
7	0.50	1.45	1.04	0.21	0.50	0.30	1.31	0.89	0.26	0.02
8	0.51	1.59	1.06	0.16	0.35	0.31	1.37	0.88	0.25	0.01

	Output						Inflation					
	RMSFE			DM: Model-DSGE			RMSFE			DM: Model-DSGE		
h	FC	BVAR	RW	FC	BVAR	RW	FC	BVAR	RW	FC	BVAR	RW
1	0.95	1.50	1.06	0.67	0.01	0.51	1.03	2.66	1.06	0.69	0.00	0.44
2	1.10	1.58	1.40	0.25	0.00	0.00	0.95	2.25	1.06	0.36	0.01	0.50
3	1.14	1.41	1.37	0.29	0.03	0.01	0.81	1.95	0.92	0.00	0.02	0.31
4	1.14	1.31	1.43	0.17	0.02	0.02	0.82	1.80	0.83	0.00	0.03	0.07
5	1.00	1.33	1.48	0.98	0.06	0.00	0.87	1.70	0.85	0.01	0.04	0.15
6	1.11	1.09	1.40	0.21	0.51	0.02	0.94	1.69	0.88	0.43	0.09	0.34
7	1.07	1.05	1.37	0.34	0.63	0.01	0.92	1.63	0.92	0.17	0.12	0.53
8	1.13	1.11	1.45	0.20	0.33	0.00	0.91	1.56	0.84	0.15	0.13	0.17

Note: The columns labeled RMSFE in the table report the RMSFE of the DSGE model forecasts (labeled "DSGE") as well as the RMSFE of the autoregressive ("AR"), forecast combinations ("FC"), VAR ("VAR"), Bayesian VAR ("BVAR") and the random walk ("RW") relative to that of the DSGE model (e.g., the column labeled "VAR" reports the RMSFE of the VAR model divided by the RMSFE of the DSGE model). The columns labeled "DM" in the table report the p-value of the Diebold and Mariano (1996) test (calculated using Giacomini and White's (2006) critical values) for comparing the specified forecasts (e.g., the column labeled "DM: Model-DSGE" and "VAR" reports the p-value for comparing the forecast of the VAR with those of Smets and Wouters' (2007) model).

model at most horizons, except for short ones, and the differences are statistically significant. Note that, overall, the performance of the BVAR is worse than that of most models; in particular it is worse than the VAR and DSGE models: the latter is statistically more accurate than the BVAR in forecasting both inflation and output growth at horizons up to five quarters. Note also that the random walk forecasts inflation better than the DSGE model at medium to long horizons, very much like the AR.

The models' relative forecasting ability has, however, changed significantly over time. Fig. 3 reports results based on the fluctuation test, which is a rolling Diebold–Mariano test with Giacomini and Rossi's (2010) critical values. Regarding output growth, Fig. 3(a) shows that the VAR performs similarly to or better than the DSGE model up to the mid-1990s, after which the DSGE model performs significantly better. The BVAR, instead, is typically worse than the DSGE model over most time periods. Interestingly, the performance of forecast combinations (labeled "Combin.") and the AR models changes drastically across both time and forecast horizons: at long horizons, they perform significantly worse than the DSGE model, especially at the end of the sample; at shorter horizons, however, the DSGE model seems to outperform AR and forecast combinations only later in the sample.

Regarding inflation, we observe the opposite pattern. The forecasting performance of the VAR, AR, and forecast combination models mostly improve over time relative to the DSGE model at most forecast horizons; the DSGE model is significantly better only in the early 1990s for short horizons. Regarding the performance of the BVAR, it performs significantly worse than the DSGE model over most of the sample and for most forecast horizons.

These empirical findings are related to those in Edge and Gürkaynak (2010) and Giacomini and Rossi (2012). Edge and Gürkaynak (2010) show that the DSGE model forecasts are better than those of the BVAR (in the sense that the former have lower RMSFE). What we find is that the large-scale BVAR may be not be a good benchmark to compare the DSGE model forecasts, as a reduced-form AR model (estimated by unrestricted least squares) as well as other reduced-form time series models perform significantly better. Giacomini and Rossi (2012) compare a DSGE model estimated for the Euro area with a BVAR, and find that the in-sample fit of the DSGE model worsens toward the end of the sample. In the most recent period they have available, the DSGE model fits the data significantly worse than the reduced-form model. However, there are two important differences between this article and Giacomini and Rossi's (2012) paper. The first and the conceptually important difference is that the latter consider the in-sample performance of the models, whereas this article considers out-of-sample forecasting performance. The second is that the latter estimates the DSGE model for the Euro area, whereas we consider U.S. data.

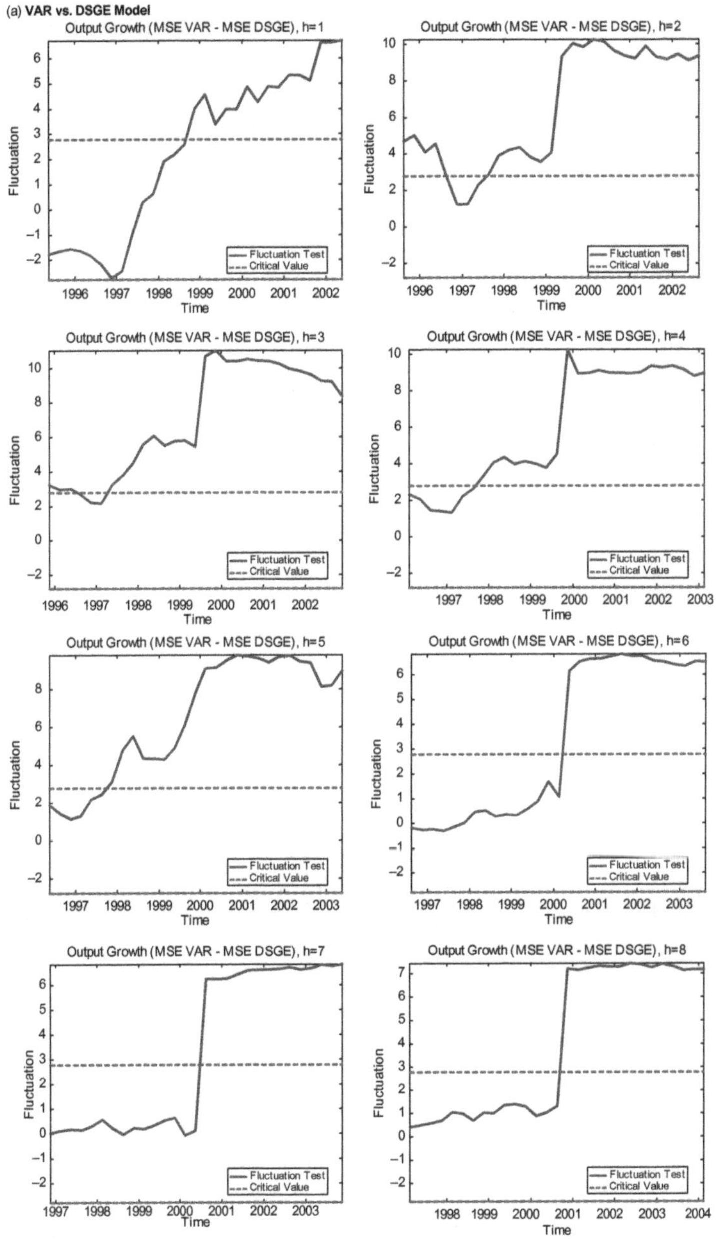

Fig. 3. (a) Fluctuation test: Output. (b) Fluctuation test: Inflation. *Notes*: The figure reports Giacomini and Rossi's (2010) fluctuation test. The figure reports results for forecasts at various horizons and several models (see the title). The target variable is output growth in panel (a) and inflation in panel (b).

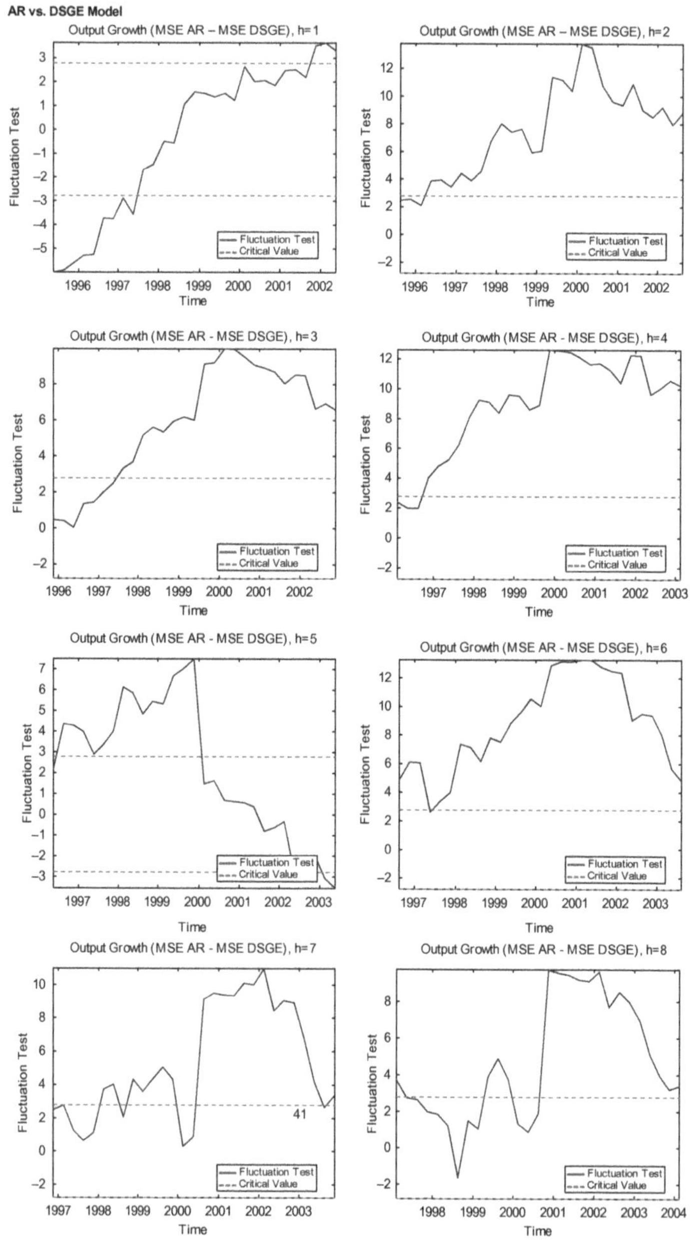

Fig. 3. Continued.

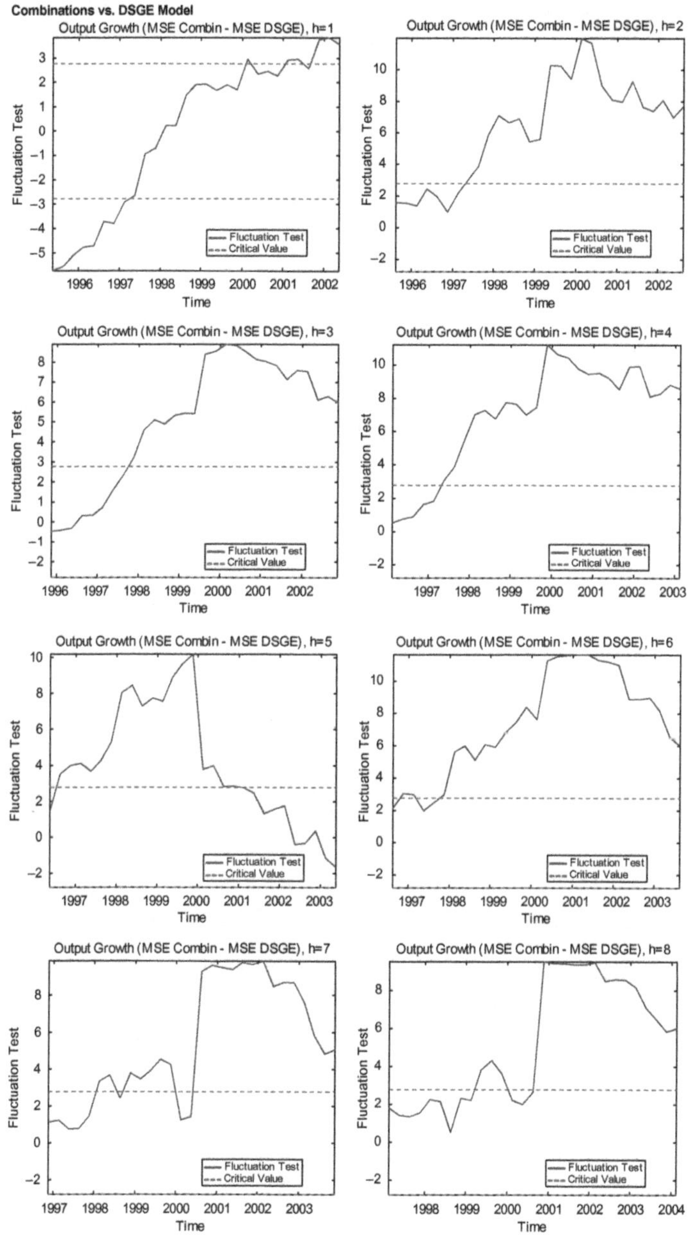

Fig. 3. Continued.

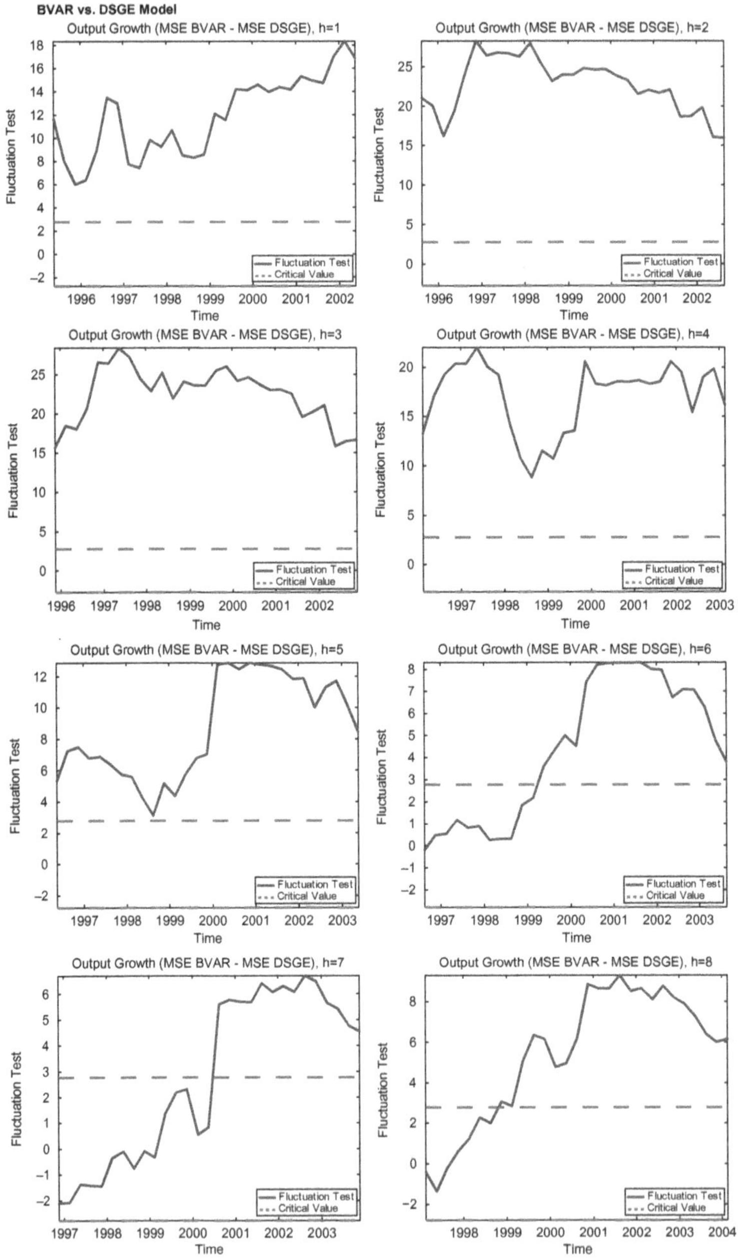

Fig. 3. *Continued.*

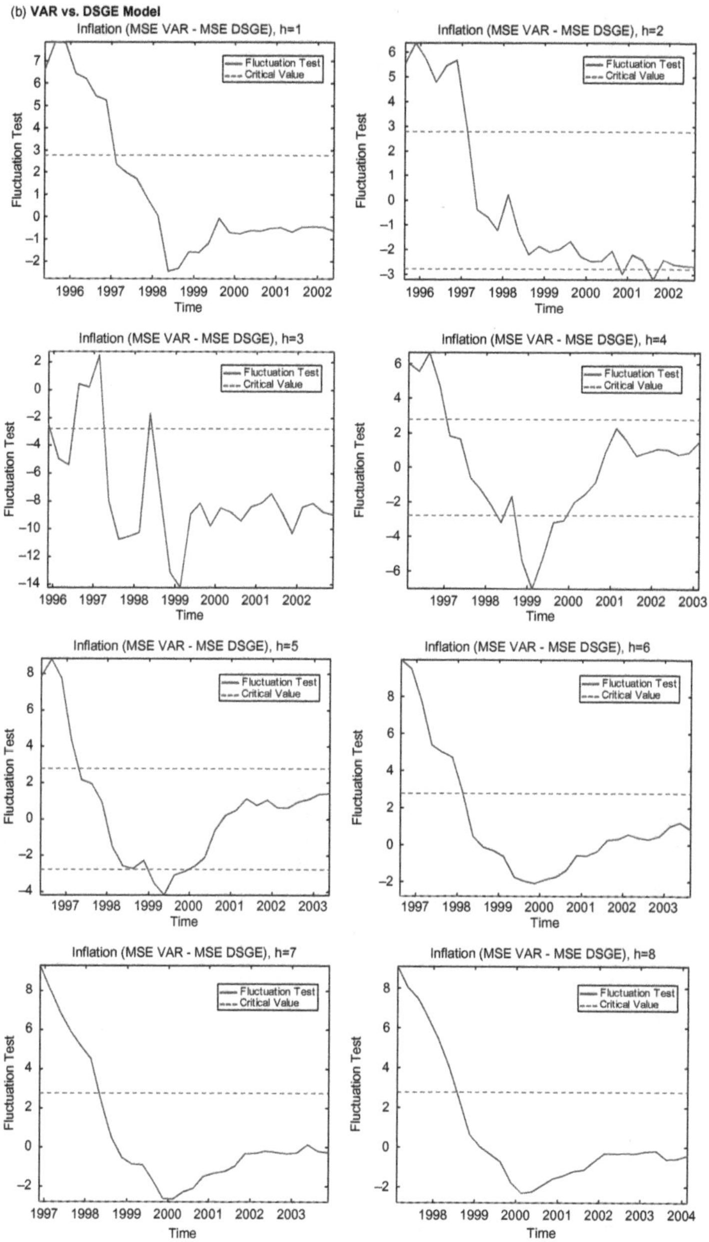

Fig. 3. Continued.

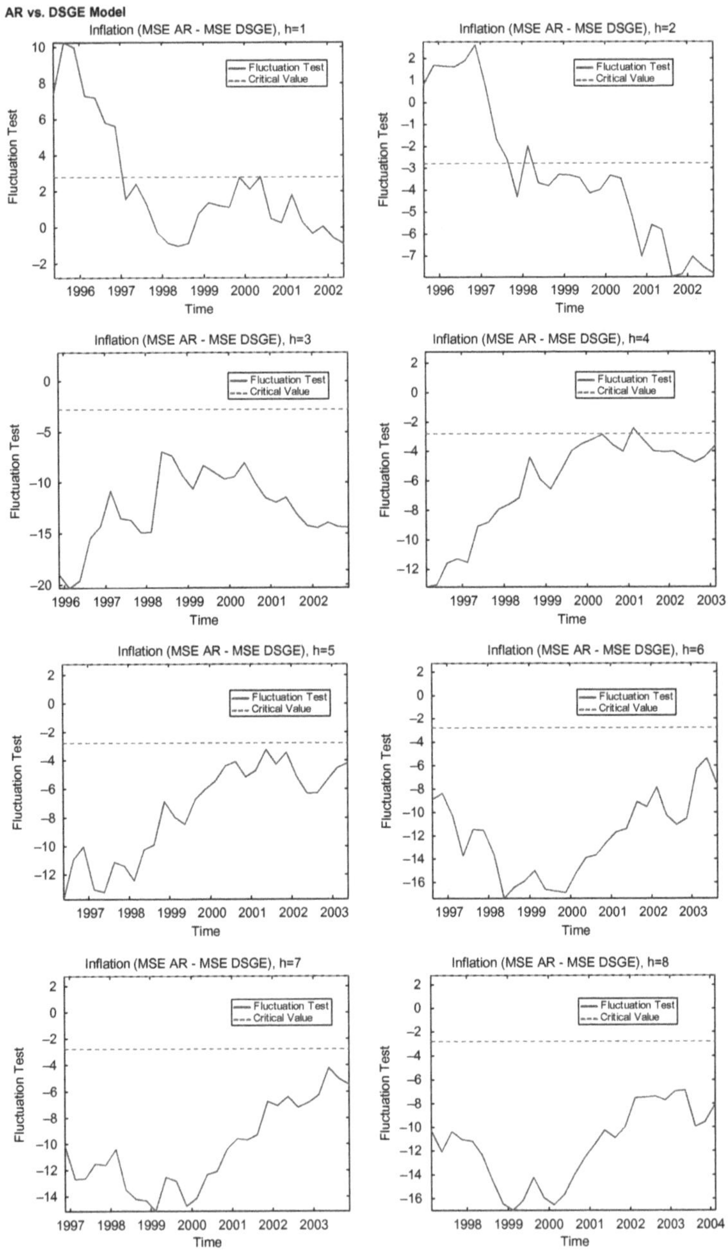

Fig. 3. Continued.

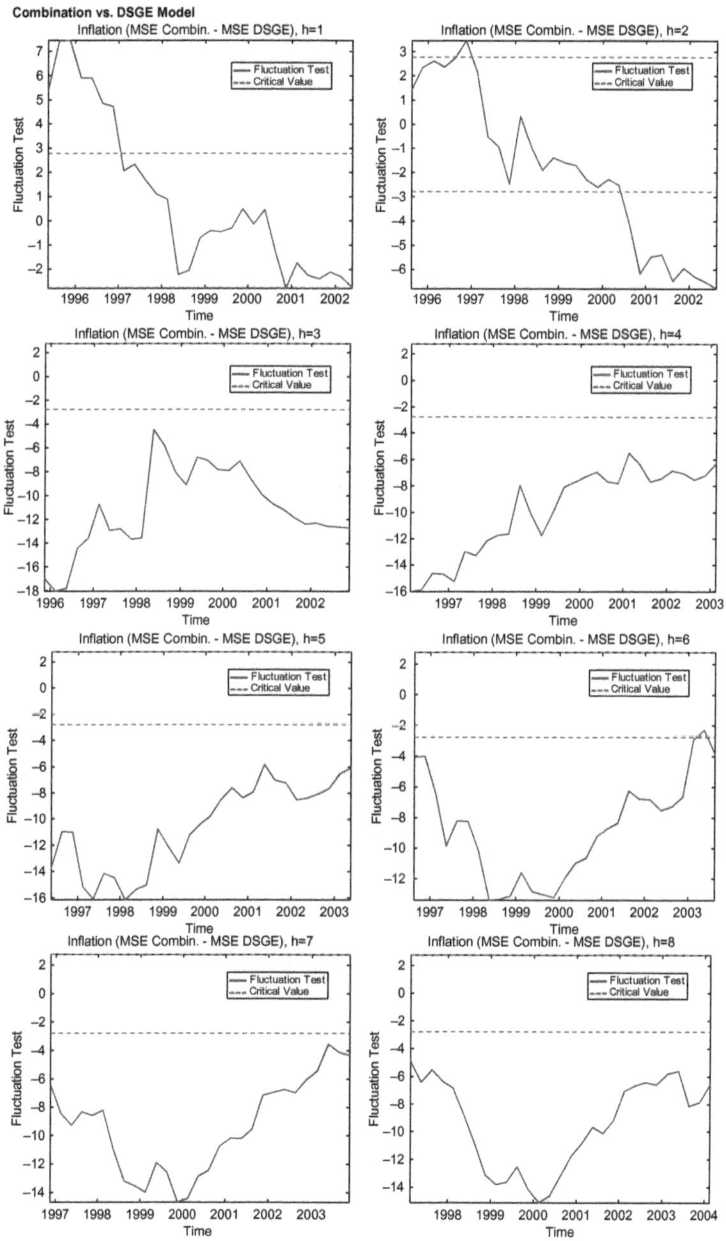

Fig. 3. *Continued.*

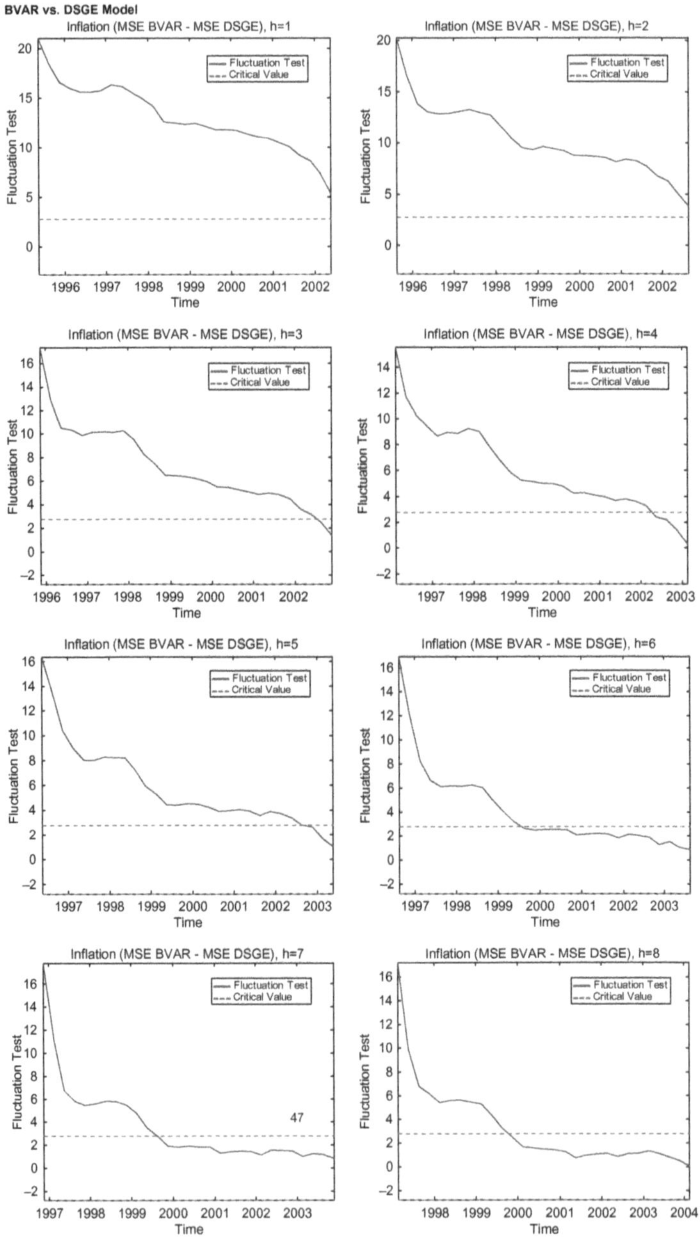

Fig. 3. Continued.

EMPIRICAL RESULTS ON FORECASTING INTEREST RATES

The results reported in the previous subsections show that none of the models produce efficient forecasts for output growth and inflation, with rare exceptions. Do these results carry over to interest rate forecasts?

We begin by showing interest rate forecasts of various models together with the realized interest rates in Fig. 4, which is the interest rate analogue of Fig. 1 for selected models.[14] To judge the quality of these forecasts, we then do the same tests we did for output growth and inflation. Table 3

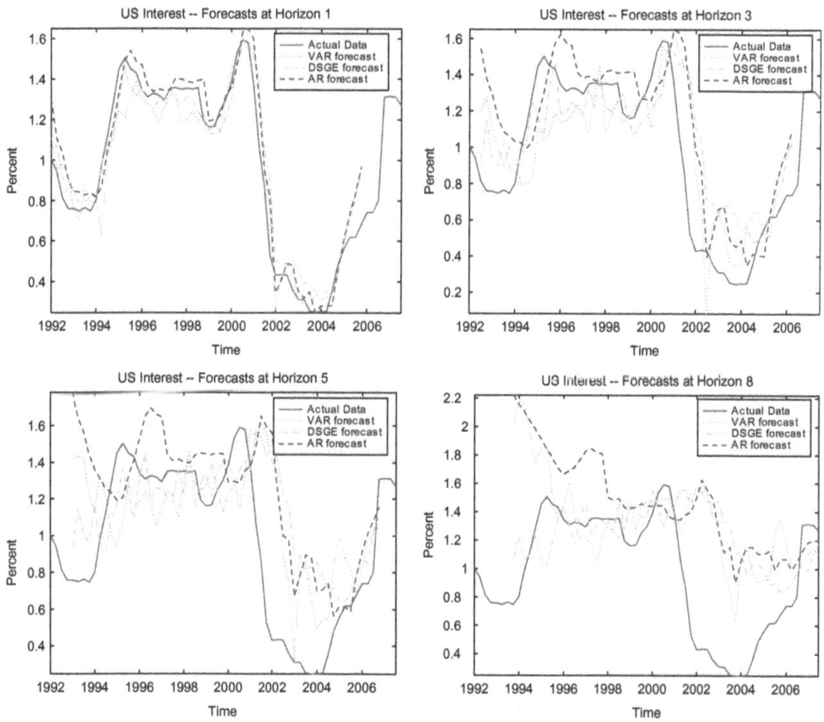

Fig. 4. Interest rate forecasts. *Notes*: The figure reports the realized value of the target variable (interest rates), labeled "Actual Data," together with forecasts of the VAR (labeled "VAR"), AR (labeled "AR"), Smets and Wouters (2007, labeled "DSGE"), Bayesian VAR (labeled "UCSV"), forecast combinations (labeled "Combin. Forecast") and the random walk (labeled "RW") models. Each graph reports results for different horizons, reported in each graph's title.

Table 3. Forecast Optimality Tests: Forecasting Interest Rates.

Panel A: VAR Model					
h	MZ p-value	$\hat{\alpha}$		$\hat{\beta}$	
1	0.66	−0.03	(0.68)	1.04	(0.69)
2	0.97	−0.02	(0.97)	1.02	(0.97)
3	0.76	0.00	(1.00)	0.95	(0.94)
4	0.36	0.01	(1.00)	0.89	(0.84)
5	0.17	0.05	(0.98)	0.81	(0.74)
6	0.05	0.20	(0.84)	0.65	(0.48)
7	0.01	0.39	(0.63)	0.47	(0.23)
8	0.00	0.43	(0.56)	0.42	(0.12)
Panel B: DSGE Model					
h	MZ p-value	$\hat{\alpha}$		$\hat{\beta}$	
1	0.00	−0.14	(0.00)	1.11	(0.03)
2	0.01	−0.24	(0.03)	1.18	(0.21)
3	0.04	−0.28	(0.26)	1.18	(0.62)
4	0.07	−0.26	(0.60)	1.12	(0.90)
5	0.11	−0.17	(0.89)	1.00	(1.00)
6	0.11	−0.03	(1.00)	0.85	(0.93)
7	0.08	0.19	(0.92)	0.66	(0.71)
8	0.05	0.53	(0.63)	0.37	(0.38)
Panel C: AR Model					
h	MZ p-value	$\hat{\alpha}$		$\hat{\beta}$	
1	0.01	−0.03	(0.48)	0.97	(0.58)
2	0.01	−0.05	(0.59)	0.94	(0.54)
3	0.01	−0.06	(0.83)	0.90	(0.56)
4	0.01	−0.05	(0.96)	0.85	(0.56)
5	0.00	−0.00	(1.00)	0.78	(0.53)
6	0.00	0.06	(0.98)	0.69	(0.43)
7	0.00	0.12	(0.94)	0.61	(0.33)
8	0.00	0.12	(0.96)	0.60	(0.32)
Panel D: Random Walk Model					
h	MZ p-value	$\hat{\alpha}$		$\hat{\beta}$	
1	0.68	0.02	(0.85)	0.97	(0.69)
2	0.38	0.08	(0.49)	0.90	(0.38)
3	0.20	0.17	(0.26)	0.81	(0.21)
4	0.12	0.27	(0.15)	0.70	(0.13)
5	0.06	0.40	(0.07)	0.58	(0.07)
6	0.03	0.51	(0.04)	0.48	(0.04)
7	0.01	0.62	(0.02)	0.38	(0.02)
8	0.01	0.71	(0.01)	0.29	(0.01)

Table 3. *(Continued)*

Panel E: Forecast Combinations Model

h	MZ p-value	$\hat{\alpha}$		$\hat{\beta}$	
1	0.01	−0.03	(0.44)	0.98	(0.74)
2	0.01	−0.05	(0.51)	0.96	(0.75)
3	0.02	−0.06	(0.75)	0.92	(0.70)
4	0.01	−0.05	(0.94)	0.87	(0.64)
5	0.01	−0.03	(0.99)	0.82	(0.66)
6	0.00	0.02	(1.00)	0.74	(0.57)
7	0.00	0.13	(0.94)	0.62	(0.38)
8	0.00	0.15	(0.93)	0.59	(0.32)

Panel F: Bayesian VAR Model

h	MZ p-value	$\hat{\alpha}$		$\hat{\beta}$	
1	0.00	0.38	(0.40)	0.33	(0.00)
2	0.00	0.38	(0.46)	0.33	(0.00)
3	0.00	0.38	(0.51)	0.34	(0.00)
4	0.00	0.39	(0.53)	0.34	(0.01)
5	0.00	0.44	(0.47)	0.32	(0.01)
6	0.00	0.50	(0.39)	0.30	(0.01)
7	0.00	0.56	(0.28)	0.27	(0.00)
8	0.00	0.61	(0.17)	0.24	(0.00)

Note: The table reports results for forecasting interest rates. Each panel reports, for several horizons (h, reported in the first column), the values of the Mincer and Zarnowitz (1969) forecast efficiency test (labeled "MZ p-value"), the estimate of $\hat{\alpha}$ in regression (5) together with the p-value of the test that the constant equals zero (in parentheses), and the estimate of $\hat{\beta}$ in regression (5) together with the p-value of the test that the slope equals zero (in parentheses).

shows that interest rate forecasts from the VAR and RW models are efficient at short horizons; the DSGE model interest rate forecasts are instead efficient at long but not at short horizons, according to the joint Mincer–Zarnowitz (1969) test. Fig. 5, however, shows that VAR, forecast combinations, and DSGE forecasts were efficient in the early 2000 but became less efficient toward the end of the sample. In contrast, BVAR forecasts were, instead, essentially never efficient.

Which model forecasts interest rates best? Table 4 shows that RW and combinations forecast interest rates more accurately than the DSGE model at the shortest horizons, whereas the DSGE model forecasts better at longer horizons (although results are not statistically significant).

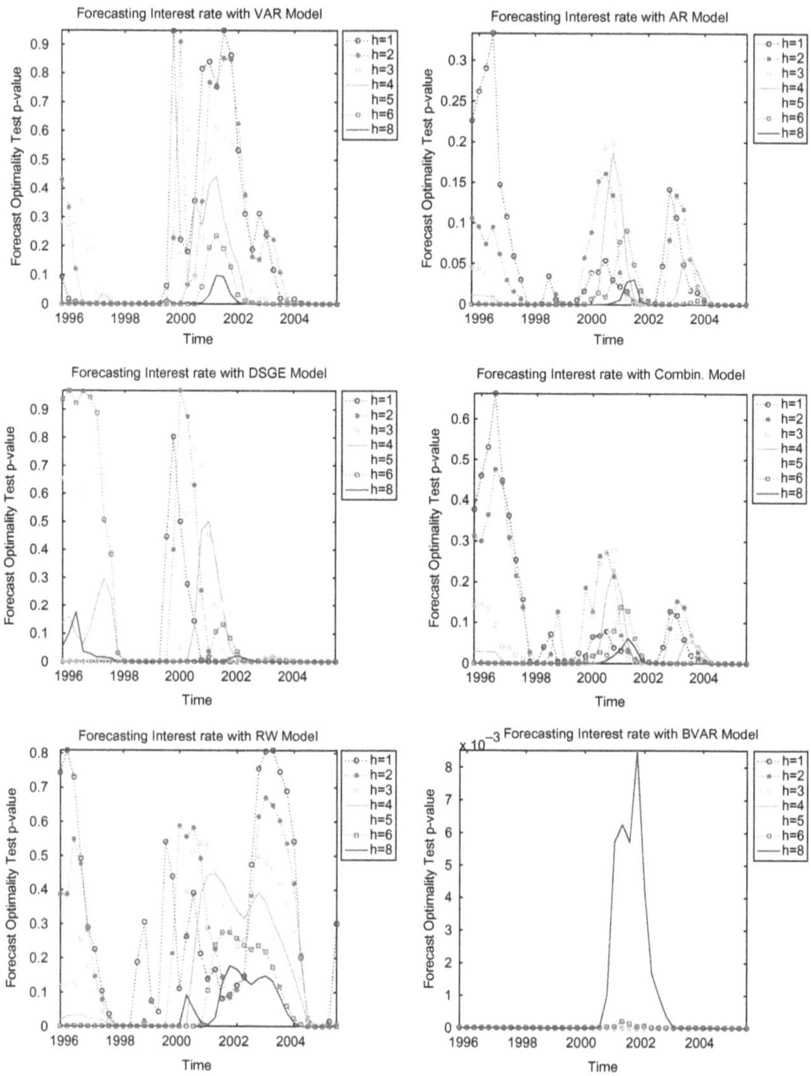

Fig. 5. Forecast optimality test over time: Interest rate. *Notes*: The figure reports the *p*-value of the Mincer and Zarnowitz (1969) test in Eq. (??) over time. The figure report results for forecasts at various horizons (see the legend) and several models (see the title). The target variable is the interest rate.

Table 4. Diebold and Mariano Test: Forecasting Interest Rates.

	RMSFE			DM: Model-DSGE	
h	DSGE	VAR	AR	VAR	AR
1	0.11	1.15	1.02	0.35	0.87
2	0.20	1.07	1.06	0.61	0.71
3	0.28	1.02	1.09	0.85	0.54
4	0.34	1.01	1.13	0.93	0.37
5	0.39	1.02	1.15	0.87	0.31
6	0.43	1.05	1.19	0.71	0.28
7	0.47	1.10	1.23	0.56	0.24
8	0.51	1.12	1.22	0.50	0.22

	RMSFE			DM: Model-DSGE		
h	FC	BVAR	RW	FC	BVAR	RW
1	0.95	8.99	0.93	0.64	0.02	0.40
2	0.99	4.88	0.95	0.93	0.03	0.54
3	1.03	3.41	0.97	0.84	0.04	0.77
4	1.08	2.68	0.98	0.62	0.06	0.90
5	1.09	2.25	1.01	0.57	0.08	0.93
6	1.12	1.96	1.02	0.49	0.10	0.89
7	1.17	1.76	1.02	0.38	0.13	0.91
8	1.18	1.72	1.01	0.34	0.14	0.96

Note: The columns labeled RMSFE in the table report the RMSFE of the DSGE model forecasts (labeled "DSGE") as well as the RMSFE of the autoregressive ("AR"), forecast combinations ("FC"), VAR ("VAR"), Bayesian VAR ("BVAR") and the random walk ("RW") relative to that of the DSGE model (e.g., the column labeled "VAR" reports the RMSFE of the VAR model divided by the RMSFE of the DSGE model). The columns labeled "DM" in the table report the *p*-value of the Diebold and Mariano (1996) test (calculated using Giacomini and White's (2006) critical values) for comparing the specified forecasts (e.g., the column labeled "DM: Model-DSGE" and "VAR" reports the *p*-value for comparing the forecast of the VAR with those of Smets and Wouters' (2007) model).

Fig. 6(a) shows that the interest rates' MSFEs of the DSGE model tends to be higher than those of the VAR model toward the beginning of the sample, signaling a worsening of forecast accuracy of the DSGE model forecasts. The differences in the MSFEs are statistically significant. Unreported results show that similar findings hold for the autoregressive model.[15] The BVAR instead is always significantly worse than the DSGE model (see Fig. 6(b)).

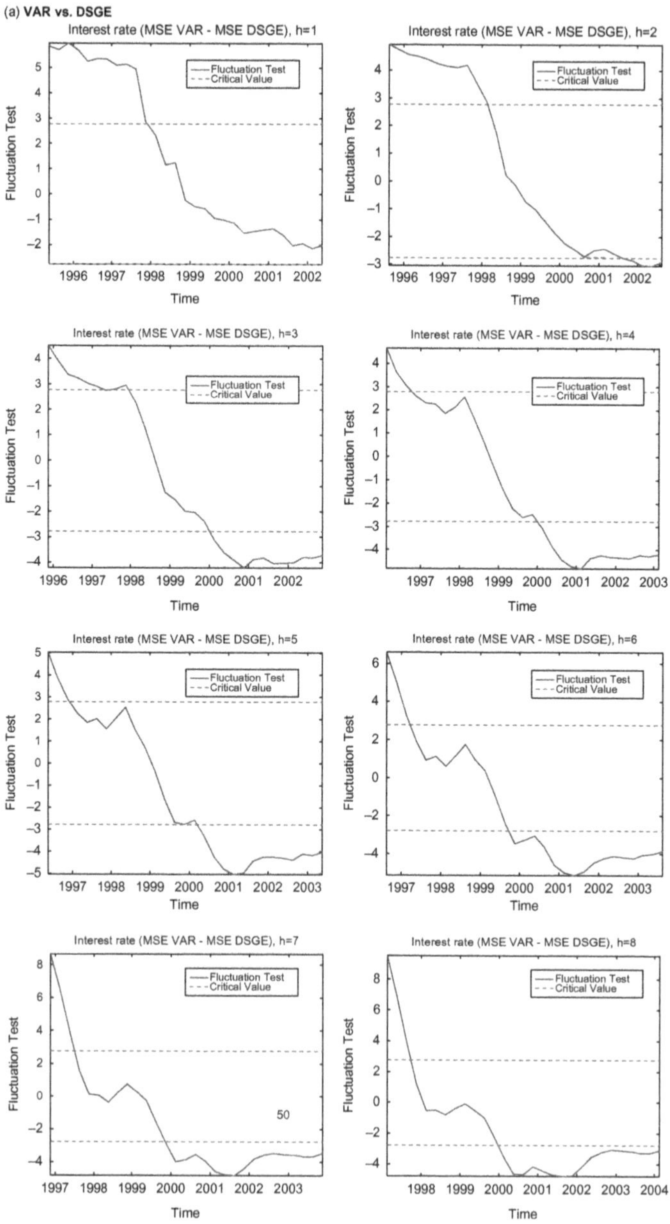

Fig. 6. Fluctuation test. Interest rate. *Notes*: The figure reports Giacomini and Rossi's (2010) fluctuation test. The figure reports results for forecasts at various horizons and several models (see the title). The target variable is the interest rate.

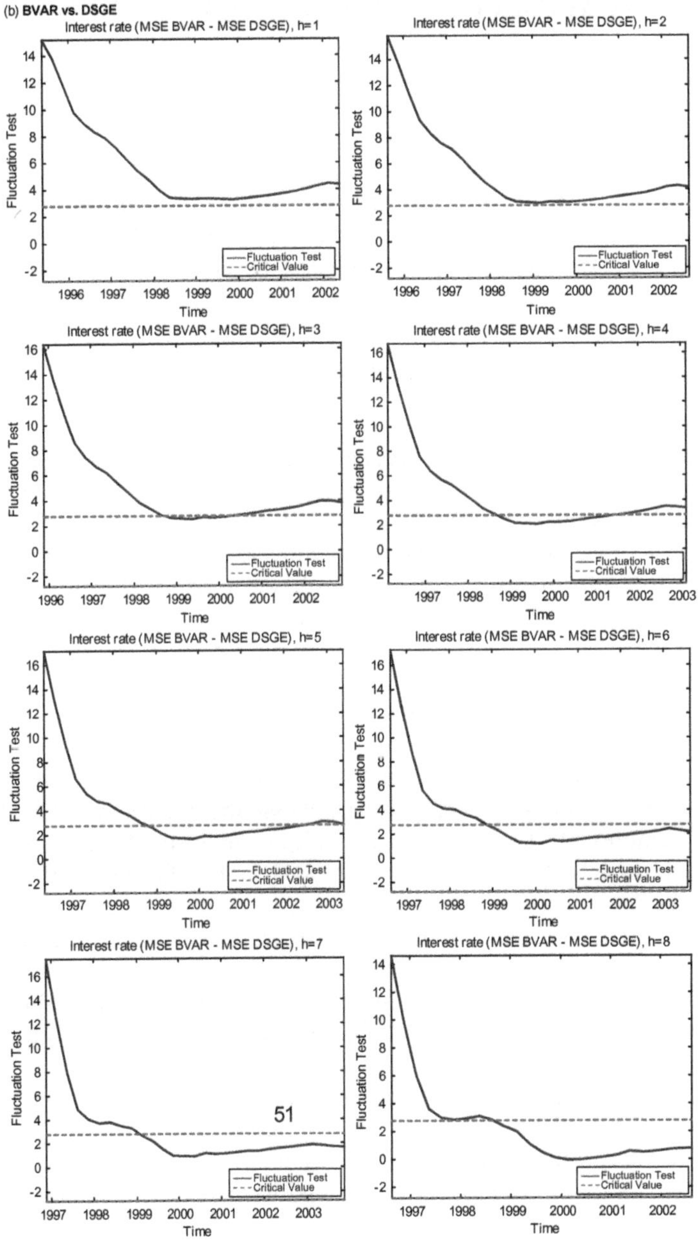

Fig. 6. Continued.

SUMMARY OF EMPIRICAL FINDINGS

In general, most models' forecasts are not efficient, no matter which model and time period we consider. When considering the models' relative forecasting performance, however, several interesting patterns emerge from the data. When forecasting output growth, DSGE models forecast better than the AR model and forecast combinations at longer horizons but slightly worse at shorter horizons, although the differences are not significant. When forecasting inflation the DSGE model forecasts better only at the shortest horizon, with AR, random walk, and the combined forecast all outperforming the DSGE model forecasts at longer horizons, significantly so at intermediate horizons. VAR and BVAR are always outperformed by the DSGE model for both output and inflation forecasting.

Interestingly, the relative forecasting performance does change over time. ARs forecast output growth better than the DSGE model in the late 1990s at shorter horizons, and significantly so; the DSGE model instead performs significantly better at longer horizons most of the time, although its performance worsens toward the end of the sample. Regarding inflation, we observe the opposite pattern. The forecasting performance of the VAR, AR, and forecast combination models improves over time relative to the DSGE model at most forecast horizons, and the DSGE model is significantly better only in the late 1990s, but not in the most recent period.

The finding that relative forecasting performance changes over time is related to Stock and Watson (2003) and Rossi and Sekhposyan (2010). The former demonstrate the existence of widespread instabilities in the parameters of models describing output growth and inflation in the United States; the latter show that most predictors for output growth lost their predictive ability in the mid-1970s, whereas, when forecasting inflation, fewer predictors are significant, and their predictive ability significantly worsened around the time of the Great Moderation. This article shows that the instabilities significantly affect DSGE and VAR models' performance as well.

Table 5 investigates how the empirical results would change using a recursive (rather than a rolling) window estimation scheme. The table shows that the results are qualitatively similar, using rolling or recursive data sets to estimate the models and make forecasts do not change the relative abilities of the models much.

Table 5. Comparisons of MSFEs: Rolling vs. Recursive Estimation.

Forecasting Output

Panel A: Recursive

h	DSGE	VAR	AR	FC	BVAR	RW
1	0.54	1.18	0.85	0.858	1.39	1.05
2	0.51	1.43	0.97	0.97	1.29	1.25
3	0.49	1.42	1.00	0.99	1.26	1.47
4	0.48	1.43	1.03	1.02	1.26	1.45
5	0.46	1.38	1.04	1.00	1.36	1.50
6	0.44	1.51	1.12	1.12	1.32	1.67
7	0.44	1.52	1.11	1.11	1.32	1.53
8	0.44	1.58	1.12	1.13	1.36	1.60

Panel B: Rolling

h	DSGE	VAR	AR	FC	BVAR	RW
1	0.54	1.09	0.94	0.95	1.50	1.06
2	0.46	1.31	1.11	1.10	1.58	1.40
3	0.46	1.41	1.16	1.14	1.41	1.37
4	0.46	1.46	1.15	1.14	1.31	1.43
5	0.48	1.24	0.98	1.00	1.33	1.48
6	0.48	1.39	1.09	1.11	1.09	1.40
7	0.50	1.45	1.04	1.07	1.05	1.37
8	0.51	1.59	1.06	1.13	1.11	1.45

Forecasting Inflation

Panel A: Recursive

h	DSGE	VAR	AR	FC	BVAR	RW
1	0.19	1.15	1.12	1.12	2.78	1.06
2	0.21	1.27	1.11	1.15	2.56	1.14
3	0.22	1.20	1.09	1.13	2.45	1.02
4	0.23	1.27	1.18	1.18	2.36	1.24
5	0.23	1.48	1.33	1.35	2.37	1.22
6	0.23	1.80	1.53	1.56	2.47	1.14
7	0.26	1.73	1.47	1.47	2.29	1.08
8	0.26	1.85	1.55	1.54	2.24	1.11

Panel B: Rolling

h	DSGE	VAR	AR	FC	BVAR	RW
1	0.19	1.12	1.06	1.03	2.66	1.06
2	0.23	1.05	0.93	0.95	2.25	1.06
3	0.26	0.95	0.80	0.81	1.95	0.92
4	0.26	1.09	0.83	0.82	1.80	0.83
5	0.28	1.20	0.85	0.87	1.70	0.85
6	0.29	1.38	0.90	0.94	1.69	0.88
7	0.30	1.31	0.89	0.92	1.63	0.92
8	0.31	1.37	0.88	0.91	1.56	0.84

Table 5. (*Continued*)

Forecasting Interest Rate

Panel A: Recursive

h	DSGE	VAR	AR	FC	BVAR	RW
1	0.14	0.72	0.90	0.81	6.94	0.72
2	0.21	0.90	1.16	1.07	4.44	1.28
3	0.28	0.94	1.20	1.11	3.15	1.39
4	0.33	1.00	1.24	1.17	2.51	1.38
5	0.38	1.01	1.29	1.21	2.14	1.39
6	0.42	1.07	1.37	1.28	1.94	1.42
7	0.45	1.13	1.42	1.33	1.78	1.41
8	0.48	1.16	1.42	1.34	1.68	1.40

Panel B: Rolling

h	DSGE	VAR	AR	FC	BVAR	RW
1	0.11	1.15	1.02	0.95	8.99	0.93
2	0.20	1.07	1.06	0.99	4.88	0.95
3	0.28	1.02	1.09	1.03	3.41	0.97
4	0.34	1.01	1.13	1.08	2.68	0.98
5	0.39	1.02	1.15	1.09	2.25	1.01
6	0.43	1.05	1.19	1.12	1.96	1.02
7	0.47	1.10	1.23	1.17	1.76	1.02
8	0.51	1.12	1.22	1.18	1.72	1.01

Note: The columns labeled RMSFE in the table report the RMSFE of the DSGE model forecasts (labeled "DSGE") as well as the RMSFE of the autoregressive ("AR"), forecast combinations ("FC"), VAR ("VAR"), Bayesian VAR ("BVAR") and the random walk ("RW") relative to that of the DSGE model (e.g., the column labeled "VAR" reports the RMSFE of the VAR model divided by the RMSFE of the DSGE model). Panel A reports results for the recursive estimation scheme and panel B for the rolling scheme.

CONCLUSIONS

In this article we studied the forecasting performance of several standard reduced-form models and the current baseline DSGE model for macroeconomic aggregates. While forecasting has gained importance in the applied literature, no comprehensive study of the merits of DSGE model forecasts and reduced-form model forecasts exists, nor have existing studies examined how the forecast accuracy of these models has evolved over time. Preliminary results in the literature suggested that, during the 1990s and the early 2000s, the DSGE model outperformed large-scale BVAR models.

This has led to the DSGE model forecast to be taken more seriously and prompted Del Negro and Schorfheide (2013) to call for the DSGE model forecast to be used as the benchmark for other forecasts' evaluation.

This article uses a variety of standard reduced-form models as well as the DSGE model to generate forecasts for the 1992–2007 sample period and studies how their forecast efficiency as well as their relative forecasting performance changed with sample periods and forecast horizons. We showed that all forecasting methods fail the "efficiency" test – their forecast errors are forecastable with current information at some forecast horizon or at some point in time. This means judgmental forecasters can always adjust these forecasts and have better outcomes, which is important to know for practitioners and policymakers who use these models' forecasts. For example, in principle, by adjusting the mean of the forecast with current quarter's information, judgmental forecasters might be able to improve the forecast bias; they might also have access to other information that is available in real time, although it is not included in the models that we consider (e.g., flash estimates of current economic conditions). However, it is not clear whether this approach would succeed in practice, as judgmental forecasts might be biased themselves and may add variability to the forecast.

The relative forecast performance, as measured by the relative RMSFE of various forecasts, is important to establish a benchmark non-judgmental forecast. We look at whether one of the forecasting methods routinely outperforms others as we change sample periods and forecasting horizons. We find that there is no single best forecasting method. The large-scale BVAR is almost never the best forecast, calling into question its use as the standard competitor for the estimated DSGE model forecast.[16] On the other hand, forecasts generated by the AR model outperform the DSGE model forecasts at different horizons for different macroeconomic aggregates and to varying degrees of statistical significance over different sample periods. This makes us conclude that among the current batch of standard forecasting methods, none qualifies as the forecast benchmark alone. Typically, AR models forecast better than VAR models, although not always.

A series of caveats are in order. First, we only consider a particular VAR with inflation, output growth, and the interest rate. Second, we use traditional forecast evaluation tests that do not take into account the real-time nature of the data. Third, we investigate large dimensional BVAR models; it might be that small dimensional BVAR models forecast better than large dimensional ones. However, we focus on the latter because they are currently used in the literature; in addition, BVARs have been introduced exactly to address the parameter proliferation in large-dimensional VARs.[17] We leave it to future work to study why certain reduced-form models' forecasts are better than the DSGE model forecast for some sample periods. It would be valuable to learn what features of the data favor what forecasting method. Perhaps it would be possible to select the most accurate forecasting method in real time. Finally, it may be that there is another forecasting method that consistently outperforms all others but is omitted in this study.

NOTES

1. See also Wieland and Wolters (2012) for an analysis of forecasting in policy making settings, Bache, Jore, Mitchell, and Vahey (2011) for an analysis of DSGE models' forecast densities, and Lees, Matheson, and Smith (2011) for an analysis of the forecasting performance of an open economy DSGE model for New Zealand.

2. For a discussion of iterated versus direct forecasts, see Marcellino, Stock, and Watson (2005), Ravazzolo and Rothman (2010), and Kilian and Vigfusson (2013).

3. Note, however, that even though both the Bayesian VAR and the DSGE models are estimated with Bayesian methods, their priors might be quite different. Note also that, if DSGE model satisfies the invertibility condition, it may be interpreted as a VAR model subject to cross-equation restrictions; however, it is not clear whether a VAR with a finite lag structure is a good approximation for the true underlying VAR.

4. That is, when the trend estimate does not use information from periods to be forecasted. This is an issue especially for models that use detrended data as the filter for detrending may be two sided and impart ex-post information into real-time data.

5. Recall that the out-of-sample output growth forecasts from the DSGE model are per capita. We transform them in aggregate values by adjusting for realized population growth. Note that the reduced-form models' forecast of output growth are aggregate, instead.

6. The number of lags used in the VAR and the autoregressive specifications is determined by the Bayesian information criterion (BIC), recursively applied each time the model is estimated. See the next subsection for details. Appendix A evaluates the robustness of the results when the AIC is used instead of the BIC.

7. Direct forecasts for the AR(p) model are obtained as follows. We estimate $\Delta y_t = a_0 + a_1 \Delta y_{t-h} + \ldots + a_p \Delta y_{t-p-h+1} + u_{AR,t,t+h}$ using information up to time t, and then use the parameter estimate and the realized values of $\Delta y_t, \ldots, \Delta y_{t-p}$ to forecast directly Δy_{t+h}.

8. It is possible that the empirical results might be different if another window size is chosen. See, for example, Inoue and Rossi (2012) and Hansen and Timmermann (2012) for a discussion and methods for preventing data mining over the estimation window size.

9. As we show in the section "Summary of Empirical Findings", results based on the recursive estimation scheme (i.e., using observations $1, \ldots, t$) are qualitatively similar.

10. That is, each time a VAR or an AR specification is estimated, the number of lags is estimated as well, according to the BIC.

11. Note that while the BIC does not necessarily select the best model, nor does the AIC. Under Inoue and Kilian's (2006) assumptions, the AIC is dominated by the BIC for one-step-ahead direct forecasts.

12. We select $m = P/2$, i.e., approximately 7 years for most forecast horizons.

13. Note that we always use a rolling window of 20 years in estimating the model and generating forecasts. But these forecasts themselves may be evaluated as a single sample or in different sub-samples using a rolling window. The window size of 7 years refers to the size of the rolling window used to smooth the forecast errors when we study the forecasts in sub-samples to see how the relative and absolute forecasting abilities of models changed over time.

14. Due to space constraints, we report results only for the VAR, DSGE, and AR models.

15. See Appendix C for detailed results.

16. An interesting avenue for research is to think about the priors used for the BVAR estimation. The Smets and Wouters priors shrink toward random walks, which may not be helping in inflation and output growth forecasting. As we show in this article, random walk is itself a poor forecast, especially for output growth.

17. Note also that Smets and Wouters (2007) shrink the priors toward a random walk. This may not be appropriate for inflation and GDP growth in the Great Moderation period. Another option would be to shrink toward a white noise prior, which might be better suited for this period.

18. Since the VAR model contains, by construction, more parameters than the AR model, we selected a more parsimonious maximum number of lags.

ACKNOWLEDGMENTS

We are grateful to Gergely Ganics and Lutz Kilian for comments on an earlier draft and to Yıldız Akkaya and Gülserim Özcan for research assistance.

REFERENCES

Adolfson, M., Linde, J., & Villani, M. (2007). Forecasting performance of an open economy dynamic stochastic general equilibrium model. *Econometric Reviews*, *26*(2–4), 289–328.

Bache, I. W., Jore, A. S., Mitchell, J., & Vahey, S. P. (2011). Combining VAR and DSGE forecast densities. *Journal of Economic Dynamics and Control*, *35*(10), 1659–1670.

Christoffel, K., Coenen, G., & Warne, A. (2011). Forecasting with DSGE Models. In M. P. Clements & D. F. Hendry (Eds.), *Handbook of economic forecasting* (pp. 89–127). Oxford, UK: Oxford University Press.

Del Negro, M., & Schorfheide, F. (2013). Forecasting with DSGE models: Theory and practice. In G. Elliott & A. Timmermann (Eds.), *Handbook of forecasting* (Vol. 2). Amsterdam, The Netherlands: Elsevier.

Diebold, F. X., & Mariano, R. S. (1995). Comparing predictive accuracy. *Journal of Business and Economic Statistics*, *13*, 253–263.

Edge, R. M., & Gürkaynak, R. S. (2010). How useful are estimated DSGE model forecasts for central bankers? *Brookings Papers on Economic Activity*, *41*(2), 209–259.

Edge, R. M., Gürkaynak, R. S., & Kısacıkoğlu, B. (2013). *Judging the DSGE model by its forecast*. mimeo, Bilkent University.

Edge, R. M., Kiley, M. T., & Laforte, J. P. (2010). A comparison of forecast performance between federal reserve staff forecasts, simple reduced-form models, and a DSGE model. *Journal of Applied Econometrics*, *25*(4), 720–754.

Giacomini, R., & Rossi, B. (2010). Forecast comparisons in unstable environments. *Journal of Applied Econometrics*, *25*(4), 595–620.

Giacomini, R., & Rossi, B. (2012). *Model comparisons in unstable environments*. mimeo, Univ. PompeuFabra.

Giacomini, R., & White, H. (2006). Tests of conditional predictive ability. *Econometrica*, *74*, 1545–1578.

Hansen, P. R., & Timmermann, A. (2012). *Choice of sample split in out-of-sample forecast evaluation*. mimeo, EUI.

Inoue, A., & Kilian, L. (2006). On the selection of forecasting models. *Journal of Econometrics*, *130*(2), 273–306.

Inoue, A., & Kilian, L. (2008). How useful is bagging in forecasting economic time series? A case study of U.S. consumer price inflation. *Journal of the American Statistical Association*, *103*, 511–522.

Inoue, A., & Rossi, B. (2012). Out-of-sample forecast tests robust to the choice of window size. *Journal of Business and Economic Statistics*, *30*(3), 432–453.

Kilian, L., & Vigfusson, R. J. (2013). Do oil prices help forecast U.S. real GDP? The role of nonlinearities and asymmetries. *Journal of Business and Economic Statistics*, *31*(1), 78–93.

Lees, K., Matheson, T., & Smith, C. (2011). Open economy forecasting with a DSGE-VAR: Head to head with the RBNZ published forecasts, *International Journal of Forecasting*, *27*(2), 512–528.

Marcellino, M., Stock, J. H., & Watson, M. W. (2005). A comparison of direct and iterated multistep AR methods for forecasting macroeconomic time series. *Journal of Econometrics*, *135*(1–2), 499–526.

Mincer, J., & Zarnowitz, V. (1969). The evaluation of economic forecasts. In J. Mincer (Ed.), *Economic forecasts and expectations* (pp. 81–111). New York, NY: National Bureau of Economic Research.

Ravazzolo, F., & Rothman, P. (2010). *Oil and U.S. GDP: A real-time out-of-sample examination*. Norges Bank Working Paper 2010/18.

Rossi, B. (2013). Advances in forecasting under model instability. In G. Elliott & A. Timmermann (Eds.), *Handbook of economic forecasting* (Vol. 2), Amsterdam: Elsevier.

Rossi, B., & Sekhposyan, T. (2010). Has models' forecasting performance changed over time, and when? *International Journal of Forecasting*, *26*(4).

Rossi, B., & Sekhposyan, T. (2012). *Forecast optimality tests in the presence of instabilities*: mimeo, Univ. PompeuFabra.

Smets, F., & Wouters, R. (2007). Shocks and frictions in U.S. business cycles: A Bayesian DSGE approach. *American Economic Review*, *97*(3), 586–607.

Stock, J. H., & Watson, M. W. (2001). Vector autoregressions. *Journal of Economic Perspectives*, *15*(4), 101–115.

Stock, J. H., & Watson, M. W. (2003). Forecasting output and inflation: The role of asset prices. *Journal of Economic Literature*, *41*(3), 788–829.

Stock, J. H., & Watson, M. W. (2004). Combination forecasts of output growth in a seven country data set. *Journal of Forecasting*, *23*, 405–430.

Stock, J. H., & Watson, M. W. (2007). Why has inflation become harder to forecast? *Journal of Money, Credit and Banking*, *39*(1), 3–34.

Timmermann, A. (2006). Forecast combinations. In G. Elliott, C. Granger, & A. Timmermann (Eds.), *Handbook of economic forecasting* (Vol. 1). Amsterdam, The Netherlands: North Holland Elsevier.

Wieland, V., & Wolters, M. H. (2012). Forecasting and policy making. In G. Elliott & A. Timmermann (Eds.), *Handbook of forecasting* (Vol. 2). Amsterdam, The Netherlands: Elsevier.

APPENDIX A

This appendix investigates how robust are the results in the main article when models estimated using the AIC rather than the BIC. The maximum lag length used in the information criterion procedure is the same for AIC and BIC and is three lags for the AR model and forecast combinations from AR models and two lags for the VAR model; the minimum lag length is one in every model.[18] We report several tables (Tables A1–A4) corresponding to those in the main text. To save space, we do not report the figures corresponding to those in the text, as they are very similar to those reported in the article.

Table A1. Forecast Optimality Tests: Forecasting Output Growth and Inflation.

Panel A: VAR Model												
	Output Growth Forecasts						Inflation Forecasts					
h	MZ p-value	$\hat{\alpha}$		$\hat{\beta}$			MZ p-value	$\hat{\alpha}$		$\hat{\beta}$		
1	0.00	0.69	(0.00)	0.14	(0.00)		0.00	0.34	(0.00)	0.28	(0.00)	
2	0.00	0.91	(0.00)	−0.07	(0.00)		0.00	0.42	(0.00)	0.13	(0.00)	
3	0.00	0.97	(0.00)	−0.12	(0.00)		0.00	0.35	(0.00)	0.26	(0.00)	
4	0.00	0.85	(0.00)	−0.02	(0.00)		0.00	0.41	(0.00)	0.13	(0.00)	
5	0.00	0.53	(0.00)	0.27	(0.00)		0.00	0.49	(0.00)	−0.01	(0.00)	
6	0.00	0.75	(0.00)	0.08	(0.00)		0.00	0.54	(0.00)	−0.09	(0.00)	
7	0.00	0.94	(0.00)	−0.11	(0.00)		0.00	0.51	(0.00)	−0.04	(0.00)	
8	0.00	1.24	(0.00)	−0.39	(0.00)		0.00	0.50	(0.00)	−0.02	(0.00)	
Panel B: AR Model												
	Output Growth Forecasts						InflationForecasts					
h	MZ p-value	$\hat{\alpha}$		$\hat{\beta}$			MZ p-value	$\hat{\alpha}$		$\hat{\beta}$		
1	0.00	0.53	(0.05)	0.33	(0.01)		0.03	0.27	(0.06)	0.42	(0.04)	
2	0.00	1.06	(0.00)	−0.25	(0.00)		0.01	0.38	(0.01)	0.20	(0.01)	
3	0.00	1.36	(0.00)	−0.59	(0.00)		0.00	0.35	(0.00)	0.27	(0.00)	
4	0.00	1.22	(0.00)	−0.44	(0.00)		0.00	0.38	(0.00)	0.21	(0.00)	
5	0.08	0.62	(0.12)	0.23	(0.09)		0.00	0.57	(0.00)	−0.17	(0.00)	
6	0.00	0.91	(0.00)	−0.10	(0.00)		0.00	0.65	(0.00)	−0.30	(0.00)	
7	0.00	0.90	(0.00)	−0.08	(0.00)		0.00	0.66	(0.00)	−0.29	(0.00)	
8	0.00	0.98	(0.00)	−0.16	(0.00)		0.00	0.62	(0.00)	−0.22	(0.00)	

Table A1. *(Continued)*

Panel C: Forecast Combinations Model

	Output Growth Forecasts					Inflation Forecasts				
h	MZ p-value	$\hat{\alpha}$		$\hat{\beta}$		MZ p-value	$\hat{\alpha}$		$\hat{\beta}$	
1	0.00	0.53	(0.05)	0.33	(0.00)	0.00	0.26	(0.04)	0.44	(0.01)
2	0.00	0.98	(0.00)	−0.16	(0.00)	0.00	0.41	(0.01)	0.14	(0.00)
3	0.00	1.18	(0.00)	−0.37	(0.00)	0.00	0.35	(0.00)	0.26	(0.00)
4	0.00	1.06	(0.00)	−0.26	(0.00)	0.00	0.35	(0.00)	0.26	(0.00)
5	0.03	0.51	(0.12)	0.35	(0.05)	0.00	0.50	(0.00)	−0.03	(0.00)
6	0.00	0.85	(0.00)	−0.03	(0.00)	0.00	0.68	(0.00)	−0.33	(0.00)
7	0.00	0.92	(0.00)	−0.10	(0.00)	0.00	0.67	(0.00)	−0.30	(0.00)
8	0.00	1.06	(0.00)	−0.25	(0.00)	0.00	0.57	(0.00)	−0.15	(0.00)

Note: The table reports results for forecasting output (on the left) and inflation (on the right). Each panel reports, for several horizons (h, reported in the first column), the values of the Mincer and Zarnowitz (1969) forecast efficiency test (labeled "MZ p-value"), the estimate of $\hat{\alpha}$ in regression (5) together with the p-value of the test that the constant equals zero (in parentheses), and the estimate of $\hat{\beta}$ in regression (5) together with the p-value of the test that the slope equals zero (in parentheses).

Table A2. The Diebold and Mariano Test: Forecasting Output Growth and Inflation.

	Output					Inflation				
	RMSFE			DM: Model-DSGE		RMSFE			DM: Model-DSGE	
h	DSGE	VAR	AR	VAR	AR	DSGE	VAR	AR	VAR	AR
1	0.54	1.04	0.90	0.78	0.34	0.19	1.13	1.03	0.25	0.72
2	0.46	1.35	1.15	0.11	0.10	0.23	1.06	0.92	0.55	0.15
3	0.46	1.47	1.17	0.20	0.23	0.26	0.93	0.82	0.22	0.00
4	0.46	1.61	1.21	0.19	0.08	0.26	1.09	0.83	0.41	0.00
5	0.48	1.33	1.04	0.16	0.69	0.28	1.20	0.87	0.28	0.01
6	0.48	1.41	1.17	0.21	0.09	0.29	1.38	0.91	0.19	0.16
7	0.50	1.41	1.18	0.02	0.10	0.30	1.30	0.90	0.27	0.02
8	0.51	1.40	1.22	0.01	0.13	0.31	1.37	0.89	0.25	0.02

	Output						Inflation					
	RMSFE			DM: Model-DSGE			RMSFE			DM: Model-DSGE		
h	FC	BVAR	RW	FC	BVAR	RW	FC	BVAR	RW	FC	BVAR	RW
1	0.95	1.50	1.06	0.64	0.01	0.51	1.03	2.66	1.06	0.64	0.00	0.44
2	1.11	1.58	1.40	0.25	0.00	0.00	0.95	2.25	1.06	0.39	0.01	0.50

Table A2. (*Continued*)

	Output						Inflation					
	RMSFE			DM: Model-DSGE			RMSFE			DM: Model-DSGE		
h	FC	BVAR	RW	FC	BVAR	RW	FC	BVAR	RW	FC	BVAR	RW
3	1.14	1.41	1.37	0.31	0.03	0.01	0.83	1.95	0.92	0.00	0.02	0.31
4	1.15	1.31	1.43	0.17	0.02	0.02	0.83	1.80	0.83	0.00	0.03	0.07
5	1.00	1.33	1.48	1.00	0.06	0.00	0.87	1.70	0.85	0.01	0.04	0.15
6	1.12	1.09	1.40	0.21	0.51	0.02	0.95	1.69	0.88	0.51	0.09	0.34
7	1.08	1.05	1.37	0.32	0.63	0.01	0.92	1.63	0.92	0.20	0.12	0.53
8	1.13	1.11	1.45	0.20	0.33	0.00	0.91	1.56	0.84	0.14	0.13	0.17

Note: The columns labeled RMSFE in the table report the RMSFE of the DSGE model forecasts (labeled "DSGE") as well as the RMSFE of the autoregressive ("AR"), forecast combinations ("FC"), VAR ("VAR"), Bayesian VAR ("BVAR"), and the random walk ("RW") relative to that of the DSGE model (e.g., the column labeled "VAR" reports the RMSFE of the VAR model divided by the RMSFE of the DSGE model). The columns labeled "DM" in the table report the *p*-value of the Diebold and Mariano (1996) test (calculated using Giacomini and White's (2006) critical values) for comparing the specified forecasts (e.g., the column labeled "DM: Model-DSGE" and "VAR" reports the *p*-value for comparing the forecast of the VAR with those of Smets and Wouters' (2007) model).

Table A3. Forecast Optimality Tests: Forecasting Interest Rates.

	Panel A: VAR Model				
h	MZ *p*-value	$\hat{\alpha}$		$\hat{\beta}$	
1	0.57	−0.04	(0.59)	1.04	(0.61)
2	0.95	−0.02	(0.95)	1.02	(0.96)
3	0.69	0.01	(1.00)	0.94	(0.88)
4	0.32	0.02	(0.99)	0.87	(0.78)
5	0.16	0.05	(0.98)	0.81	(0.72)
6	0.04	0.20	(0.84)	0.65	(0.46)
7	0.00	0.39	(0.63)	0.47	(0.23)
8	0.00	0.44	(0.55)	0.41	(0.12)
	Panel B: Forecast Combinations Model				
h	MZ *p*-value	$\hat{\alpha}$		$\hat{\beta}$	
1	0.01	−0.03	(0.44)	0.98	(0.75)
2	0.02	−0.05	(0.51)	0.96	(0.76)
3	0.02	−0.06	(0.76)	0.92	(0.70)
4	0.01	−0.05	(0.94)	0.87	(0.65)
5	0.01	−0.03	(0.99)	0.82	(0.66)

Table A3. *(Continued)*

Panel B: Forecast Combinations Model					
h	MZ *p*-value	$\hat{\alpha}$		$\hat{\beta}$	
6	0.00	0.02	(1.00)	0.74	(0.57)
7	0.00	0.14	(0.93)	0.62	(0.37)
8	0.00	0.16	(0.92)	0.58	(0.31)

Note: The table reports results for forecasting interest rates. Each panel reports, for several horizons (*h*, reported in the first column), the values of the Mincer and Zarnowitz (1969) forecast efficiency test (labeled "MZ *p*-value"), the estimate of $\hat{\alpha}$ in regression (5) together with the *p*-value of the test that the constant equals zero (in parentheses), and the estimate of $\hat{\beta}$ in regression (5) together with the *p*-value of the test that the slope equals zero (in parentheses).

Table A4. Diebold and Mariano Test: Forecasting Interest Rates.

	RMSFE			DM: Model-DSGE		RMSFE			DM: Model-DSGE		
h	DSGE	VAR	AR	VAR	AR	FC	BVAR	RW	FC	BVAR	RW
1	0.11	1.09	1.00	0.63	0.99	0.95	8.99	0.93	0.64	0.02	0.40
2	0.20	1.05	1.05	0.75	0.75	0.99	4.88	0.95	0.93	0.03	0.54
3	0.28	1.03	1.09	0.84	0.55	1.03	3.41	0.97	0.84	0.04	0.77
4	0.34	1.02	1.14	0.88	0.37	1.08	2.68	0.98	0.62	0.06	0.90
5	0.39	1.02	1.16	0.88	0.29	1.09	2.25	1.01	0.57	0.08	0.93
6	0.43	1.06	1.19	0.70	0.27	1.12	1.96	1.02	0.49	0.10	0.89
7	0.47	1.10	1.23	0.55	0.24	1.17	1.76	1.02	0.38	0.13	0.91
8	0.51	1.12	1.22	0.49	0.22	1.18	1.72	1.01	0.35	0.14	0.96

Note: The columns labeled RMSFE in the table report the RMSFE of the DSGE model forecasts (labeled "DSGE") as well as the RMSFE of the autoregressive ("AR"), forecast combinations ("FC"), VAR ("VAR"), Bayesian VAR ("BVAR") and the random walk ("RW") relative to that of the DSGE model (e.g., the column labeled "VAR" reports the RMSFE of the VAR model divided by the RMSFE of the DSGE model). The columns labeled "DM" in the table report the *p*-value of the Diebold and Mariano (1996) test (calculated using Giacomini and White's (2006) critical values) for comparing the specified forecasts (e.g., the column labeled "DM: Model-DSGE" and "VAR" reports the *p*-value for comparing the forecast of the VAR with those of Smets and Wouters' (2007) model.

APPENDIX B

This appendix investigates how robust the AR and VAR results in the main article are to obtaining the forecasts by iteration. We report several tables corresponding to those in the main text. To save space, we do not report the figures corresponding to those in the text, as they are very similar to those reported in the article.

Overall, qualitatively, forecasts from iterated VAR and AR models behave similarly: typically, when one has a MSFE lower than that of the DSGE model, the other does as well. In addition, iterated VAR and AR forecasts, like their direct counterparts, are not efficient.

Specifically, Tables B1–B3 show that, when forecasting output growth, the iterated VAR forecasts have higher MSFE than the direct forecast (except for $h = 8$) and the iterated AR forecasts have a higher MSFE than the direct AR forecasts at all horizons. When forecasting inflation, the iterated VAR produces forecasts that have higher MSFEs for short to medium horizons (up to four quarters) and vice versa for medium to long horizons (longer than four quarters); similarly, iterated AR forecasts have higher MSFEs for short to medium horizons (up to two quarters) and vice versa for medium to long horizons (three quarters or longer). When forecasting interest rates, the results depend on the forecast horizon.

Table B1. Iterated VAR Forecasts: Forecast Optimality Tests.

	Output Growth Forecasts					Inflation Forecasts				
h	MZ p-value	$\hat{\alpha}$		$\hat{\beta}$		MZ p-value	$\hat{\alpha}$		$\hat{\beta}$	
1	0.00	0.55	(0.10)	0.24	(0.00)	0.00	0.33	(0.00)	0.30	(0.00)
2	0.00	1.05	(0.00)	−0.18	(0.00)	0.00	0.43	(0.00)	0.10	(0.00)
3	0.00	1.01	(0.00)	−0.14	(0.00)	0.00	0.41	(0.00)	0.15	(0.00)
4	0.00	0.80	(0.00)	0.03	(0.00)	0.00	0.43	(0.00)	0.10	(0.00)
5	0.00	0.31	(0.06)	0.42	(0.00)	0.00	0.50	(0.00)	−0.02	(0.00)
6	0.00	0.62	(0.00)	0.16	(0.00)	0.00	0.59	(0.00)	−0.17	(0.00)
7	0.00	0.74	(0.00)	0.08	(0.00)	0.00	0.60	(0.00)	−0.18	(0.00)
8	0.00	0.83	(0.00)	0.01	(0.00)	0.00	0.56	(0.00)	−0.12	(0.00)

	Interest Rate Forecasts				
h	MZ p-value	$\hat{\alpha}$		$\hat{\beta}$	
1	0.66	−0.03	(0.68)	1.04	(0.69)
2	0.92	−0.03	(0.93)	1.02	(0.97)
3	0.82	0.03	(0.97)	0.93	(0.88)
4	0.53	0.12	(0.80)	0.82	(0.62)

Table B1. (*Continued*)

	Interest Rate Forecasts				
h	MZ p-value	$\hat{\alpha}$		$\hat{\beta}$	
5	0.32	0.22	(0.64)	0.70	(0.42)
6	0.17	0.34	(0.47)	0.57	(0.26)
7	0.08	0.45	(0.35)	0.47	(0.16)
8	0.04	0.54	(0.25)	0.38	(0.09)

Table B2. Iterated AR Forecasts: Forecast Optimality Tests.

	Output Growth Forecasts					Inflation Forecasts				
h	MZ p-value	$\hat{\alpha}$		$\hat{\beta}$		MZ p-value	$\hat{\alpha}$		$\hat{\beta}$	
1	0.00	0.63	(0.00)	0.22	(0.00)	0.01	0.30	(0.03)	0.36	(0.01)
2	0.00	0.90	(0.00)	−0.08	(0.00)	0.01	0.40	(0.02)	0.15	(0.01)
3	0.00	1.18	(0.00)	−0.40	(0.00)	0.07	0.26	(0.18)	0.43	(0.11)
4	0.00	1.10	(0.00)	−0.31	(0.00)	0.01	0.22	(0.19)	0.48	(0.05)
5	0.33	0.66	(0.34)	0.20	(0.37)	0.00	0.42	(0.00)	0.12	(0.00)
6	0.00	1.11	(0.00)	−0.34	(0.00)	0.00	0.64	(0.00)	−0.27	(0.00)
7	0.00	1.16	(0.00)	−0.39	(0.00)	0.00	0.60	(0.00)	−0.18	(0.00)
8	0.00	1.12	(0.00)	−0.33	(0.00)	0.00	0.42	(0.03)	0.10	(0.00)

	Interest rate Forecasts				
h	MZ p-value	$\hat{\alpha}$		$\hat{\beta}$	
1	0.01	−0.03	(0.48)	0.97	(0.58)
2	0.01	−0.04	(0.73)	0.94	(0.51)
3	0.02	−0.03	(0.95)	0.89	(0.41)
4	0.01	0.00	(1.00)	0.83	(0.36)
5	0.01	0.05	(0.96)	0.77	(0.31)
6	0.01	0.11	(0.87)	0.70	(0.24)
7	0.00	0.17	(0.79)	0.65	(0.20)
8	0.00	0.21	(0.74)	0.60	(0.16)

Table B3. Iterated AR and VAR Forecasts: The Diebold and Mariano Test.

	Output					Inflation				
	RMSFE			DM: Model-DSGE		RMSFE			DM: Model-DSGE	
h	DSGE	VAR	AR	VAR	AR	DSGE	VAR	AR	VAR	AR
1	0.54	1.09	0.94	0.45	0.61	0.19	1.12	1.06	0.31	0.46
2	0.46	1.40	1.12	0.02	0.18	0.23	1.07	0.94	0.44	0.40
3	0.46	1.44	1.16	0.04	0.25	0.26	0.97	0.79	0.61	0.00
4	0.46	1.48	1.16	0.02	0.17	0.26	1.01	0.78	0.83	0.00
5	0.48	1.31	1.02	0.01	0.89	0.28	1.07	0.81	0.57	0.00
6	0.48	1.45	1.12	0.07	0.24	0.29	1.15	0.86	0.35	0.09
7	0.50	1.45	1.10	0.12	0.22	0.30	1.13	0.81	0.39	0.01
8	0.51	1.47	1.08	0.17	0.32	0.31	1.11	0.78	0.44	0.00

	Interest Rate				
	RMSFE			DM	
h	DSGE	VAR	AR	VAR-DSGE	AR-DSGE
1	0.11	1.15	1.02	0.35	0.87
2	0.20	1.04	1.00	0.73	0.97
3	0.28	1.04	1.03	0.74	0.78
4	0.34	1.03	1.05	0.82	0.66
5	0.39	1.01	1.05	0.91	0.64
6	0.43	1.00	1.04	0.97	0.66
7	0.47	0.98	1.02	0.84	0.78
8	0.51	0.96	1.00	0.60	0.98

APPENDIX C

Fig. C1 reports the fluctuation test for comparing the AR model with the DSGE model when forecasting interest rates.

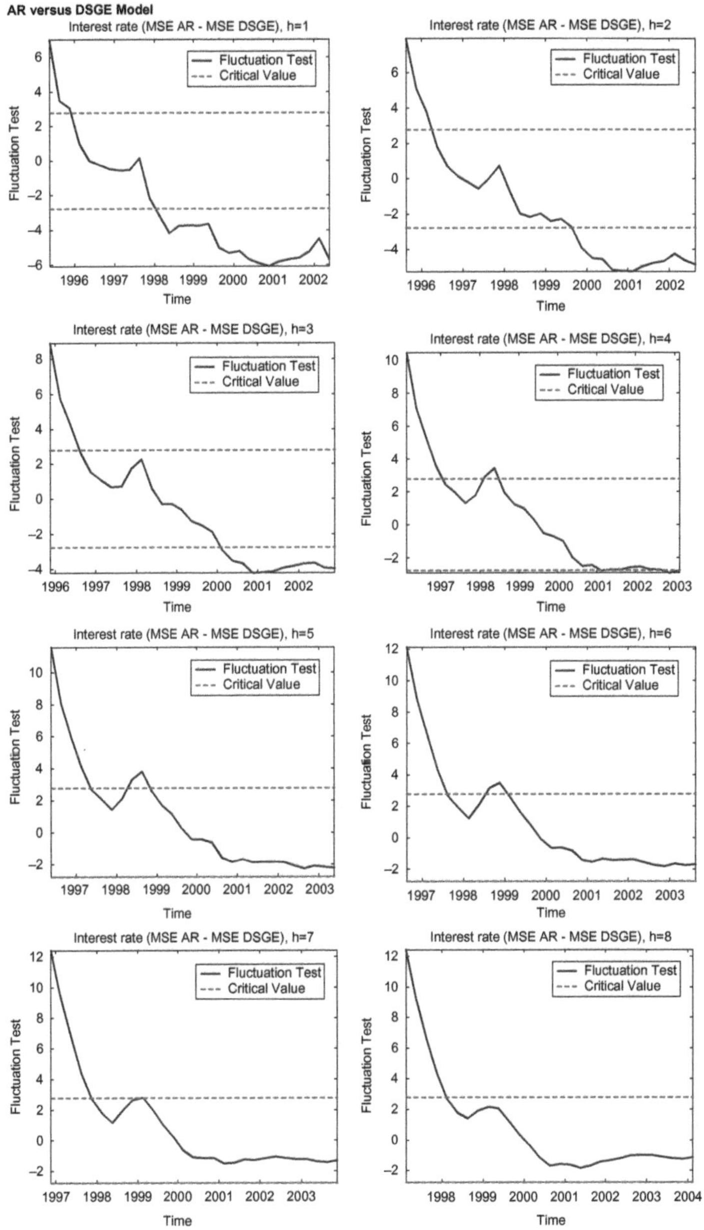

Fig. C1. Fluctuation test: Interest rate.

UNIT ROOTS, COINTEGRATION, AND PRETESTING IN VAR MODELS ☆

Nikolay Gospodinov, Ana María Herrera and Elena Pesavento

ABSTRACT

This article investigates the robustness of impulse response estimators to near unit roots and near cointegration in vector autoregressive (VAR) models. We compare estimators based on VAR specifications determined by pretests for unit roots and cointegration as well as unrestricted VAR specifications in levels. Our main finding is that the impulse response estimators obtained from the levels specification tend to be most robust when the magnitude of the roots is not known. The pretest specification works well only when the restrictions imposed by the model are satisfied. Its performance deteriorates even for small deviations from the exact unit root for one or more model variables. We illustrate the practical

☆The views expressed here are the authors and not necessarily those of the Federal Reserve Bank of Atlanta or the Federal Reserve System.

VAR Models in Macroeconomics – New Developments and Applications: Essays in Honor of Christopher A. Sims
Advances in Econometrics, Volume 32, 81–115

ISSN: 0731-9053/doi:10.1108/S0731-9053(2013)0000031003

relevance of our results through simulation examples and an empirical application.

Keywords: Impulse response functions; structural VAR; short/long-run identification; pretesting; unit roots; cointegration

JEL classifications: C32; C51; E52

INTRODUCTION

Most economic variables that are used for policy analysis and forecasting are characterized by high persistence and possibly nonstationary behavior. It is common practice in applied work to subject these series to pretests for unit roots and cointegration prior to the vector autoregressive (VAR) analysis to determine the appropriate transformations that render the data stationary. Incorporating information about the integration and cointegration properties of the data in VAR models reduces the estimation uncertainty and the degree of small-sample bias of impulse response estimates. Such pretests, however, suffer from low power in the neighborhood of the null hypothesis. In fact unit root pretests lack robustness to small deviations from unit roots and cointegration as they do not reject the null with probability approaching one for near-integrated processes that are parameterized as local-to-unity. For example, Gospodinov, Maynard, and Pesavento (2011) demonstrate that arbitrarily small deviations from an exact unit root may produce impulse response estimators that are highly distorted and misleading. The main goal of this article is to study and quantify the role of pretests for unit roots and cointegration for VAR specifications and identification schemes for impulse response analysis. We contrast the pretest approach to the specification of VAR models with the alternative of expressing all VAR variables in levels. The latter practice, which is also common in applied work, is susceptible to problems of small-sample bias, especially when the data are highly persistent. Moreover, it requires appropriate choices of bootstrap or asymptotic methods for inference that are developed to accommodate near-integrated processes. However, the levels specification makes the estimates of structural impulse responses, identified by short-run restrictions, asymptotically robust to the unknown order of integration of the model variables. The relative merits of the pretest VAR approach and the levels approach for VAR impulse response analysis are not well understood. The objective of this article is to examine this issue and to provide users of VAR models with advice on how to proceed in practice.

The limitations of the pretest approach to VAR specification have been discussed previously in the literature. Elliott (1998) illustrated the possibly large size distortions of the cointegration methods that arise in systems with near unit roots. It is widely documented that similar distortions tend to characterize the properties of sequential modeling and specification procedures that are based on pretests for unit roots. These tests are known to have low power in rejecting the null of a unit root when the data are highly persistent and nearly integrated. Nevertheless, Elliott's (1998) warning has been ignored by many applied researchers who continue to rely on pretests, perhaps because of the perception that the levels specification is not without its own potential drawbacks. This article examines how severe the problem documented by Elliott (1998) is for estimators of VAR impulse responses and which of the commonly used approaches to determining the VAR specification is most accurate in practice. We conduct a Monte Carlo study that assesses the mean squared errors (MSEs) of selected impulse response estimates and quantify how dangerous the pretest strategy can be in practice. The pretests include individual unit root tests such as the augmented Dickey–Fuller test with GLS detrending (ADF-GLS) and tests for cointegrating rank, which determine whether the model should be estimated as a VAR in differences, a vector error-correction model (VECM) or a VAR in levels. Our simulation results provide support for the robustness of the level VAR specification to departures from exact unit roots and cointegration and to the possible presence of unmodeled low frequency co-movement among the variables of interest.

Our evidence in favor of the levels specification highlights the importance of valid inference on impulse response estimators when there is uncertainty about the integration and cointegration properties of the data. We provide a brief review of several asymptotic and bootstrap approaches and methods for inference in levels models including local-to-unity asymptotics (Gospodinov, 2004, 2010; Mikusheva, 2012; Pesavento & Rossi, 2006; Phillips, 1998), delta method asymptotics (Sims, Stock, & Watson, 1990), and bootstrap (Inoue & Kilian, 2002; Kilian, 1998a). We show that the best approach may differ depending on the properties of the data, the dimensionality of the system, the forecast horizon, the identification scheme, etc.

The remainder of this article is structured as follows. The second section provides a short description of various representations of multivariate integrated processes. The third section introduces the common trend decomposition for nearly integrated process and discusses some theoretical results for impulse response analysis with nearly integrated (and possibly cointegrated) processes based on short-run and long-run identifying restrictions. The fourth section reviews the theoretical literature on inference for impulse response functions which is robust to the magnitude of the largest roots of the process. The fifth section presents the results of several simulation exercises. The sixth section concludes.

A BRIEF REVIEW OF REPRESENTATIONS OF MULTIVARIATE INTEGRATED PROCESSES

Let y_t be an $I(1)$ multivariate $(m \times 1)$ process. Since Δy_t is $I(0)$ by assumption, Δy_t has a moving average (MA) representation of the form

$$\Delta y_t = C(L)u_t \tag{1}$$

where $C(L) = C_0 + C_1 L + C_2 L^2 + \ldots$ is a $m \times m$ matrix lag polynomial and $u_t \sim iid(0, \Omega)$. Then, Δy_t has a long-run covariance matrix $C(1)\Omega C(1)'$.

For any $\beta \neq 0_{m\times r}$ $(r = 1, \ldots, m)$, the long-run variance of $\beta' \Delta y_t$ is $\beta' C(1) \Omega C(1)'\beta$. If β are r cointegrating vectors of dimension $m \times 1$, then $\beta' y_t$ is $I(0)$ and $\beta' \Delta y_t$ is overdifferenced with a zero long-run variance. Therefore, for β to be cointegrating vectors, we must have

$$\beta' C(1)\Omega C(1)'\beta = 0_{r\times r}$$

This equality cannot hold if $C(1)$ has full rank. Thus, cointegration requires $C(1)$ to have rank less than m and β must lie in the null space of $C(1)$, i.e., so that $\beta' C(1) = 0_{r\times m}$.

Assuming that $C(L)$ is invertible, model (1) has the VAR representation

$$A(L)y_t = u_t \tag{2}$$

where $A(L) = (1 - L)C(L)^{-1}$ is a pth order lag polynomial. Evaluating $(1 - L)I_m = A(L)C(L)$ at $L = 1$ yields

$$A(1)C(1) = 0_{m\times m}$$

Therefore, $A(1)$ lies in the null space of $C(1)$ and, from the previous results for the MA representation, must be a linear combination of the cointegrating vectors β. More precisely, $A(1) = -\alpha\beta'$, where α is an $m \times r$ matrix of rank r.

Finally, model (2) admits an VECM representation

$$\Delta y_t = \Pi y_{t-1} + \Gamma(L)\Delta y_t + u_t$$

where $\Gamma(L) = \Gamma_1 L + \Gamma_2 L^2 + \ldots + \Gamma_{p-1} L^{p-1}$, $\Gamma_i = -\sum_{j=i+1}^{p} A_j$ and $\Pi = -A(1) = \alpha\beta'$. In this representation, rank(Π)$= r < m$, i.e., Π has a reduced rank if there are r cointegrating relations. If rank(Π)$= m$, all elements of y_t are $I(0)$ and the appropriate model is an unrestricted VAR in levels, whereas if rank(Π)$= 0$, all elements of y_t are $I(1)$ and not cointegrated, and the appropriate model is VAR in first differences. Practitioners are commonly faced with the decision of which of these three models (VAR in levels, VAR in first differences, or VECM) to estimate, and this decision is often made on the basis of pretests for unit root and cointegration.

A second issue faced by practitioners interested in identifying the effect of a structural shock of interest (e.g., monetary policy shocks, oil price shocks, or technological innovations) is the choice of identifying restrictions. In the applied literature, the most commonly used identification strategies are short-run and long-run restrictions. Short-run identification schemes can be characterized as follows. Let B_0 denote an $m \times m$ invertible matrix with ones on the main diagonal. Pre-multiplying both sides of Eq. (2) by B_0 yields the structural VAR (SVAR) model

$$B(L)y_t = \varepsilon_t$$

where $B(L) = B_0 A(L)$ and $\varepsilon_t = B_0 u_t$ denotes an $m \times 1$ vector of structural shocks which are assumed to be orthogonal with a diagonal covariance matrix Σ. Hence

$$E(u_t u_t') \equiv \Omega = B_0^{-1} \Sigma B_0^{-1\prime}$$

Given that Ω is symmetric with $m(m+1)/2$ elements, in order to identify all of the free parameters of B_0 we need to impose $m(m-1)/2$ restrictions. One possibility is to restrict B_0 to be lower triangular which is equivalent to a Choleski decomposition of Ω. Note that the short-run identifying restrictions do not depend on the specification of the reduced-form VAR model (e.g., Lütkepohl & Reimers, 1992).

Alternatively, one can impose long-run identifying restrictions that render the MA matrix $A(1)^{-1} B_0^{-1}$ lower triangular. The use of long-run restrictions is less general in that it requires some model variables to be $I(1)$ and others to be $I(0)$, making this approach particularly susceptible to any misspecification of the integration properties of the individual series. For example, Gospodinov et al. (2011) demonstrate that even arbitrary small deviations from the exact unit, when combined with long run identification restrictions, can produce impulse response estimates that are highly distorted.

In the fourth section we study how alternative choices of model specification and identification restrictions can affect the MSE of the structural impulse responses.

SOME THEORETICAL RESULTS FOR NEAR-INTEGRATED PROCESSES

In this article, our main interest lies in data generating processes (DGPs) in which the variables in the VAR are highly persistent with roots that are either close to one or equal to one.[1] This is the typical situation a practitioner estimating a VAR with macroeconomic variables would face. Theoretically, these variables are well approximated by near-integrated processes. Thus, to understand the effect of persistent variables on the estimation of the VARs, we start by reviewing some theoretical results for near-intergrated processes.

Common Trend Decomposition of Near-Integrated Processes

Let y_t denote a multivariate $m \times 1$ process

$$(I_m - \Phi L)y_t = \Psi(L)u_t$$

where $\Phi = I_m + C/T$ and C is a matrix of fixed constants. It is often convenient to define $C = \text{diag}(c_1, c_2, ..., c_m)$ in order to rule out $I(2)$ processes when the diagonal elements are nonzero. We discuss the effect of nonzero off-diagonal elements later. Assume that $y_0 = 0$; $\Psi(L) = \sum_{i=0}^{\infty} \Psi_i L^i$ with $\Psi_0 = I_m$, $\sum_{i=0}^{\infty} i|\Psi_i| < \infty$ and $\sum_{i=0}^{\infty} \Psi_i \neq 0_{m \times m}$, and u_t is a homoskedastic martingale difference sequence with a covariance matrix Ω and finite fourth moments. Deterministic terms are assumed away for notational convenience.

Let $e_t = \Psi(L)u_t$. Then, $e_t = [\Psi(1) + (1-L)\Psi^*(L)]u_t$ by the Beveridge–Nelson (BN) decomposition, where $\Psi(1) = \sum_{i=0}^{\infty} \Psi_i$, $\Psi^*(L) = \sum_{i=1}^{\infty} \Psi_i^* L^{i-1}$ and $\Psi_i^* = -\sum_{j=i}^{\infty} \Psi_j$. Denoting $S_t = \sum_{j=1}^{t} u_t$ and using recursive substitution and summation by parts

$$y_t = \sum_{j=1}^{t} \Phi^{t-j}\Psi(L)u_j = \Psi(1)S_t + v_t + \frac{C}{T}\sum_{j=1}^{t-1} \Psi(1)\Phi^{t-j}S_{j-1} + \frac{C}{T}\sum_{j=1}^{t-1} \Phi^{t-j}v_{j-1} \tag{3}$$

where $v_t = \Psi^*(L)u_t$.

Expression (3) is an algebraic decomposition (or factorization) of the process y_t that contains the standard BN decomposition for exact unit root processes as a special case ($C = 0_{m\times m}$). While the standard BN decomposition is given by the permanent, $\Psi(1)S_t$, and transitory, v_t, components, the decomposition in Eq. (3) contains two additional terms. The fourth term $\frac{C}{T}\sum_{j=1}^{t-1} \Phi^{t-j}v_{j-1}$ is asymptotically negligible, whereas the term $\frac{C}{T}\sum_{j=1}^{t-1} \Psi(1) \Phi^{t-j}S_{j-1}$ is not and is of the same order $O_p(T^{1/2})$ as $\Psi(1)S_t$. Therefore, Eq. (3) contains two permanent component terms (i.e., two terms of order $O_p(T^{1/2})$). Omitting one of the them (by wrongly assuming an exact unit root, for example) will give rise to a misspecification bias and undermine the size of hypothesis tests.

Consider, for example, the cointegration model (Elliott, 1998)

$$y_{1,t} = \beta y_{2,t} + e_{1,t}$$

$$y_{2,t} = (1 + c/T)y_{2,t-1} + e_{2,t}$$

which can be cast in a similar representation as the model above. In this case, the restriction $\beta\Psi(1) = 0$ does not annihilate the permanent component. As a result, reduced rank regressions that impose only this restriction are not valid and would cause misspecification bias and test size distortions as shown by Elliott (1998). Similar misspecification biases are expected to play a role in impulse response analysis based on long-run identifying restrictions on $\Psi(1)$, that is, restrictions imposed on the "wrong" permanent component when $C \neq 0_{m\times m}$.

In brief, wrongly assuming that a multivariate process y_t has one or more unit roots, when the roots are close but not equal to unity, leads to distortions in reduced rank regressions, as well as in the BN decomposition and impulse response analysis.

Impulse Response Analysis in VAR Models With Near Unit Roots Identified by Short-Run Restrictions

Let the DGP be given by

$$\tilde{A}(L)\tilde{y}_t = \tilde{u}_t \tag{4}$$

or

$$(I_m - \Lambda L) = \Psi(L)\tilde{u}_t$$

where $\tilde{A}(L) = I_m - \tilde{A}_1 L - \ldots - \tilde{A}_p L^p$ has roots outside or on the unit circle, and Λ can be expressed in terms of its eigenvalues and eigenvectors as $\Lambda = V^{-1}\Phi V$ with Φ as defined above (Pesavento & Rossi, 2006). This model allows for unit roots and possible cointegration and several representations in other parts of the article can be regarded as specifications in the rotated variables $y_t = V\tilde{y}_t$ and $u_t = V\tilde{u}_t$.

Despite the possible presence of unit roots and cointegration, the estimates of $\tilde{A}_1, \ldots, \tilde{A}_p$ from the levels VAR (4) are root-T consistent and individually normally distributed (Sims et al., 1990). Using these estimates and a short-run (recursive) identification scheme, one can construct l-period ahead orthogonalized impulse responses Θ_l. Provided that the response horizon is fixed with respect to the sample size, the impulse response estimators are, under weak regularity conditions, consistent and asymptotically normally distributed (see Lütkepohl & Reimers, 1992). This explains the robustness of impulse response estimators based on the level specification to possible deviations from exact unit roots and cointegration, as long as the structural parameters and shocks are identified using short-run restrictions. In contrast, inference becomes nonstandard if we rely on long-run restrictions for identification, as discussed below. A similar problem would arise, regardless of the identification, if we allow the response horizon to grow fast enough with respect to the sample size. The latter result casts doubt on impulse response estimates for long horizons when the data are persistent. Indeed, Kilian and Chang (2000) document that the reliability of impulse response estimates from standard levels specifications may deteriorate substantially at longer horizons.

Impulse Response Analysis in VAR Models With Near Unit Roots Identified by Long-Run Restrictions

This case is best illustrated in the context of the Blanchard and Quah (1989) model for which $y_t = (\Delta Y_t, U_t)'$, where Y_t is the output (log real

GNP) and U_t is the unemployment rate, $A(L) = \Psi(L)^{-1}(I_2 - \Phi L)$ and $B_0 = \begin{bmatrix} 1 & -b_{12}^{(0)} \\ -b_{21}^{(0)} & 1 \end{bmatrix}$. The structural shocks $\varepsilon_t = B_0 u_t$ can be interpreted as aggregate supply and demand shocks $(\varepsilon_t^s, \varepsilon_t^d)'$. The long-run identifying restriction that the demand shocks ε_t^d have no long-run effect on output imposes a lower triangular structure on the MA matrix $A(1)^{-1}B_0^{-1}$. Hence, under this identifying restriction, the matrix of long-run multipliers in the structural model, $B(1) = B_0 A(1)$, is also lower triangular.

The robustness of the impulse response analysis depends crucially on the persistence of U_t and, hence, on the parameterization of the matrix Φ. We follow Gospodinov et al. (2011) and express Φ as $\Phi = \begin{bmatrix} 1 & \delta \\ 0 & \rho \end{bmatrix}$, where $\delta = -\gamma(1-\rho)$ is the parameter that determines the low frequency co-movement between the variables, and $\rho = 1 + c/T$ (for $c < 0$) denotes the largest root of the unemployment rate. The nonzero off-diagonal element $\gamma(1-\rho)$ allows for the possibility that a small low frequency component of unemployment affects output growth.

To gain some further insight into this parameterization, let us assume for simplicity that $\Psi(L) = I_2$ and note that the vector error-correction form of the model is given by

$$\Delta y_t = (\rho - 1)\begin{bmatrix} 0 & -\gamma \\ 0 & 1 \end{bmatrix} y_{t-1} + u_t$$

which can be written as

$$\Delta y_t = \alpha\beta' y_{t-1} + u_t$$

where $\alpha = (\rho - 1)\begin{pmatrix} \gamma \\ 1 \end{pmatrix}$ and $\beta = (0, 1)'$. Therefore, the model has a trivial cointegration vector $(0, 1)'$ and the speed of adjustment α is a function of γ which carries important information from the levels of the series, provided that $\rho < 1$.

With this parameterization of Φ, the long-run multiplier matrix is given by

$$B(1) = \begin{bmatrix} \psi_{11}(1) - b_{12}^{(0)}\psi_{21}(1) & (1-\rho)\left([\gamma\psi_{11}(1) + \psi_{12}(1)] - b_{12}^{(0)}[\gamma\psi_{21}(1) + \psi_{22}(1)]\right) \\ \psi_{21}(1) - b_{21}^{(0)}\psi_{11}(1) & (1-\rho)\left([\gamma\psi_{21}(1) + \psi_{22}(1)] - b_{21}^{(0)}[\gamma\psi_{11}(1) + \psi_{12}(1)]\right) \end{bmatrix}$$

where $\psi_{ij}(1)$ denotes the corresponding elements of the matrix $\Psi(1)^{-1}$. Imposing the long-run restriction that aggregate demand shocks have no permanent effect on output implies that $b_{12}^{(0)} = [\gamma\psi_{11}(1) + \psi_{12}(1)]/[\gamma\psi_{21}(1) + \psi_{22}(1)]$. In contrast, imposing a unit root in unemployment ($\rho = 1$) and differencing both output and unemployment renders the restriction $b_{12}^{(0)} = \psi_{12}(1)/\psi_{22}(1)$. Thus, the differenced VAR ignores any information contained in the levels of output and unemployment.

The main message to the practitioner is that the differenced VAR specification is not robust to small low frequency co-movements. While the levels specification (in which some of the variables are in first differences and some are in levels) preserves the information on the low-frequency co-movement, the estimates of the impulse response functions are inconsistent and their distribution is fat-tailed in the presence of local-to-unity processes (Gospodinov, 2010). In summary, the applied researchers should exert extreme caution when working with near-integrated variables and the SVAR is identified by long-run restrictions.

Unit Root Pretest VAR Specification

Similarly to the case of the differenced VAR, lack of robustness is expected to characterize the behavior of specifications based on pretest for a unit root given that this pretest will select the differenced specification with probability approaching one when the process is near integrated. Hence, the unit root pretest VAR specification will inherit the non-robustness properties of the differenced VAR specification for slowly co-moving, near-integrated variables.

ROBUST INFERENCE FOR IMPULSE RESPONSE FUNCTIONS

Given the low power of most pretests for unit root and cointegration, and costs from estimating a VAR that erroneously imposes unit roots, the current literature has moved in the direction of methods for inference that are robust to the possible presence of unit roots. These robust methods are designed for VAR models based on short-run identifying restrictions only, of course, as departures from exact unit roots immediately invalidate the use of long-run identifying restrictions. Consider the model

$$(I_m - \Phi L)y_t = \Psi(L)u_t$$

where $\Phi = I_m + C/T$ and $C = \text{diag}(c_1, c_2, \ldots, c_m)$. Note that y_t could denote rotated variables to account for the possibility of cointegration with a known cointegration vector (for more details, see Pesavento & Rossi, 2006). Suppose that the interest lies in inference on the impulse responses of y_t at long response horizons l, y_{t+l}, to the structural shocks $\varepsilon_t = B_0 u_t$.

To better capture the uncertainty in estimating the impulse responses at long horizons, it is convenient to parameterize l as a function of the sample size. More specifically, let $l = [\delta T]$ for some fixed $\delta > 0$. Under this parameterization, we have that $C^l \rightarrow e^{C\delta}$ as $T \rightarrow \infty$, where $e^{\delta C}$ is a diagonal matrix with $(e^{\delta c_1}, e^{\delta c_2}, \ldots, e^{\delta c_m})$ on the main diagonal (Gospodinov, 2004; Pesavento & Rossi, 2006). Furthermore

$$\Theta_l \equiv \frac{\partial y_{t+l}}{\partial u_t} = \Phi^l (I_m + \Phi^{-1}\Psi_1 + \Phi^{-1}\Psi_2 + \ldots) + o(1) \simeq \Phi^l \Psi(1) \tag{5}$$

and the l-period impulse response function of the jth variable in y_t to the kth structural shock in ε_t is given by (Pesavento & Rossi, 2006)

$$\frac{\partial y_{j,t+l}}{\partial \varepsilon_{k,t}} \simeq \iota_j' \Phi^l \Psi(1) B_0^{-1} \iota_k \rightarrow \iota_j' e^{C\delta} \Psi(1) B_0^{-1} \iota_k \tag{6}$$

as $T \rightarrow \infty$, where ι_j and ι_k are the corresponding columns of the identity matrix I_m. Inference on the impulse response function can then proceed by plugging a consistent estimate of $\Psi(1)$ and obtaining the confidence interval limits for the localizing constant c_j of the jth variable $y_{j,t}$, which is typically done by inverting a unit root test (see, e.g., Stock, 1991). In this sense, this method is a two-step procedure, which involves the construction of a confidence interval for c_j in the first step and using this confidence interval and the expression (6) to construct a confidence interval for the impulse response function $\frac{\partial y_{j,t+l}}{\partial \varepsilon_{k,t}}$.

Gospodinov (2004) proposes a one-step test-inversion method for directly constructing confidence bands for the impulse response function. Let $\theta = (\text{vec}(\Phi)', \text{vec}(\Psi_1)', \text{vec}(\Psi_2)', \ldots)'$, where in practice $\Psi_1, \Psi_2, \ldots$ are approximated by a finite-order VAR, and $h_l(\theta) = \Theta_l(\theta) - \Theta_{0,l}$ denotes the constraint that the expression in Eq. (5) is equal to a particular value $\Theta_{0,l}$ under the null hypothesis $H_0 : \Theta_l(\theta) = \Theta_{0,l}$. Finally, let $LR_T = T\text{tr}[\hat{\Omega}^{-1}(\tilde{\Omega} - \hat{\Omega})]$ denote the likelihood ratio statistic of the hypothesis $H_0 : h_l(\theta) = 0$, where $\tilde{\Omega}$ and $\hat{\Omega}$ denote the variance-covariance matrices of the restricted and the estimated residuals, respectively, from the approximating finite-order VAR model. Then, under the null $H_0 : h_l(\theta) = 0$, $\Phi = I_m + C/T$ and $l = [\delta T]$

$$LR_T \xrightarrow{d} \mathrm{tr}\left[\left(\int_0^1 \mathbf{J}_c(s)d\mathbf{J}_c(s)'\right)'\left(\int_0^1 \mathbf{J}_c(s)\mathbf{J}_c(s)'ds\right)^{-1}\left(\int_0^1 \mathbf{J}_c(s)d\mathbf{J}_c(s)'\right)\right]$$

where $\mathbf{J}_c(s)$ is an Ornstein–Uhlenbeck vector process. Confidence sets can be obtained by inverting the LR_T test on a grid of possible values of $\Theta_{0,l}$ and using projection methods for constructing confidence bands for individual impulse responses.

One drawback of the methods proposed by Gospodinov (2004) and Pesavento and Rossi (2006) is that they provide accurate approximations only for near-integrated processes ($\Phi = I_m + C/T$) and long response horizons ($l = [\delta T]$). While these methods continue to remain asymptotically valid over the other parts of the parameter space and horizon space, they tend to be conservative. One solution to this problem is to adopt an approximation procedure which is uniform over the whole parameter space. Mikusheva (2012) shows that the grid bootstrap of Hansen (1999) adapted to the *LR* test inversion method of Gospodinov (2004) delivers the desirable uniform approximation regardless of the persistence of the underlying process. Unfortunately, this approximation method suffers from the curse of dimensionality and can be computationally demanding (or even infeasible) as the dimensionality of the VAR model increases, due to the multidimensional grid required to perform the test inversion.

Other practically appealing bootstrap-based methods for inference on impulse response functions are also available. For example, Inoue and Kilian (2002), building on the work of Sims et al. (1990), show that the conventional bootstrap (Runkle, 1987) is asymptotically valid for inference on typical nonlinear combinations of VAR parameters even in the presence of unit roots or near unit roots, provided the VAR model includes more than one lag. However, despite their asymptotic validity in the presence of highly persistent variables, standard bootstrap methods for higher-order VAR models may have poor coverage accuracy for impulse responses in small samples. One source of the unsatisfactory behavior of the conventional bootstrap is that it further exacerbates the bias that characterizes the least-squares estimator of VAR models with highly persistent variables. For this reason, Kilian (1998a, 1998b) develops a two-stage bootstrap method that explicitly estimates and removes the bias in the VAR parameters before approximating the impulse response distribution. However, this bias-correction method is not designed for unit root processes, unless the unit root is imposed in the estimation, and its theoretical validity to date has been established only for stationary VAR models. We consider these bootstrap methods in our simulation section.

MONTE CARLO SIMULATIONS

Bivariate DGP With Near Unit Roots, But No Near Cointegration

Our first simulation experiment follows closely the structure of the models used for quantifying the contribution of aggregate demand (non-technology) and supply (technology) shocks to business cycle fluctuations (Blanchard & Quah, 1989; Christiano, Eichenbaum, & Vigfusson, 2006; Galí, 1999, among others). Blanchard and Quah (1989) use output and unemployment to identify the demand and supply shocks while Galí (1999) and Christiano et al. (2006) model the dynamic behavior of labor productivity and hours worked to study the prediction of the real business cycle theory that technology shocks have a positive effect on hours worked. These papers impose a unit root on output or technology, that is, the first variable (output or labor productivity) in the SVAR is expressed in first differences. The second variable (unemployment or hours worked) is highly persistent which has generated a debate about the appropriate empirical specification of this variable; the question is whether this second variable should be included in levels or in first differences (see Gospodinov et al., 2011). While the two variables are not cointegrated, they appear to be driven by some low-frequency co-movement which is preserved in the levels VAR specification but is eliminated after differencing.

In this simulation exercise, we investigate the robustness of the levels VAR specification (first variable in differenced form and second variable in levels) and a specification based on a pretest of a unit root for the second variable to various degrees of persistence and low-frequency co-movement. We employ the ADF-GLS test at 5% significance level to pretest for a unit root in the second variable. If the test rejects the null, we model the second variable in levels; instead, if the null is not rejected, we model it in first differences. We also follow Gospodinov et al. (2011) in allowing for a small low-frequency co-movement when the root of the second variable is strictly less than unity. Finally, we consider both long-run identification (based on the assumption that the shocks to the second variable have no permanent effect on the level of the first variable) and short-run (recursive) identification (based on the assumption that the second variable has no contemporaneous effect on the first variable). More specifically, the data are generated from

$$\begin{bmatrix} \Delta y_{1,t} \\ y_{2,t} \end{bmatrix} = \begin{bmatrix} 0 & (1-\rho) \\ 0 & \rho \end{bmatrix} \begin{bmatrix} \Delta y_{1,t-1} \\ y_{2,t-1} \end{bmatrix} + \begin{bmatrix} u_{1,t} \\ u_{2,t} \end{bmatrix} \tag{7}$$

where $\begin{bmatrix} u_{1,t} \\ u_{2,t} \end{bmatrix} \sim iid\, N\left(\begin{bmatrix} 0 \\ 0 \end{bmatrix}, \begin{bmatrix} 1 & 0.3 \\ 0.3 & 1 \end{bmatrix}\right)$ The sample size is $T = 200$ (50 years of quarterly data that matches our empirical application) and $\rho \in \{0.92, 0.95, 0.98, 1\}$. The two variables are demeaned prior to estimation. The number of Monte Carlo replications is 10,000. Note that in what follows, the first variable is always modeled in first differences ($\Delta y_{1,t}$) and the second variable is either in levels $y_{2,t}$ (we refer to this as levels VAR specification) or in first differences $\Delta y_{2,t}$.

The simulation results are summarized by plotting the coverage rates of 95% confidence intervals of the estimated impulse responses and the MSEs of the impulse response estimator for horizons $1, 2, \ldots, 24$. The confidence bands for the impulse responses are constructed using the bias-corrected bootstrap method proposed by Kilian (1998a, 1998b). To illustrate the advantages of Kilian's (1998a, 1998b) method over the conventional bootstrap (Runkle, 1987), Figs. 1 and 2 report the coverage rates of the 95% confidence intervals of the (i,j) impulse response $(i = 1, 2; j = 1, 2)$, constructed using Kilian's (1998a, 1998b) and Runkle's (1987) bootstrap methods, for the levels VAR specification using long-run identification ($\rho = 0.95$ and 0.98). Overall, the coverage rates of Kilian's (1998a, 1998b) bootstrap method are very close to the nominal level while Runkle's (1987) conventional bootstrap method tends to undercover. In what follows, we use Kilian's (1998a, 1998b) bias-corrected bootstrap method to construct confidence bands for the impulse response of interest.

We start with the case of long-run identification and consider both coverage rates and MSEs of the (2,1) and (2,2) impulse response functions. These impulse responses trace the responses of the second variable, for which there is uncertainty about the value of its largest AR root, to both shocks. Recall that the levels VAR specification always models the first variable in first differences and the second variable in levels. Instead, the pretest VAR specification is based on a sequential procedure. In the first step, the ADF-GLS test is used to determine if the second variable is

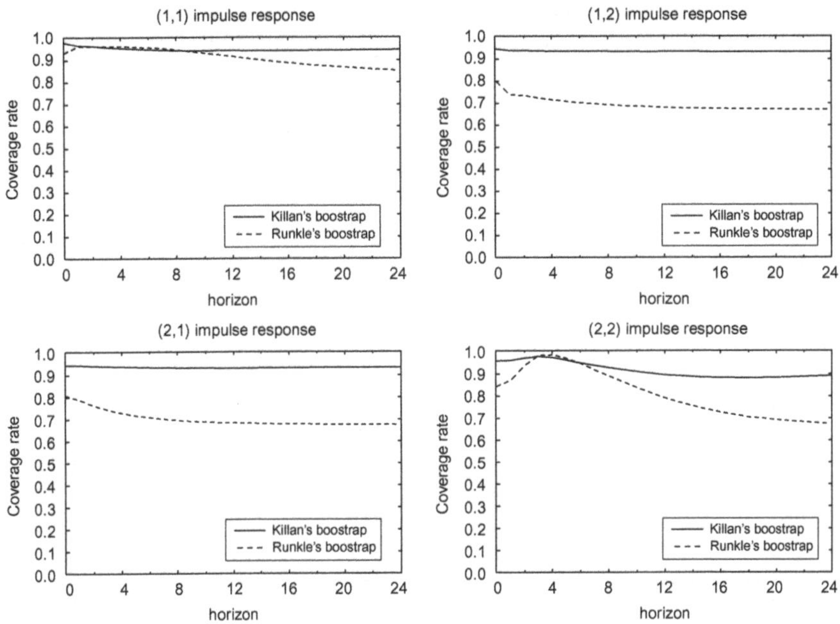

Fig. 1. Coverage rates of 95% confidence intervals for the (i,j) impulse response (response of the ith $(i=1,2)$ variable to the jth $(j=1,2)$ shock) constructed by Kilian's (1998a, 1998b) bias-corrected bootstrap and Runkle's (1987) conventional bootstrap methods. The data are generated from model (7) with $\rho=0.95$. The impulse responses are identified using the long-run identification scheme.

integrated of order one or zero. If the null hypothesis is not rejected, the second variable is modeled in first differences; otherwise, the second variable enters the model in levels.

Figs. 3–5 plot the coverage rates and MSEs for $\rho=0.92, 0.95$, and 0.98. The coverage rates and the MSEs of the (2,1) and (2,2) impulse responses from the pretest model deteriorate when the root ρ approaches one. This is due to the presence of a low frequency component which is differenced away by the pretest specification when the null of a unit root in not rejected. In the absence of this low-frequency co-movement, the coverage rates and MSEs of the pretest specification should improve. Gospodinov et al. (2011) illustrate how imposing a unit root in the presence of an arbitrarily small low-frequency co-movement leads to estimated impulse responses that are vastly different from the true ones. In the simulation setup (7), the VAR model in first differences (except for $\rho=1$) is

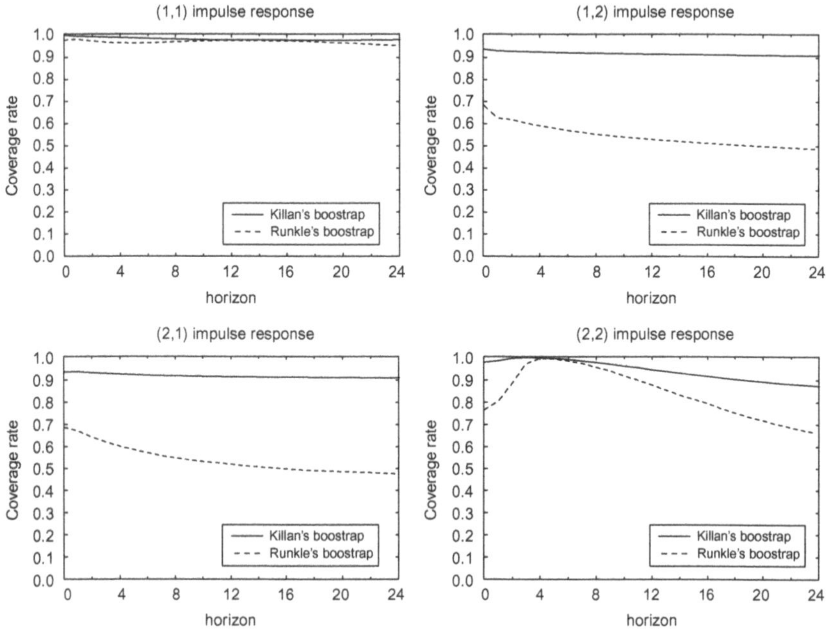

Fig. 2. Coverage rates of 95% confidence intervals for the (i,j) impulse response (response of the ith $(i=1,2)$ variable to the jth $(j=1,2)$ shock) constructed by Kilian's (1998a, 1998b) bias-corrected bootstrap and Runkle's (1987) conventional bootstrap methods. The data are generated from model (7) with $\rho=0.98$. The impulse responses are identified using the long-run identification scheme.

misspecified and the estimated impulse responses from this model perform worse than the ones from the pretest specification. In contrast, the levels specification performs well and appears to be robust to the different degrees of persistence and the presence of a low-frequency co-movement. However, as the root approaches unity, the structural parameters are only weakly identified and the estimation uncertainty of the impulse responses is large (see Gospodinov, 2010).

The corresponding short-run identification scheme seems to be immune to some of the problems that arise under the long-run identifying restrictions (Christiano et al., 2006; Gospodinov et al., 2011). Figs. 6–9 plot the coverage rates and MSEs for $\rho=0.92$, 0.95, 0.98, and 1 under the short-run identification scheme. Again, the pretest specification is not robust to deviations from the unit root and exhibits distorted coverage rates and

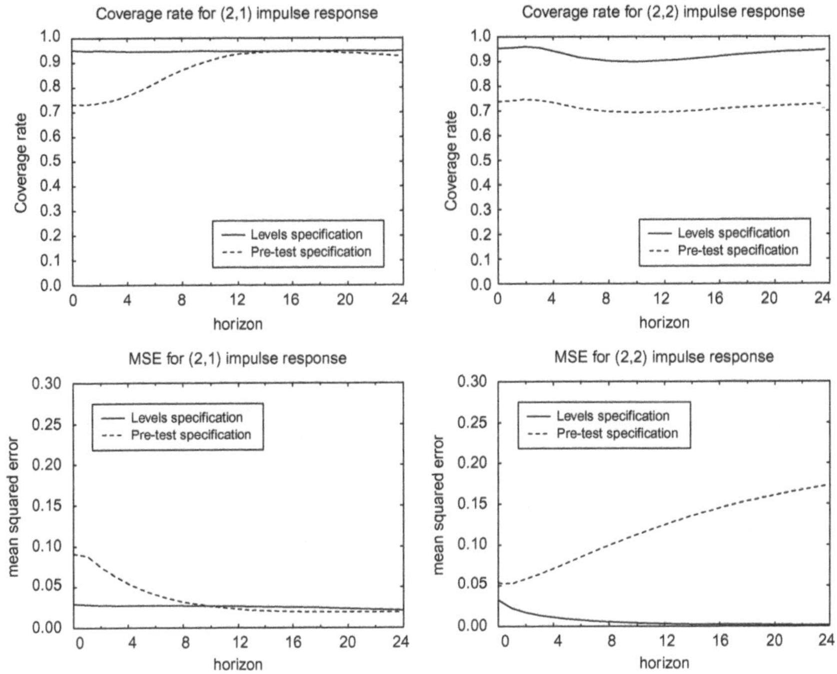

Fig. 3. Coverage rates of 95% bias-corrected bootstrap confidence intervals and MSEs for the (2, 1) and (2, 2) impulse responses for the levels (solid line) and pretest (dashed line) VAR specifications. The data are generated from model (7) with $\rho=0.92$. The impulse responses are identified using the long-run identification scheme.

increased MSEs. When $\rho=1$, the differenced model is correct and the pretest VAR specification dominates the levels VAR. But even in this case, the costs from estimating the levels VAR are not large especially if one employs inference procedures that are asymptotically valid in the case when the root lies on or near the unit circle.

To summarize, both for the long-run and the short-run identification schemes, the levels VAR specification fares well in terms of coverage rates and MSEs. The sampling uncertainty associated with the estimated impulse response functions tends to be well approximated by the bias-corrected bootstrap procedure of Kilian (1998a, 1998b). The coverage rates and the MSEs of the levels specification continue to be satisfactory even when the largest root is very close to unity ($\rho=0.95$). The specification based on

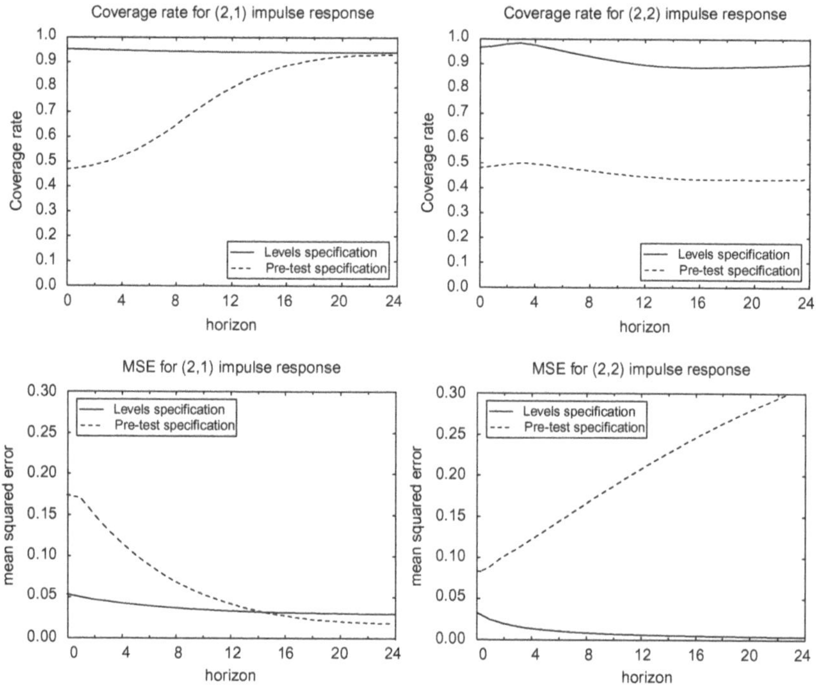

Fig. 4. Coverage rates of 95% bias-corrected bootstrap confidence intervals and MSEs for the (2, 1) and (2, 2) impulse responses for the levels (solid line) and pretest (dashed line) VAR specifications. The data are generated from model (7) with $\rho = 0.95$. The impulse responses are identified using the long-run identification scheme.

the pretest for a unit root exhibits finite-sample distortions and inflated MSEs. Overall, the levels specification emerges as the preferred specification for this simulation experiment.

Bivariate DGP With Near Unit Roots and Near Cointegration

The DGPs so far did not allow for cointegration among the model variables. In the next simulation experiment, we consider a bivariate system with possible cointegration. More specifically, we simulate data from the following model

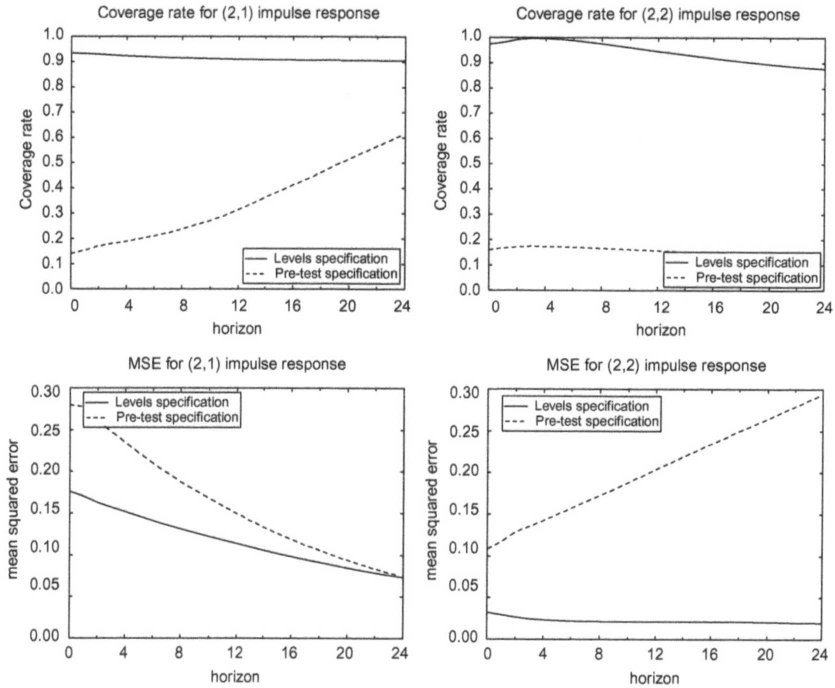

Fig. 5. Coverage rates of 95% bias-corrected bootstrap confidence intervals and MSEs for the (2, 1) and (2, 2) impulse responses for the levels (solid line) and pretest (dashed line) VAR specifications. The data are generated from model (7) with $\rho=0.98$. The impulse responses are identified using the long-run identification scheme.

$$y_{1,t}=\beta y_{2,t}+u_{1,t}$$

$$y_{2,t}=\rho y_{2,t-1}+u_{2,t}$$

$$u_{1,t}=\rho_u u_{1,t-1}+\xi_t$$

where $\beta=1-\rho_u$, and ξ_t and $u_{2,t}$ are mutually uncorrelated *iid* $N(0,1)$ random variables. The variable $y_{2,t}$ is generated as a (near-)integrated process with $\rho\in\{1,0.98,0.95,0.90,0.85\}$. The autoregressive parameter ρ_u determines the persistence of $y_{1,t}$ and the cointegration properties between $y_{1,t}$ and $y_{2,t}$. In our numerical experiment, we set ρ_u to 0.4. Note that when

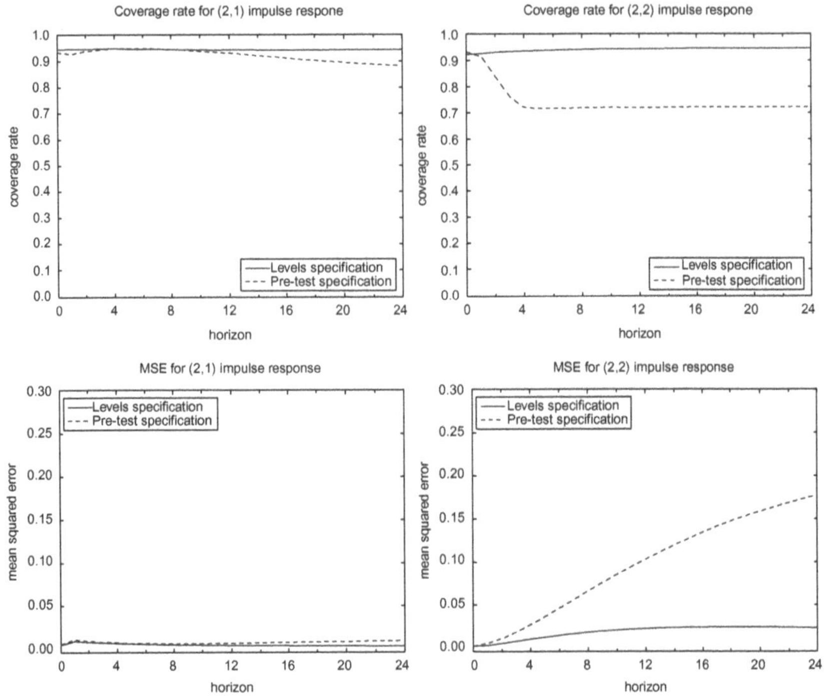

Fig. 6. Coverage rates of 95% bias-corrected bootstrap confidence intervals and MSEs for the (2, 1) and (2, 2) impulse responses for the levels (solid line) and pretest (dashed line) VAR specifications. The data are generated from model (7) with $\rho = 0.92$. The impulse responses are identified using the short-run (recursive) identification scheme.

$\rho_u = 1$, $y_{1,t}$ is an exact unit root process and there is no cointegration between the two variables, whereas for $\rho_u = 0.4$, $u_{1,t}$ is stationary and there is cointegration between the two variables.

Consider first the sequential pretest procedure. The two series are first subjected to a unit root test. If the null of a unit root is not rejected for both series, we test for cointegration between the two series. Depending on the outcome of the test, the model is estimated as a differenced VAR (if no cointegration is detected) or as a VECM using the estimated cointegrated vector (if cointegration is detected). If the null of a unit root is rejected, the levels VAR is estimated. We contrast this sequential modeling strategy to an unrestricted VAR in levels or first differences regardless of the pretest

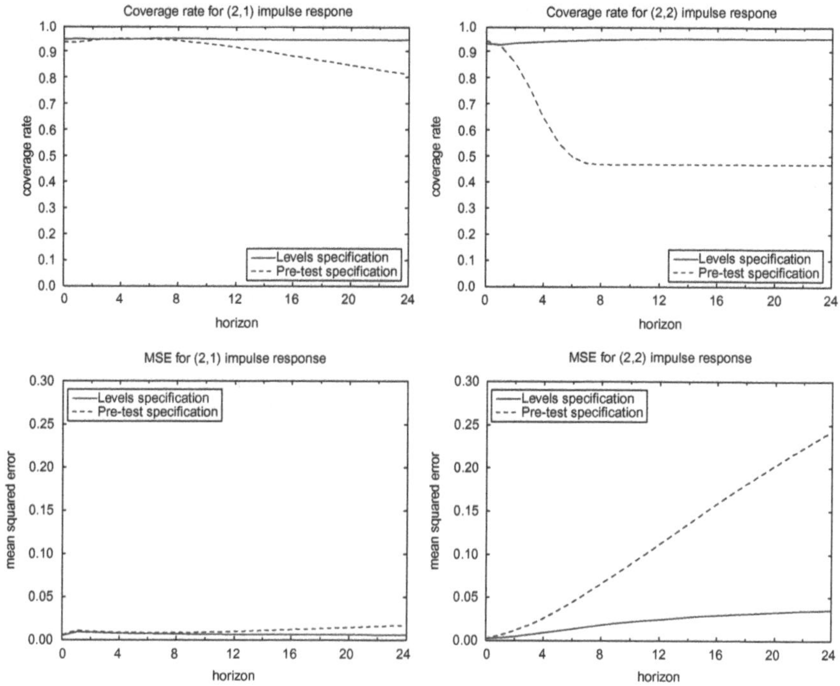

Fig. 7. Coverage rates of 95% bias-corrected bootstrap confidence intervals and MSEs for the (2, 1) and (2, 2) impulse responses for the levels (solid line) and pretest (dashed line) VAR specifications. The data are generated from model (7) with $\rho = 0.95$. The impulse responses are identified using the short-run (recursive) identification scheme.

results. We also consider the case in which the practitioner only tests for unit roots but no cointegration, and then uses the results of these pretests to decide whether to include each series in levels or in first differences. These models are used to estimate impulse responses, identified by short-run (recursive) restrictions, which are compared to the true impulse responses from the model. Unit root pretests are performed using the ADF-GLS test at 5% significance level, with the number of lags chosen by the modified information criterion of Ng and Perron (2001). For the VAR and VECM estimation, we select the number of lags using the BIC (the maximum lag order is set to 4). We run 10,000 replications for simulated samples of 200 observations. The results are presented in Figs. 10–14.

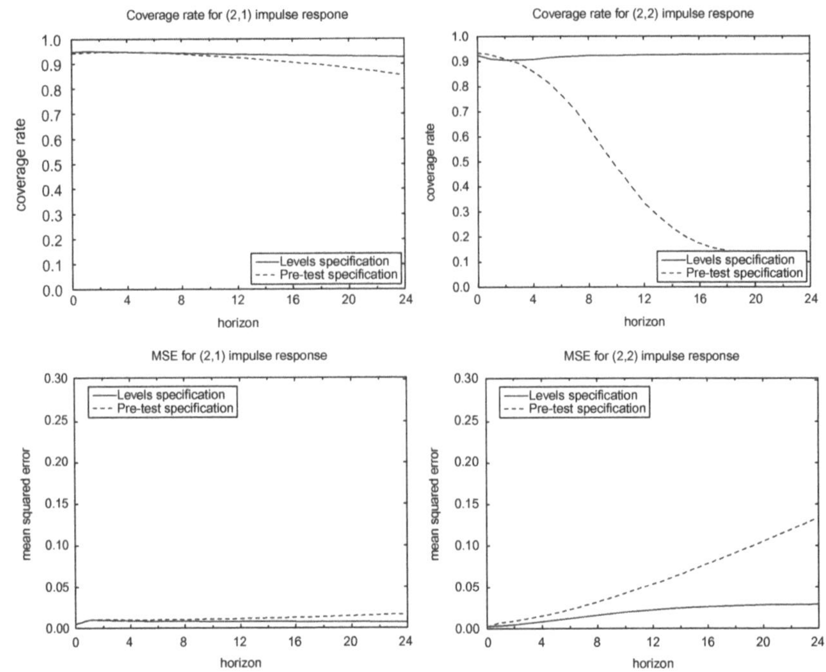

Fig. 8. Coverage rates of 95% bias-corrected bootstrap confidence intervals and MSEs for the (2, 1) and (2, 2) impulse responses for the levels (solid line) and pretest (dashed line) VAR specifications. The data are generated from model (7) with $\rho = 0.98$. The impulse responses are identified using the short-run (recursive) identification scheme.

Figs. 10–14 show the MSE of the (i, j) impulse response ($i = 1, 2; j = 1, 2$) for the case in which there is cointegration in the DGP ($\rho_u = 0.4$). Similar to the results in the previous section, the levels VAR performs better in terms of MSE than a VAR in first differences except in the case in which we know that the root is equal to one. Pretesting for a unit root has a similar effect as in the no-cointegration case: pretesting provides very little gain when the roots are large and it is not better than estimating the impulse response from the levels VAR in most cases. For the (2,1) response, when the root is large, pretesting is actually worse in term of MSE than simply running the VAR in first differences. As expected, when ρ gets smaller, pretests have good power and they are able to correctly suggest to estimate the VAR in levels.

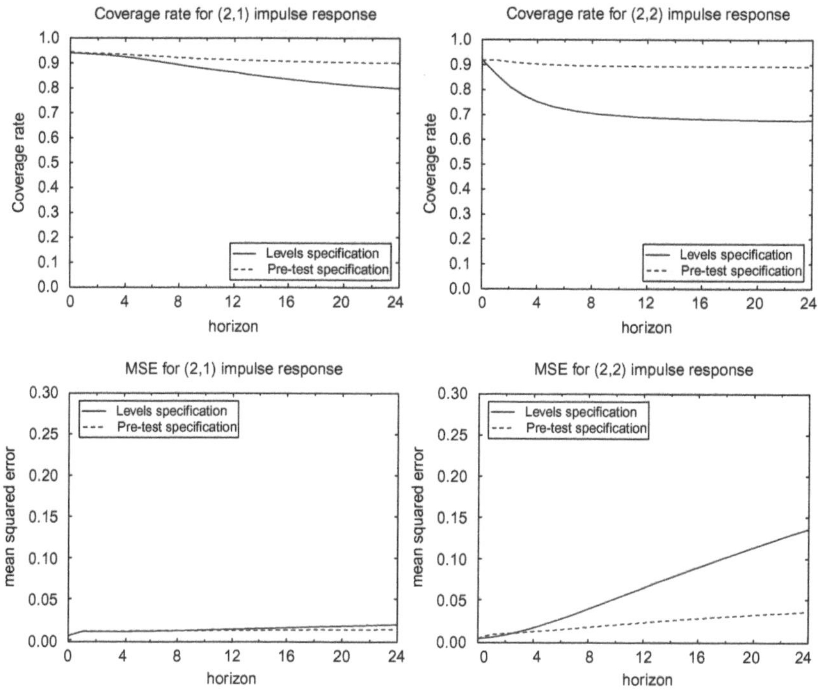

Fig. 9. Coverage rates of 95% bias-corrected bootstrap confidence intervals and MSEs for the (2, 1) and (2, 2) impulse responses for the levels (solid line) and pretest (dashed line) VAR specifications. The data are generated from model (7) with $\rho = 1$. The impulse responses are identified using the short-run (recursive) identification scheme.

Comparing the results from estimating a VAR in levels ignoring and not ignoring cointegration reveals some interesting results: the levels VAR provides a smaller MSE for the (1,1) and (2,1) impulse responses compared to the specification based on a pretest for cointegration even if cointegration is actually present ($\rho = 1$). For the (1,2) and (2,2) impulse responses there is a gain from estimating a VECM when indeed there is cointegration ($\rho = 1$). When ρ is less than one, technically there is no cointegration but we would still expect the variables to behave similarly to the unit root case. For roots equal to 0.98 and smaller, the levels VARs performs better for all four impulse responses. Overall, the results seem to suggest that, in most cases, the levels VAR dominates the other specifications in terms of MSE

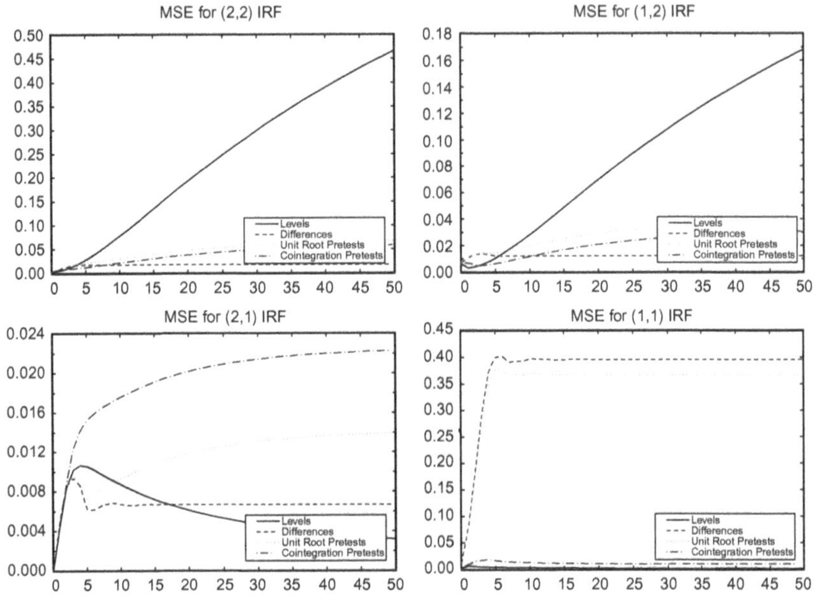

Fig. 10. MSEs for $\rho = 1$.

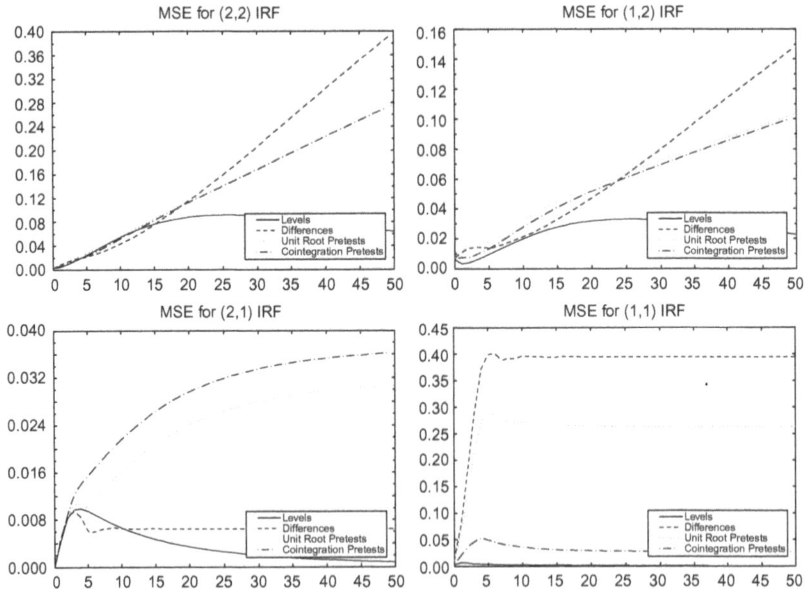

Fig. 11. MSEs for $\rho = 0.98$.

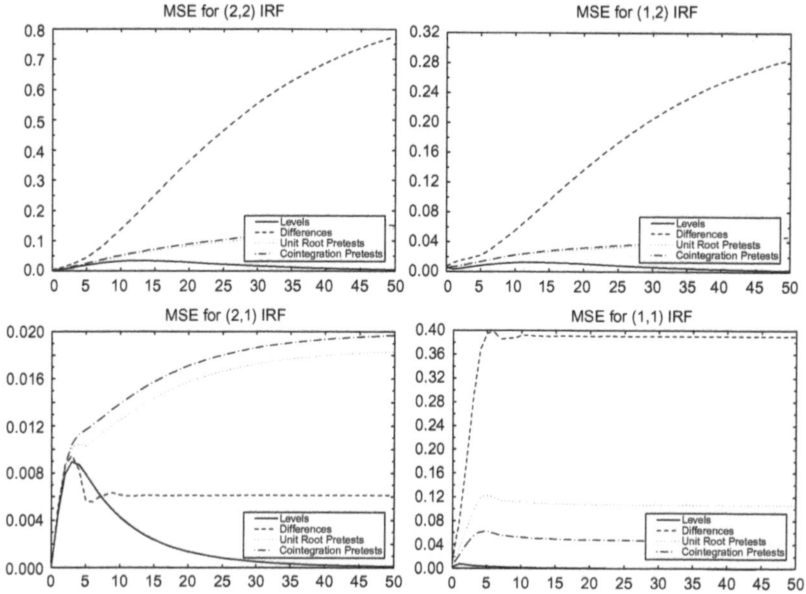

Fig. 12. MSEs for $\rho = 0.95$.

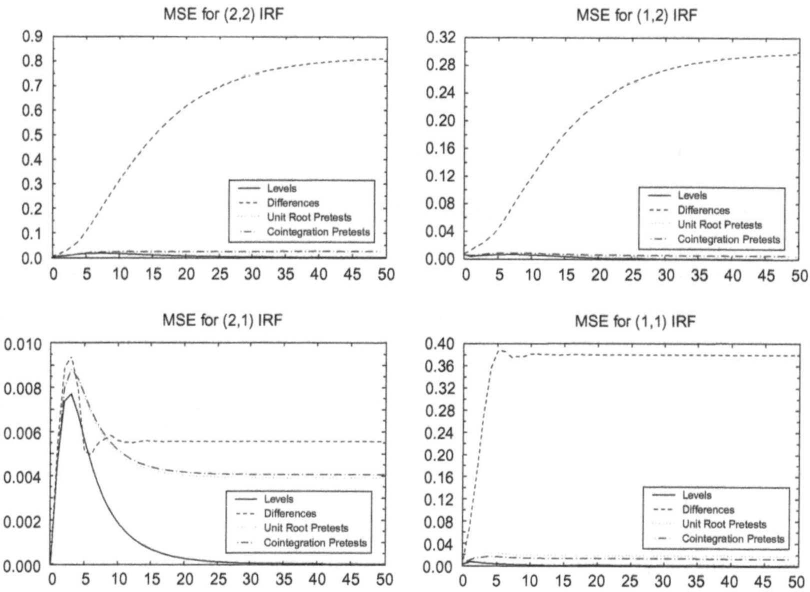

Fig. 13. MSEs for $\rho = 0.90$.

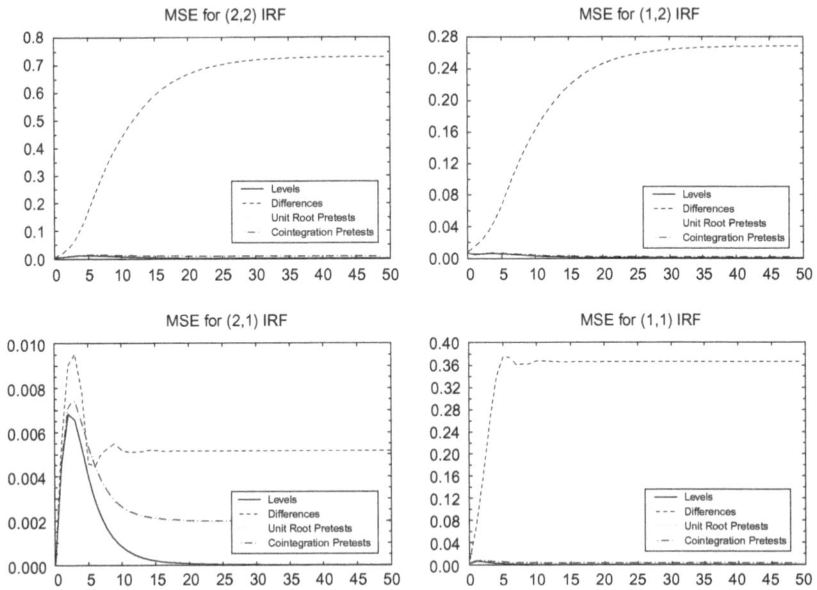

Fig. 14. MSEs for $\rho = 0.85$.

although this depends on the impulse responses of interest and the magnitude of the largest roots in the model.

A Multivariate Model Used in Monetary Policy Analysis

In the previous section, we considered a bivariate DGP and evaluated the options faced by the practitioner regarding model specification. When only one unit root is suspected, these options include: (a) running a VAR with $y_{2,t}$ in levels ignoring a possible unit root, (b) imposing a unit root and running a VAR in first differences, or (c) carrying out a pretest for a unit root in $y_{2,t}$ and specifying the VAR based on the outcome for the pretest. Instead, when two unit roots are suspected, the practitioner would proceed with a sequential procedure that could lead to a VAR model in first differences, a VECM, a VAR model in levels, or a mixed VAR model with some variables in levels and some in first differences.

A similar problem is faced by the practitioner when estimating the multivariate VARs that are commonly employed in analyzing the effect of monetary policy shocks. Below, we use the benchmark model of Christiano, Eichenbaum, and Evans (1999) – hereafter CEE – to study the effect of exogenous monetary policy shocks on economic activity. In this model, the economy is described by a 7-dimensional VAR(4) given by

$$A(L)y_t = c + u_t \tag{8}$$

where $y_t = [Y_t, P_t, \text{PCOM}_t, \text{FF}_t, \text{TR}_t, \text{NBR}_t, M_t]'$, Y_t is the log of real GDP, P_t is the log of the implicit GDP deflator, PCOM_t is the log of the Conference Board commodity price index, FF_t is the federal funds rate, TR_t is the log of total reserves, NBR_t is the log of nonborrowed reserves, $M1_t$ is the log of M1, and $u_t \sim iid(0,\Omega)$.

Motivated by the literature on the transmission of monetary policy shocks, we consider a short-run (recursive) identification scheme based on the lower triangular decomposition of the variance-covariance matrix. As in Bernanke and Blinder (1992) and Sims (1986, 1992), among others, we measure the monetary policy instrument by the federal funds rate.[2] The variables in y_t are ordered as described above such that Y_t, P_t, and PCOM_t do not react on impact to monetary policy shocks.

In the subsequent analysis, we treat the 7-dimensional VAR(4) in Eq. (8) as the DGP. We follow the standard practice to estimate the model in levels and with an intercept included. The estimation method is ordinary least squares.

The data are quarterly and, with the exception of the commodity price index, are obtained from the Federal Reserve Bank of St. Louis (FRED) database. The commodity price index corresponds to the close of the Reuters/Jefferies-CRB Index at the end of the quarter. FF_t equals the average of the monthly federal funds rate over the quarter. The monetary aggregates are measured in millions of dollars at the end of the quarter. The data span the period between 1959:1 and 2006:4. We exclude the data prior to the onset of the Great Recession as well as subsequent data, given that the conduct of monetary policy during this period has been markedly different from earlier years. In particular, nonborrowed reserves took on negative numbers during almost all of 2008 – something that has never been observed in the documented history of this series – and then increased at a very fast pace in the following year. Total reserves, on the other hand, exploded during 2008 and have continued to increase at a very rapid pace. As can be seen in Fig. 15, all of the variables appear to exhibit a high degree of persistence.

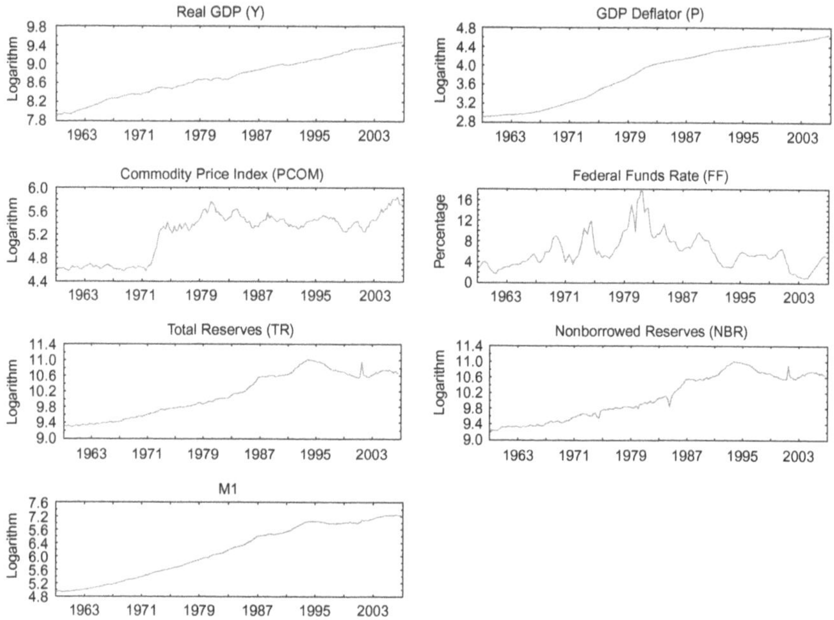

Fig. 15. Data used in the monetary policy VAR.

Multivariate Model Without Considering Cointegration

In this section, we consider the approach followed by a practitioner who would first pretest for unit roots in the individual series before estimating the VAR. Then, based on the test results, the practitioner would include the variables either in levels or first differences. As we mentioned before, we treat the estimated 7-variable VAR(4) model in levels as the DGP.

To simulate the data, we draw randomly from the vector y_t to obtain the starting values $y_1^{(i)}, \ldots, y_4^{(i)}$, and draw with replacement from the estimated vector of residuals $\hat{u}_t$ to obtain $u_t^{(i)}$. We generate 10,000 samples, $y_t^{(i)}$ for $i = 1, 2, \ldots, 10,000$, each consisting of a sequence of data of the same length as that used to estimate the 7-variable VAR(4).

Then, for each of the 10,000 trials we pretest the individual series in the vector $y_t^{(i)}$ ($Y_t^{(i)}$, $P_t^{(i)}$, $\text{PCOM}_t^{(i)}$, $\text{FF}_t^{(i)}$, $\text{TR}_t^{(i)}$, $\text{NBR}_t^{(i)}$, $M1_t^{(i)}$) using an ADF-GLS test at a 5% significance level. A time trend is included in the GLS detrending step for all variables except for $\text{FF}_t^{(i)}$. As in the previous section, the lag order for the unit root test is chosen by the modified information criterion of Ng and Perron (2001). If the pretest rejects the null, we keep the simulated variable in levels. Instead, if we fail to reject, we first difference the simulated variable. In all of the 10,000 replications, we estimate a VAR(4) model with each series either in levels or first differences depending on the outcome of the individual ADF-GLS test. Impulse response functions are then computed using the recursive identification scheme (Choleski decomposition) described in the previous section. Finally, we compute the responses to a federal funds shock and report the MSEs with respect to the impulse responses from the DGP.

The solid and dashed lines in Fig. 16 represent the MSEs of the impulse responses for the levels specification and the ADF-GLS pretesting strategy, respectively. The numbers on the horizontal axis represent quarters since the monetary policy shock. Note that for the variables that exhibit a large degree of persistence, the MSE deteriorates rapidly as the horizon

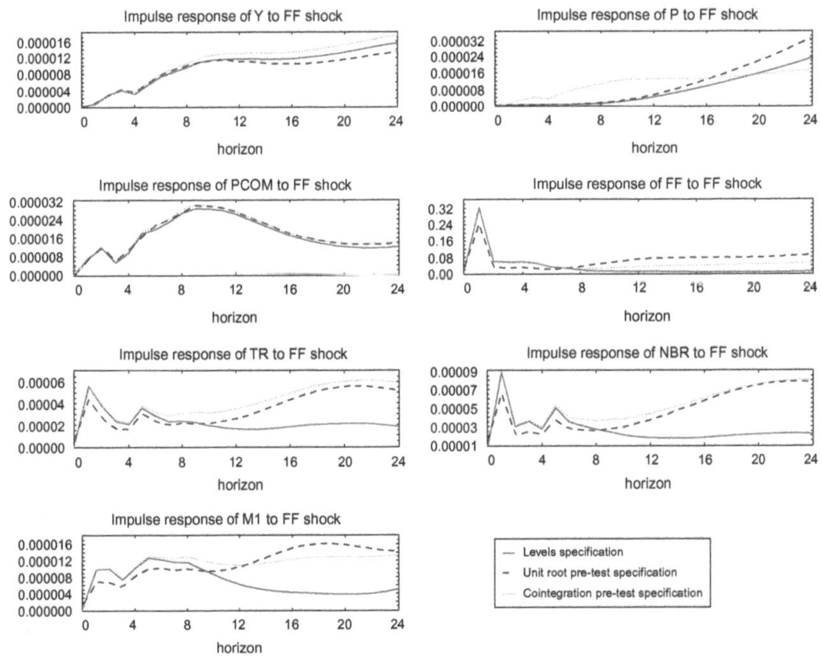

Fig. 16. MSEs for the impulse responses to a federal funds rate shock in CEE model in Eq. (8). The data are generated as described in the Section "A multivariate model used in monetary policy analysis." The impulse responses are identified using short-run (recursive) restrictions.

increases. This is particularly the case for the log of the GDP deflator, P_t. Instead, for the federal funds rate, the MSE increases a quarter after the shock but it drops in the following quarter staying about the same level in the long run.

Comparing the results for the levels specification and the DF-GLS pretest specification provides some interesting insights. First, pretesting for unit roots has the effect of increasing the MSE in the long run for all impulse response functions but real GDP. The increase in the MSE, relative to the levels specification, is larger for the monetary variables, which are characterized by lower persistence. As it can be seen from the scale of the MSEs, the cost associated with pretesting for unit roots is not large.[3]

Multivariate Model With Possible Cointegration

An alternative model selection strategy faced by a practitioner involves considering the possibility that some of the variables in the system might be cointegrated. In this case, a possible avenue would be to pretest for unit roots in each of the series in y_t, and then consider cointegration among the subset of variables that have a unit root. Although this strategy might be employed in practice, the most commonly used approach is to directly tests for the cointegration rank, without pretesting for unit roots. We thus follow the bulk of the literature and address the issue of cointegration in the full set of variables.

Thus, consider the case where y_t might have a VECM representation

$$\Delta y_t = \Pi y_{t-1} + \Gamma_1 \Delta y_{t-1} + \dots + \Gamma_{p-1} \Delta y_{t-p+1} + u_t$$

For instance, a number of possible cointegrating relationships in a model of monetary transmission is explored by Juselius (1998a, 1998b). One possibility is that equilibrium in the money market is attained via stationarity of the liquidity ratio $M_t - P_t - Y_t$, so that a cointegration relationship among these three variables exists. In addition, inflation and the nominal interest rate could be cointegrated given the stationarity of the real interest rate. Moreover, the IS relationship would suggest that the trend-adjusted real GDP and inflation (or the interest rate) are cointegrated. Finally, Strongin (1995) argues that, initially, monetary policy shocks lead only to changes in the composition of total reserves between borrowed and nonborrowed reserves. Thus, one could conjecture that changes in the ratio of nonborrowed to total reserves are only short-lived and that a long-run relationship exists so that TR_t and NBR_t are cointegrated.

In this section, we evaluate the effect of Johansen's (1988) method, used to pretest for the cointegration rank, on the MSE of the impulse response functions. In order to do this, we simulate 10,000 samples of the DGP as described in the previous section. For each of these samples, we follow a sequential procedure to determine the cointegrating rank of the 7-variable system (see Lütkepohl, 2005, and the references therein). That is, using the trace statistic, we test the sequence of null hypotheses

$$H_0 : \text{rank}(\Pi) = 0; H_0 : \text{rank}(\Pi) = 1, \ldots; H_0 : \text{rank}(\Pi) = 6 \quad (9)$$

and stop the test procedure when the null hypothesis cannot be rejected for the first time. The cointegrating rank is selected to be the value when the test is stopped. That is, if we cannot reject the null that $\text{rank}(\Pi) = 0$, the cointegrating rank is taken to be 0 and we estimate a model in first differences. If the null is rejected, we proceed to test $H_0 : \text{rank}(\Pi) = 1$. This continues until we are not able to reject the null. If the first time we reject the null is for $0 < \text{rank}(\Pi) = r < 7$, then the analysis proceeds with a cointegrating rank of r and a VECM is estimated. Finally, if we cannot reject the null that $\text{rank}(\Pi) = 7$, then we estimate a VAR in levels. We use the trace statistic at 5% significance level to test the sequence of null hypothesis describe in Eq. (9). The critical values for the trace test are taken from MacKinnon, Haug, and Michelis (1999).

Before we proceed to describe the effects of pretesting on the impulse response functions, it is worth mentioning that in 55% percent of the simulations, the trace statistic leads us to conclude that the rank $\text{rank}(\Pi) = 7$. Hence, in more than a half of the simulations we use a VAR in levels when computing the impulse response functions. This result is not surprising because our DGP is given by the estimated VAR in levels (Eq. (8)).

The comparison of the dashed and the dotted lines in Fig. 16 suggests some differences in the MSEs for the impulse responses when we use the ADF-GLS pretest specification and the Johansen pretest specification. Only slight differences are observed in the MSEs on impact and in the short run. As the horizon increases, the MSE for the impulse responses of Y_t, P_t, TR_t, and NBR_t are larger under the Johansen pretest strategy than if we pretest for unit roots.

The cost associated with pretesting for cointegrating rank, as measured by the MSEs, is not large. This finding is consistent with our previous findings based on short-run (recursive) identification scheme. Nevertheless, in the long run, pretesting for cointegration results in larger MSEs for all impulse responses – except for the log of the GDP deflator – than the levels specification.

CONCLUDING REMARKS AND PRACTICAL RECOMMENDATIONS

The practical relevance of structural VARs has been recently questioned (Cooley & Dwyer, 1998; Chari, Kehoe, & McGratten, 2008, among others) with criticisms targeting their theoretical underpinnings, identification assumptions, and statistical methodology. Despite these criticisms, the SVARs prove to be an indispensable tool for policy analysis and evaluation of dynamic economic models. However, their robustness to uncertainty about the magnitude of the largest roots in the system and possible co-movements between the variables has not been fully investigated. For example, the costs from erroneously imposing restrictions (by differencing the data and/or incorporating cointegrating relationships) on the models or from estimating unrestricted VAR models in levels (when the true model is a VECM) have not been quantified for the purpose of impulse response analysis under different identification schemes. In this article, we evaluated the robustness of alternative VAR specifications to deviations from exact unit roots and cointegration.

The main findings and practical recommendations that emerge from our results are the following. First, under the long-run identification scheme, specification strategies based on pretests and restricted VAR models are not robust to uncertainty about the largest roots of the process and may lead to highly distorted inference. On the other hand, impulse response estimates from VAR models using long-run restrictions are inconsistent when the non-unit root variables are expressed in levels, but have near-unit roots. Thus, applied researchers should exert caution in using models based on long-run restrictions. Nevertheless, we showed that in practice, the VAR specification with the non-unit root variable in levels is more robust to possible low-frequency co-movements and departures from exact unit roots.

Second, under the short-run identification scheme, the restricted (based on pretests for unit roots and cointegration) and unrestricted VAR specifications do not exhibit substantial differences in their computed impulse responses. In other words, the costs from imposing restrictions on the variables of the model (first differences or cointegration), when these restrictions do not hold exactly in the true model, do not tend to be too large when the structural impulse responses are identified through short-run restrictions. However, the unrestricted VAR in levels appears to be the most robust specification when there is uncertainty about the magnitude of the largest roots and the co-movement between the variables. We conclude that estimating VAR models in levels and identifying the structural impulse responses through short-run restrictions emerges as the most reliable strategy for applied work.

NOTES

1. Persistent variables can also be approximated by fractionally integrated processes. Allowing for this possibility is outside the scope of this article, although some of the results in our simulations may generalize to the case in which the DGP contains fractionally intergrated variables.
2. The results obtained when shocks to monetary policy are identified with shocks to nonborrowed reserves (e.g., Christiano & Eichenbaum, 1992; Eichenbaum, 1992) are virtually the same.
3. Simulation results (not reported here) suggest that using nonborrowed reserves to identify the monetary policy shock does not alter our conclusions. In fact, when the data are subject to an ADF-GLS pretest specification, the federal funds rate model and the nonborrowed reserves model exhibit MSEs of similar magnitudes for the main variables of interest (i.e., Y_t and P_t).

ACKNOWLEDGMENT

We would like to thank the participants at the 12th Annual Advances in Econometrics Conference for useful feedback. We are particularly indebted to Lutz Kilian for numerous detailed comments and insightful suggestions that substantially improved the presentation of the article.

REFERENCES

Bernanke, B. S., & Blinder, A. S. (1992). The federal funds rate and the channels of monetary transmission. *American Economic Review*, *82*, 901–921.

Blanchard, O., & Quah, D. (1989). The dynamic effects of aggregate demand and supply disturbances. *American Economic Review*, *79*, 655–673.

Chari, V. V., Kehoe, P. J., & McGrattan, E. R. (2008). Are structural VARs with long-run restrictions useful in developing business cycle theory? *Journal of Monetary Economics*, *55*, 1337–1352.

Christiano, L. J., & Eichenbaum, M. (1992). Identification and the liquidity effect of a monetary policy shock. In A. Cukierman, Z. Hercowitz, & L. Leiderman (Eds.), *Political economy, growth and business cycles* (pp. 335–370). London: MIT Press.

Christiano, L. J., Eichenbaum, M., & Evans, C. (1999). Monetary policy shocks: What have we learned and to what end? In M. Woodford & J. Taylor (Eds.), *Handbook of macroeconomics* (Vol. 1A), Amsterdam: New York, NY.

Christiano, L. J., Eichenbaum, M., & Vigfusson, R. (2006). Assessing structural VARs. In D. Acemoglu, K. Rogoff, & M. Woodford (Eds.), *NBER macroeconomics annual*. Cambridge, MA: MIT Press.

Cooley, T., & Dwyer, M. (1998). Business cycle analysis without much theory: A look at structural VARs. *Journal of Econometrics*, *83*, 57–88.

Eichenbaum, M. (1992). Comment on interpreting the macroeconomic time series facts: The effects of monetary policy. *European Economic Review*, *36*, 1001–1011.

Elliott, G. (1998). On the robustness of cointegration methods when regressors almost have unit roots. *Econometrica*, *66*, 149–158.

Gali, J. (1999). Technology, employment, and the business cycle: Do technology shocks explain aggregate fluctuations? *American Economic Review*, *89*, 249–271.

Gospodinov, N. (2004). Asymptotic confidence intervals for impulse responses of near-integrated processes. *Econometrics Journal*, *7*, 505–527.

Gospodinov, N. (2010). Inference in nearly nonstationary SVAR models with long-run identifying restrictions. *Journal of Business and Economic Statistics*, *28*, 1–12.

Gospodinov, N., Maynard, A., & Pesavento, E. (2011). Sensitivity of impulse responses to small low frequency co-movements: Reconciling the evidence on the effects of technology shocks. *Journal of Business and Economic Statistics*, *29*, 455–467.

Hansen, B. E. (1999). The grid bootstrap and the autoregressive model. *Review of Economics and Statistics*, *81*, 594–607.

Inoue, A., & Kilian, L. (2002). Bootstrapping autoregressive processes with possible unit roots. *Econometrica*, *70*, 377–391.

Johansen, S. (1988). Statistical analysis of cointegration vectors. *Journal of Economic Dynamics and Control*, *12*, 231–254.

Juselius, K. (1998a). A structured VAR for Denmark under changing monetary regimes. *Journal of Business and Economic Statistics*, *16*, 400–411.

Juselius, K. (1998b). Changing monetary transmission mechanisms within the EU. *Empirical Economics*, *23*, 455–481.

Kilian, L. (1998a). Small-sample confidence intervals for impulse response functions. *Review of Economics and Statistics*, *80*, 218–230.

Kilian, L. (1998b). Confidence intervals for impulse responses under departures from normality. *Econometric Reviews*, *17*, 1–29.

Kilian, L., & Chang, P. L. (2000). How accurate are confidence intervals for impulse responses in large VAR models? *Economics Letters*, *69*, 299–307.

Lütkepohl, H. (2005). *New introduction to multiple time series analysis*. Berlin: Springer-Verlag.

Lütkepohl, H., & Reimers, H. E. (1992). Impulse response analysis of cointegrated systems. *Journal of Economic Dynamics and Control*, *16*, 53–78.

MacKinnon, J. G., Haug, A. A., & Michelis, L. (1999). Numerical distribution functions of likelihood ratio tests for cointegration. *Journal of Applied Econometrics*, *14*, 563–577.

Mikusheva, A. (2012). One-dimensional inferences in autoregressive models in a potential presence of a unit root. *Econometrica*, *80*, 173–212.

Ng, S., & Perron, P. (2001). Lag length selection and the construction of unit root tests with good size and power. *Econometrica*, *69*, 1519–1554.

Pesavento, E., & Rossi, B. (2006). Small sample confidence intervals for multivariate impulse response functions at long horizons. *Journal of Applied Econometrics*, *21*, 1135–1155.

Phillips, P. C. B. (1998). Impulse response and forecast error variance asymptotics in nonstationary VARs. *Journal of Econometrics*, *83*, 21–56.

Runkle, D. E. (1987). Vector autoregressions and reality. *Journal of Business and Economic Statistics*, *5*, 437–442.

Sims, C. A. (1986). Are forecasting models usable for policy analysis? *Federal Reserve Bank of Minneapolis Quarterly Review*, *10*, 2–16.

Sims, C. A., Stock, J. H., & Watson, M. W. (1990). Inference in linear time series with some unit roots. *Econometrica*, *58*, 113–144.

Sims, C. A. (1992). Interpreting the macroeconomic time series facts: The effects of monetary policy. *European Economic Review*, 36, 975–1000.

Stock, J. H. (1991). Confidence intervals for the largest autoregressive root in U.S. macroeconomic time series. *Journal of Monetary Economics*, *28*, 435–459.

Strongin, S. (1995). The identification of monetary policy disturbances explaining the liquidity puzzle. *Journal of Monetary Economics*, *35*, 463–497.

EVALUATING THE ACCURACY OF FORECASTS FROM VECTOR AUTOREGRESSIONS☆

Todd E. Clark and Michael W. McCracken

ABSTRACT

This article surveys recent developments in the evaluation of point and density forecasts in the context of forecasts made by vector autoregressions. Specific emphasis is placed on highlighting those parts of the existing literature that are applicable to direct multistep forecasts and those parts that are applicable to iterated multistep forecasts. This literature includes advancements in the evaluation of forecasts in population (based on true, unknown model coefficients) and the evaluation of forecasts in the finite sample (based on estimated model coefficients). The article then examines in Monte Carlo experiments the finite-sample properties of some tests of equal forecast accuracy, focusing on the comparison of VAR forecasts to AR forecasts. These experiments show the tests to

☆The views expressed herein are solely those of the authors and do not necessarily reflect the views of the Federal Reserve Bank of Cleveland, Federal Reserve Bank of St. Louis, Federal Reserve System, or any of its staff.

VAR Models in Macroeconomics – New Developments and Applications: Essays in Honor of Christopher A. Sims
Advances in Econometrics, Volume 32, 117–168

ISSN: 0731-9053/doi:10.1108/S0731-9053(2013)0000031004

behave as should be expected given the theory. For example, using critical values obtained by bootstrap methods, tests of equal accuracy in population have empirical size about equal to nominal size.

Keywords: Prediction; forecasting; out-of-sample

JEL classifications: C53; C52; C12; C32

INTRODUCTION

Since the seminal work of Doan, Litterman, and Sims (1984), Litterman (1986), and Sims (1980), vector autoregressions (VARs) have become an essential tool for out-of-sample forecasting in macroeconomics. Their key role is evident in the long list of central bank research studies that deploy VARs, sometimes explicitly indicating VARs are in use at a particular central bank. While a complete listing is beyond the scope of the article, recent examples of such studies include Coletti and Murchison (2002), Kapetanios, Labhard, and Price (2008), Bjornland, Gerdrup, Jore, Smith, and Thorsrud (2012), Andersson, Karlsson, and Svensson (2007), Beauchemin and Zaman (2011), and Baumeister and Kilian (2012b, 2012c). By now, VARs are commonly used for not only point forecasts but also density forecasts, sometimes in the form of fan charts.

Research on VAR forecasting has almost always included comparisons of the accuracy of VAR forecasts to the accuracy of forecasts from univariate benchmark models. Early work, such as Litterman (1986), commonly used random walk forecasts as a benchmark for comparison, reporting Theil's U statistic, without a formal measure of the statistical significance of differences in forecast accuracy. It has also become common to compare forecasts from VARs to forecasts from autoregressive (AR) models, since, for many (not all) variables of interest, an AR model forecast is more accurate than a random walk. In recent work, it has become increasingly common to provide some measure of the statistical significance of differences in forecast accuracy, sometimes with caveats about a need to interpret the results with some caution because existing asymptotic theory does not necessarily apply to forecasts from VARs.[1] Indeed, while much is now known about proper inference for forecasts from a univariate model (summarized in surveys such as West (2006) and Clark and McCracken (2011b, 2012b)), much less is known about proper methods for testing equal accuracy of forecasts from VARs.

In this article, we review possible methods for evaluating the accuracy of out-of-sample forecasts from VARs. As we detail below, what methods are appropriate hinges on three criteria. The first is the model specification used to produce multistep forecasts. If the forecasts are generated with a direct multistep (DMS) model specification, existing results on the evaluation of forecasts from a single model can be applied directly.[2] But if multistep forecasts are produced by iterating forward 1-step ahead projections based on VAR models, some of the existing results either become much more complicated to use or cannot be used.

A second criterion is the null hypothesis of interest – whether the relevant null is equal accuracy in the finite sample or equal accuracy in population.[3] The distinction between these two hypotheses is whether parameter estimation error is captured under the null hypothesis (finite sample) or whether the parameters are treated as known under the null (population). If the null of interest is the former, the results of Giacomini and White (2006) permit the use of t-tests of equal accuracy and standard normal critical values, but only for forecasts produced with a rolling window of data for model estimation and not for forecasts produced with an expanding window. Conceptually, this poses some challenge with the evaluation of VAR forecasts, since recursive estimation schemes are commonly used (although the rolling scheme is also regularly used). For a recursive scheme, at this point in time, there are no theoretically justified methods for testing equal accuracy of VAR forecasts. From our Monte Carlo analysis, it seems as though the same tests that are theoretically appropriate under a rolling scheme and the null of equal accuracy in the finite sample may have reasonable size and power properties under a recursive scheme, but these tests lack a theoretical basis for the recursive scheme.

If the null hypothesis of interest is equal accuracy in population, the third criterion is the relationship of the VAR to the benchmark forecasting model. If the VAR does not nest the benchmark model, it is possible to rely on the results of West (1996) for conducting formal forecast inference, using standard asymptotic distributions. (If the forecasts are obtained by iteration, West's (1996) results apply, but can be computationally difficult.) However, if the VAR does nest the benchmark model of interest, as is often the case, inference becomes more complicated. At the 1-step horizon, the results of Clark and McCracken (2001, 2005a, 2012a) and McCracken (2007) provide a basis for inference, using asymptotic critical values obtained by Monte Carlo simulation or bootstrap. For forecasts at longer horizons obtained by iteration, it becomes difficult to derive the relevant asymptotic distributions. Still, in Monte Carlo experiments, some of the simulation or bootstrap methods that work well at the 1-step horizon or with DMS forecasts also seem to work well with iterated forecasts from VARs for the purpose of assessing predictability in population (as noted above, however, this finding does not pertain to the null of equal accuracy in the finite sample).

We begin the article by detailing a brief summary of possible methods for testing equal accuracy of out-of-sample forecasts from VARs, focusing on the relevant econometric theory. Throughout our analysis we focus on evaluating accuracy variable-by-variable, rather than on evaluating accuracy for the system of variables as a whole. Although a few studies have considered multivariate measures of the accuracy of point forecasts (see, e.g., Komunjer & Owyang (2012)), most research on point forecasts focuses on univariate measures of accuracy. In the case of density evaluation, a measure of accuracy for the system of variables – the log predictive score for the vector of variables in the model – is perhaps more natural than in the case of point forecasts, but even studies of density accuracy commonly consider univariate measures. We leave as a subject for future research the further development of tests of the accuracy of point and density forecasts for a vector of variables.

We follow the theory treatment with a Monte Carlo analysis of a small range of tests of equal accuracy, using data generating processes that allow us to impose equal accuracy in population, but not in the finite sample. In the interest of brevity, we focus the Monte Carlo study on nested models, comparing VAR forecasts to AR forecasts, both point and density. In the Monte Carlo experiments, the tests of equal mean square error and forecast encompassing considered have size and power properties that seem reasonable. Perhaps most importantly, when the true model is a set of AR equations that are nested by the VAR model, and therefore the null hypothesis of no predictive accuracy in population is satisfied, using bootstrap methods to generate critical values yields rejection rates reasonably close to the nominal size of 10 percent. As would be expected and appropriate, tests of equal accuracy in the finite sample have lower rejection rates than tests of equal accuracy in population.

Throughout we emphasize tests of predictive ability based on forecasts generated by stationary VARs. We do so because the literature on tests of predictive ability is almost completely silent on inference when unit roots are present. The sole exception is Corradi, Swanson, and Olivetti (2001), who extend the analysis of West (1996) to permit cointegrating relationships but do so only for one-step ahead forecasts from univariate models. Their results indicate that in certain special cases, tests of predictive ability are asymptotically normal as in West (1996). More generally, these tests have nonstandard asymptotic distributions due to the presence of unit roots. They argue that many of their results extend directly to longer horizon, DMS forecasts but do not show this explicitly. Regardless, there exist no results that are specifically applicable to VARs with unit roots and hence we focus our attention on stationary VARs.

In addition, we focus exclusively on analytical results that ignore real-time vintage data issues. We do so in order to focus attention on the primary results within the literature without having to introduce the additional notation needed to keep track of data revisions. This should not be interpreted as meaning that real-time data issues are unimportant for out-of-sample inference. To the contrary, Clark and McCracken (2009) extend the results in West (1996) allowing for real-time data and find that while many test statistics remain asymptotically normal they do so with a distinct asymptotic variance structure due to the presence of data revisions. In the context of this article, the topic is of particular importance since VARs are a popular tool for forecasting at many central banks which, by their very nature, must construct forecasts in real-time using sequences of vintage data. Even so, for clarity we omit the issue within the remainder.

The rest of the article proceeds as follows. The second section gives a very brief overview of estimation and iterated multistep (IMS) forecasting with stationary VARs, with an emphasis on how these forecasts differ from those based on DMS models. The third section discusses the existing literature in the context of tests of population predictive ability as compared to tests of finite sample predictive ability. The fourth section presents Monte Carlo results on the performance of some basic testing procedures. The fifth section concludes.

CONSTRUCTING THE FORECASTS

Before examining options for testing the predictive ability of VAR-based forecasts, it is helpful to review methods for generating the forecasts. Researchers and practitioners use a range of approaches to estimating VAR models and generating forecasts. As we explain below, how the forecasts are generated bears on how the forecasts can be evaluated. This section proceeds by first distinguishing IMS and DMS approaches to producing forecasts and then detailing the most common VAR estimation approaches.

DMS versus IMS Forecasts

In the VAR forecasting literature, it is most common to generate multistep point forecasts with the IMS approach (detailed below). However, it is sometimes the case that multistep point forecasts are obtained with a DMS approach (e.g., Koop, 2012). As Schorfheide (2005) and Marcellino, Stock, and Watson (2006) note, the DMS approach may yield more accurate forecasts than the IMS approach if the VAR is sufficiently misspecified. In the VAR results of Carriero, Clark, and Marcellino (2012), IMS and DMS forecast accuracy is comparable, with a little advantage to the IMS approach, more so for density forecasts than point forecasts. Likewise, Marcellino et al. (2006) report evidence in favor of the IMS approach, provided enough lags are included in the VAR model.

While IMS forecasts are more common than DMS forecasts in VAR-based work, the testing literature tends to, either theoretically or by example, focus on procedures designed to accommodate forecasts constructed using the DMS approach. In some parts of the literature, like that of Clark and McCracken (2001, 2005a), theoretical results on tests of equal forecast accuracy between two nested models only apply to DMS forecasts. In other parts of the literature, like West (1996) or McCracken (2000), both IMS and DMS forecasts are allowed but the theory is much more easily implemented when DMS rather than IMS forecasts are used and hence the empirical applications focus on DMS forecasts.

To get a feel for the difference between the two forecasting procedures and for why results for out-of-sample tests of predictive ability may differ when IMS versus DMS forecasts are used, suppose we are interested in forecasting a scalar y_1 variable τ-periods in the future using the current, time t value of y_1 as a predictor. A DMS approach to point forecasting would use OLS to estimate the parameters in a simple linear regression of the form

$$y_{1,t} = \beta_0 + \beta_1 y_{1,t-\tau} + \varepsilon_{1,t} = \boldsymbol{\beta}' \boldsymbol{x}_{t-\tau} + \varepsilon_{1,t} \tag{1}$$

and then construct the point forecast as $\hat{y}_{1,t+\tau} = \hat{\beta}_{0,t} + \hat{\beta}_{1,t} y_{1,t} = \hat{\boldsymbol{\beta}}_t' \boldsymbol{x}_t$ where $\hat{\boldsymbol{\beta}}_t = (\hat{\beta}_{0,t}, \hat{\beta}_{1,t})' = (t^{-1} \sum_{s=1}^{t-\tau} \boldsymbol{x}_s \boldsymbol{x}_s')^{-1} (t^{-1} \sum_{s=1}^{t-\tau} \boldsymbol{x}_s y_{1,s+\tau})$. In contrast an IMS approach to forecasting would use OLS to estimate the parameters in a slightly different simple linear regression model of the form

$$y_{1,t} = \beta_0 + \beta_1 y_{1,t-1} + \varepsilon_{1,t} = \boldsymbol{\beta}' \boldsymbol{x}_{t-1} + \varepsilon_{1,t} \tag{2}$$

With these parameters in hand, and with an implicit assumption that the errors in Eq. (2) form a martingale difference sequence, the IMS point forecast of $y_{1,t+\tau}$ is $\hat{y}_{1,t+\tau}=\hat{\beta}_{0,t}(1+\sum_{j=1}^{\tau-1}\hat{\beta}_{1,t}^{j})+\hat{\beta}_{1,t}^{\tau}y_{1,t}$, where $\hat{\boldsymbol{\beta}}_t=(\hat{\beta}_{0,t},\hat{\beta}_{1,t})'=(t^{-1}\sum_{s=1}^{t-1}\boldsymbol{x}_s\boldsymbol{x}_s')^{-1}(t^{-1}\sum_{s=1}^{t-1}\boldsymbol{x}_s y_{1,s+1})$. The DMS and IMS forecasts are clearly related to one another. They are both linear functions of $\boldsymbol{x}_t=(1,y_{1,t})'$. Moreover, if the errors in Eq. (2) form a martingale difference sequence, it is straightforward to show that (the population parameters) β_0 and β_1 in Eq. (1) equal $\beta_0(1+\sum_{j=1}^{\tau-1}\beta_1^j)$ and β_1^τ in Eq. (2) and hence in large enough samples the forecasts are identical.

That said, as a practical matter the point forecasts made using the IMS approach are a more complicated function of the parameter estimates $\hat{\boldsymbol{\beta}}_t$ than they are when using the DMS approach. The complexity of the problem only increases as we move from IMS forecasts from AR models to IMS forecasts from VARs. To see this consider extending the models in Eqs. (1) and (2) to allow information from other time t predictors. In particular suppose we are interested in forecasting a scalar y_1 variable τ-periods in the future using the current, time t value of y_1 as a predictor and the time t values of predictors y_2 and y_3. A DMS approach to point forecasting would use OLS to estimate the parameters in the linear regression

$$y_{1,t}=\beta_0+\beta_1 y_{1,t-\tau}+\beta_2 y_{2,t-\tau}+\beta_3 y_{3,t-\tau}+\varepsilon_{1,t}=\boldsymbol{\beta}'\boldsymbol{x}_{t-\tau}+\varepsilon_{1,t} \tag{3}$$

and then, construct the point forecast as $\hat{y}_{1,t+\tau}=\hat{\beta}_{0,t}+\hat{\beta}_{1,t}y_{1,t}+\hat{\beta}_{2,t}y_{2,t}+\hat{\beta}_{3,t}y_{3,t}=\hat{\boldsymbol{\beta}}_t\boldsymbol{x}_t$ where $\boldsymbol{x}_t=(1,y_{1,t},y_{2,t},y_{3,t})'$ and $\hat{\boldsymbol{\beta}}_t=(t^{-1}\sum_{s=1}^{t-\tau}\boldsymbol{x}_s\boldsymbol{x}_s')^{-1}(t^{-1}\sum_{s=1}^{t-\tau}\boldsymbol{x}_s y_{1,s+\tau})$. In contrast an IMS approach to forecasting would use OLS to estimate the parameters in the three equation linear system

$$\begin{aligned}\begin{pmatrix} y_{1,t}\\ y_{2,t}\\ y_{3,t}\end{pmatrix} &= \begin{pmatrix} c_1\\ c_2\\ c_3\end{pmatrix}+\begin{pmatrix} A_{11} & A_{12} & A_{13}\\ A_{21} & A_{22} & A_{23}\\ A_{31} & A_{32} & A_{33}\end{pmatrix}\begin{pmatrix} y_{1,t-1}\\ y_{2,t-1}\\ y_{3,t-1}\end{pmatrix}+\begin{pmatrix} \varepsilon_{1,t}\\ \varepsilon_{2,t}\\ \varepsilon_{3,t}\end{pmatrix}\\ &= \boldsymbol{C}+\mathbf{A}\boldsymbol{Y}_{t-1}+\boldsymbol{\varepsilon}_t\end{aligned} \tag{4}$$

which is equivalent to

$$\begin{aligned}\boldsymbol{Y}_t &= \boldsymbol{\Lambda}\boldsymbol{x}_{t-1}+\boldsymbol{\varepsilon}_t\\ &= (\boldsymbol{x}_{t-1}'\otimes\mathbf{I}_3)\boldsymbol{\beta}+\boldsymbol{\varepsilon}_t\end{aligned}$$

if we define $\boldsymbol{Y}_t = (y_{1,t}, y_{2,t}, y_{3,t})'$, $\boldsymbol{\beta} = \text{vec}(\boldsymbol{\Lambda})$, and

$$\boldsymbol{\Lambda} = \begin{pmatrix} c_1 & A_{11} & A_{12} & A_{13} \\ c_2 & A_{21} & A_{22} & A_{23} \\ c_3 & A_{31} & A_{32} & A_{33} \end{pmatrix}$$

With these parameters in hand, and with an implicit assumption that the errors in Eq. (4) form a martingale difference sequence, the IMS point forecast of $y_{1,t+\tau}$ is $\hat{y}_{1,t+\tau} = \boldsymbol{\iota}'((\mathbf{I} + \sum_{j=0}^{\tau-1} \hat{\mathbf{A}}_t^j)\hat{\boldsymbol{C}}_t + \hat{\mathbf{A}}_t^\tau \boldsymbol{Y}_t)$ where $\boldsymbol{\iota}' = (1, 0, 0)$ and $\hat{\boldsymbol{\Lambda}}_t = (t^{-1} \sum_{s=1}^{t-1} \boldsymbol{Y}_{s+1} \boldsymbol{x}_s')(t^{-1} \sum_{s=1}^{t-1} \boldsymbol{x}_s \boldsymbol{x}_s')^{-1}$. The DMS and IMS forecasts continue to be related to one another. They are both linear functions of $\boldsymbol{x}_t$ and as was the case above, if the errors in Eq. (4) form a martingale difference, it is straightforward to show that (the population) $\boldsymbol{\beta}$ in Eq. (3) equals $(\boldsymbol{\iota}'(\mathbf{I} + \sum_{j=0}^{\tau-1} \mathbf{A}^j)\boldsymbol{C}, \boldsymbol{\iota}'\mathbf{A}^\tau)$ and hence in large enough samples the forecasts are identical. Regardless, the IMS forecasts remain a much more complicated function of $\boldsymbol{\beta}$ for the IMS forecasts in Eq. (4) than for the DMS forecasts in Eq. (3).

Estimating and Forecasting with VARs

Of course the comparison between DMS and IMS forecasts can be generalized further to allow more lags and more predictors in the system. Accordingly, it will be useful to generalize the notation for VAR-based IMS forecasts for use throughout the remainder of the text. To do so suppose that y_1 is the scalar variable we wish to predict at the τ-step horizon but now assume that there exist $n-1$ other predictors and that we allow for $m \geq 1$ lags. The VAR model can be written as

$$\boldsymbol{Y}_t = \boldsymbol{C} + \mathbf{A}(L)\boldsymbol{Y}_{t-1} + \boldsymbol{\varepsilon}_t$$

where $\boldsymbol{Y} = (y_1, y_2, ..., y_n)'$, $\varepsilon = (\varepsilon_1, \varepsilon_2, ..., \varepsilon_n)'$, and $\mathbf{A}(L) = \sum_{j=1}^{m} \mathbf{A}_j L^j$ for $n \times 1$ and $n \times n$ parameter matrices $\boldsymbol{C}$ and $\mathbf{A}_j$, $j = 1, ..., m$, respectively. This is again equivalent to

$$Y_t = \boldsymbol{\Lambda}\boldsymbol{x}_{t-1} + \boldsymbol{\varepsilon}_t$$

$$= (\boldsymbol{x}_{t-1}' \otimes \mathbf{I}_n)\boldsymbol{\beta} + \boldsymbol{\varepsilon}_t$$

if we define $\boldsymbol{x}_t = (1, \boldsymbol{Y}_t', ..., \boldsymbol{Y}_{t-m+1}')'$, $\boldsymbol{\beta} = vec(\boldsymbol{\Lambda})$, and $\boldsymbol{\Lambda} = (\boldsymbol{C}, \mathbf{A}_1, ..., \mathbf{A}_m)$.

Under the maintained assumption that the model errors $\boldsymbol{\varepsilon}_t$ form a martingale difference sequence, the IMS τ-step ahead forecasting function $\hat{y}_{1,t+\tau}=\boldsymbol{\iota}'\hat{\boldsymbol{Y}}_{t+\tau}$, with $\boldsymbol{\iota}=(1,0,...,0)'$, can be inferred from the recursion $\hat{\boldsymbol{Y}}_{t+\tau}=\boldsymbol{\iota}'(\boldsymbol{C}+\mathbf{A}_1\hat{\boldsymbol{Y}}_{t+\tau-1}+...+\mathbf{A}_m\hat{\boldsymbol{Y}}_{t+\tau-m})$, where for the moment we abstract from how the regression parameters are estimated. But of course the regression parameters do need to be estimated before the forecasts can be constructed. While there are many ways to do so, we focus on OLS and Bayesian estimation since they appear to be the most common in applied work.

OLS Estimation

As noted in the previous section, the OLS estimates of $\boldsymbol{\Lambda}$ take the simple closed form $\hat{\boldsymbol{\Lambda}}_t=(t^{-1}\sum_{s=1}^{t-1}\boldsymbol{Y}_{s+1}\boldsymbol{x}_s')(t^{-1}\sum_{s=1}^{t-1}\boldsymbol{x}_s\boldsymbol{x}_s')^{-1}$. For reference in the following section it is worth noting that if we're willing to assume that $\boldsymbol{\varepsilon}_t$ forms a martingale difference sequence and the observables are stationary, the vector of parameter estimates can be represented as equivalent to the true parameter values plus a term driven by estimation error

$$\hat{\boldsymbol{\Lambda}}_t=\boldsymbol{\Lambda}+(t^{-1}\sum_{s=1}^{t-1}\boldsymbol{\varepsilon}_{s+1}\boldsymbol{x}_s')(t^{-1}\sum_{s=1}^{t-1}\boldsymbol{x}_s\boldsymbol{x}_s')^{-1}$$

Equivalently, if we map this into the definition of $\boldsymbol{\beta}=\text{vec}(\boldsymbol{\Lambda})$, this takes the form

$$\hat{\boldsymbol{\beta}}_t=\boldsymbol{\beta}+\mathbf{B}(t)\mathbf{H}(t)$$

where $\mathbf{B}=((t^{-1}\sum_{s=1}^{t-1}\boldsymbol{x}_s\boldsymbol{x}_s')^{-1}\otimes\mathbf{I}_n)$ and $\boldsymbol{H}(t)=\text{vec}(t^{-1}\sum_{s=1}^{t-1}\boldsymbol{\varepsilon}_{s+1}\boldsymbol{x}_s')$.

Bayesian Estimation

By now, in the VAR forecasting literature, it is probably more common to use models estimated by Bayesian methods than models estimated by least squares, particularly with larger models. One key reason is that studies such as Clark and McCracken (2008), Banbura, Giannone, and Reichlin (2010), and Wright (2012) have shown forecasts from BVARs to often be more accurate than forecasts from OLS-estimated VARs, with gains that increase in the size of the model (however, there are examples, such as Baumeister and Kilian (2012a), in which an OLS-estimated VAR forecasts better than a BVAR). In no small part this is due to the number of degrees of freedom used up when estimating a VAR. Bayesian methods, like ridge regression techniques, use shrinkage of the parameters toward a given set of priors in order to increase the (finite-sample) precision of the parameters.

As detailed in resources such as Kadiyala and Karlsson (1997), Geweke and Whiteman (2006), and Karlsson (2012), BVARs can be estimated with a wide array of approaches, depending in part on the priors that are used.[4] Due to computational constraints, Litterman (1986) developed an approach that is equivalent to applying ridge regression on an equation-by-equation basis, based on an assumption of an error covariance matrix that is fixed and diagonal. With current computational power, Monte Carlo simulation methods are now often seen as more or less the standard for BVAR forecasting (except with large models). The simulation method required hinges on the particular prior used. We will proceed to briefly describe some approaches and then conclude with a discussion of how Bayesian estimation fits within the methods admitted in the theory underlying forecast evaluation.

Among possible prior specifications or approaches, the simplest case is the Normal-Wishart prior, which yields a posterior of the same form. Under a Normal-Wishart prior, the posterior mean of the VAR coefficients can be computed directly, without simulation. As a consequence, the posterior mean of 1-step ahead forecasts, which is a linear function of the VAR coefficients, can also be computed directly, without simulation. Studies such as Carriero et al. (2012) and Koop (2012) use a Normal-Wishart prior with VARs specified in DMS form to similarly obtain multistep forecasts without simulation.[5] In the IMS case, because the multistep forecasts are nonlinear functions of the VAR coefficients, computing the mean of the posterior distribution of forecasts requires simulation of the distribution. In practice, though, Carriero et al. (2012) find that the properly simulated mean is essentially the same as an approximate mean computed by using the posterior mean coefficients to compute multistep forecasts by iteration, as outlined above.

More specifically, consider the implementation of a Normal-Wishart prior and posterior, abstracting for now from the details of the prior. By grouping the coefficients and explanatory variables, the VAR can be written as

$$\boldsymbol{Y}_t = \boldsymbol{\Lambda x}_{t-1} + \boldsymbol{\varepsilon}_t \tag{5}$$

or, even more compactly, as

$$\mathbf{Y} = \mathbf{X}\boldsymbol{\Lambda}' + \mathbf{E} \tag{6}$$

where $\mathbf{Y}=[\boldsymbol{Y}_1,\ldots,\boldsymbol{Y}_T]'$, $\mathbf{X}=[\boldsymbol{x}_1,\ldots,\boldsymbol{x}_T]'$, and $\mathbf{E}=[\boldsymbol{\varepsilon}_1,\ldots,\boldsymbol{\varepsilon}_T]'$ are, respectively, $T\times n$, $T\times(nm+1)$, and $T\times n$ matrices. The conjugate Normal-Wishart prior takes the form

$$\mathbf{\Lambda}'|\Sigma \sim N(\mathbf{\Lambda}_0', \Sigma\otimes\Omega_0),\quad \Sigma\sim IW(\mathbf{S}_0, \nu_0) \tag{7}$$

which yields a conditional posterior distribution that is also Normal-Wishart (Zellner, 1971)

$$\mathbf{\Lambda}'|\Sigma,\mathbf{Y} \sim N(\overline{\mathbf{\Lambda}}', \Sigma\otimes\overline{\Omega}),\quad \Sigma|\mathbf{Y}\sim IW(\overline{\mathbf{S}}, \overline{\nu}) \tag{8}$$

Defining $\hat{\mathbf{\Lambda}}'$ and $\hat{\mathbf{E}}$ as the OLS estimates, we have that

$$\begin{aligned} \overline{\mathbf{\Lambda}}' &= (\Omega_0^{-1}+\mathbf{X}'\mathbf{X})^{-1}(\Omega_0^{-1}\mathbf{\Lambda}_0'+\mathbf{X}'\mathbf{Y}) \\ \overline{\Omega} &= (\Omega_0^{-1}+\mathbf{X}'\mathbf{X})^{-1} \\ \overline{\nu} &= \nu_0+T \end{aligned} \tag{9}$$

$$\overline{\mathbf{S}} = \mathbf{\Lambda}_0'+\hat{\mathbf{E}}'\hat{\mathbf{E}}+\hat{\mathbf{\Lambda}}\mathbf{X}'\mathbf{X}\hat{\mathbf{\Lambda}}'+\mathbf{\Lambda}_0\Omega_0^{-1}\mathbf{\Lambda}_0'-\overline{\mathbf{\Lambda}}\overline{\Omega}^{-1}\overline{\mathbf{\Lambda}}'$$

Karlsson (2012) details the steps involved in simulating the posterior distribution.

This Normal-Wishart approach has both advantages and disadvantages. One advantage is that the posterior mean of the coefficients can be computed without simulation, which means that 1-step ahead forecasts can also be computed without simulation and multistep point forecasts can at least be approximated very well without simulation. These considerations have led to the use of the Normal-Wishart prior and posterior (without simulation) in large model studies such as Banbura et al. (2010) and Koop (2012). A second advantage is that the posterior distribution can be quickly and easily simulated. The simulations are quick because the posterior variance of the coefficients has a Kronecker structure, which greatly speeds the necessary matrix computations. The simulations are easy because the distributions can be directly simulated, without resorting to full Markov chain methods such as Gibbs sampling. The disadvantage of the Normal-Wishart approach is that it does not permit one of the key components of the Minnesota prior as developed by Litterman (1986): it does not permit one to make the prior on lags of variable j in equation i tighter than the prior on lags of variable i in the same equation. The Normal-Wishart specification instead requires the prior variances to be symmetric across equations.

Avoiding this disadvantage of the Normal-Wishart specification requires using Markov chain methods such as Gibbs sampling to estimate the BVAR. For example, to implement a Minnesota prior that includes a different rate of shrinkage on "own" lags versus "other" lags, the most common approach rests on the Normal-diffuse prior detailed in Kadiyala and Karlsson (1997), among others. The model is then estimated by Gibbs sampling, using conditional distributions for the VAR coefficients and the error variance–covariance matrix. Karlsson (2012) details the steps involved in this sampling. While readily tractable with smaller models, the computations become more costly as the model size increases, due to the need for repeated computations with a large variance–covariance matrix of coefficients that lacks a Kronecker structure.[6]

Regardless of the particular approach used to estimate a VAR by Bayesian methods, Bayesian estimation can be viewed as falling within the scheme of model estimation approaches assumed in the evaluation theory reviewed in the next section. Most immediately, because the asymptotic approach of Giacomini and White (2006) does not involve any assumptions about model estimation, forecasts from Bayesian VARs are allowed in their framework. While the asymptotic approaches of West (1996), West and McCracken (1998), and Clark and McCracken (2001, 2005a) each make assumptions about model estimation methods, Bayesian estimation can be seen as fitting within these assumptions. Consider, for example, the nested model theory of Clark and McCracken (2001, 2005a), which requires some form of least squares estimation, but uses asymptotics that treat the estimation sample size as diverging to infinity. As the sample size used for estimation expands, the importance of the prior used in Bayesian estimation declines, reaching zero in the limit. This is clear in the posterior mean of the VAR coefficients given in Eq. (9). As the sample size grows to infinity, the data terms $\mathbf{X}'\mathbf{X}$ and $\mathbf{X}'\mathbf{Y}$ will overwhelm the prior terms Ω_0 and $\Omega_0^{-1}\mathbf{\Lambda}_0$, and the posterior mean of the VAR coefficients will converge to the OLS estimates and, in turn, the population value. Therefore, asymptotically, Bayesian estimation can be seen as equivalent to the least squares estimation admitted by the asymptotic theory. Admittedly, this interpretation is not as straightforward under priors that require Gibbs sampling or other alternatives to the Normal-Wishart prior and posterior. But even in these cases, it will generally be the case that, asymptotically, the posterior mean of the VAR coefficient estimates will converge to the OLS estimates.

THEORY

In this section we provide an overview of asymptotically valid tests of predictability with an eye toward distinguishing between inference when DMS or IMS forecasts are being used. For each test, the statistics are sample averages of functions of a sequence of forecasts and other observables available at the time of the forecast or when the forecast error is observed.

Examples and Notation

Before proceeding to the theoretical results we lay out some notation in the context of the example of a test of zero mean prediction error when forecasts are made at a four-step ahead horizon. We do so once assuming DMS forecasts akin to that in Eq. (3) and once assuming IMS forecasts using a quarterly frequency trivariate VAR(1) akin to that in Eq. (4). We then generalize this very simple testing procedure to pairwise tests of equally accurate point and density forecasts.

Suppose we have quarterly data from 1947:Q1 to 2012:Q2 for a total of $T = 262$ observations. The sample is split into an in-sample period and an out-of-sample period, respectively. The former is used to estimate the parameters associated with the first forecast while the latter denotes the number of forecast origins from which four quarter-ahead ($\tau = 4$) forecasts can be constructed. Let 1947:Q1−1979:Q4 denote the in-sample period and 1980: Q1−2011:Q2 denote the out-of-sample period, giving us $R = 132$ in-sample observations and $P - 3 = 127$ out-of-sample periods. Note that data from 2011:Q3 to 2012:Q2 are not treated as forecast origins but will be used to construct the forecast errors.

How we construct the test of zero mean prediction error depends on whether DMS or IMS forecasts are being used.

1. *DMS forecast*: At each forecast origin $t = R, \ldots, T-4$, the linear regression model in Eq. (3) is estimated by OLS using all available data and hence $\hat{\boldsymbol{\beta}}_t = (t^{-1}\sum_{s=1}^{t-4}\boldsymbol{x}_s\boldsymbol{x}_s')^{-1}(t^{-1}\sum_{s=1}^{t-4}\boldsymbol{x}_s y_{1,s+4})$. Together with the predictors $\boldsymbol{x}_t$ these parameter estimates yield a sequence of forecast errors of the form $\hat{\varepsilon}_{1,t+4} = y_{1,t+4} - \hat{y}_{1,t+4}$, where $\hat{y}_{1,t+4} = \boldsymbol{x}_t'\hat{\boldsymbol{\beta}}_t$. The test of zero mean prediction error is based on a statistic of the form $(P-3)^{-1/2}\sum_{t=R}^{T-4}\hat{\varepsilon}_{1,t+4}/\hat{\Omega}^{1/2}$, where $\hat{\Omega}$ is a consistent estimate of an unknown long-run variance Ω (characterized later in section "Tests of Predictive Ability in Population: Non-Nested Models"). Note that the numerator of the statistic

can be written in the generic form $(P-3)^{-1/2}\sum_{t=R}^{T-4} f(\boldsymbol{X}_{t+4},\hat{\boldsymbol{\beta}}_t)$, where $\boldsymbol{X}_{t+4}$ is the vector of observables $(y_{1,t+4},\boldsymbol{x}_t^{'})^{'}$ and $f(\boldsymbol{X}_{t+4},\hat{\boldsymbol{\beta}}_t)=\hat{\varepsilon}_{1,t+4}$. In the next section we describe how to consistently estimate Ω and conduct inference.

2. *IMS forecast*: At each forecast origin $t=R,...,T-4$, the VAR(1) model in Eq. (4) is estimated by OLS using all available data and hence $\hat{\mathbf{\Lambda}}_t=(t^{-1}\sum_{s=1}^{t-1}\mathbf{Y}_{s+1}\boldsymbol{x}_s^{'})(t^{-1}\sum_{s=1}^{t-1}\boldsymbol{x}_s\boldsymbol{x}_s^{'})^{-1}$. With these parameter estimates in hand, forecasts of $y_{1,t+4}$ take the form $\hat{y}_{1,t+4}=\boldsymbol{\iota}^{'}((\mathbf{I}+\sum_{j=0}^{3}\hat{\mathbf{\Lambda}}_t^{j})\hat{\boldsymbol{C}}_t+\hat{\mathbf{\Lambda}}_t^{4}\boldsymbol{Y}_t))=\boldsymbol{\iota}^{'}((\mathbf{I}+\sum_{j=0}^{3}\hat{\mathbf{\Lambda}}_t^{j})\hat{\boldsymbol{C}}_t,\hat{\mathbf{\Lambda}}_t^{4})\boldsymbol{x}_t$, with subsequent forecast errors $\hat{\varepsilon}_{1,t+4}=y_{1,t+4}-\hat{y}_{1,t+4}$. Based on these forecast errors a test of zero mean prediction error is constructed using the same statistic $(P-3)^{-1/2}\sum_{t=R}^{T-4}\hat{\varepsilon}_{1,t+4}/\hat{\Omega}^{1/2}$, where $\hat{\Omega}$ is a consistent estimate of an unknown long-run variance Ω which may differ from that above associated with DMS forecasts. As was the case for the DMS forecast above, the numerator of the statistic can be written in the generic form $(P-3)^{-1/2}\sum_{t=R}^{T-4} f(\boldsymbol{X}_{t+4},\hat{\boldsymbol{\beta}}_t)$, where $\boldsymbol{X}_{t+4}$ is the vector of observables $(y_{1,t+4},\boldsymbol{x}_t^{'})^{'}$ and $f(\mathbf{X}_{t+4},\hat{\boldsymbol{\beta}}_t)=\hat{\varepsilon}_{1,t+4}$, but only after defining $\hat{\boldsymbol{\beta}}_t=\mathrm{vec}(\hat{\mathbf{\Lambda}}_t)$. It's also worth noting that since $\hat{\varepsilon}_{1,t+4}$ is a fourth order polynomial in the elements of $\hat{\mathbf{\Lambda}}_t$, the function $f(\mathbf{X}_{t+4},\hat{\boldsymbol{\beta}}_t)$ is a far more complex function of $\hat{\boldsymbol{\beta}}_t$ when IMS forecasts are used than it is when DMS forecasts are used.

This simple example of testing for zero mean prediction error is easily generalized to a wider range of null hypotheses of the form $H_0: Ef(\boldsymbol{X}_{t+4},\boldsymbol{\beta}^{*})=0$. Perhaps the most common of these tests are ones designed to compare the relative accuracy of two models. For example, suppose that in the zero mean prediction error example we actually had two models, indexed as $i=1,2$, that were used to construct forecasts (either both DMS or both IMS) rather than just one model. Both models could be used to construct a sequence of forecast errors $\hat{\varepsilon}_{1,t+4}^{(i)}$, $t=R,...,T-4$, and, for a given loss function $L(\cdot)$, the relative accuracy of the two models could be evaluated. By far the most common of these loss functions is the quadratic, for which $L(\varepsilon)=\varepsilon^2$. If this loss function is used the relative accuracy of the two models can be tested using the same framework described above if we define $f(\boldsymbol{X}_{t+4},\hat{\boldsymbol{\beta}}_t)=(\hat{\varepsilon}_{1,t+4}^{(1)})^2-(\hat{\varepsilon}_{1,t+4}^{(2)})^2$, where $\boldsymbol{X}_{t+4}$ denotes the vector of observables used to construct both forecast errors and $\hat{\boldsymbol{\beta}}_t$ is defined as the stacked vector of parameter estimates $\hat{\boldsymbol{\beta}}_t=(\hat{\boldsymbol{\beta}}_t^{(1)'},\hat{\boldsymbol{\beta}}_t^{(2)'})^{'}$ from the two models.

While quadratic loss is the most common among all loss functions, density forecasts are increasingly common. Most typically, the density forecasts are obtained by Monte Carlo simulation, using an IMS approach to horizons greater than one. However, studies such as Koop (2012) obtain density forecasts directly (i.e., without simulation) from the t-distribution implied by a Normal-Wishart posterior and a DMS specification. Regardless, suppose that two distinct VARs are used, each of which implies a distinct predictive density $g_i(\boldsymbol{X}_{t+\tau}, \hat{\boldsymbol{\beta}}_t^{(i)})$. Amisano and Giacomini (2007) consider comparing the accuracy of the two VARs based on their relative log predictive scores. In our notation, this is equivalent to testing the (population) null hypothesis that $H_0: E[\ln g_1(\boldsymbol{X}_{t+\tau}, \boldsymbol{\beta}^{(1)^*}) - \ln g_2(\boldsymbol{X}_{t+\tau}, \boldsymbol{\beta}^{(2)^*})] = 0$ using the function $f(X_{t+\tau}, \hat{\boldsymbol{\beta}}_t) = \ln g_1(X_{t+\tau}, \hat{\boldsymbol{\beta}}_t^{(1)}) - \ln g_2(\boldsymbol{X}_{t+\tau}, \hat{\boldsymbol{\beta}}_t^{(2)})$ to construct the test statistic $(P-\tau+1)^{-1/2}\sum_{t=R}^{T-\tau} f(\boldsymbol{X}_{t+\tau}, \hat{\boldsymbol{\beta}}_t)/\hat{\Omega}^{1/2}$.

Tests of Predictive Ability in Population: Non-Nested Models

In reviewing existing theory for VAR forecast evaluation, we begin with inference based on West (1996), which, for comparisons of models, can be applied with non-nested models only. In subsequent sections we take up nested model comparisons and then testing based on Giacomini and White (2006) asymptotics.

Regardless of whether DMS or IMS forecasts are used, the theory in West (1996) can be used to conduct asymptotically valid inference for many (though not all) statistics that are scaled sample averages of the form

$$(P-\tau+1)^{-1/2}\sum_{t=R}^{T-\tau} f(\boldsymbol{X}_{t+\tau}, \hat{\boldsymbol{\beta}}_t) \tag{10}$$

under modest assumptions on (i) the observables $\boldsymbol{X}_{t+\tau}$, (ii) how the parameter estimates $\hat{\boldsymbol{\beta}}_t$ are constructed, and (iii) smoothness conditions on the function $f(\boldsymbol{X}_{t+\tau}, \boldsymbol{\beta})$ as a function of $\boldsymbol{\beta}$. The major building block of this theory is Lemma 4.3 of West (1996), in which it is shown that under the null hypothesis $H_0: Ef(\boldsymbol{X}_{t+\tau}, \boldsymbol{\beta}^*) = 0$, the out of sample average of the function $f(\boldsymbol{X}_{t+\tau}, \hat{\boldsymbol{\beta}}_t)$ can be decomposed into two parts

$$\begin{aligned}&(P-\tau+1)^{-1/2}\sum_{t=R}^{T-\tau} f(\boldsymbol{X}_{t+\tau}, \hat{\boldsymbol{\beta}}_t)\\ &= (P-\tau+1)^{-1/2}\sum_{t=R}^{T-\tau} f(\boldsymbol{X}_{t+\tau}, \boldsymbol{\beta}^*) + \mathbf{FB}((P-\tau+1)^{-1/2}\sum_{t=R}^{T-\tau} \boldsymbol{H}(t)) + o_p(1)\end{aligned} \tag{11}$$

where $\mathbf{F} = E\partial f(\boldsymbol{X}_{t+\tau}, \boldsymbol{\beta})/\partial\boldsymbol{\beta}|_{\beta=\beta^*}$, $\mathbf{B}$ is a non-stochastic matrix satisfying $\hat{\boldsymbol{\beta}}_t - \boldsymbol{\beta}^* = \mathbf{B}\boldsymbol{H}(t) + o_{a.s.}(1)$, and $\boldsymbol{H}(t) = t^{-1}\sum_{s=1}^{t-\tau} \boldsymbol{h}_{s+\tau}$ for a sequence of mean zero increments $\boldsymbol{h}_{s+\tau}$. The first right-hand side term in Eq. (11) denotes that part of the statistic associated with the forecast had the population values of the parameters been known while the second right-hand side term denotes that part of the statistic associated with the fact that the model parameters cannot be observed but instead must be estimated.

Building on Lemma 4.3, West (1996) shows that the statistic $(P-\tau+1)^{-1/2}\sum_{t=R}^{T-\tau} f(\boldsymbol{X}_{t+\tau},\hat{\boldsymbol{\beta}}_t)/\hat{\Omega}^{1/2}$ is asymptotically standard normal so long as Ω is positive and $\hat{\Omega}$ is consistent for its population counterpart. The details of how to estimate Ω are perhaps the main technical developments in West (1996). Before providing this result, some additional notation and assumptions are needed.[7]

(A1) $\hat{\boldsymbol{\beta}}_t=\boldsymbol{\beta}^*+\mathbf{B}\boldsymbol{H}(t)+o_{a.s.}(1)$, where for some mean zero process $\boldsymbol{h}_{t+\tau}=\boldsymbol{h}_{t+\tau}(\boldsymbol{\beta}^*)$ (with $\boldsymbol{h}$ denoting the orthogonality conditions used to estimate parameters, such as $\boldsymbol{h}_{t+\tau}=\boldsymbol{x}_t u_{t+\tau}$ for a single linear regression), $\boldsymbol{H}(t)$ equals $t^{-1}\sum_{s=1}^{t-\tau}\boldsymbol{h}_{s+\tau}$, $R^{-1}\sum_{s=t-R+1}^{t-\tau}\boldsymbol{h}_{s+\tau}$, and $R^{-1}\sum_{s=1}^{R-\tau}\boldsymbol{h}_{s+\tau}$ for the recursive, rolling, and fixed schemes, respectively, and $\mathbf{B}$ denotes a non-stochastic matrix.

(A2) The vector $(f(\boldsymbol{X}_{t+\tau},\boldsymbol{\beta}^*),\boldsymbol{h}'_{t+\tau})'$ is covariance stationary and satisfies mild mixing and moment conditions.

(A3) $\lim_{P,R\to\infty}P/R=\pi$, a constant that is finite for the rolling and fixed schemes but can be infinite for the recursive scheme.

(A4) The vector $\mathbf{F}=E[\partial f(\boldsymbol{X}_{t+\tau},\boldsymbol{\beta})/\partial\boldsymbol{\beta}]_{\beta=\beta^*}$ is finite.[8]

(A5) Ω is positive definite.

Given these assumptions, West (1996) shows that the asymptotic variance Ω can take a variety of forms depending on how the parameters are estimated

$$\Omega=S_{ff}+2\lambda_{fh}\mathbf{FBS}'_{fh}+\lambda_{hh}\mathbf{FBS}_{hh}\mathbf{B}'\mathbf{F}' \tag{12}$$

where $S_{ff}=\lim_{T\to\infty}\mathrm{Var}(T^{-1/2}\sum_{s=1}^{T-\tau}f(\boldsymbol{X}_{s+\tau},\boldsymbol{\beta}^*))$, $\mathbf{S}_{hh}=\lim_{T\to\infty}\mathrm{Var}(T^{-1/2}\sum_{s=1}^{T-\tau}\boldsymbol{h}_{s+\tau})$, $S_{fh}=\lim_{T\to\infty}\mathrm{Cov}(T^{-1/2}\sum_{s=1}^{T-\tau}f(\boldsymbol{X}_{s+\tau},\boldsymbol{\beta}^*),T^{-1/2}\sum_{s=1}^{T-\tau}\boldsymbol{h}_{s+\tau})$, and

	$\lambda_{fh}=$	$\lambda_{hh}=$
Recursive	$1-\pi^{-1}\ln(1+\pi)$	$2(1-\pi^{-1}\ln(1+\pi))$
Rolling, $\pi\leq 1$	$\pi/2$	$\pi-\pi^2/3$
Rolling, $1<\pi<\infty$	$1-(2\pi)^{-1}$	$1-(3\pi)^{-1}$
Fixed	0	π

In Eq. (12) we see that Ω consists of three terms. The first, S_{ff}, is the long-run variance of the measure of accuracy when the parameters are known. The third term, $\lambda_{hh}\mathbf{FBS}_{hh}\mathbf{B}'\mathbf{F}'$, captures the contribution of the variance due purely to the fact that we do not observe $\boldsymbol{\beta}^*$ but must estimate it instead. The second term, $2\lambda_{fh}\mathbf{FBS}'_{fh}$, captures the covariance between the measure of accuracy and the estimation error associated with $\hat{\boldsymbol{\beta}}_t$. Because the parameter estimates can be constructed using three different observation windows (recursive, rolling, and fixed) it is not surprising that the terms that arise due to estimation error depend on that choice via the terms λ_{fh} and λ_{hh}.

Estimating Ω is reasonably straightforward when either IMS or DMS forecasts are used but can be considerably simpler when DMS forecasts are used. Regardless of which approach is taken, since $\hat{\pi} = P/R \rightarrow \pi$ and both λ_{fh} and λ_{hh} are continuous in π, substituting $\hat{\pi}$ for π is sufficient for estimating both λ_{fh} and λ_{hh}. For the remaining components one needs to know whether the IMS or DMS approach is used to construct multistep forecasts. The distinction matters for how $\mathbf{B}$, $\boldsymbol{h}_{t+\tau}$, and $\mathbf{F}$ are defined and, in turn, how each is to be estimated.

To see this consider the test of zero mean prediction error described above with $f(\boldsymbol{X}_{t+4}, \hat{\boldsymbol{\beta}}_t) = \hat{\varepsilon}_{1,t+4}$. In the following we delineate how each of the remaining components of Ω can be estimated depending on whether DMS or IMS forecasts are used.

- $\mathbf{B}$: For the DMS case $\mathbf{B} = (E\boldsymbol{x}_t\boldsymbol{x}'_t)^{-1}$ and hence $\hat{\mathbf{B}} = (T^{-1}\sum_{s=1}^{T-4}\boldsymbol{x}_s\boldsymbol{x}'_s)^{-1}$ is a consistent estimator of $\mathbf{B}$. For the IMS case $\mathbf{B} = ((E\boldsymbol{x}_t\boldsymbol{x}'_t)^{-1} \otimes \boldsymbol{I}_n)$ and hence $\hat{\mathbf{B}} = ((T^{-1}\sum_{s=1}^{T-4}\boldsymbol{x}_s\boldsymbol{x}'_s)^{-1} \otimes \mathbf{I}_n)$ is a consistent estimator of $\mathbf{B}$.[9]
- S_{ff}, $\mathbf{S}_{fh}$, and $\mathbf{S}_{hh}$: For the long-run variances and covariances needed to compute the test statistic, West (1996) shows that standard kernel-based estimators, which are averages of $f(\boldsymbol{X}_{t+4}, \hat{\boldsymbol{\beta}}_t)$ and $\boldsymbol{h}_{t+4}(\hat{\boldsymbol{\beta}}_t)$, are consistent. We know $f(\boldsymbol{X}_{t+4}, \hat{\boldsymbol{\beta}}_t) = \hat{\varepsilon}_{1,t+4}$ for both the IMS and DMS cases. For the DMS case it suffices to define $\boldsymbol{h}_{t+4}(\hat{\boldsymbol{\beta}}_t) = \boldsymbol{x}_t\hat{\varepsilon}_{1,t+4}$ while for the IMS case it suffices to define $\boldsymbol{h}_{t+4}(\hat{\boldsymbol{\beta}}_t) = vec(\hat{\boldsymbol{\varepsilon}}_{t+4}\boldsymbol{x}'_t)$. With these in hand if we define $\bar{f} = (P-3)^{-1}\sum_{t=R}^{T-4} f(\boldsymbol{X}_{t+4}, \hat{\boldsymbol{\beta}}_t)$, $\hat{\Gamma}_{ff}(j) = (P-3)^{-1} \sum^{t=R+jT-\tau} (f(\boldsymbol{X}_{t+4}, \hat{\boldsymbol{\beta}}_t) - \bar{f})(f(\boldsymbol{X}_{t+4-j}, \hat{\boldsymbol{\beta}}_{t-j}) - \bar{f})'$, $\hat{\boldsymbol{\Gamma}}_{hh}(j) = T^{-1}\sum_{t=j+1}^{T-4} \boldsymbol{h}_{t+4}(\hat{\boldsymbol{\beta}}_t)\boldsymbol{h}'_{t+4-j}(\hat{\boldsymbol{\beta}}_{t-j})$ and $\hat{\boldsymbol{\Gamma}}_{fh}(j) = (P-3)^{-1}\sum_{t=R+j}^{T-\tau} f(\boldsymbol{X}_{t+4}, \hat{\boldsymbol{\beta}}_t)\boldsymbol{h}'_{t+4-j}(\hat{\boldsymbol{\beta}}_{t-j})$, with $\hat{\boldsymbol{\Gamma}}_{ff}(j) = \hat{\boldsymbol{\Gamma}}_{ff}(-j)$, $\hat{\boldsymbol{\Gamma}}_{hh}(j) = \hat{\boldsymbol{\Gamma}}'_{hh}(-j)$, and $\hat{\boldsymbol{\Gamma}}_{fh}(j) = \hat{\boldsymbol{\Gamma}}'_{fh}(-j)$, the long-run variance estimates $\hat{S}_{ff}$, $\hat{\mathbf{S}}_{hh}$, and $\hat{\mathbf{S}}_{fh}$ are then constructed by weighting the relevant leads and lags of these covariances, as in HAC estimators such as the one developed by Newey and West (1987).[10]

- **F**: This term poses probably the greatest complication in estimating Ω, particularly with the IMS approach to generating multistep forecasts. To see this recall that $\mathbf{F} = E[\partial f(\boldsymbol{X}_{t+4}, \boldsymbol{\beta})/\partial\boldsymbol{\beta}]_{\beta=\beta^*}$.[11] In the DMS case this is trivially equal to $-E\boldsymbol{x}_t$ and can easily be estimated using $\hat{\mathbf{F}} = -(P-3)^{-1}\sum_{t=R}^{T-4}\boldsymbol{x}_t$. But in the IMS case this term is significantly more complicated but still feasible using derivations from Section 3.5.2 of Lutkepohl (1991). Specifically if we define the $(nm+1)\times(nm+1)$ matrix

$$\mathbf{W} = \begin{pmatrix} 1 & \boldsymbol{0} & \ldots & \boldsymbol{0} \\ & \boldsymbol{\Lambda} & & \\ \boldsymbol{0} & \mathbf{I}_{n(m-1)} & & \boldsymbol{0} \end{pmatrix}$$

the $n\times(nm+1)$ selection matrix $\mathbf{J}_1 = (\mathbf{0}_{n\times 1}, \mathbf{I}_n, \mathbf{0}_{n\times n(m-1)})$, and $\Phi_i = \mathbf{J}_1\mathbf{W}^i\mathbf{J}_1'$ we obtain

$$\mathbf{F} = -\iota'\sum_{i=0}^{\tau-1}(E\boldsymbol{x}_t'(\mathbf{W}')^{\tau-1-i}\otimes\Phi_i)$$

If we use the estimator $\hat{\boldsymbol{\Lambda}}_T$ to obtain $\hat{\mathbf{W}}$ and $\hat{\Phi}_i$ and define $\bar{\boldsymbol{x}} = (P-3)^{-1}\sum_{t=R}^{T-4}\boldsymbol{x}_t$, a straightforward estimator of **F** is simply

$$\hat{\mathbf{F}} = -\iota'\sum_{i=0}^{\tau-1}(\bar{\boldsymbol{x}}'(\hat{\mathbf{W}}')^{\tau-1-i}\otimes\hat{\Phi}_i)$$

Interestingly, for three special cases estimating Ω is as simple as using the estimate $\hat{\Omega} = \hat{S}_{ff}$. This arises when the second and third terms in Eq. (12), those due to estimation error, cancel and hence we say the estimation error is asymptotically irrelevant.

Case 1. If $\pi = 0$, then both λ_{fh} and λ_{hh} are zero and hence $\Omega = S_{ff}$. This case arises naturally when the sample split is chosen so that the number of out-of-sample observations is small relative to the number of in-sample observations. Chong and Hendry (1986) first observed that parameter estimation error is irrelevant if P is small relative to R.

Case 2. If $\mathbf{F} = \mathbf{0}$, then $\Omega = S_{ff}$. This case arises under certain very specific circumstances but arises most naturally when the measure of "accuracy" is explicitly used when estimating the model parameters and holds regardless of whether DMS or IMS forecasts are used. The canonical example is the use of a quadratic loss function (MSE) to evaluate the accuracy of forecasts from two non-nested models estimated by OLS. In this situation, the **F** term equals zero and estimation error is asymptotically irrelevant.

Case 3. There are instances where $-S_{fh}\mathbf{B}'\mathbf{F}' = \mathbf{F}\mathbf{B}\mathbf{S}_{hh}\mathbf{B}'\mathbf{F}'$ and hence under the recursive scheme, estimation error is asymptotically irrelevant. In this case, it is not so much that any particular term equals zero but that the sum of the components just happens to cancel to zero. One such example is our test of zero mean prediction error so long as that model contains an intercept. In this case, estimation error is asymptotically irrelevant regardless of whether DMS or IMS forecasts are used. See West (1996, 2006) and West and McCracken (1998) for other examples.

But more generally, all three components of Ω will have to be estimated and in particular $\mathbf{F}$ will have to be estimated. Fortunately, the formula for $\mathbf{F}$ associated with a test of zero mean prediction error can be generalized fairly easily in many useful cases. Specifically, if we assume that the moment function of interest satisfies $f(\boldsymbol{X}_{t+\tau}, \hat{\boldsymbol{\beta}}_t) = \tilde{f}(\boldsymbol{X}_{t+\tau}, y_{1,t+\tau} - \hat{y}_{1,t+\tau}) = \tilde{f}_{t+\tau}(\hat{\varepsilon}_{1,t+\tau})$, and hence the parameter estimate only affects the function through the forecast error (as it does for tests of zero mean prediction error, efficiency, and serial correlation), we obtain

$$\mathbf{F} = -\iota' \sum_{i=0}^{\tau-1} \left(E\left(\frac{\partial \tilde{f}_{t+\tau}(\varepsilon_{1,t+\tau})}{\partial \varepsilon_1} \boldsymbol{x}_t'\right)(\mathbf{W}')^{\tau-1-i} \otimes \Phi_i\right)$$

For this scenario the same formula for $\hat{\mathbf{F}}$ applies if we continue to use the estimator $\hat{\boldsymbol{\Lambda}}_T$ to obtain $\hat{\mathbf{W}}$ and $\hat{\Phi}_i$ and redefine $\bar{\boldsymbol{x}}$ as $(P-3)^{-1} \sum_{t=R}^{T-4} \frac{\partial \tilde{f}_{t+\tau}(\hat{\varepsilon}_{1,t+\tau})}{\partial \varepsilon_1} \boldsymbol{x}_t'$.

Tests of Predictive Ability in Population: Nested Models

In the theoretical discussion above the statistic $(P-\tau+1)^{-1/2} \sum_{t=R}^{T-\tau} f(\boldsymbol{X}_{t+\tau}, \hat{\boldsymbol{\beta}}_t)$ was asymptotically normal with variance Ω. To achieve this result, however, it must be the case that Ω is positive – a point emphasized directly in West (1996) through his assumptions. Unfortunately Ω is not trivially positive in certain practical situations including tests of equal forecast accuracy or encompassing when the two nested models are being compared. To see the issue consider a simple example in which DMS forecasts from the AR model

$$y_{1,t} = \beta_0^{(1)} + \beta_1^{(1)} y_{1,t-\tau} + \varepsilon_{1,t}^{(1)} = \boldsymbol{\beta}^{(1)'} \boldsymbol{x}_{t-\tau}^{(1)} + \varepsilon_{1,t}^{(1)}$$

are being compared to those from

$$y_{1,t} = \beta_0^{(2)} + \beta_1^{(2)} y_{1,t-\tau} + \beta_2^{(2)} y_{2,t-\tau} + \beta_3^{(2)} y_{3,t-\tau} + \varepsilon_{1,t}^{(2)} = \beta^{(2)\prime} x_{t-\tau}^{(2)} + \varepsilon_{1,t}^{(2)}$$

in terms of their mean squared errors. Specifically suppose that both models are estimated by OLS and we define $(\hat{\varepsilon}_{1,t+\tau}^{(j)})^2 = (y_{1,t+\tau} - \hat{\beta}_t^{(j)\prime} x_{t-\tau}^{(j)})^2$ for the restricted and unrestricted models $j = 1, 2$, respectively. Under the null hypothesis that $E((\varepsilon_{1,t+\tau}^{(1)})^2 - (\varepsilon_{1,t+\tau}^{(2)})^2) = 0$, Clark and McCracken (2001, 2005a) and McCracken (2007) show that the statistic $(P - \tau + 1)^{-1/2} \sum_{t=R}^{T-\tau} f(X_{t+\tau}, \hat{\beta}_t)$ with $f(X_{t+\tau}, \hat{\beta}_t) = (\hat{\varepsilon}_{1,t+\tau}^{(1)})^2 - (\hat{\varepsilon}_{1,t+\tau}^{(2)})^2$ converges in probability to zero and is therefore not asymptotically normal is any useful sense.

To get around this problem they extend the results in West (1996) to a framework that allows for nested model comparisons but do so only when the models being compared are OLS estimated linear models and the forecasts are formed using the DMS approach. Specifically, under assumptions close to those considered in West (1996) they extend West's Lemma 4.3 by showing

$$\sum_{t=R}^{T-\tau} f(X_{t+\tau}, \hat{\beta}_t) = \sum_{t=R}^{T-\tau} ((\hat{\varepsilon}_{1,t+\tau}^{(1)})^2 - (\hat{\varepsilon}_{1,t+\tau}^{(2)})^2)$$

$$= 2 \sum_{t=R}^{T-\tau} h_{1,t+\tau}^{(2)\prime} (-\mathbf{J}\mathbf{B}^{(1)}\mathbf{J}' + \mathbf{B}^{(2)}) H^{(2)}(t)$$

$$- \sum_{t=R}^{T-\tau} H^{(2)}(t)' (-\mathbf{J}\mathbf{B}^{(1)}\mathbf{J}' + \mathbf{B}^{(2)}) H^{(2)}(t) + o_p(1)$$

$$= O_p(1)$$

where $\mathbf{B}^{(j)} = (E x_t^{(j)} x_t^{(j)\prime})^{-1}$, $h_{1,t+\tau}^{(2)} = \varepsilon_{1,t+\tau}^{(2)} x_t^{(2)}$, $H^{(2)}(t) = t^{-1} \sum_{s=1}^{t-\tau} h_{1,s+\tau}^{(2)}$, and $\mathbf{J} = (\mathbf{I}_{\dim(x_t^{(1)})}, \mathbf{0})'$.

Building upon that expansion they derive the asymptotic distributions of the $\text{MSE} - t = (P - \tau + 1)^{-1/2} \sum_{t=R}^{T-\tau} f(X_{t+\tau}, \hat{\beta}_t) / \hat{S}_{ff}^{1/2}$ and $\text{MSE} - F = \sum_{t=R}^{T-\tau} f(X_{t+\tau}, \hat{\beta}_t) / ((P - \tau + 1)^{-1} \sum_{t=R}^{T-\tau} (\hat{\varepsilon}_{1,t+\tau}^{(2)})^2)$ tests of equal forecast accuracy as well as the $\text{ENC} - t$ and $\text{ENC} - F$ tests of forecast encompassing, for which $f(X_{t+\tau}, \hat{\beta}_t) = \hat{\varepsilon}_{1,t+\tau}^{(1)} (\hat{\varepsilon}_{1,t+\tau}^{(1)} - \hat{\varepsilon}_{1,t+\tau}^{(2)})$. (Note that, in the Monte Carlo section, we also write out these test statistics using just forecast errors.)

Each of the four statistics have asymptotic distributions that can be represented using stochastic integrals of quadratics of Brownian motion and are, in nearly all instances, not asymptotically normal. While useful, the analytics are derived after restricting attention to DMS forecasts since

$(\hat{\varepsilon}^{(1)}_{1,t+\tau})^2 - (\hat{\varepsilon}^{(2)}_{1,t+\tau})^2$ (and $\hat{\varepsilon}^{(1)}_{1,t+\tau}(\hat{\varepsilon}^{(1)}_{1,t+\tau} - \hat{\varepsilon}^{(2)}_{1,t+\tau})$) are then only quadratic functions of the parameter estimates and the necessary second order Taylor expansion of $\sum_{t=R}^{T-\tau}((\hat{\varepsilon}^{(1)}_{1,t+\tau})^2 - (\hat{\varepsilon}^{(2)}_{1,t+\tau})^2)$ or $\sum_{t=R}^{T-\tau}\hat{\varepsilon}^{(1)}_{1,t+\tau}(\hat{\varepsilon}^{(1)}_{1,t+\tau} - \hat{\varepsilon}^{(2)}_{1,t+\tau})$, as a function of $\boldsymbol{\beta}_1$ and $\boldsymbol{\beta}_2$, is straightforward. As detailed in such sources as Clark and McCracken (2005a, 2012a,b), critical values from these asymptotic distributions can easily be obtained by bootstrap methods.

In contrast, if the forecasts were constructed using an IMS approach, the second order Taylor expansion of $\sum_{t=R}^{T-\tau}((\hat{\varepsilon}^{(1)}_{1,t+\tau})^2 - (\hat{\varepsilon}^{(2)}_{1,t+\tau})^2)$ as a function of $\boldsymbol{\beta}_1$ and $\boldsymbol{\beta}_2$ would be much more complicated. In particular, if $\tau=4$ we know that $\sum_{t=R}^{T-\tau}((\hat{\varepsilon}^{(1)}_{1,t+\tau})^2 - (\hat{\varepsilon}^{(2)}_{1,t+\tau})^2)$ is an 8th order polynomial in both $\boldsymbol{\beta}_1$ and $\boldsymbol{\beta}_2$. That is not to say that a similar collection of asymptotics would not be applicable when IMS methods are used to construct forecasts, rather that the results have not been formally derived anywhere in the literature, in large part because to do so would be algebraically tedious and difficult. For that reason, in the fourth section we provide Monte Carlo evidence on these tests when the competing model is a VAR that nests a baseline AR model and all forecasts are generated using an IMS approach. We will compare the tests against bootstrapped critical values.

Tests of Finite-Sample Predictive Ability

The theoretical results in West (1996) and other studies such as Clark and McCracken (2001, 2005a) focus on testing a null hypothesis of the form $H_0 : Ef(\boldsymbol{X}_{t+\tau}, \boldsymbol{\beta}^*) = 0$. In words, this hypothesis is designed to evaluate the predictive ability of a model (or models) if the true parameter values of the model(s) are known. But as a practical matter, forecasting agents never know the true parameter values of a model and must estimate them. Hence, intuitively, one might prefer to evaluate the predictive content of the model accounting for the fact that the model parameters must be estimated and will never be known. Doing so changes the null hypothesis to something akin to $H_0 : Ef(\boldsymbol{X}_{t+\tau}, \hat{\boldsymbol{\beta}}_t) = 0$, where the estimation error associated with the parameter estimates is now captured under the null hypothesis.

While such a hypothesis could be considered for any function $f(\boldsymbol{X}_{t+\tau}, \hat{\boldsymbol{\beta}}_t)$, including that associated with zero mean prediction error, the hypothesis arises most clearly when comparing two ostensibly nested models. In the theory of Clark and McCracken (2001, 2005a) the null of equal population predictive ability maps directly into testing whether the additional predictors in the unrestricted model are associated with coefficients that are zero and hence the hypothesis is $H_0 : Ef(\boldsymbol{X}_{t+\tau}, \boldsymbol{\beta}^*) = E((\varepsilon^{(1)}_{1,t+\tau})^2 - (\varepsilon^{(2)}_{1,t+\tau})^2) = 0$.

In contrast, a null of equal finite sample predictive ability between two nested models is akin to the hypothesis $H_0 : Ef(\boldsymbol{X}_{t+\tau},\hat{\boldsymbol{\beta}}_t) = E((\hat{\varepsilon}^{(1)}_{1,t+\tau})^2 - (\hat{\varepsilon}^{(2)}_{1,t+\tau})^2) = 0$. For this latter hypothesis, the issue isn't so much whether certain coefficients are zero or not, but whether they are nonzero *and* whether they can be estimated sufficiently precisely in a finite sample to be useful for forecasting.

Even though the null hypothesis $H_0 : Ef(\boldsymbol{X}_{t+\tau},\hat{\boldsymbol{\beta}}_t) = 0$ is reasonably intuitive, the results in West (1996) or Clark and McCracken (2001, 2005a) do not apply under this null and hence their theory cannot be used to infer the asymptotic distribution of $(P-\tau+1)^{-1/2}\sum_{t=R}^{T-\tau} f(\boldsymbol{X}_{t+\tau},\hat{\boldsymbol{\beta}}_t)$ under this null regardless of whether DMS or IMS forecasts are used. Instead, theoretical results in Giacomini and White (2006) can be used to conduct inference. Perhaps most importantly from the standpoint of this article, their results are applicable regardless of whether DMS or IMS forecasts are being used. Specifically, Giacomini and White (2006) show that so long as all parameter estimates are estimated using a small rolling window of observations (small in the sense that R is finite and P diverges to infinity), $(P-\tau+1)^{-1/2}\sum_{t=R}^{T-\tau} f(\boldsymbol{X}_{t+\tau},\hat{\boldsymbol{\beta}}_t) \to^d N(0,S_{\hat{f}\hat{f}})$ under the null hypothesis

$$\lim_{P\to\infty} (P-\tau+1)^{-1/2}\sum\nolimits_{t=R}^{T-\tau} Ef(\boldsymbol{X}_{t+\tau},\hat{\boldsymbol{\beta}}_t) = 0$$

where $S_{\hat{f}\hat{f}} = \lim_{P\to\infty}\mathrm{Var}(P^{-1/2}\sum_{t=R}^{T-\tau} f(\boldsymbol{X}_{t+\tau},\hat{\boldsymbol{\beta}}_t))$.

The theoretical results in Giacomini and White (2006) can be used for a wide range of applications, including tests of equal forecast accuracy between nested or non-nested models, zero mean prediction error, forecast encompassing, equal log predictive densities, etc. So long as the statistic takes the form $(P-\tau+1)^{-1/2}\sum_{t=R}^{T-\tau} f(\boldsymbol{X}_{t+\tau},\hat{\boldsymbol{\beta}}_t)$, and all parameters are estimated using a small rolling window of observations, normal critical values can be used to conduct inference on the null hypothesis $\lim_{P\to\infty}(P-\tau+1)^{-1/2}\sum_{t=R}^{T-\tau} Ef(\boldsymbol{X}_{t+\tau},\hat{\boldsymbol{\beta}}_t) = 0$. Building on Giacomini and White (2006), Amisano and Giacomini (2007) formally develop such tests of equal accuracy of density forecasts, using a logarithmic scoring rule (i.e., the log predictive score). Their tests encompass a t-test applied to the mean score differential and weighted likelihood ratio tests that permit concentration on particular areas of the predictive density.

It's important to also note that, under the asymptotics of Giacomini and White (2006), there are almost no restrictions on how the parameters are estimated. The parameters can be estimated by OLS but can also be estimated using Bayesian methods so long as a small rolling window of R

observations is used to estimate the parameters at each forecast origin. Univariate as well as VAR models are permitted.

The only other set of results that apply to tests of the null of finite sample predictive ability are those considered in Clark and McCracken (2011a). There they develop a procedure for testing the null hypothesis $\lim_{P,R\rightarrow\infty}\sum_{t=R}^{T-\tau}E((\hat{\varepsilon}_{1,t+\tau}^{(1)})^2-(\hat{\varepsilon}_{1,t+\tau}^{(2)})^2)=0$ against an alternative in which the unrestricted model 2 is more accurate than the restricted model 1. As was the case for their previous work on tests of equal population predictive ability between nested models, the asymptotic distribution is nonstandard and in particular is not asymptotically normal. Instead a bootstrap approach to inference is shown to be asymptotically valid when OLS estimated, linear DMS forecasts are being compared. IMS forecasts from univariate or VAR models are not allowed again due to the fact that the forecast errors are a complicated function of the parameter estimates.

MONTE CARLO EVIDENCE

In this section we examine the finite sample properties of some of the inference approaches described above. In the interest of brevity and computational tractability, we focus our evaluation on the subset of inference problems that seem to be most common in the VAR forecasting literature. In particular, we examine tests that compare forecasts from VARs and BVARs to baseline, nested AR models, with all forecasts obtained by iteration. We consider both point and density forecasts, generated under both recursive and rolling estimation schemes.

In these experiments, we use trivariate data-generating processes (DGPs) parameterized on the basis of estimates of AR(4) and VAR(4) models in U.S. data on GDP growth, inflation in the price index for personal consumption expenditures (PCE), and the 3-month Treasury bill rate, for an estimation sample of 1961–2007.[12] We consider one set of experiments in which the DGP is a set of AR(4) models for each variable, such that forecasts from the AR models should be at least as accurate as forecasts from a VAR or BVAR. We consider another set of experiments in which the DGP is a VAR(4), such that, with a long enough data sample, forecasts from the VAR or BVAR should be more accurate than forecasts from the AR models. While we would like to be able to include experiments in which the forecasts from the AR and VAR models are equally accurate in the finite sample, there is no easy way of parameterizing a VAR-based DGP to make the forecasts equally accurate in the finite sample. Such a finite-sample evaluation is an important topic for future research. In this article, we focus on the more tractable DGP settings just described.

In all experiments, we take forecasts from AR(4) models as benchmarks for each variable, and compare against them forecasts from a VAR(4) and a BVAR(4). In the evaluation of point forecasts, to make bootstrap evaluation computationally tractable, we generate the forecasts without simulation. The AR and VAR point forecasts are obtained by using OLS estimates of coefficients and simple iteration of forecasts based on the OLS coefficients. In the case of the BVAR, we also obtain forecasts by iteration, using just the posterior mean coefficient estimates obtained with the analytical solution for the Normal-Wishart posterior. At horizons greater than 1 period, proper Bayesian methods would require simulation of the posterior distribution, but Carriero et al. (2012) find that multistep point forecasts based on just the posterior mean coefficients are essentially the same as properly simulated point forecasts.

In the evaluation of density forecasts, because the density forecasts are obtained by simulation, computational constraints lead us to avoid bootstrapping and instead limit our analysis to tests compared against asymptotic critical values. We use Bayesian simulation methods to produce the density forecasts, using an extremely loose prior to effectively implement OLS for AR and VAR forecasts and an informative prior to obtain BVAR forecasts.

We proceed by detailing the test statistics and critical values used, our implementation of Bayesian methods, and the DGPs and other aspects of experiment design. We then present the results.

Test Statistics and Critical Values

For the evaluation of point forecasts, we consider four test statistics, consisting of both F-type and t-tests of equal MSE and forecast encompassing. To define these tests for a given variable y_j of the VAR of interest, let model 1 denote the benchmark model for the given variable and model 2 denote the alternative model, $\hat{d}_{t+\tau} = (\hat{\varepsilon}^{(1)}_{j,t+\tau})^2 - (\hat{\varepsilon}^{(2)}_{j,t+\tau})^2$, $\hat{c}_{t+\tau} = \hat{\varepsilon}^{(1)}_{j,t+\tau}(\hat{\varepsilon}^{(1)}_{j,t+\tau} - \hat{\varepsilon}^{(2)}_{j,t+\tau})$, and $\text{MSE}^{(i)} = (P-\tau+1)^{-1}\sum_{t=R}^{T-\tau}(\hat{\varepsilon}^{(i)}_{j,t+\tau})^2$. Let $\hat{S}_{dd}$ and $\hat{S}_{cc}$ denote the long-run variances of $\hat{d}_{t+\tau}$ and $\hat{c}_{t+\tau}$, computed using a rectangular kernel, $\tau-1$ lags, and the small-sample adjustment of Harvey, Leybourne, and

Newbold (1997). For each variable, the four test statistics for point forecasts are computed as

$$\text{MSE}-F=\left(\sum\nolimits_{t=R}^{T-\tau}\hat{d}_{t+\tau}\right)/\text{MSE}^{(2)}$$

$$\text{MSE}-t=((P-\tau+1)^{-1/2}\sum\nolimits_{t=R}^{T-\tau}\hat{d}_{t+\tau})/\hat{S}_{dd}^{1/2}$$

$$\text{ENC}-F=\left(\sum\nolimits_{t=R}^{T-\tau}\hat{c}_{t+\tau}\right)/\text{MSE}^{(2)}$$

$$\text{ENC}-t=((P-\tau+1)^{-1/2}\sum\nolimits_{t=R}^{T-\tau}\hat{c}_{t+\tau})/\hat{S}_{cc}^{1/2}$$

(note that the ENC-t test corresponds to the MSE-adjusted test of Clark and West (2006, 2007)). We compare the t-tests against standard normal critical values and all four tests against bootstrapped critical values. We consider two different bootstrap methods, one we refer to as a restricted VAR bootstrap (VARBS) and the other a fixed regressor bootstrap (FRBS), both detailed below.

For the evaluation of density forecasts, we consider just one test statistic, a t-test for equal average log predictive score (based on the results of Amisano and Giacomini (2007)), which we will identify as the score-t test. While we focus on this test because the log score is the broadest measure of density accuracy, we certainly recognize that other density-based tests – such as Berkowitz's (2001) test based on normalized forecast errors, Christofferson's (1998) test for interval forecasts, and a test for equality of the mean cumulative ranked probability score of Gneiting and Raftery (2007) – are also of interest. The development and evaluation of such tests for application to VAR forecasts is an important topic for future research.

For variable y_j, letting $\hat{s}_{t+\tau}^{(i)}$ denote the log predictive score of model i (as described below, computed from a simulated forecast distribution), $\hat{d}_{t+\tau}=\hat{s}_{t+\tau}^{(2)}-\hat{s}_{t+\tau}^{(1)}$, and $\hat{S}_{dd}$ = the long-run variance of $\hat{d}_{t+\tau}$ computed using a rectangular kernel, $\tau-1$ lags, and the small-sample adjustment of Harvey et al. (1997), the test is computed as

$$\text{score-}t=((P-\tau+1)^{-1/2}\sum\nolimits_{t=R}^{T-\tau}\hat{d}_{t+\tau})/\hat{S}_{dd}^{1/2}$$

We compare the t-test against standard normal critical values. As noted above, our simulation-based approach to computing density forecasts makes the evaluation of bootstrap-based inference infeasible from a computational perspective.

Based on 2,000 Monte Carlo draws, we report the percentage of Monte Carlo trials in which the null of no predictive content is rejected – the percentage of trials in which the sample test statistics exceed the critical values (conducting one-sided tests in all cases). In the reported results, the tests are compared against 10% critical values. Using 5% critical values yields similar findings.

In most cases, our Monte Carlo results apply to tests of equal accuracy in population. For point forecasts, both the restricted VAR and fixed regressor bootstraps generate critical values for tests of the null of equal population predictive ability. Similarly, comparing the ENC-*t* test against standard normal critical values should be seen as an assessment of predictive content in the population, since, at least in the DMS forecasting case, the normal critical values seem to be fairly close approximations of the quantiles of an asymptotic distribution for the null of no predictive content in population (Clark & West, 2007).

However, comparing the MSE-*t* test against standard normal critical values may be seen as a test of equal accuracy in the finite sample, since, under the asymptotics of Giacomini and White (2006) and a null of equal accuracy in the finite sample, the *t*-test has a standard normal distribution. The same applies to the score-*t* test used to test the equality of average log predictive scores. That said, as noted above, the asymptotic normality results of Giacomini and White (2006) and Amisano and Giacomini (2007) require that forecasts be generated with a rolling sample approach to estimation. We will nonetheless evaluate the efficacy of MSE-*t* and score-*t* tests compared against normal critical values for both recursive and rolling estimation schemes. We leave as a subject for future research the further investigation of inference of equal accuracy in the finite sample. As we indicated above, this investigation will require finding a way to simulate data to ensure the model-based forecasts are equally accurate in the finite sample. It may also involve developing other tests or bootstrap approaches for testing equal accuracy in the finite sample.

Restricted VAR Bootstrap

One bootstrap method we consider is patterned on the approach first used in Kilian (1999) and applied in the univariate forecast evaluation of such studies as Clark and McCracken (2001, 2005a). With the null forecasting model for each variable taken to be an AR model, the bootstrap model is a set of AR(4) equations for each variable, with coefficients estimated by

OLS using the full sample of data. Bootstrapped time series on the vector $\boldsymbol{y}_t$ are generated by drawing with replacement from the vector of OLS residuals and using the autoregressive structures of the AR models to iteratively construct data.[13] In each of 499 bootstrap replications, the bootstrapped data are used to recursively estimate AR, VAR, and BVAR models and generate forecasts. In each replication, the resulting forecasts are then used to calculate forecast test statistics. Critical values are simply computed as percentiles of the bootstrapped test statistics.

Fixed Regressor Bootstrap

The other bootstrap method we consider is patterned on the fixed regressor wild bootstrap developed in Goncalves and Kilian (2004) and extended to out-of-sample tests by Clark and McCracken (2012a). We generate artificial data on the predictands using (a) fitted values from AR(4) models for each variable, which correspond to the restricted forecasting models, and (b) residuals from the relevant unrestricted forecasting model, which is either the VAR or BVAR. For the data simulation, all of the models used are estimated with the full sample of data. We construct artificial time series $\boldsymbol{y}_t^*$ by adding the vector of fitted values from the AR models to a vector of innovations equal to the innovations from the relevant unrestricted model times a draw from the standard normal distribution.

More specifically, for each variable $y_{i,t}$, we use the full sample of data to estimate an AR(4) model to obtain a vector of coefficients $\hat{\boldsymbol{\beta}}_i$. We also estimate the VAR and BVAR models, to obtain vectors of residuals $\hat{\boldsymbol{\varepsilon}}_t^{(\mathrm{VAR})}$ and $\hat{\boldsymbol{\varepsilon}}_t^{(\mathrm{BVAR})}$. For comparing VAR forecasts against the AR benchmark, we repeatedly construct a vector of simulated time series $y_{i,t}^*$ using

$$y_{i,t}^* = \hat{\beta}_{i,0} + \sum_{j=1}^{4} \hat{\beta}_{i,j} y_{i,t-j} + \eta_t \hat{\varepsilon}_{i,t}^{(\mathrm{VAR})}$$

where η_t is a draw from an $N(0,1)$ distribution, and the same draw is used for each variable of the VAR. To obtain a bootstrapped distribution of test statistics, for each data draw we obtain pseudo-AR model forecasts by regressing $y_{i,t}^*$ on lags of $y_{i,t}$, using lagged values of $y_{i,t}$ to iteratively construct forecasts of horizons 1 through τ, and forming forecast errors as $y_{i,t+\tau}^*$ less the forecast. We obtain pseudo-VAR forecasts by regressing each variable $y_{i,t}^*$ on lags of the vector $\boldsymbol{y}_t$, using lagged values of $\boldsymbol{y}_t$ to iteratively construct forecasts of horizons 1 through τ, and forming forecast errors for each variable as $y_{i,t+\tau}^*$ less the forecast. For comparing BVAR forecasts against the AR benchmark, the procedure is the same, except that residuals from the BVAR are used in generating artificial data $y^*_{i,t}$, and simulated forecasts are computed for the BVAR rather than the VAR. For each draw of simulated forecast errors, we compute the four test statistics of interest, to obtain a bootstrapped distribution against which we compare the sample test statistics from the Monte Carlo-generated data set.

Details on Implementation of Bayesian Methods

The BVAR takes the form described in section "Bayesian Estimation":

$$\begin{aligned} \boldsymbol{Y}_t &= \boldsymbol{C} + \mathbf{A}(L)\boldsymbol{Y}_{t-1} + \boldsymbol{\varepsilon}_t, \quad \boldsymbol{\varepsilon}_t \sim N(\mathbf{0},\Sigma) \\ &= \boldsymbol{\Lambda x}_{t-1} + \boldsymbol{\varepsilon}_t \end{aligned}$$

For this model, we use the Normal-Wishart prior and posterior detailed in such studies as Kadiyala and Karlsson (1997) and Banbura et al. (2010).

The prior takes a conventional Minnesota form, without cross-variable shrinkage, in which the prior for the VAR coefficients is normally distributed

$$\mathrm{vec}(\boldsymbol{\Lambda}) \sim N(\mathrm{vec}(\underline{\boldsymbol{\mu}}_{\Lambda}), \underline{\Omega}_{\Lambda})$$

Letting $\boldsymbol{\Lambda}_l^{(ij)}$ denote the coefficient on lag l of variable j in equation i and $\boldsymbol{\Lambda}_l^{(ij)}$ with l=0 denote the intercept of equation i, we define the prior mean and variance as follows:

$$\underline{\boldsymbol{\mu}}_{\Lambda} \text{ such that } E[\boldsymbol{\Lambda}_l^{(ij)}] = 0 \forall \; i,j,l \tag{13}$$

$$\underline{\Omega}_{\Lambda} \text{ such that } V\left[\boldsymbol{\Lambda}_l^{(ij)}\right] = \begin{cases} \dfrac{\theta^2}{l^2}\dfrac{\sigma_i^2}{\sigma_j^2} & \text{for} \quad l > 0 \\ \varepsilon^2 \sigma_i^2 & \text{for} \quad l = 0 \end{cases} \tag{14}$$

In our BVAR implementation, we set the tightness hyperparameter θ at 0.5 and the intercept hyperparameter ε to 1,000, and we set the scale parameters σ_i^2 at estimates of residual variances from AR(4) models from the estimation sample available at each forecast origin. In using the same simulation code for producing density forecasts from AR and VAR models, we make the prior uninformative by setting $\theta = 1{,}000$ and dropping the l^2 in the denominator of the prior variance.

For the evaluation of point forecasts from the BVAR, we obtain forecasts using the posterior mean value of the coefficients and the iterative approach to forecasting at horizons greater than 1 period. For the evaluation of density forecasts, we simulate 2,000 draws of forecasts from the posterior distribution as follows: for each draw of coefficients and Σ from their posterior distributions, we (1) draw shocks to $\boldsymbol{y}_t$ over the forecast horizon with variance–covariance matrix equal to the draw of Σ, and (2) iterate forward using the AR structure of the VAR and the draw of the VAR coefficients to obtain artificial $\boldsymbol{y}_{t+\tau}$. At each forecast horizon, for each variable $y_{i,t+\tau}$, we compute the log score based on the actual outcome for $y_{i,t+\tau}$ and an empirical estimate of the forecast density obtained with a Gaussian kernel and the draws of forecasts from the posterior distribution.

Monte Carlo Design

For two different trivariate DGPs – one in which the true model consists of AR(4) processes for each variable and the other in which the true model is a VAR(4) – we generate data using independent draws of innovations from the normal distribution and the AR structure of the DGP. The initial observations necessitated by the lag structure of each DGP are generated with draws from the unconditional normal distribution implied by the DGP. With quarterly data in mind, we report results for forecast horizons of 1, 2, 4, and 8 periods ahead, with sample sizes of $R, P = 80{,}40$; 80,80; and 80,120. We obtained qualitatively similar results (available upon request) for sample sizes of $R, P = 40{,}80$; 40,120; 120,40; and 120,80.[14]

The parameters of the DGPs are set on the basis of 1961–2007 model estimates for U.S. data on GDP growth, PCE inflation, and the 3-month Treasury bill rate. In DGP 1, the true model consists of AR(4) processes for each variable, with coefficients given in Table 1 and an error variance–covariance matrix of

$$\text{var}(\boldsymbol{\varepsilon}_t) = \begin{pmatrix} 9.81 & & \\ -0.05 & 1.35 & \\ 0.32 & 0.16 & 0.48 \end{pmatrix}$$

Table 1. Monte Carlo DGP 1 Coefficients.

Explanatory Variable	$y_{1,t}$ Equation	$y_{2,t}$ Equation	$y_{3,t}$ Equation
$y_{1,t-1}$	0.24	0.00	0.00
$y_{1,t-2}$	0.16	0.00	0.00
$y_{1,t-3}$	−0.05	0.00	0.00
$y_{1,t-4}$	0.03	0.00	0.00
$y_{2,t-1}$	0.00	0.61	0.00
$y_{2,t-2}$	0.00	0.13	0.00
$y_{2,t-3}$	0.00	0.30	0.00
$y_{2,t-4}$	0.00	−0.14	0.00
$y_{3,t-1}$	0.00	0.00	1.30
$y_{3,t-2}$	0.00	0.00	−0.64
$y_{3,t-3}$	0.00	0.00	0.59
$y_{3,t-4}$	0.00	0.00	−0.30
Intercept	2.08	0.34	0.30

Notes:
1. The table provides the coefficients of Monte Carlo DGP 1, in the form of a VAR with 0 restrictions to make the model correspond to an AR.
2. The variance–covariance matrix of innovations and other aspects of the Monte Carlo design are described in section "Monte Carlo Design."

In DGP 2, the true model is a VAR(4), with coefficients given in Table 2 and an error variance–covariance matrix of

$$\mathrm{var}(\boldsymbol{\varepsilon}_t) = \begin{pmatrix} 8.35 & & \\ -0.06 & 1.16 & \\ 0.36 & 0.13 & 0.41 \end{pmatrix} \tag{15}$$

Results

Tables 3 and 4 provide empirical rejection rates from Monte Carlo experiments with DGP 1 based on a recursive estimation scheme, and Tables 5 and 6 provide rejection rates from DGP 1 experiments using a rolling estimation scheme. Tables 7 and 8 provide recursive scheme results from experiments with DGP 2. Each table provides results for three different sample size combinations, for either VAR versus AR forecasts or BVAR versus AR forecasts. Each table covers all three variables being forecast and forecast horizons of 1, 2, 4, and 8 periods ahead. Because the BVAR

Table 2. Monte Carlo DGP 2 Coefficients.

Explanatory Variable	$y_{1,t}$ Equation	$y_{2,t}$ Equation	$y_{3,t}$ Equation
$y_{1,t-1}$	0.18	0.04	0.05
$y_{1,t-2}$	0.19	−0.01	0.04
$y_{1,t-3}$	−0.04	−0.03	−0.01
$y_{1,t-4}$	0.05	0.12	0.01
$y_{2,t-1}$	−0.15	0.59	0.06
$y_{2,t-2}$	0.09	0.17	0.10
$y_{2,t-3}$	−0.19	0.28	−0.01
$y_{2,t-4}$	0.14	−0.06	−0.03
$y_{3,t-1}$	0.36	0.17	1.15
$y_{3,t-2}$	−1.73	−0.20	−0.63
$y_{3,t-3}$	1.24	−0.06	0.65
$y_{3,t-4}$	−0.06	0.07	−0.26
Intercept	3.47	−0.13	−0.23

Notes:
1. The table provides the coefficients of Monte Carlo DGP 2, in the form of a VAR.
2. The variance–covariance matrix of innovations and other aspects of the Monte Carlo design are described in section "Monte Carlo Design."

and VAR results are broadly similar, in the ensuing discussion we focus on the VAR versus AR comparisons, in the interest of brevity.[15] The test statistic and source of critical values are indicated in the first column of each table.

In the DGP 1 experiments, in which the true model is a set of AR equations, and therefore the null hypothesis of no predictive accuracy in population is satisfied, the bootstrap-based tests yield rejection rates reasonably close to the nominal size of 10 percent. However, the size results seem to be more consistently close to 10 percent when critical values are obtained from the restricted VAR bootstrap (VARBS) than when critical values are obtained from the fixed regressor bootstrap (FRBS). While the FRBS rejection rates are typically close to 10 percent, there are a number of instances in which the rejection rates are somewhat above or below 10 percent. Consider, for example, the VAR versus AR results in Table 3 for a sample size combination of R,P=80,80. For most variables and forecast horizons, the empirical size of the MSE-F test based on the FRBS is 10 or 11 percent. However, with variable 1, the FRBS-based size is about 13 percent at horizons of 4 and 8, while with variable 3, the FRBS-based size is about 7 percent at horizons of 4 and 8. Using critical values from the VARBS more consistently yields empirical rejection rates close to 10

Table 3. Monte Carlo Rejection Rates, 10% Critical Values: DGP 1, Recursive Scheme, VAR versus AR.

Test, Critical Values	Forecasts of $y_{1,t}$				Forecasts of $y_{2,t}$				Forecasts of $y_{3,t}$			
	$\tau=1$	$\tau=2$	$\tau=4$	$\tau=8$	$\tau=1$	$\tau=2$	$\tau=4$	$\tau=8$	$\tau=1$	$\tau=2$	$\tau=4$	$\tau=8$
R,*P* = 80,40												
MSE-*F*, VARBS	0.107	0.115	0.107	0.130	0.100	0.106	0.097	0.098	0.105	0.110	0.111	0.108
MSE-*F*, FRBS	0.113	0.142	0.149	0.148	0.112	0.139	0.118	0.095	0.126	0.120	0.104	0.085
MSE-*t*, VARBS	0.110	0.114	0.113	0.111	0.105	0.105	0.101	0.100	0.104	0.105	0.110	0.096
MSE-*t*, FRBS	0.114	0.136	0.124	0.139	0.111	0.115	0.081	0.086	0.123	0.095	0.060	0.071
ENC-*F*, VARBS	0.096	0.105	0.109	0.134	0.088	0.091	0.091	0.098	0.106	0.111	0.115	0.120
ENC-*F*, FRBS	0.113	0.132	0.133	0.111	0.113	0.158	0.132	0.104	0.117	0.155	0.134	0.096
ENC-*t*, VARBS	0.107	0.104	0.104	0.116	0.096	0.096	0.095	0.096	0.105	0.104	0.103	0.104
ENC-*t*, FRBS	0.112	0.107	0.091	0.106	0.101	0.093	0.073	0.084	0.108	0.079	0.065	0.085
MSE-*t*, normal	0.013	0.020	0.051	0.088	0.015	0.029	0.050	0.077	0.013	0.021	0.043	0.072
ENC-*t*, normal	0.110	0.121	0.137	0.143	0.105	0.118	0.128	0.146	0.113	0.124	0.142	0.154
score-*t*, normal	0.015	0.026	0.046	0.063	0.011	0.030	0.052	0.084	0.012	0.024	0.048	0.087
R,*P* = 80,80												
MSE-*F*, VARBS	0.126	0.110	0.106	0.103	0.116	0.101	0.108	0.115	0.092	0.093	0.101	0.104
MSE-*F*, FRBS	0.101	0.116	0.134	0.132	0.106	0.117	0.107	0.101	0.092	0.083	0.074	0.072
MSE-*t*, VARBS	0.118	0.108	0.105	0.100	0.113	0.101	0.108	0.113	0.091	0.095	0.102	0.097
MSE-*t*, FRBS	0.103	0.121	0.122	0.121	0.109	0.117	0.076	0.073	0.099	0.073	0.043	0.053
ENC-*F*, VARBS	0.084	0.081	0.086	0.098	0.096	0.088	0.104	0.107	0.088	0.097	0.102	0.121
ENC-*F*, FRBS	0.105	0.122	0.115	0.092	0.120	0.159	0.140	0.104	0.098	0.138	0.115	0.093
ENC-*t*, VARBS	0.090	0.098	0.088	0.101	0.104	0.096	0.104	0.109	0.086	0.098	0.098	0.108
ENC-*t*, FRBS	0.090	0.095	0.075	0.087	0.109	0.084	0.061	0.071	0.088	0.061	0.052	0.067
MSE-*t*, normal	0.002	0.005	0.019	0.044	0.003	0.005	0.022	0.052	0.004	0.005	0.022	0.046
ENC-*t*, normal	0.085	0.094	0.092	0.121	0.098	0.098	0.111	0.127	0.081	0.095	0.106	0.126
score-*t*, normal	0.005	0.010	0.025	0.042	0.003	0.011	0.025	0.045	0.003	0.007	0.015	0.047
R,*P* = 80,120												
MSE-*F*, VARBS	0.131	0.122	0.113	0.112	0.121	0.113	0.117	0.119	0.108	0.110	0.105	0.096
MSE-*F*, FRBS	0.093	0.110	0.135	0.147	0.098	0.105	0.108	0.091	0.104	0.082	0.063	0.060
MSE-*t*, VARBS	0.119	0.117	0.116	0.104	0.113	0.111	0.119	0.111	0.110	0.110	0.105	0.092
MSE-*t*, FRBS	0.096	0.127	0.129	0.133	0.103	0.123	0.077	0.066	0.111	0.084	0.041	0.038

ENC-*F*, VARBS	0.085	0.097	0.106	0.115	0.099	0.099	0.106	0.113	0.101	0.106	0.111	0.108
ENC-*F*, FRBS	0.106	0.131	0.135	0.136	0.115	0.149	0.138	0.098	0.119	0.147	0.120	0.080
ENC-*t*, VARBS	0.097	0.100	0.107	0.100	0.107	0.099	0.104	0.100	0.106	0.107	0.103	0.095
ENC-*t*, FRBS	0.095	0.092	0.088	0.082	0.109	0.085	0.057	0.051	0.111	0.072	0.047	0.054
MSE-*t*, normal	0.001	0.003	0.010	0.038	0.000	0.005	0.014	0.034	0.001	0.004	0.011	0.027
ENC-*t*, normal	0.084	0.091	0.102	0.111	0.096	0.090	0.099	0.110	0.098	0.100	0.097	0.102
score-*t*, normal	0.001	0.004	0.013	0.032	0.002	0.008	0.013	0.022	0.002	0.003	0.010	0.021

Notes:
1. The data generating process is a set of AR(4) equations, with coefficients given in Table 1 and error variance–covariance matrix given in section "Monte Carlo Design."

2. For each artificial data set, point forecasts of the trivariate vector of variables are formed recursively using AR(4) and VAR(4) models estimated by OLS. Density forecasts from AR(4) and VAR(4) models are obtained (with recursive estimation) using the Bayesian MCMC estimation approach described in section "Details on Implementation of Bayesian methods," under an extremely loose (Normal-inverted Wishart) prior. At each forecast origin and horizon, we compute the log score using an empirical estimate of the forecast density obtained with a Gaussian kernel and 2,000 draws of forecasts from the posterior distribution. The point forecasts and log scores are then used to form the indicated test statistics. *R* and *P* refer to the number of in-sample observations and 1-step ahead forecasts, respectively.

3. The test statistics MSE-*F*, MSE-*t*, ENC-*F*, ENC-*t*, and score-*t* are defined in section "Test Statistics and Critical Values." All tests are conducted on a one-sided basis.

4. VARBS refers to a VAR-based bootstrap, described in section "Test Statistics and Critical Values." FRBS refers to a fixed-regressor bootstrap, described in section "Test Statistics and Critical Values." "Normal" refers to critical values from the standard normal distribution. The bootstraps used 499 replications. The number of Monte Carlo simulations is 2,000.

percent. In the same example, across the three variables and four horizons, the size of the MSE-F test based on the VARBS ranges from 9.2 to 12.6 percent. Similarly, as indicated in Table 3, for many variables and horizons, the empirical size of the ENC-F test based on the FRBS is 10 or 11 percent. However, with variable 2 and a 2-step-ahead horizon, the FRBS-based size spikes to almost 16 percent. Again, size performance is more even with the VARBS, yielding ENC-F rejection rates that range from 8.1 to 12.1 percent (compared to a range of 9.2–15.9 percent with the FRBS).

Given the bootstrap method, size results are broadly similar for the MSE-F, MSE-t, ENC-F, and ENC-t tests. The most noticeable difference is that, when the FRBS is used to generate critical values, the t-tests are less prone to oversizing and more prone to undersizing than their F-test counterparts. Consider, for example, the FRBS-based encompassing test results in the Table 3 experiments with R, P=80,80. The size of the ENC-F test ranges from 9.2 to 15.9 percent; the size of the ENC-t test varies from 5.2 to 10.9 percent. In corresponding results in the experiments with R, P=80,40, the empirical rejection rate of the ENC-F test ranges from 9.6 to 15.8 percent; the size of the ENC-t test varies from 6.5 to 11.2 percent.

When the t-tests for point forecasts are compared against standard normal critical values, the MSE-t and ENC-t tests behave fairly differently, in line with expectations. As noted above, the MSE-t test compared against normal critical values is in effect (subject to the conditions of Giacomini and White (2006)) a test of the null of equal accuracy in the finite sample, rather than in population. Accordingly, when the AR model is the true model and therefore likely to be more accurate than a VAR in the finite sample, the (one-sided) MSE-t test compared against standard normal critical values usually yields a rejection rate well below 10%. How much below depends on the forecast horizon: the rejection rate tends to rise with the forecast horizon, particularly with smaller P, probably due to challenges in estimating precisely the HAC standard deviation in the denominator of the test statistic. For example, as shown in Table 3's results with R, P=80,120, the rejection rate of the MSE-t test ranges from 0.0 to 3.8 percent. In results with R, P=80,40, the rejection rate varies from 1.3 to 8.8 percent.

In contrast, the ENC-t test compared against normal critical values is approximately a test of the null of no predictive content in population and may therefore be expected to yield rejection rates in the DGP 1 experiments that are close to the nominal size of 10 percent. The test is indeed about correctly sized when P is relatively large, but for smaller P, it tends to be oversized at longer forecast horizons. For instance, in Table 3's results with R, P=80,120, the rejection rate of the ENC-t test based on standard normal critical values ranges from 8.4 to 11.1 percent. In results with R, P=80,40, the rejection rate varies from 10.5 to 15.4 percent.

Turning to the density forecast results, comparing the score-t test against standard normal critical values yields rejection rates broadly similar to those obtained for the MSE-t test for point forecasts. Like the MSE-t test, the score-t test compared against normal critical values is in effect (subject to the conditions of Amisano and Giacomini (2007)) a test of the null of equal (density) accuracy in the finite sample, rather than in population. Accordingly, when the AR model is the true model and therefore likely to be more accurate than a VAR in the finite sample, the (one-sided) score-t test compared against standard normal critical values usually yields a rejection rate well below 10%, more so at shorter horizons than longer horizons. As an example, in Table 3's results with R, P=80,120, the rejection rate of the score-t test ranges from 0.1 to 3.2 percent. Taken at face value, these findings suggest that tests for equal accuracy in density forecasts are likely to yield results similar to tests for equal accuracy of point forecasts. In practice, however, results could differ due to complications outside the scope of our experiments: non-Gaussian errors, time-varying volatility (particularly if one model captures it and the other does not), or model misspecification.

The results in Tables 5 and 6 indicate that, with DGP 1, using a rolling scheme for estimation and forecasting yields results qualitatively similar to those obtained using a recursive scheme. It is again the case that the bootstrap-based tests yield rejection rates reasonably close to the nominal size of 10 percent, with VARBS being more consistent across specifications than FRBS. As an example, in Table 5's results for the VAR versus the AR with R, P=80,40, the size of the MSE-F test ranges from 9.3 to 15.8 percent under the FRBS and from 9.7 to 12.0 percent under the VARBS. In addition, size results are similar across the different tests, with the biggest difference being that, when critical values are obtained from the FRBS, the t-tests are less prone to oversizing and more prone to undersizing than their F-test counterparts. In the same example, the size of the MSE-t test compared against standard normal critical values ranges from 8.0 to 14.6 percent. Finally, under the rolling scheme, it is also the case that, when the t-tests are compared against standard normal critical values, the MSE-t test usually yields a rejection rate well below 10 percent, while the ENC-t test has a rejection rate ranging from about 9 to 16 percent. As in the recursive case, the score-t test for density forecasts yields a rejection rate very similar to the normal critical values-based rejection rate of the MSE-t test for equal accuracy of point forecasts.

Table 4. Monte Carlo Rejection Rates, 10% Critical Values: DGP 1, Recursive Scheme, BVAR versus AR.

Test, Critical Values	Forecasts of $y_{1,t}$				Forecasts of $y_{2,t}$				Forecasts of $y_{3,t}$			
	$\tau=1$	$\tau=2$	$\tau=4$	$\tau=8$	$\tau=1$	$\tau=2$	$\tau=4$	$\tau=8$	$\tau=1$	$\tau=2$	$\tau=4$	$\tau=8$
$R,P=80,40$												
MSE-*F*, VARBS	0.112	0.105	0.111	0.131	0.098	0.095	0.088	0.100	0.102	0.097	0.103	0.102
MSE-*F*, FRBS	0.129	0.131	0.137	0.134	0.104	0.129	0.115	0.088	0.103	0.112	0.093	0.065
MSE-*t*, VARBS	0.115	0.107	0.107	0.105	0.104	0.100	0.101	0.097	0.105	0.104	0.102	0.102
MSE-*t*, FRBS	0.123	0.117	0.099	0.120	0.104	0.104	0.069	0.073	0.104	0.085	0.055	0.071
ENC-*F*, VARBS	0.104	0.106	0.115	0.137	0.082	0.089	0.084	0.105	0.083	0.104	0.110	0.112
ENC-*F*, FRBS	0.114	0.132	0.121	0.113	0.104	0.145	0.130	0.096	0.086	0.134	0.115	0.077
ENC-*t*, VARBS	0.115	0.104	0.105	0.101	0.093	0.093	0.087	0.096	0.093	0.093	0.099	0.107
ENC-*t*, FRBS	0.108	0.095	0.088	0.094	0.096	0.083	0.058	0.073	0.088	0.068	0.051	0.070
MSE-*t*, normal	0.036	0.045	0.073	0.091	0.026	0.030	0.068	0.098	0.010	0.026	0.066	0.107
ENC-*t*, normal	0.153	0.154	0.163	0.144	0.130	0.128	0.147	0.163	0.117	0.141	0.171	0.185
score-*t*, normal	0.029	0.050	0.070	0.085	0.026	0.046	0.071	0.111	0.005	0.021	0.052	0.104
$R,P=80,80$												
MSE-*F*, VARBS	0.110	0.110	0.102	0.100	0.107	0.097	0.097	0.107	0.103	0.092	0.091	0.096
MSE-*F*, FRBS	0.108	0.129	0.124	0.118	0.106	0.110	0.103	0.090	0.075	0.074	0.062	0.056
MSE-*t*, VARBS	0.113	0.113	0.100	0.091	0.105	0.096	0.102	0.106	0.099	0.093	0.092	0.096
MSE-*t*, FRBS	0.104	0.121	0.105	0.098	0.105	0.092	0.061	0.060	0.079	0.068	0.034	0.050
ENC-*F*, VARBS	0.087	0.087	0.092	0.100	0.095	0.091	0.098	0.111	0.061	0.085	0.092	0.107
ENC-*F*, FRBS	0.099	0.117	0.115	0.091	0.111	0.141	0.139	0.104	0.072	0.114	0.090	0.071
ENC-*t*, VARBS	0.097	0.101	0.091	0.096	0.100	0.095	0.101	0.110	0.075	0.085	0.096	0.109
ENC-*t*, FRBS	0.090	0.083	0.070	0.076	0.097	0.066	0.053	0.058	0.072	0.052	0.032	0.052
MSE-*t*, normal	0.011	0.018	0.040	0.050	0.005	0.011	0.036	0.068	0.002	0.009	0.029	0.065
ENC-*t*, normal	0.128	0.136	0.114	0.123	0.130	0.112	0.132	0.146	0.096	0.114	0.131	0.153
score-*t*, normal	0.013	0.027	0.038	0.059	0.009	0.018	0.042	0.058	0.002	0.009	0.026	0.066
$R,P=80,120$												
MSE-*F*, VARBS	0.116	0.116	0.116	0.106	0.104	0.101	0.108	0.117	0.121	0.096	0.095	0.083
MSE-*F*, FRBS	0.102	0.127	0.134	0.134	0.091	0.097	0.089	0.075	0.080	0.067	0.050	0.046
MSE-*t*, VARBS	0.115	0.116	0.110	0.100	0.102	0.100	0.115	0.106	0.109	0.095	0.096	0.085
MSE-*t*, FRBS	0.102	0.120	0.107	0.110	0.092	0.090	0.048	0.037	0.084	0.062	0.027	0.029

ENC-*F*, VARBS	0.095	0.105	0.112	0.110	0.091	0.095	0.103	0.111	0.077	0.093	0.097	0.089
ENC-*F*, FRBS	0.105	0.135	0.134	0.102	0.107	0.137	0.128	0.091	0.093	0.121	0.090	0.062
ENC-*t*, VARBS	0.102	0.108	0.102	0.097	0.094	0.098	0.099	0.097	0.089	0.093	0.093	0.088
ENC-*t*, FRBS	0.094	0.087	0.071	0.069	0.090	0.067	0.038	0.041	0.082	0.046	0.028	0.038
MSE-*t*, normal	0.005	0.011	0.019	0.039	0.003	0.009	0.022	0.043	0.001	0.005	0.014	0.042
ENC-*t*, normal	0.131	0.133	0.119	0.115	0.127	0.111	0.112	0.120	0.113	0.113	0.120	0.134
score-*t*, normal	0.005	0.009	0.026	0.044	0.005	0.013	0.021	0.037	0.002	0.006	0.009	0.033

Notes:

1. The data generating process is a set of AR(4) equations, with coefficients given in Table 1 and error variance–covariance matrix given in section "Monte Carlo Design."

2. For each artificial data set, point forecasts of the trivariate vector of variables are formed recursively using AR(4) models estimated by OLS and a BVAR(4) estimated under a Normal-Wishart prior (using iteration and the posterior mean coefficients). Density forecasts from AR(4) and BVAR(4) models are obtained using the Bayesian MCMC estimation approach described in section "Details on Implementation of Bayesian Methods," with an extremely loose prior for the AR models. At each forecast origin and horizon, we compute the log score using an empirical estimate of the forecast density obtained with a Gaussian kernel and 2,000 draws of forecasts. The point forecasts and log scores are then used to form the indicated test statistics. *R* and *P* refer to the number of in-sample observations and 1-step ahead forecasts, respectively.

3. The test statistics MSE-*F*, MSE-*t*, ENC-*F*, ENC-*t*, and score-*t* are defined in section "Test Statistics and Critical Values." All tests are conducted on a one-sided basis.

4. VARBS refers to a VAR-based bootstrap, described in section "Test Statistics and Critical Values." FRBS refers to a fixed-regressor bootstrap, described in section "Test Statistics and Critical Values." "Normal" refers to critical values from the standard normal distribution. The bootstraps used 499 replications. The number of Monte Carlo simulations is 2,000.

Table 5. Monte Carlo Rejection Rates, 10% Critical Values: DGP 1, Rolling Scheme, VAR versus AR.

Test, Critical Values	Forecasts of $y_{1,t}$				Forecasts of $y_{2,t}$				Forecasts of $y_{3,t}$			
	$\tau=1$	$\tau=2$	$\tau=4$	$\tau=8$	$\tau=1$	$\tau=2$	$\tau=4$	$\tau=8$	$\tau=1$	$\tau=2$	$\tau=4$	$\tau=8$
R,P = 80,40												
MSE-*F*, VARBS	0.098	0.103	0.107	0.120	0.097	0.097	0.104	0.115	0.106	0.117	0.115	0.114
MSE-*F*, FRBS	0.106	0.127	0.158	0.147	0.111	0.133	0.134	0.112	0.123	0.125	0.115	0.093
MSE-*t*, VARBS	0.098	0.102	0.105	0.103	0.097	0.095	0.108	0.100	0.106	0.111	0.109	0.096
MSE-*t*, FRBS	0.107	0.129	0.146	0.143	0.111	0.124	0.106	0.110	0.122	0.107	0.080	0.092
ENC-*F*, VARBS	0.100	0.105	0.118	0.119	0.099	0.104	0.106	0.114	0.112	0.116	0.111	0.118
ENC-*F*, FRBS	0.104	0.127	0.122	0.084	0.115	0.153	0.139	0.097	0.121	0.161	0.144	0.094
ENC-*t*, VARBS	0.092	0.103	0.107	0.103	0.096	0.098	0.096	0.105	0.113	0.114	0.110	0.108
ENC-*t*, FRBS	0.095	0.101	0.100	0.098	0.103	0.091	0.084	0.100	0.118	0.098	0.083	0.093
MSE-*t*, normal	0.009	0.015	0.040	0.069	0.004	0.015	0.036	0.072	0.009	0.024	0.046	0.072
ENC-*t*, normal	0.099	0.120	0.139	0.136	0.105	0.120	0.135	0.159	0.121	0.137	0.146	0.155
score-*t*, normal	0.008	0.015	0.036	0.045	0.009	0.017	0.044	0.080	0.005	0.020	0.043	0.064
R,P = 80,80												
MSE-*F*, VARBS	0.096	0.094	0.092	0.095	0.101	0.101	0.097	0.103	0.096	0.097	0.112	0.121
MSE-*F*, FRBS	0.095	0.118	0.143	0.164	0.110	0.125	0.113	0.115	0.100	0.088	0.092	0.102
MSE-*t*, VARBS	0.099	0.095	0.096	0.088	0.101	0.104	0.098	0.096	0.095	0.099	0.107	0.114
MSE-*t*, FRBS	0.110	0.137	0.151	0.159	0.116	0.153	0.105	0.099	0.105	0.104	0.084	0.104
ENC-*F*, VARBS	0.097	0.098	0.093	0.107	0.107	0.108	0.113	0.121	0.094	0.101	0.111	0.126
ENC-*F*, FRBS	0.108	0.116	0.102	0.069	0.128	0.161	0.137	0.101	0.110	0.149	0.127	0.093
ENC-*t*, VARBS	0.096	0.099	0.091	0.086	0.107	0.114	0.106	0.100	0.092	0.100	0.102	0.120
ENC-*t*, FRBS	0.099	0.091	0.081	0.074	0.115	0.105	0.085	0.078	0.095	0.075	0.072	0.099
MSE-*t*, normal	0.000	0.003	0.014	0.026	0.000	0.002	0.018	0.036	0.000	0.003	0.013	0.044
ENC-*t*, normal	0.096	0.096	0.097	0.104	0.106	0.119	0.122	0.132	0.093	0.105	0.122	0.149
score-*t*, normal	0.001	0.003	0.015	0.018	0.002	0.003	0.018	0.040	0.000	0.001	0.015	0.036
R,P = 80,120												
MSE-*F*, VARBS	0.092	0.085	0.091	0.106	0.097	0.091	0.097	0.100	0.104	0.097	0.097	0.104
MSE-*F*, FRBS	0.092	0.119	0.151	0.197	0.092	0.103	0.116	0.128	0.104	0.081	0.084	0.108
MSE-*t*, VARBS	0.097	0.091	0.096	0.092	0.095	0.093	0.099	0.095	0.100	0.102	0.097	0.095
MSE-*t*, FRBS	0.101	0.148	0.180	0.195	0.104	0.170	0.128	0.114	0.117	0.128	0.084	0.110

ENC-*F*, VARBS	0.093	0.092	0.106	0.109	0.092	0.097	0.113	0.121	0.102	0.109	0.110	0.121
ENC-*F*, FRBS	0.105	0.115	0.118	0.072	0.116	0.151	0.140	0.095	0.122	0.159	0.127	0.087
ENC-*t*, VARBS	0.093	0.094	0.100	0.092	0.098	0.091	0.108	0.107	0.108	0.104	0.102	0.101
ENC-*t*, FRBS	0.099	0.092	0.086	0.088	0.104	0.092	0.086	0.082	0.115	0.092	0.083	0.089
MSE-*t*, normal	0.000	0.001	0.003	0.019	0.000	0.000	0.003	0.013	0.000	0.001	0.007	0.029
ENC-*t*, normal	0.091	0.093	0.104	0.111	0.099	0.106	0.121	0.130	0.107	0.111	0.114	0.130
score-*t*, normal	0.001	0.000	0.004	0.013	0.000	0.000	0.005	0.018	0.000	0.001	0.006	0.027

Notes:

1. The data generating process is a set of AR(4) equations, with coefficients given in Table 1 and error variance-covariance matrix given in section "Monte Carlo Design."

2. For each artificial data set, point forecasts of the trivariate vector of variables are formed on a rolling basis (holding the estimation sample size at R observations) using AR(4) and VAR(4) models estimated by OLS. Density forecasts from AR(4) and VAR(4) models are obtained (with rolling estimation) using the Bayesian MCMC estimation approach described in section "Details on Implementation of Bayesian Methods," under an extremely loose (Normal-inverted Wishart) prior. At each forecast origin and horizon, we compute the log score using an empirical estimate of the forecast density obtained with a Gaussian kernel and 2,000 draws of forecasts from the posterior distribution. The point forecasts and log scores are then used to form the indicated test statistics. R and P refer to the number of in-sample observations and 1-step ahead forecasts, respectively.

3. The test statistics MSE-*F*, MSE-*t*, ENC-*F*, ENC-*t*, and score-*t* are defined in section "Test Statistics and Critical Values." All tests are conducted on a one-sided basis.

4. VARBS refers to a VAR-based bootstrap, described in section "Test Statistics and Critical Values." FRBS refers to a fixed-regressor bootstrap, described in section "Test Statistics and Critical Values." "Normal" refers to critical values from the standard normal distribution. The bootstraps used 499 replications. The number of Monte Carlo simulations is 2,000.

Table 6. Monte Carlo Rejection Rates, 10% Critical Values: DGP 1, Rolling Scheme, BVAR versus AR.

Test, Critical Values	Forecasts of $y_{1,t}$				Forecasts of $y_{2,t}$				Forecasts of $y_{3,t}$			
	$\tau=1$	$\tau=2$	$\tau=4$	$\tau=8$	$\tau=1$	$\tau=2$	$\tau=4$	$\tau=8$	$\tau=1$	$\tau=2$	$\tau=4$	$\tau=8$
$R,P=80,40$												
MSE-*F*, VARBS	0.097	0.101	0.107	0.120	0.102	0.101	0.105	0.117	0.098	0.103	0.103	0.111
MSE-*F*, FRBS	0.109	0.136	0.139	0.122	0.103	0.130	0.120	0.088	0.104	0.112	0.097	0.065
MSE-*t*, VARBS	0.102	0.101	0.096	0.102	0.104	0.104	0.104	0.098	0.100	0.102	0.091	0.097
MSE-*t*, FRBS	0.114	0.119	0.110	0.126	0.100	0.104	0.074	0.083	0.106	0.085	0.065	0.071
ENC-*F*, VARBS	0.102	0.105	0.117	0.128	0.093	0.096	0.106	0.119	0.103	0.107	0.116	0.121
ENC-*F*, FRBS	0.101	0.119	0.115	0.088	0.096	0.143	0.123	0.089	0.092	0.132	0.111	0.063
ENC-*t*, VARBS	0.100	0.100	0.104	0.098	0.096	0.097	0.093	0.092	0.103	0.102	0.102	0.108
ENC-*t*, FRBS	0.096	0.085	0.085	0.090	0.091	0.071	0.057	0.082	0.090	0.066	0.053	0.071
MSE-*t*, normal	0.029	0.040	0.057	0.088	0.018	0.030	0.065	0.098	0.007	0.029	0.061	0.102
ENC-*t*, normal	0.151	0.164	0.167	0.139	0.144	0.142	0.157	0.178	0.133	0.161	0.168	0.187
score-*t*, normal	0.025	0.044	0.064	0.069	0.018	0.030	0.075	0.132	0.006	0.030	0.056	0.101
$R,P=80,80$												
MSE-*F*, VARBS	0.101	0.103	0.098	0.090	0.107	0.107	0.104	0.112	0.081	0.095	0.103	0.120
MSE-*F*, FRBS	0.115	0.118	0.130	0.117	0.107	0.111	0.094	0.080	0.075	0.076	0.072	0.060
MSE-*t*, VARBS	0.102	0.095	0.097	0.073	0.109	0.107	0.102	0.101	0.083	0.095	0.099	0.111
MSE-*t*, FRBS	0.115	0.118	0.114	0.118	0.103	0.102	0.062	0.065	0.082	0.075	0.046	0.067
ENC-*F*, VARBS	0.096	0.104	0.096	0.104	0.112	0.111	0.115	0.122	0.086	0.102	0.114	0.126
ENC-*F*, FRBS	0.105	0.120	0.100	0.061	0.127	0.147	0.127	0.091	0.085	0.126	0.090	0.059
ENC-*t*, VARBS	0.100	0.089	0.095	0.080	0.110	0.108	0.105	0.104	0.087	0.100	0.102	0.119
ENC-*t*, FRBS	0.094	0.080	0.071	0.065	0.099	0.075	0.054	0.066	0.081	0.052	0.046	0.063
MSE-*t*, normal	0.006	0.014	0.026	0.040	0.005	0.007	0.029	0.060	0.001	0.009	0.027	0.074
ENC-*t*, normal	0.165	0.154	0.135	0.115	0.168	0.141	0.151	0.161	0.111	0.139	0.154	0.190
score-*t*, normal	0.007	0.011	0.029	0.049	0.008	0.015	0.035	0.065	0.001	0.006	0.025	0.062
$R,P=80,120$												
MSE-*F*, VARBS	0.100	0.100	0.097	0.096	0.093	0.089	0.096	0.100	0.101	0.104	0.101	0.103
MSE-*F*, FRBS	0.099	0.118	0.132	0.139	0.075	0.085	0.078	0.068	0.088	0.066	0.058	0.053
MSE-*t*, VARBS	0.101	0.100	0.088	0.088	0.092	0.087	0.093	0.096	0.114	0.103	0.095	0.098
MSE-*t*, FRBS	0.101	0.122	0.127	0.135	0.082	0.101	0.051	0.058	0.101	0.072	0.041	0.049

ENC-*F*, VARBS	0.097	0.102	0.115	0.110	0.099	0.104	0.114	0.119	0.119	0.109	0.114	0.118
ENC-*F*, FRBS	0.101	0.119	0.111	0.060	0.115	0.142	0.116	0.074	0.113	0.115	0.089	0.046
ENC-*t*, VARBS	0.099	0.092	0.099	0.090	0.107	0.094	0.099	0.103	0.107	0.107	0.111	0.111
ENC-*t*, FRBS	0.095	0.074	0.068	0.070	0.097	0.070	0.046	0.051	0.099	0.052	0.043	0.051
MSE-*t*, normal	0.002	0.005	0.010	0.025	0.001	0.002	0.009	0.028	0.000	0.004	0.013	0.049
ENC-*t*, normal	0.177	0.167	0.145	0.116	0.173	0.138	0.146	0.153	0.141	0.167	0.158	0.185
score-*t*, normal	0.004	0.006	0.014	0.030	0.002	0.004	0.018	0.027	0.000	0.004	0.009	0.052

Notes:

1. The data generating process is a VAR(4), with coefficients given in Table 2 and error variance–covariance matrix given in section "Monte Carlo Design."

2. For each artificial data set, point forecasts of the trivariate vector of variables are formed on a rolling basis (holding the estimation sample size at R observations) using AR(4) models estimated by OLS and a BVAR(4) estimated with a Normal-inverted Wishart prior and posterior (using iteration and the posterior mean coefficients). Density forecasts from AR(4) and BVAR(4) models are obtained (with rolling estimation) using the Bayesian MCMC estimation approach described in section "Details on Implementation of Bayesian Methods," using an extremely loose (Normal-inverted Wishart) prior for the AR models. At each forecast origin and horizon, we compute the log score using an empirical estimate of the forecast density obtained with a Gaussian kernel and 2,000 draws of forecasts from the posterior distribution. The point forecasts and log scores are then used to form the indicated test statistics. R and P refer to the number of in-sample observations and 1-step ahead forecasts, respectively.

3. The test statistics MSE-*F*, MSE-*t*, ENC-*F*, ENC-*t*, and score-*t* are defined in section "Test Statistics and Critical Values." All tests are conducted on a one-sided basis.

4. VARBS refers to a VAR-based bootstrap, described in section "Test Statistics and Critical Values." FRBS refers to a fixed-regressor bootstrap, described in section "Test Statistics and Critical Values." "Normal" refers to critical values from the standard normal distribution. The bootstraps used 499 replications. The number of Monte Carlo simulations is 2,000.

Table 7. Monte Carlo Rejection Rates, 10% Critical Values: DGP 2, Recursive Scheme, VAR versus AR.

Test, Critical Values	Forecasts of $y_{1,t}$				Forecasts of $y_{2,t}$				Forecasts of $y_{3,t}$			
	$\tau=1$	$\tau=2$	$\tau=4$	$\tau=8$	$\tau=1$	$\tau=2$	$\tau=4$	$\tau=8$	$\tau=1$	$\tau=2$	$\tau=4$	$\tau=8$
R,P = 80,40												
MSE-*F*, VARBS	0.708	0.707	0.489	0.432	0.636	0.488	0.414	0.220	0.646	0.681	0.519	0.415
MSE-*F*, FRBS	0.710	0.736	0.515	0.462	0.634	0.537	0.456	0.249	0.660	0.680	0.490	0.380
MSE-*t*, VARBS	0.690	0.664	0.392	0.263	0.606	0.454	0.328	0.162	0.623	0.625	0.420	0.272
MSE-*t*, FRBS	0.676	0.676	0.373	0.292	0.601	0.475	0.284	0.159	0.631	0.476	0.193	0.180
ENC-*F*, VARBS	0.889	0.875	0.598	0.517	0.786	0.610	0.472	0.235	0.834	0.862	0.645	0.509
ENC-*F*, FRBS	0.913	0.910	0.645	0.502	0.828	0.737	0.599	0.280	0.870	0.913	0.726	0.532
ENC-*t*, VARBS	0.835	0.796	0.425	0.282	0.720	0.566	0.381	0.172	0.778	0.780	0.528	0.328
ENC-*t*, FRBS	0.839	0.784	0.349	0.256	0.735	0.554	0.355	0.173	0.789	0.683	0.372	0.270
MSE-*t*, normal	0.263	0.322	0.207	0.202	0.217	0.188	0.182	0.130	0.234	0.308	0.255	0.228
ENC-*t*, normal	0.841	0.823	0.507	0.364	0.740	0.613	0.462	0.257	0.798	0.814	0.633	0.469
score-*t*, normal	0.285	0.340	0.250	0.249	0.213	0.202	0.214	0.146	0.210	0.293	0.261	0.290
R,P = 80,80												
MSE-*F*, VARBS	0.901	0.903	0.680	0.583	0.826	0.689	0.573	0.301	0.846	0.868	0.705	0.578
MSE-*F*, FRBS	0.893	0.904	0.692	0.612	0.814	0.699	0.589	0.327	0.843	0.822	0.625	0.489
MSE-*t*, VARBS	0.918	0.909	0.632	0.445	0.841	0.683	0.531	0.267	0.862	0.866	0.649	0.470
MSE-*t*, FRBS	0.905	0.903	0.589	0.441	0.826	0.671	0.433	0.221	0.863	0.685	0.236	0.210
ENC-*F*, VARBS	0.983	0.983	0.796	0.700	0.944	0.830	0.688	0.335	0.969	0.973	0.833	0.692
ENC-*F*, FRBS	0.988	0.990	0.840	0.693	0.957	0.890	0.765	0.389	0.975	0.987	0.871	0.684
ENC-*t*, VARBS	0.976	0.970	0.708	0.486	0.928	0.812	0.642	0.288	0.955	0.951	0.779	0.577
ENC-*t*, FRBS	0.974	0.967	0.606	0.405	0.931	0.790	0.581	0.273	0.956	0.899	0.547	0.400
MSE-*t*, normal	0.393	0.447	0.252	0.255	0.294	0.230	0.202	0.118	0.298	0.410	0.302	0.270
ENC-*t*, normal	0.975	0.972	0.725	0.556	0.928	0.816	0.666	0.349	0.954	0.957	0.803	0.647
score-*t*, normal	0.376	0.448	0.339	0.333	0.285	0.229	0.227	0.118	0.283	0.373	0.288	0.296
R,P = 80,120												
MSE-*F*, VARBS	0.971	0.968	0.793	0.706	0.928	0.824	0.689	0.372	0.938	0.957	0.821	0.691
MSE-*F*, FRBS	0.965	0.966	0.788	0.724	0.915	0.801	0.673	0.365	0.937	0.926	0.722	0.561
MSE-*t*, VARBS	0.979	0.978	0.780	0.612	0.944	0.834	0.675	0.342	0.955	0.963	0.804	0.622
MSE-*t*, FRBS	0.976	0.979	0.727	0.573	0.938	0.819	0.535	0.263	0.957	0.840	0.267	0.242
ENC-*F*, VARBS	0.997	0.997	0.911	0.828	0.986	0.923	0.800	0.409	0.994	0.995	0.939	0.819

ENC-*F*, FRBS	0.999	0.997	0.932	0.818	0.988	0.957	0.863	0.449	0.995	0.996	0.939	0.780
ENC-*t*, VARBS	0.994	0.996	0.868	0.679	0.981	0.920	0.780	0.390	0.992	0.992	0.915	0.772
ENC-*t*, FRBS	0.995	0.995	0.796	0.543	0.984	0.906	0.722	0.332	0.991	0.979	0.711	0.530
MSE-*t*, normal	0.532	0.597	0.344	0.327	0.409	0.290	0.245	0.139	0.429	0.554	0.383	0.326
ENC-*t*, normal	0.995	0.995	0.865	0.715	0.980	0.919	0.785	0.416	0.991	0.992	0.918	0.796
score-*t*, normal	0.506	0.589	0.419	0.417	0.386	0.297	0.277	0.137	0.393	0.515	0.400	0.370

Notes:

1. The data generating process is VAR(4), with coefficients given in Table 2 and error variance–covariance matrix given in section "Monte Carlo Design."

2. For each artificial data set, point forecasts of the trivariate vector of variables are formed recursively using AR(4) and VAR(4) models estimated by OLS. Density forecasts from AR(4) and VAR(4) models are obtained (with recursive estimation) using the Bayesian MCMC estimation approach described in section "Details on Implementation of Bayesian Methods," under an extremely loose (Normal-inverted Wishart) prior. At each forecast origin and horizon, we compute the log score using an empirical estimate of the forecast density obtained with a Gaussian kernel and 2,000 draws of forecasts from the posterior distribution. The point forecasts and log scores are then used to form the indicated test statistics. R and P refer to the number of in-sample observations and 1-step ahead forecasts, respectively.

3. The test statistics MSE-*F*, MSE-*t*, ENC-*F*, ENC-*t*, and score-*t* are defined in section "Test Statistics and Critical Values." All tests are conducted on a one-sided basis.

4. VARBS refers to a VAR-based bootstrap, described in section "Test Statistics and Critical Values." FRBS refers to a fixed-regressor bootstrap, described in section "Test Statistics and Critical Values." "Normal" refers to critical values from the standard normal distribution. The bootstraps used 499 replications. The number of Monte Carlo simulations is 2,000.

Table 8. Monte Carlo Rejection Rates, 10% Critical Values: DGP 2, Recursive Scheme, BVAR versus AR.

Test, Critical Values	Forecasts of $y_{1,t}$				Forecasts of $y_{2,t}$				Forecasts of $y_{3,t}$			
	$\tau=1$	$\tau=2$	$\tau=4$	$\tau=8$	$\tau=1$	$\tau=2$	$\tau=4$	$\tau=8$	$\tau=1$	$\tau=2$	$\tau=4$	$\tau=8$
R,P = 80,40												
MSE-*F*, VARBS	0.740	0.742	0.512	0.457	0.627	0.457	0.371	0.184	0.610	0.684	0.507	0.401
MSE-*F*, FRBS	0.749	0.774	0.543	0.464	0.650	0.535	0.453	0.220	0.598	0.678	0.466	0.338
MSE-*t*, VARBS	0.664	0.625	0.369	0.237	0.579	0.419	0.307	0.151	0.596	0.599	0.392	0.251
MSE-*t*, FRBS	0.681	0.640	0.303	0.227	0.591	0.435	0.272	0.145	0.570	0.410	0.124	0.148
ENC-*F*, VARBS	0.889	0.889	0.619	0.547	0.692	0.460	0.355	0.165	0.762	0.837	0.645	0.485
ENC-*F*, FRBS	0.914	0.917	0.667	0.542	0.743	0.619	0.512	0.225	0.805	0.894	0.684	0.451
ENC-*t*, VARBS	0.808	0.769	0.398	0.275	0.635	0.456	0.317	0.149	0.713	0.720	0.468	0.290
ENC-*t*, FRBS	0.816	0.745	0.279	0.219	0.649	0.430	0.269	0.144	0.725	0.527	0.192	0.181
MSE-*t*, normal	0.387	0.416	0.258	0.206	0.289	0.235	0.228	0.157	0.188	0.340	0.289	0.266
ENC-*t*, normal	0.864	0.851	0.541	0.366	0.712	0.538	0.439	0.245	0.767	0.807	0.653	0.502
score-*t*, normal	0.415	0.458	0.313	0.281	0.317	0.262	0.270	0.189	0.163	0.333	0.285	0.341
R,P = 80,80												
MSE-*F*, VARBS	0.913	0.917	0.719	0.596	0.842	0.695	0.573	0.273	0.800	0.867	0.707	0.569
MSE-*F*, FRBS	0.919	0.924	0.721	0.606	0.840	0.719	0.602	0.299	0.767	0.832	0.579	0.439
MSE-*t*, VARBS	0.897	0.887	0.618	0.434	0.831	0.682	0.526	0.249	0.814	0.843	0.624	0.437
MSE-*t*, FRBS	0.899	0.884	0.494	0.342	0.829	0.666	0.416	0.177	0.778	0.592	0.127	0.156
ENC-*F*, VARBS	0.982	0.984	0.817	0.715	0.905	0.706	0.567	0.237	0.926	0.962	0.830	0.659
ENC-*F*, FRBS	0.988	0.990	0.861	0.709	0.923	0.813	0.709	0.328	0.942	0.981	0.842	0.601
ENC-*t*, VARBS	0.975	0.963	0.693	0.492	0.882	0.730	0.562	0.252	0.909	0.929	0.743	0.527
ENC-*t*, FRBS	0.974	0.960	0.516	0.340	0.888	0.694	0.471	0.216	0.916	0.791	0.271	0.249
MSE-*t*, normal	0.547	0.589	0.326	0.272	0.447	0.308	0.271	0.161	0.240	0.445	0.345	0.329
ENC-*t*, normal	0.981	0.980	0.763	0.572	0.909	0.766	0.635	0.328	0.928	0.955	0.832	0.669
score-*t*, normal	0.538	0.595	0.414	0.375	0.452	0.327	0.302	0.138	0.239	0.443	0.330	0.354
R,P = 80,120												
MSE-*F*, VARBS	0.977	0.976	0.818	0.727	0.942	0.829	0.705	0.351	0.914	0.960	0.826	0.683
MSE-*F*, FRBS	0.976	0.976	0.816	0.726	0.937	0.824	0.699	0.341	0.894	0.925	0.668	0.516
MSE-*t*, VARBS	0.976	0.970	0.770	0.605	0.943	0.822	0.680	0.326	0.939	0.960	0.788	0.586

10% Critical Values

MSE-*t*, FRBS	0.977	0.968	0.653	0.475	0.933	0.802	0.524	0.220	0.919	0.733	0.125	0.163
ENC-*F*, VARBS	0.998	0.998	0.920	0.843	0.967	0.844	0.699	0.315	0.981	0.994	0.934	0.791
ENC-*F*, FRBS	0.999	0.998	0.939	0.839	0.977	0.910	0.808	0.376	0.985	0.996	0.923	0.720
ENC-*t*, VARBS	0.994	0.996	0.866	0.670	0.961	0.857	0.711	0.332	0.976	0.984	0.901	0.713
ENC-*t*, FRBS	0.995	0.994	0.729	0.466	0.967	0.824	0.612	0.260	0.979	0.923	0.372	0.341
MSE-*t*, normal	0.685	0.709	0.433	0.356	0.572	0.402	0.337	0.161	0.331	0.602	0.416	0.395
ENC-*t*, normal	0.997	0.997	0.885	0.724	0.972	0.873	0.745	0.381	0.982	0.990	0.930	0.809
score-*t*, normal	0.673	0.722	0.521	0.447	0.572	0.426	0.370	0.169	0.342	0.584	0.432	0.423

Notes:

1. The data generating process is a VAR(4), with coefficients given in Table 2 and error variance–covariance matrix given in section "Monte Carlo Design."

2. For each artificial data set, point forecasts of the trivariate vector of variables are formed recursively using AR(4) models estimated by OLS and a BVAR(4) estimated under a Normal-Wishart prior (using iteration and the posterior mean coefficients). Density forecasts from AR(4) and BVAR(4) models are obtained using the Bayesian MCMC estimation approach described in section "Details on Implementation of Bayesian Methods," with an extremely loose prior for the AR models. At each forecast origin and horizon, we compute the log score using an empirical estimate of the forecast density obtained with a Gaussian kernel and 2,000 draws of forecasts. The point forecasts and log scores are then used to form the indicated test statistics. R and P refer to the number of in-sample observations and 1-step ahead forecasts, respectively.

3. The test statistics MSE-*F*, MSE-*t*, ENC-*F*, ENC-*t*, and score-*t* are defined in section "Test Statistics and Critical Values." All tests are conducted on a one-sided basis.

4. VARBS refers to a VAR-based bootstrap, described in section "Test Statistics and Critical Values." FRBS refers to a fixed-regressor bootstrap, described in section "Test Statistics and Critical Values." "Normal" refers to critical values from the standard normal distribution. The bootstraps used 499 replications. The number of Monte Carlo simulations is 2,000.

Returning to results based on the recursive estimation and forecasting scheme, in the DGP 2 experiments of Tables 7 and 8, in which the true model is a VAR, the bootstrap-based tests yield reasonably high rejection rates, particularly at shorter forecast horizons. Broadly, rejection rates are similar for the VAR-based and fixed regressor bootstraps. It is also the case that rejection rates are broadly similar for tests of VAR accuracy versus the AR and for tests of BVAR accuracy versus the AR. Accordingly, in the interest of brevity, in the following discussion we focus on test results for VAR forecasts obtained with the VAR-based bootstrap and with standard normal critical values.

Consider, for example, results in Table 7 for variable 2 and R,P=80,80 obtained with VARBS. At the 1-step horizon, the rejection rates are 82.6 percent for MSE-F, 84.1 percent for MSE-t, 94.4 percent for ENC-F, and 92.8 for ENC-t. At the 4-step horizon, the corresponding rejection rates are, respectively, 57.3, 53.1, 68.8, and 64.2 percent. As might be expected, power systematically rises as the size of the forecast sample (P) rises and falls as the forecast horizon falls. For instance, using the VARBS, the rejection rate for the MSE-F test applied to 1-step ahead forecasts of variable 2 rises from 63.6 percent for P=40 to 82.6 percent for P=80 to 92.8 percent for P=120. Using the VARBS, the rejection rate for the MSE-F test applied to forecasts of variable 2 falls from 82.6 percent for $h=1$ to, respectively, 68.9, 57.3, and 30.1 percent for horizons 2, 4, and 8.

For the different tests compared to bootstrap critical values, their power rankings broadly line up with the results of Clark and McCracken (2001, 2005a,b) on univariate forecasts. The ENC-F test rejects the null of equal accuracy in population more frequently than does the MSE-F test. Consider VARBS-based results for forecasts of the second variable in experiments with R,P=80,80 (Table 7). The ENC-F rejection rate is 94.4 percent at the 1-step horizon and 68.8 percent at the 4-step horizon; the corresponding MSE-F rejection rates are 82.6 and 57.3 percent, respectively. It is also the case that the ENC-t test rejects the null more frequently than does the MSE-t test. In the same example, the ENC-t rejection rate is 92.8 and 64.2 percent at horizons of 1 and 4 periods, respectively, compared to MSE-t rejection rates of 84.1 and 53.1 percent. As these examples suggest, it is also usually the case that power is at least as great with the F-type tests as with t-type tests, with larger differences at longer horizons than shorter horizons.

In the DGP 2 experiments, as in the DGP 1 experiments, the MSE-t and ENC-t tests compared against standard normal critical values behave fairly differently. The ENC-t test compared against normal critical values rejects the null of equal accuracy in population at least as often as does the same test compared against bootstrapped critical values. Consider again forecasts of the second variable in experiments with R, P=80,80 (Table 7). Comparing the ENC-t test against VARBS critical values yields a rejection rate of 92.8 and 64.2 percent at horizons of 1 and 4 periods, respectively, while comparing the test against standard normal critical values yields corresponding rejection rates of 92.8 and 66.6 percent, respectively. Because the MSE-t test compared against normal critical values is effectively a test of the null of equal accuracy in the finite sample, rather than in population, it systematically rejects less often than do any of the other tests, all of which are more appropriately thought of as tests of equal accuracy in population. Continuing with the same example, the rejection rate for the MSE-t test compared against standard normal critical values is 29.4 percent at the 1-step horizon and 20.2 percent at the 4-step horizon.

Another parallel to the DGP 1 results is that, in the DGP 2 experiments, rejection rates for the score-t test applied to density forecasts are similar to those for the MSE-t test applied to point forecasts and compared to standard normal critical values. For instance, in forecasts of the second variable in experiments with R, P=80,80, the rejection rate of the score-t test is 28.5 percent at the 1-step horizon and 22.7 percent at the 4-step horizon. The only notable differences in the rejection rates of the two tests occur with variable 1 and longer forecast horizons, for which the score-t test has a rejection rate as much as 10 percentage points higher than the MSE-t test.

To summarize the implications of the Monte Carlo analysis, based on our results for testing VAR forecasts versus AR forecasts, what we refer to as the VAR-based bootstrap (in which the null of no predictive content is imposed on the bootstrap DGP) appears to be reasonably reliable for testing equal accuracy in population. It is harder to know much about the reliability of tests for equal accuracy in the finite sample, based on standard normal critical values, because there is no simple way of parameterizing a Monte Carlo DGP to make AR and VAR forecasts equally accurate in the finite sample. Our results for such tests, based on DGPs in which either the AR or VAR is the model generating the data, may be seen as weakly suggestive that the finite-sample tests will behave as desired. We draw that inference based on our finding that, when the data are generated by AR models, such that there is no predictability in population, rejection rates for the finite sample tests are low.

CONCLUSION

This article reviews recent developments in the evaluation of point and density forecasts in the context of VARs. In particular we focus on testing issues specific to IMS forecasts made from VARs as compared to DMS forecasts generated by single equation models. We discussed these issues as they relate to the evaluation of forecasts in population (based on true, unknown model coefficients) as well as the evaluation of forecasts in the finite sample (based on estimated model coefficients).

For evaluation in finite-samples, the results of Giacomini and White (2006) broadly apply with VAR forecasts, subject to the requirement that the forecasts be generated with a fixed or rolling window estimation approach. In a VAR context, it would be very difficult to define a data-generating process that satisfied the null hypothesis and thereby permitted a Monte Carlo analysis of the performance of the tests of equal finite-sample predictive ability. However, we provide Monte Carlo analysis to show that tests of equal accuracy in the finite sample behave as expected when applied under a DGP designed for the null of equal accuracy in population. In particular, the finite-sample tests are much more conservative (in rejection rates) than the tests of equal accuracy in population. These differences are important. For example, Alquist, Kilian, and Vigfusson (2013) demonstrate by example that conventional tests of equal population predictability are prone to overstating the finite-sample forecasting ability of VAR models. This problem is especially severe for large-dimension VAR models.

For evaluation in population, the distinction between IMS and DMS approaches becomes important. For DMS forecasts, existing results from sources such as West (1996), Clark and McCracken (2001, 2005a), and McCracken (2007) go through directly and simply. For IMS forecasts and evaluations that do not involve nested models, the results of West (1996) still apply, but the correction to standard errors necessitated by parameter estimation becomes more complicated. For IMS forecasts and evaluations of nested models, the asymptotic results of Clark and McCracken (2001a, 2005) and McCracken (2007) no longer apply, because the multistep forecasts are not linear functions of least-squares estimated parameters.

Still, even in this case, the bootstrap methods considered in Clark and McCracken (2001a, 2005) for testing equal predictability in population tend to be reasonably accurate when applied to nested model forecasts from VAR models. Our Monte Carlo experiments bear this out. Overall, our Monte Carlo analysis suggests that bootstrap-based tests of equal accuracy have reasonable properties under the null of equal accuracy in population.

NOTES

1. Some recent examples include Clark and McCracken (2010), Clark (2011), and Baumeister and Kilian (2012a).

2. Some might prefer that a system of equations all taking a direct multistep form not be referred to as a vector autoregression. However, because other studies, such as Koop (2012) and Schorfheide (2005), have referred to these models as direct multistep VARs, we follow their precedent.

3. See Inoue and Kilian (2004) for an early example of a study drawing a distinction between equal accuracy in the finite sample and in population.

4. Giannone, Lenza, and Primiceri (2012) extend the common Normal-Wishart setup to incorporate a hierarchical prior that covers the usual hyperparameters of the Minnesota prior.

5. However, such an approach ignores the serial correlation in the errors of a VAR in DMS form.

6. Carriero et al. (2012) investigate the performance of a Litterman-style approach that is more tractable for simulation because it treats each equation separately, under the assumption of a fixed and diagonal error covariance matrix. They don't find such an approach to improve on a Normal-Wishart setup, partly because the negative consequences of the assumption of a fixed and diagonal error covariance matrix offset the positive effects of imposing cross-variable shrinkage.

7. These assumptions are intended to be expository, not complete. See West (1996) for more detail.

8. McCracken (2000) weakens this assumption to $\mathbf{F} = \partial E[f(\boldsymbol{X}_{t+\tau}, \boldsymbol{\beta})]/\partial \boldsymbol{\beta}_{\beta=\beta^*}$ so that the function $f(\boldsymbol{X}_{t+\tau}, \boldsymbol{\beta})$ need not be differentiable in $\boldsymbol{\beta}$.

9. If more than one model is being used to construct $f(\boldsymbol{X}_{t+\tau}, \hat{\boldsymbol{\beta}}_t)$ (so that $\hat{\boldsymbol{\beta}}_t = (\hat{\boldsymbol{\beta}}_t^{(1)'}, \hat{\boldsymbol{\beta}}_t^{(2)'})'$), then $\mathbf{B}$ is the block diagonal matrix $\text{diag}(\mathbf{B}^{(1)}, \mathbf{B}^{(2)})$ and hence a consistent estimate is $\hat{\mathbf{B}} = \text{diag}(\hat{\mathbf{B}}^{(1)}, \hat{\mathbf{B}}^{(2)})$.

10. If more than one model is being used to construct $f_{t+\tau}(\hat{\boldsymbol{\beta}}_t)$ (so that $\hat{\boldsymbol{\beta}}_t = (\hat{\boldsymbol{\beta}}_t^{(1)'}, \hat{\boldsymbol{\beta}}_t^{(2)'})'$), then $\boldsymbol{h}_{t+\tau}$ is the stacked vector $(\boldsymbol{h}_{t+\tau}^{(1)'}, \boldsymbol{h}_{t+\tau}^{(2)'})'$ and hence the appropriate estimator is $\hat{\boldsymbol{h}}_{t+\tau} = (\hat{\boldsymbol{h}}_{t+\tau}^{(1)'}, \hat{\boldsymbol{h}}_{t+\tau}^{(2)'})'$.

11. If $f(\boldsymbol{X}_{t+\tau}, \boldsymbol{\beta})$ is non-differentiable see McCracken (2004) for an alternative estimator.

12. We deliberately end the sample before the financial crisis to avoid some of the sizable effects the severe recession and slow recovery had on VAR coefficient estimates.

13. The initial observations are selected by sampling from the actual data as in Stine (1987).

14. In results for these other sample sizes, the most notable difference is that, with DGP experiments using R, P=40,80, the tests sometimes display slightly larger size distortions.

15. The largest difference seems to be that, in the DGP 1 results, the ENC-t test compared against standard normal critical values yields a higher rejection rate when applied to BVAR versus AR forecasts than when applied to VAR versus AR forecasts.

ACKNOWLEDGMENT

The authors gratefully acknowledged helpful comments from Lutz Kilian.

REFERENCES

Alquist, R., Kilian, L., & Vigfusson, R. J. (2013). Forecasting the price of oil. In G. Elliott & A. Timmermann (Eds.), *Handbook of economic forecasting* (Vol. 2). Amsterdam: North-Holland.

Amisano, G., & Giacomini, R. (2007). Comparing density forecasts via weighted likelihood ratio tests. *Journal of Business and Economic Statistics*, *25*, 177–190.

Andersson, M. K., Karlsson, G., & Svensson, J. (2007). *The Riksbank's forecasting performance*. Sveriges Riksbank Working Paper No. 218.

Banbura, M., Giannone, D., & Reichlin, L. (2010). Large Bayesian vector autoregressions. *Journal of Applied Econometrics*, *25*, 71–92.

Baumeister, C., & Kilian, L. (2012a). Real-time forecasts of the real price of oil. *Journal of Business and Economic Statistics*, *30*, 326–336.

Baumeister, C., & Kilian, L. (2012b). *Real-time analysis of oil price risks using forecast scenarios*. Mimeo. University of Michigan.

Baumeister, C., & Kilian, L. (2012c). *What central bankers need to know about forecasting oil prices*. Forthcoming, International Economic Review.

Beauchemin, K. R. & Zaman, S. (2011). *A medium scale forecasting model for monetary policy*. Federal Reserve Bank of Cleveland, Working Paper No. 11-28.

Berkowitz, J. (2001). Testing density forecasts, with applications to risk management. *Journal of Business and Economic Statistics*, *19*, 465–474.

Bjornland, H. C., Gerdrup, K., Jore, A. S., Smith, C., & Thorsrud, L. A. (2012). Does forecast combination improve Norges Bank inflation forecasts? *Oxford Bulletin of Economics and Statistics*, *74*, 163–179.

Carriero A., Clark, T. E., & Marcellino, M. (2012). Bayesian VARs: Specification choices and forecast accuracy. *Journal of Applied Econometrics*, Forthcoming.

Chong, Y. Y., & Hendry, D. F. (1986). Econometric evaluation of linear macroeconomic models. *Review of Economic Studies*, *53*, 671–690.

Christoffersen, P. F. (1998). Evaluating interval forecasts. *International Economic Review*, *39*, 841–862.

Clark, T. E. (2011). Real-time density forecasts from BVARs with stochastic volatility. *Journal of Business and Economic Statistics*, *29*, 327–341.

Clark, T. E., & McCracken, M. W. (2001). Tests of equal forecast accuracy and encompassing for nested models. *Journal of Econometrics*, *105*, 85–110.

Clark, T. E., & McCracken, M. W. (2005a). Evaluating direct multistep forecasts. *Econometric Reviews*, *24*, 369–404.

Clark, T. E., & McCracken, M. W. (2005b). The power of tests of predictive ability in the presence of structural breaks. *Journal of Econometrics*, *124*, 1–31.

Clark, T. E., & McCracken, M. W. (2008). Forecasting with small macroeconomic VARs in the presence of instabilities. In D. E. Rapach & M. E. Wohar (Eds.), *Forecasting in the presence of structural breaks and model uncertainty*. Amsterdam: Elsevier.

Clark, T. E., & McCracken, M. W. (2009). Tests of equal predictive ability with real-time data. *Journal of Business and Economic Statistics*, *27*, 441–454.

Clark, T. E., & McCracken, M. W. (2010). Averaging forecasts from VARs with uncertain instabilities. *Journal of Applied Econometrics*, *25*, 5–29.

Clark, T. E., & McCracken, M. W. (2011a). *Nested forecast model comparisons: A new approach to testing equal accuracy*. Manuscript. Federal Reserve Bank of St. Louis.

Clark, T. E., & McCracken, M. W. (2011b). Testing for unconditional predictive ability. In Michael P. Clements & David F. Hendry (Eds.), *Oxford handbook of economic forecasting*. Oxford, UK: Oxford University Press.

Clark, T. E., & McCracken, M. W. (2012a). Reality checks and comparisons of nested predictive models. *Journal of Business and Economic Statistics*, *30*, 53–66.

Clark, T. E., & McCracken, M. W. (2013). Advances in forecast evaluation. In G. Elliott & A. Timmermann (Eds.), *Handbook of economic forecasting* (Vol. 2). Amsterdam: North-Holland.

Clark, T. E., & West, K. D. (2006). Using out-of-sample mean squared prediction errors to test the martingale difference hypothesis. *Journal of Econometrics*, *135*, 155–186.

Clark, T. E., & West, K. D. (2007). Approximately normal tests for equal predictive accuracy in nested models. *Journal of Econometrics*, *138*, 291–311.

Coletti, D., & Murchison, S. (2002). Models in policy-making. *Bank of Canada Review*, Summer 2002, 19–26.

Corradi, V., Swanson, N. R., & Olivetti, C. (2001). Predictive ability with cointegrated variables. *Journal of Econometrics*, *105*, 315–358.

Doan, T., Litterman, R. B., & Sims, C. A. (1984). Forecasting and conditional projection using realistic prior distributions. *Econometric Reviews*, *3*, 1–100.

Geweke, J., & Whiteman, C. (2006). Bayesian forecasting. In G. Elliott, C. Granger, & A. Timmermann (Eds.), *Handbook of economic forecasting*. Amsterdam: North-Holland.

Giacomini, R., & White, H. (2006). Tests of conditional predictive ability. *Econometrica*, *74*, 1545–1578.

Giannone, D., Lenza, M., & Primiceri, G. (2012). *Prior selection for vector autoregressions*. Mimeo. Free University of Brussels.

Gneiting, T., & Raftery, A. E. (2007). Strictly proper scoring rules, prediction, and estimation. *Journal of the American Statistical Association*, *102*, 359–378.

Goncalves, Silvia, & Kilian, L. (2004). Bootstrapping autoregressions with conditional heteroskedasticity of unknown form. *Journal of Econometrics*, *123*, 89–120.

Harvey, D. I., Leybourne, S. J., & Newbold, P. (1997). Testing the equality of prediction mean squared errors. *International Journal of Forecasting*, *13*, 281–91.

Inoue, A., & Kilian, L. (2004). In-sample or out-of-sample tests of predictability: Which one should we use? *Econometric Reviews*, *23*, 371–402.

Kadiyala, K. R., & Karlsson, S. (1997). Numerical methods for estimation and inference in Bayesian VAR-models. *Journal of Applied Econometrics*, *12*, 99–132.

Kapetanios, G., Labhard, V., & Price, S. (2008). Forecast combination and the bank of England's suite of statistical forecasting models. *Economic Modelling*, *25*, 772–792.

Karlsson, S. (2013). Forecasting with Bayesian VAR models. In G. Elliott & A. Timmermann (Eds.), *Handbook of economic forecasting* (Vol. 2). Amsterdam: North-Holland.

Kilian, L. (1999). Exchange rates and monetary fundamentals: What do we learn from long-horizon regressions? *Journal of Applied Econometrics*, *14*, 491–510.

Komunjer, I., & Owyang, M. T. (2012). Multivariate forecast evaluation and rationality testing. *Review of Economics and Statistics*, *94*, 1066–1080.

Koop, G. (2013). Forecasting with medium and large Bayesian VARs. *Journal of Applied Econometrics*, *28*, 177–203.

Litterman, R. B. (1986). Forecasting with Bayesian vector autoregressions – Five years of experience. *Journal of Business and Economic Statistics*, *4*, 25–38.

Lutkepohl, H. (1991). *Introduction to multiple time series analysis*. Berlin: Springer-Verlag.

Marcellino, M., Stock, J. H., & Watson, M. W. (2006). A comparison of direct and iterated multistep AR methods for forecasting macroeconomic time series. *Journal of Econometrics*, *127*, 499–526.

McCracken, M. W. (2000). Robust out-of-sample inference. *Journal of Econometrics*, *99*, 195–223.

McCracken, M. W. (2004). Parameter estimation error and tests of equal forecast accuracy between non-nested models. *International Journal of Forecasting*, *20*, 503–514.

McCracken, M. W. (2007). Asymptotics for out-of-sample tests of granger causality. *Journal of Econometrics*, *140*, 719–752.

Newey, W. K., & West, K. D. (1987). A simple, positive semi-definite, heteroskedasticity and autocorrelation consistent covariance matrix. *Econometrica*, *55*, 703–708.

Schorfheide, F. (2005). VAR forecasting under misspecification. *Journal of Econometrics*, *128*, 99–136.

Sims, C. A. (1980). Macroeconomics and reality. *Econometrica*, *48*, 1–48.

Stine, R. A. (1987). Estimating properties of autoregressive forecasts. *Journal of the American Statistical Association*, *82*, 1072–1078.

West, K. D. (1996). Asymptotic inference about predictive ability. *Econometrica*, *64*, 1067–1084.

West, K. D. (2006). Forecast evaluation. In G. Elliott, C. W. J. Granger, & A. Timmermann (Eds.), *Handbook of economic forecasting*. North-Holland, Amsterdam.

West, K. D., & McCracken, M. W. (1998). Regression-based tests of predictive ability. *International Economic Review*, *39*, 817–840.

Wright, J. H. (2012). Evaluating real-time VAR forecasts with an informative democratic prior. *Journal of Applied Econometrics*. Forthcoming.

Zellner, A. (1971). *An introduction to Bayesian inference in econometrics*. New York, NY: Wiley.

IDENTIFYING STRUCTURAL VECTOR AUTOREGRESSIONS VIA CHANGES IN VOLATILITY☆

Helmut Lütkepohl

ABSTRACT

Identification of shocks of interest is a central problem in structural vector autoregressive (SVAR) modeling. Identification is often achieved by imposing restrictions on the impact or long-run effects of shocks or by considering sign restrictions for the impulse responses. In a number of articles changes in the volatility of the shocks have also been used for identification. The present study focuses on the latter device. Some possible setups for identification via heteroskedasticity are reviewed and

☆This article was written while the author was a Bundesbank Professor at the Freie Universität Berlin. An earlier version of the paper was published as DIW Discussion Paper 1259 – http://www.diw.de/sixcms/detail.php?id=diw_0.1.c.412678.de

VAR Models in Macroeconomics – New Developments and Applications:
Essays in Honor of Christopher A. Sims
Advances in Econometrics, Volume 32, 169–203

ISSN: 0731-9053/doi:10.1108/S0731-9053(2013)0000031005

their potential and limitations are discussed. Two detailed examples are considered to illustrate the approach.

Keywords: Markov switching model; vector autoregression; heteroskedasticity; vector GARCH; conditional heteroskedasticity

JEL classification: C32

INTRODUCTION

In structural vector autoregressive (SVAR) analysis, typically a critical issue is the identification of economically meaningful shocks. Restrictions on the impact effects, the long-run responses of the variables, or the signs of the impulse responses are usually used to identify or characterize the shocks of interest (e.g., Amisano & Giannini, 1997; Blanchard & Quah, 1989; Canova & De Nicoló, 2002; Faust, 1998; King, Plosser, Stock, & Watson, 1991; Sims, 1980; Uhlig, 2005). Such restrictions are sometimes controversial. Taking that into account, another strand of the literature uses statistical data properties for identification. In particular, changes in the volatility of the residuals or the variables is used as additional identification information. For instance, Rigobon (2003), Rigobon and Sack (2003), and Lanne and Lütkepohl (2008) use heteroskedasticity, while Normandin and Phaneuf (2004), Bouakez and Normandin (2010), and Lanne, Lütkepohl, and Maciejowska (2010) consider conditional heteroskedasticity for the identification of shocks.

In the present study the principle of identification by heteroskedasticity is discussed and specific modeling strategies are considered. First the general modeling strategy is presented and its advantages and limitations are discussed in the second section. The central idea is that the structural shocks are extracted by transforming the reduced form residuals. The basic assumption that structural shocks should be instantaneously uncorrelated is insufficient to uniquely determine or identify the shocks. To uniquely identify the shocks, or at least to characterize them sufficiently to make them distinguishable, further restrictions must be imposed. This is typically done through exclusion or sign restrictions. The present study considers the question of how changes in volatility can be used for this purpose. The assumption that the impulse responses remain time-invariant, as the volatility of the reduced-form shocks changes, provides additional restrictions that can be used to uniquely pin down the structural shocks.

Some specific models for changes in volatility that can be used in this context are considered in the third section. In particular, a standard model of heteroskedasticity with exogenously generated shifts in variance is discussed, an approach based on a multivariate generalized autoregressive heteroskedastic (MGARCH) model for the movements in volatility is presented, and a Markov regime switching mechanism is considered for capturing changes in the residual volatility. Detailed examples are provided in the fourth and fifth sections. The first example compares different identification schemes for monetary policy shocks that have been considered in the literature. The second example is a model for the crude oil market. Finally, the sixth section concludes.

THE MODEL SETUP

The Basic Model

The point of departure is a K-dimensional reduced form VAR(p)

$$y_t = \nu + A_1 y_{t-1} + \cdots + A_p y_{t-p} + u_t \tag{1}$$

Here ν is a $(K \times 1)$ constant term, A_j $(j = 1, \ldots, p)$ are $(K \times K)$ VAR coefficient matrices, and u_t is a zero-mean white noise error term. The model may contain cointegrated variables and may in fact be set up as a vector error correction model. Also there may be more general deterministic terms such as seasonal dummies or a polynomial trend.

The standard SVAR analysis assumes that the structural shocks, say ε_t, are obtained from the reduced form residuals by a linear transformation, say $\varepsilon_t = B^{-1}u_t$ or, equivalently, $B\varepsilon_t = u_t$. In a conventional SVAR analysis B is chosen such that the structural shocks, that is, the components of ε_t are instantaneously uncorrelated. In other words, ε_t has a diagonal covariance matrix, that is, $E(\varepsilon_t\varepsilon_t') = \Sigma_\varepsilon$ is a diagonal matrix. In practice, B is often chosen such that Σ_ε is an identity matrix. This implies that $\varepsilon_t \sim (0, I_K)$ and, hence, $E(u_t u_t') = \Sigma_u = BB'$ if u_t is homoskedastic and, thus, has a time-invariant covariance matrix. Notice that by substituting $B\varepsilon_t$ for u_t in Eq. (1), the matrix B is easily recognized as the matrix of impact effects of the structural shocks.

The matrix B is not uniquely determined by the relation $\Sigma_u = BB'$ because Σ_u is symmetric and, hence, has at most $K(K+1)/2$ different elements while B has K^2 elements. Consequently, at least $K(K-1)/2$ further relations or restrictions are needed to uniquely specify the transformation matrix B and thereby characterize the shocks. As mentioned earlier, restrictions on the impact effects of the shocks, that is, on the elements of B directly, or indirect constraints on the long-run effects are often used in a conventional SVAR analysis to specify unique structural shocks. Alternatively, sign restrictions on the impulse responses may be imposed. They will not fix unique structural shocks in general but will specify a range of structural shocks. Such an approach may be sufficient for answering at least some questions of interest.

Another possibility for specifying unique shocks opens up if the variances of the reduced form and, hence, also the structural shocks change during the sample period. To see this suppose that $E(u_t u_t') = \Sigma_1$ in the first part of the sample, say for $t = 1, \ldots, T_1$, and $E(u_t u_t') = \Sigma_2$ in the second part of the sample $(t > T_1)$, where $\Sigma_1 \neq \Sigma_2$. Moreover, suppose that all other VAR parameters in Eq. (1) remain time-invariant. Then there is a matrix decomposition result that ensures the existence of a matrix B and a diagonal matrix $\Lambda = \text{diag}(\lambda_1, \ldots, \lambda_K)$ such that

$$\Sigma_1 = BB' \quad \text{and} \quad \Sigma_2 = B\Lambda B' \tag{2}$$

Using this B matrix to obtain structural shocks from the reduced form errors, that is, defining $\varepsilon_t = B^{-1}u_t$, we get

$$E(\varepsilon_t \varepsilon_t') = \begin{cases} I_K, & t = 1, \ldots, T_1 \\ \Lambda, & t > T_1 \end{cases}$$

Since Λ is diagonal, the structural errors obtained in this way satisfy the basic requirement of being instantaneously uncorrelated. Moreover, Lanne et al. (2010) show that the matrix B is unique apart from changes in the signs and permutations of the columns if the diagonal elements of Λ are all distinct. The latter condition can be checked easily, of course. If it is satisfied, we have unique shocks by just imposing the basic requirement that the structural shocks are instantaneously uncorrelated. Of course, using the same transformation matrix B for the whole sample period implies that the impact effects of the shocks are time-invariant as well and only the variances change. All other requirements for uniqueness of the joint decomposition of the two covariance matrices do not affect the shocks substantively. The possible changes in sign just mean that we may consider negative instead of positive shocks and vice versa. Also the fact that the columns of B can be permuted reflects just that the ordering of the shocks can be chosen freely. It is worth looking at an example in some detail.

An Illustrative Example

Consider a bivariate system so that

$$u_t = \begin{bmatrix} u_{1t} \\ u_{2t} \end{bmatrix} = \begin{bmatrix} b_{11} & b_{12} \\ b_{21} & b_{22} \end{bmatrix} \begin{bmatrix} \varepsilon_{1t} \\ \varepsilon_{2t} \end{bmatrix}$$

For this case the two relations in (2) are

$$\begin{bmatrix} \sigma_{1,1}^2 & \sigma_{12,1} \\ \sigma_{12,1} & \sigma_{2,1}^2 \end{bmatrix} = \begin{bmatrix} b_{11}^2 + b_{12}^2 & b_{11}b_{21} + b_{12}b_{22} \\ b_{11}b_{21} + b_{12}b_{22} & b_{21}^2 + b_{22}^2 \end{bmatrix}$$

and

$$\begin{bmatrix} \sigma_{1,2}^2 & \sigma_{12,2} \\ \sigma_{12,2} & \sigma_{2,2}^2 \end{bmatrix} = \begin{bmatrix} \lambda_1 b_{11}^2 + \lambda_2 b_{12}^2 & \lambda_1 b_{11}b_{21} + \lambda_2 b_{12}b_{22} \\ \lambda_1 b_{11}b_{21} + \lambda_2 b_{12}b_{22} & \lambda_1 b_{21}^2 + \lambda_2 b_{22}^2 \end{bmatrix}$$

Thus, we get six relations

$$\begin{aligned} \sigma_{1,1}^2 &= b_{11}^2 + b_{12}^2 \\ \sigma_{12,1} &= b_{11}b_{21} + b_{12}b_{22} \\ \sigma_{2,1}^2 &= b_{21}^2 + b_{22}^2 \\ \sigma_{1,2}^2 &= \lambda_1 b_{11}^2 + \lambda_2 b_{12}^2 \\ \sigma_{12,2} &= \lambda_1 b_{11}b_{21} + \lambda_2 b_{12}b_{22} \\ \sigma_{2,2}^2 &= \lambda_1 b_{21}^2 + \lambda_2 b_{22}^2 \end{aligned}$$

from which we can solve for the six structural parameters $b_{11}, b_{12}, b_{21}, b_{22}$, λ_1, λ_2. Here the latter two parameters are the diagonal elements of Λ. The solution is unique (up to sign) if λ_1 and λ_2 are distinct and are ordered in some specific way. For instance, $\lambda_1 < \lambda_2$. Notice that the variances of the structural shocks are normalized to one in the first part of the sample. Hence, the λ_i indicate the change in variance from the first to the second variance regime. In other words, they can be interpreted as relative variances in the second part of the sample. The uniqueness requirement just means that the change in variance is not homogeneous across both variables. In fact, only the variance of one of the variables has to change. This already suffices for a unique solution.

It should be clear, however, that this is a purely statistical approach to identifying unique shocks and does not necessarily result in economically meaningful shocks. Just like using a Choleski decomposition of the residual covariance matrix to obtain a unique set of orthogonal shocks does not lead automatically to economically meaningful shocks but requires further arguments to make them meaningful. For example, Rigobon (2003) discusses the case of a bivariate system with demand and supply shocks. Although in this case it is indeed difficult to disentangle the shocks in the absence of economic restrictions and, hence, the statistical identification approach comes in handy, it must be kept in mind that the method just normalizes the shocks in a mathematical way and the shocks identified in this way need not correspond to supply and demand shocks nor would it be clear which shock is the demand shock and which the supply shock. Such an interpretation requires further economic insights. Thus, this approach needs to be viewed with some caution. What we can say if the identification conditions are satisfied is that there are only two specific shocks that satisfy the basic requirements of structural shocks and induce time-invariant responses throughout the full sample. The latter assumption is, of course, implicit in a conventional analysis that ignores the changes in volatility or does not use them for identification purposes. On the other hand, one may question the assumption that the instantaneous effects of the shocks are not affected by the volatility changes. Indeed, also the VAR coefficients are assumed to be time-invariant and that assumption may be unrealistic in some cases. It is required, however, to make the approach work. If also the VAR coefficients change over time, then the assumption of time-invariant impulse responses does not make sense any more and without that assumption enforcing time-invariant impact effects is meaningless.

The General Model

In general there may be more than two volatility regimes. Suppose there are M such regimes and the corresponding covariance matrices are $\Sigma_1, \dots, \Sigma_M$. In that case a decomposition

$$\Sigma_1 = BB', \quad \Sigma_m = B\Lambda_m B', \quad m = 2, \dots, M \tag{3}$$

with diagonal matrices $\Lambda_m = \text{diag}(\lambda_{m1}, \dots, \lambda_{mK})$ $(m = 2, \dots, M)$ may not exist if arbitrary covariance matrices Σ_m $(m = 1, \dots, M)$ are allowed for. In fact, the decomposition (3) imposes testable restrictions on the covariance matrices. Thereby we can check whether the data are compatible with the decomposition and, hence, we can use B to transform the reduced form residuals into structural errors with time-invariant impact effects. Lanne et al. (2010, Proposition 1) show that the structural shocks obtained in this way are unique (apart from ordering and sign) if for any two subscripts $k, l \in \{1, \ldots, K\}$, $k \neq l$, there is a $j \in \{2, \ldots, M\}$ such that $\lambda_{jk} \neq \lambda_{jl}$. Although this condition for exact (local) identification is more complicated than in the two-state case, it shows that not much is needed for identification of structural shocks via heteroskedasticity. Moreover, the identification condition can be checked with statistical tests because, if there are M distinct volatility regimes, then the diagonal elements of the Λ_m matrices are identified and, hence, can be estimated consistently with a proper asymptotic distribution under common assumptions. We will discuss the issue further when we consider specific models for the volatility regimes in the third section and we will present examples in the fourth and fifth sections.

If unique structural shocks are obtained via heteroskedasticity, then any further restrictions become overidentifying. In particular, restrictions on the impact or the long-run effects imposed in conventional SVAR analyses will be overidentifying and, hence, can be tested against the data. This implication of the present framework is very convenient if there are controversial identifying restrictions in a conventional analysis. In the present framework the data can speak up against such restrictions or confirm that they are compatible with the data. Of course, it can happen that even this framework does not allow us to discriminate between competing economic theories. In other words, two competing models may both be in line with the data in the present framework. Moreover, rejecting a particular model in the present framework could be the result of the underlying assumptions of the statistical model not being compatible with the data. For example, the assumption of regime invariant impact effects may be problematic and then a particular set of restrictions may be rejected although it is actually the model framework that is too narrow for the test.

So far we have considered what is known as a B-model in the SVAR literature (see Lütkepohl, 2005, Chapter 9). In practice A-models are sometimes preferred because they allow to impose restrictions on the instantaneous interactions between the observed variables more easily. An A-model has the form

$$A_0^* y_t = \nu^* + A_1^* y_{t-1} + \cdots + A_p^* y_{t-p} + \varepsilon_t \tag{4}$$

In this model the A_i^* $(i=0,1,\ldots,p)$ are structural form VAR coefficient matrices. The matrix A_0^* represents the instantaneous relations between the observed elements of y_t. The B-model can be obtained from the A-model by pre-multiplying Eq. (4) with $B=A_0^{*-1}$. Thus, the two models are closely related. However, in the A-model the matrix A_0^* is typically assumed to have unit diagonal elements. Thereby each of the K variables can appear as dependent variable on the left-hand side of one of the equations. In that case, assuming unit variances of the structural errors is not reasonable. Thus, the error term ε_t in the A-model is usually assumed to have diagonal covariance matrix.

In that situation it may also be preferable not to normalize the variances of the structural shocks in the first regime if there are different volatility regimes. Hence, one may prefer to choose a decomposition

$$\Sigma_m = A_0^{*-1}\Lambda_m(A_0^{*-1})', \quad m=1,\ldots,M \tag{5}$$

where A_0^* has a unit main diagonal. Although this modification of our basic model is in principle easy to deal with as far as estimation is concerned, the identification conditions are more difficult to state in this model. Hence, it may be better to leave the diagonal elements of A_0^* unrestricted and normalize the variances of the structural shocks in the first regime. Such a modification leaves zero restrictions on A_0^* unaffected. Thus, they may be imposed as in a conventional analysis. Alternatively, one may set up a so-called AB-model of the form

$$A_0^* y_t = \nu^* + A_1^* y_{t-1} + \cdots + A_p^* y_{t-p} + B\varepsilon_t \tag{6}$$

and specify B to be a diagonal matrix and require the structural shocks ε_t to have unit variances throughout the sample in a conventional analysis or in the first volatility regime in a heteroskedastic model.

There are two other generalizations of the present setup that have been used in practice. First, one may include additional unmodeled right-hand side variables. As long as time-invariance of all parameters apart from those in the residual covariance structure can be justified, the extension is straightforward. The same holds if there are cointegrated variables and a vector error correction model is specified. In that case one may estimate the cointegration relations in a first step without accounting for heteroskedasticity and fix the cointegration parameters at those estimates in the subsequent analysis.

SPECIFIC MODELS FOR CHANGES IN RESIDUAL VOLATILITY

In this section different models for the changes in residual volatility are presented. The first one assumes exogenous changes in given time periods. The second model supposes that the volatility changes are generated by a vector generalized autoregressive conditional heteroskedasticity (MGARCH) process. The third approach is also based on conditional heteroskedasticity and models the volatility changes with the help of a Markov process. All three models have been used in applications in the literature. Which one of them is most suitable in a particular situation is best judged by the analyst and subject matter considerations may play a role for the choice. For instance, for higher frequency financial market data conditional heteroskedasticity is often a plausible assumption whereas in low frequency macro data models assuming that some exogenous event has caused a change in the variance at a specific point in time may be plausible. Also the standard tools for model checking can be used to confirm whether a particular model captures the volatility features of the data. In practice it may be difficult, however, to discriminate between different models purely on the basis of statistical tools because they capture similar data features.

Exogenous Changes in Volatility

The General Setup

In the original article on identifying structural shocks via heteroskedasticity in an SVAR analysis, Rigobon (2003) considers changes in the unconditional variances of the shocks at specific time points. Although he also discusses the implications of misspecifying the change points, the baseline model is one where the change in volatility occurs exogenously. Thus, it is assumed that the reduced form residual covariance matrix in time period t is

$$E(u_t u_t') = \Sigma_t = \begin{cases} \Sigma_1 & \text{for } t \in \mathcal{T}_1 \\ \vdots & \\ \Sigma_M & \text{for } t \in \mathcal{T}_M \end{cases} \tag{7}$$

where $\mathcal{T}_m = \{T_{m-1}+1, \ldots, T_m\}$ $(m=1,\ldots,M)$ are M given volatility regimes that typically consist of consecutive time periods. Here $T_0=0$ and $T_M=T$ is assumed. The T_m, for $m>0$, represent the points of volatility changes.

Inference

Assuming that the error term in Eq. (1) is Normal, that is, $u_t \sim \mathcal{N}(0, \Sigma_t)$, with covariance matrix structure as in Eq. (7), estimation of the model can be done by maximum likelihood (ML). The log-likelihood function is

$$\log l(\beta, \sigma) = -\frac{KT}{2}\log 2\pi - \frac{1}{2}\sum_{t=1}^{T}\log|\Sigma_t| - \frac{1}{2}\sum_{t=1}^{T} u_t'\Sigma_t^{-1}u_t \tag{8}$$

where σ contains all unknown covariance parameters and the time-invariant VAR parameters including deterministic terms are collected in β (see Lütkepohl (2005, Chapter 17) for further details). The log-likelihood (8) can also be used for quasi ML estimation if u_t and hence y_t is not normally distributed. Alternatively one may use a feasible GLS procedure by estimating Eq. (1) with equationwise OLS first and then using the residuals obtained, say $\hat{u}_t$, for estimating the covariance matrices as

$$\hat{\Sigma}_m = \frac{1}{T_m - T_{m-1}}\sum_{t\in\mathcal{T}_m}\hat{u}_t\hat{u}_t'$$

In the next step these estimates can then be used in a GLS estimator

$$\hat{\beta} = \left(\sum_{t=1}^{T} Z_{t-1}Z_{t-1}'\otimes\hat{\Sigma}_t^{-1}\right)^{-1}\left(\sum_{t=1}^{T}(Z_{t-1}\otimes\hat{\Sigma}_t^{-1})y_t\right) \tag{9}$$

where $\beta = \text{vec}[\nu, A_1, \ldots, A_p]$, $Z_{t-1} = (1, y_{t-1}', \ldots, y_{t-p}')'$ and $\hat{\Sigma}_t = \hat{\Sigma}_m$ for $t \in \mathcal{T}_m$. If the VAR process is stable, these estimators have standard asymptotic properties and can be used accordingly.

The structural parameters can be estimated by substituting BB' for Σ_1, $B\Lambda_m B'$ for Σ_m $(m = 2, \ldots, M)$, and $y_t - (Z_t'\otimes I_K)\beta$ for u_t in Eq. (8) and using a full maximization of the log-likelihood. Alternatively, the β parameters can be estimated in a first step and in the second step the concentrated log-likelihood is optimized only with respect to the structural parameters B and Λ_m $(m = 2, \ldots, M)$.

Again the estimators have standard asymptotic properties under common assumptions. These estimators can thus be used for checking the crucial assumptions underlying the present analysis. For example, since the estimators of the diagonal elements of the Λ_m are asymptotically Normal, they can be used to test for distinct diagonal elements and, hence, for identification of the shocks. Moreover, a likelihood ratio (LR) test for checking the null hypothesis that the covariance matrices can be decomposed as in expression (3) can be set up by comparing the log-likelihood maxima with and without the restriction. The LR statistic has an asymptotic χ^2 distribution with

$$\frac{1}{2}MK(K+1)-K^2-(M-1)K$$

degrees of freedom under the null hypothesis (see Lanne et al., 2010).

Notice, however, that the asymptotic properties of the estimators and test statistics may differ substantially from the small sample properties because we are in a dense parameter environment and macro time series are typically relatively short. Recall also that a certain volatility regime may be present only for a relatively short time span. Hence, the interpretation of the statistical results should be done cautiously. Remember also that the volatility regimes are assumed known, which is not true in practice. Instead there will be uncertainty regarding the change points and perhaps some pretesting may have occurred to determine the change points (see, e.g., Ehrmann, Fratzscher, & Rigobon, 2011). All these problems do not add to the reliability of the inference in this context. Still, the fact that the data have a chance to speak up regarding some of the crucial issues related to the identification of the shocks is an advantage over having no information on these issues.

Although this type of model requires nonlinear estimation techniques, doing a bootstrap for inference on impulse responses should not result in insurmountable problems.

Applications

Rigobon (2003) applies this approach to investigate the relationship between sovereign bonds in three Latin American countries, Argentina, Brazil and Mexico, using daily bond yields for the period January 1994 to December 2001. There is no natural instantaneous causal direction of the relation between the bond yields in these three countries. In fact, shocks hitting the system of the three series of interest are expected to have an instantaneous effect on all of them. Also, the long-term impact of shocks does not provide easily acceptable identifying restrictions. Hence, using the changes in volatility makes sense.

Rigobon and Sack (2003) point out the instantaneous nature of the relation between short-term interest rates and stock prices and use identification via heteroskedasticity to disentangle the relation between these variables. In particular, they use differences in volatility shifts in the stock market and in monetary policy for identification. More precisely, they assume that the volatility of monetary shocks is constant whereas the volatility in equity prices varies. They distinguish four different volatility regimes that are determined by 30-day rolling variance estimates. Their analysis is based on daily US data from March 1985 to December 1999. They find that monetary policy reacts to unexpected movements in the stock market.

Rigobon and Sack (2004) address related problems as in the previous application. In other words, they also study the links between monetary policy and the stock market. Using identification through heteroskedasticity they find that earlier results from event studies may be biased.

Lanne and Lütkepohl (2008) use identification through heteroskedasticity to compare different models for the US money market that have been considered in the earlier literature. Their analysis will be discussed in more detail as an example in the fourth section.

Ehrmann et al. (2011) use the approach to investigate the linkages between the US and euro area money, bond and equity markets by analyzing a system of short-term interest rates, long-term bond rates, stock returns from both regions and an exchange rate. They use two-day returns for a period from 1989 to 2008. The change in volatility is determined by a rolling window procedure that results in various volatility regimes for their dataset. Again identification through heteroskedasticity is used because there are not enough identifying exclusion restrictions for the impact effects of the shocks.

In some of the applications surveyed in this section, identification by heteroskedasticity is used to compensate for lack of economic identification conditions. From the discussion in the second section it should be clear, however, that shocks obtained in this way may not correspond to the economically interesting shocks the authors are looking for.

Vector GARCH Residuals

Alternatively, conditional heteroskedasticity may facilitate the identification of the structural VAR shocks. Normandin and Phaneuf (2004) propose to use multivariate GARCH models to capture changes in volatility. Their setup is presented in this section.

The General Setup

The reduced form errors u_t are assumed to be generated by a MGARCH process of the form

$$\Sigma_{u,t|t-1} = E(u_t u_t' | u_{t-1}, \ldots) = B\Sigma_{\varepsilon,t|t-1}B' \tag{10}$$

where $\Sigma_{\varepsilon,t|t-1} = \operatorname{diag}(\sigma^2_{1,t|t-1}, \ldots, \sigma^2_{K,t|t-1})$ is a diagonal matrix with

$$\sigma^2_{k,t|t-1} = \gamma_{k0} + \sum_{j=1}^{q} \gamma_{kj}\varepsilon^2_{t-j} + \sum_{j=1}^{s} g_{kj}\sigma^2_{k,t-j|t-j-1}, \quad k = 1, \ldots, K \tag{11}$$

In other words, the structural shocks ε_t are assumed to be instantaneously uncorrelated and have a diagonal MGARCH(q, s) structure. This is basically the generalized orthogonal GARCH (GO-GARCH) model of van der Weide (2002). A closely related model was also proposed by Vrontos, Dellaportas, and Politis (2003). These authors consider a model of the form (10) with a triangular matrix B. This choice simplifies the identification issue. Normandin and Phaneuf (2004) assume an MGARCH(1,1) model which is a standard model in practice for variables with GARCH effects. By sticking to the simple first order class they avoid a more difficult GARCH order selection step.

Sentana and Fiorentini (2001) provide general results that imply identification of the structural shocks (apart from permutation and sign changes) if all but one of the components are actually conditionally heteroskedastic and $\Gamma'\Gamma$ is invertible, where $(\sigma^2_{k,1|0}, \ldots, \sigma^2_{k,T|T-1})$ is the kth row of Γ'.[1] In other words, in the GARCH(1,1) case we can have for at most one $k \in \{1, \ldots, K\}$ that $\gamma_{k1} = g_{k1} = 0$. Invertibility of $\Gamma'\Gamma$ ensures that the conditional heteroskedasticity is not driven by a small number of processes like in a factor model where a single process may determine the development of a larger set of variables. Of course, this requirement corresponds to the condition that there must be sufficient heterogeneity in the volatility changes when there are only finitely many volatility regimes (see Section "The General Model"). Provided the identification conditions are satisfied, the setup in Eq. (10) ensures that, if we choose $\varepsilon_t = B^{-1}u_t$, the structural shocks will be unique and the impulse responses including the impact effects, are time-invariant.

Inference

For parameter estimation Normandin and Phaneuf (2004) propose a two-step procedure that fits a reduced form VAR(p) process in the first step and then estimates the GARCH and structural parameters by maximizing the corresponding Gaussian log-likelihood. In other words, B and the GARCH parameters are estimated by maximizing

$$\log l^c(B,\gamma) = -\frac{1}{2}\sum_{t=1}^{T}\left[\log \det(B\Sigma_{\varepsilon,t|t-1}B') + \hat{u}_t'(B\Sigma_{\varepsilon,t|t-1}B')^{-1}\hat{u}_t\right] \tag{12}$$

where $\hat{u}_t$, $t=1,\ldots,T$, are the reduced form residuals from the first estimation stage and γ is the vector of all GARCH parameters. For example, for a GARCH(1,1)

$$\gamma = (\gamma_{01},\ldots,\gamma_{0K},\gamma_{11},\ldots,\gamma_{1K},g_{11},\ldots,g_{1K})'$$

Obviously, the estimation problem is a highly nonlinear optimization task and Bouakez and Normandin (2010) mention that a full optimization of the function in Eq. (12) may be difficult for high dimensional systems. In that case further simplifications may be necessary. For example, one may split up the system and introduce further estimation stages, in particular, if suitable constraints on B are justifiable.

In principle, the identification conditions can be checked with statistical tools. In particular, it can be tested whether the GARCH parameters in Eq. (11) are significantly different from zero. Moreover, the nonsingularity of the matrix $\Gamma'\Gamma$ can be tested. Normandin and Phaneuf (2004) basically check only the nonsingularity of the empirical counterpart of this matrix. It would be desirable, of course, to develop a proper statistical test of the null hypothesis that the population matrix $\Gamma'\Gamma$ is singular.

Once structural shocks are identified, an impulse response analysis can be performed. Usual methods for generating confidence intervals for them with a bootstrap may be computationally demanding, given the difficulties in maximizing the likelihood function. Normandin and Phaneuf (2004) mention that they used a Bayesian method from Sims and Zha (1999) for that purpose without giving details.

Applications

Normandin and Phaneuf (2004) use their MGARCH framework to compare identification schemes for monetary policy shocks similar to the example discussed in the fourth section based on US monthly data for the 1982M11–1998M12 period. They reject some of the assumptions for identifying monetary policy shocks in a conventional framework. In the fourth section a slightly different model will be considered in a related analysis that illustrates how identification through changes in volatility can be used in this context.

Bouakez and Normandin (2010) study the effects of US monetary policy shocks on exchange rates between the United States and the G7 countries. The monetary policy shocks are identified via GARCH in the residuals. They use monthly data for 1982M11–1998M12 for one set of countries and from 1982M11 to 2004M10 for the remaining countries. They find that monetary policy shocks have a substantial impact on exchange rate fluctuations.

Markov Switching in Residual Volatility

The first proposal for modeling volatility changes assumed a finite number of states that are exogenously generated while the previous proposal of an MGARCH process generating the volatility changes amounts to assuming an endogenously generated continuum of possible volatility regimes. In the present subsection an intermediate proposal is presented where only a finite number of volatility regimes are assumed, which are, however, generated endogenously via a Markov process. In the context of structural shock identification through changes in volatility this approach was first suggested by Lanne et al. (2010). A more complete methodology for using it in the SVAR framework is presented by Herwartz and Lütkepohl (2011).

The General Setup

In this approach the distribution of the reduced form error term u_t is assumed to depend on a discrete Markov process s_t $(t=0,\pm 1,\pm 2,\ldots)$ with M states. In other words, $s_t \in \{1,\ldots,M\}$. The transition probabilities between the states are

$$p_{ij} = \Pr(s_t = j | s_{t-1} = i), \quad i, j = 1, \ldots, M$$

Lanne et al. (2010) assume that the conditional distribution of u_t, given the state s_t, is Normal

$$u_t | s_t \sim \mathcal{N}(0, \Sigma_{s_t}) \qquad (13)$$

This assumption allows them to do ML estimation but is otherwise not critical. Notice that the VAR coefficients do not depend on the Markov process s_t and are assumed time-invariant. Hence, this model is more restrictive than the Markov switching (MS) models considered by Rubio-Ramirez, Waggoner, and Zha (2005), Sims and Zha (2006), and Sims, Waggoner, and Zha (2008), for example.

Although there is only a finite number of volatility states, the model can mix these states by assigning probabilities strictly between zero and one to them in any particular period t. Thus, it may capture gradual transitions from one state to another and one may argue that it can also generate a continuum of states. They are centered at a finite number of states, however, which may simplify the interpretation. The state covariance matrices $\Sigma_1, \ldots, \Sigma_M$ are used for the identification of shocks, as before. In other words, the state covariance matrices are decomposed as in expression (3) and uniqueness of the structural shocks obtained as $\varepsilon_t = B^{-1}u_t$ is ensured if the diagonal elements of the Λ_m matrices have a fixed ordering and are sufficiently heterogeneous, as specified in Section "The General Model."

Inference

The statistical analysis of this model has to address the following issues in addition to the usual VAR inference: The number of volatility states has to be determined, the parameters of the overall model have to be estimated, the diagonal elements of the Λ_m matrices have to be compared, and possibly identifying restrictions from a conventional analysis have to be tested within the present framework. Under the assumption of a conditional Normal distribution given the states, the likelihood function can be set up easily

$$\log l(\beta, B, \boldsymbol{\lambda}, P|\mathbf{y}) = \sum_{t=1}^{T} \log\left(\sum_{m=1}^{M} \Pr(s_t = m|Y_{t-1}) f(y_t|s_t = m, Y_{t-1})\right) \tag{14}$$

where $\boldsymbol{\lambda}$ is the vector of all diagonal elements of $\Lambda_2, \ldots, \Lambda_M$, P is the matrix of transition probabilities, $\mathbf{y}$ is the full sample, $Y_{t-1} = (y'_{t-1}, \ldots, y'_1)'$, and

$$f(y_t|s_t = m, Y_{t-1}) = (2\pi)^{-K/2} \det(\Sigma_m)^{-1/2} \exp\left\{-\frac{1}{2} u'_t \Sigma_m^{-1} u_t\right\}$$

Optimizing this log-likelihood function is a highly nonlinear optimization problem that poses a number of computational challenges. A suitable EM algorithm is provided by Herwartz and Lütkepohl (2011) who adopt an algorithm from Krolzig (1997) for the present model. Unfortunately, this algorithm seems to work reliably only for reasonably small systems with a small number of variables K, a small number of volatility states M, and a moderate number of lags p. A detailed discussion of the problems associated with the optimization of the log-likelihood function is given by Herwartz and Lütkepohl (2011). In practice, if the conditional Normality of the residuals given State m is not satisfied, the procedure is a quasi ML procedure. The Normal distribution assumption is made for convenience. It is not essential for the asymptotic properties of the estimates to hold, provided some general conditions are satisfied.

Inference on the λ parameters can be conducted with standard procedures. For example, equality of some of them can be tested with standard Wald or LR tests. The latter tests have the drawback in the present situation that they involve two difficult log-likelihood maximizations. Hence, Wald tests may be preferable from a computational point of view. Unfortunately, they are known to have poor small sample properties in similar situations and, hence, may not be fully satisfactory either. In any case, in principle these tests can be used to check the identification conditions. Moreover, if the shocks can be identified via the volatility structure, other restrictions can be tested as well with conventional tests.

Standard model selection criteria can be used for choosing between models with different numbers of volatility states. Based on simulation experiments Psaradakis and Spagnolo (2003, 2006) find that they work reasonably well. This is important because testing models with different numbers of states against the data is not easy. The usual tests have nonstandard properties because of unidentified parameters under the null hypothesis (e.g., Hansen, 1992; Garcia, 1998). Of course, model selection criteria can also be used to compare models with other sets of restrictions.

When a structural model with identified shocks is specified, it can be used for structural analysis. Impulse responses are a standard tool for that purpose. In a classical ML analysis, confidence bands for impulse responses are typically generated with bootstrap methods. Clearly, such methods are problematic when a single estimation is difficult. To mitigate the computational problems, Herwartz and Lütkepohl (2011) propose a fixed design wild bootstrap procedure. Such a procedure makes sense in the present framework because it can take care of conditional heteroskedasticity, as shown by Goncalves and Kilian (2004). The bootstrap samples are constructed conditionally on the ML estimates using

$$y_t^* = \hat{\nu} + \hat{A}_1 y_{t-1} + \cdots + \hat{A}_p y_{t-p} + u_t^* \tag{15}$$

Here the right-hand side variables for a given t are always the original lagged variables and the errors $u_t^* = \eta_t \hat{u}_t$, where η_t is a binary random variable with values 1 and -1 that have equal probability. The procedure preserves heteroskedasticity and the pattern of contemporaneous dependence of the data. Herwartz and Lütkepohl (2011) propose to bootstrap parameter estimates θ^* of $\theta = \text{vec}[\nu, A_1, \ldots, A_p]$ and B^* of B, conditionally on the initially estimated transition probabilities and $\boldsymbol{\lambda}$ to limit the computational burden. Computing the bootstrap impulse responses requires nonlinear optimization of the log-likelihood even with these simplifications and is computationally demanding.

Applications

Lanne et al. (2010) illustrate their approach by considering a small US macro model from Primiceri (2005) consisting of inflation, unemployment, and an interest rate. They find that previously used identification restrictions are problematic and are not supported by the data when changes in volatility are taken into account. As a second example they consider a US model from Sims et al. (2008) for log GDP, inflation, and a short-term interest rate. Based on quarterly data from 1959Q1 to 2005Q4 they find support for conventional identification restrictions, as used by Sims et al. (2008).

In a study that advances the methodology needed for a structural MS-VAR analysis, Herwartz and Lütkepohl (2011) consider an example from Peersman (2005) that aims at investigating the causes of the early millennium economic slow down. They use a four-variable system consisting of the price of oil, output, a consumer price index, and a short-term interest rate. Peersman uses zero restrictions on the impact effects and the long-run effects of the shocks for identification. Using quarterly US data from 1980Q1 to 2002Q2 and models with two and three volatility states they find that some but not all of the identifying restrictions used by Peersman are in line with the data when changes in volatility are accounted for. They question that the model based on conventional identifying restrictions properly reflects the causes of the economic downturn at the beginning of the millennium.

Lütkepohl and Netšunajev (2013) illustrate how the MS-SVAR approach can be used for checking models identified through sign restrictions. They consider models for the crude oil market from Kilian (2009) and Kilian and Murphy (2012). That analysis is discussed as a more detailed example in the fifth section and is therefore not elaborated upon here.

Netšunajev (2013) applies the MS-SVAR methodology to reconsider alternative approaches for identifying technology shocks and sheds new light on the conflicting evidence on the impact on hours worked in the bivariate system. He does not find clear evidence of an increase in hours worked in response to a positive technology shock.

COMPARISON OF MONEY MARKET MODELS

This section is based on Lanne and Lütkepohl (2008) (henceforth LL). Alternative identification schemes for monetary shocks are reviewed in the next subsection and then an empirical comparison using identification of shocks via heteroskedasticity is performed.

Alternative Identification Schemes

The models for the US money market considered in the following use the variables:

- TR_t – total reserves;
- NBR_t – nonborrowed reserves;
- BR_t – borrowed reserves; and
- FF_t – federal funds rate.

Because TR = BR + NBR, only three of these variables are really modeled. Hence, we define $y_t = (TR_t, NBR_t, FF_t)'$.

Bernanke and Mihov (1998) and Christiano, Eichenbaum, and Evans (1999) consider the following general structural model for the market for bank reserves: The demand for total reserves is described as

$$TR_t = -\alpha FF_t + f_{TR}(\text{policy information}) + \sigma_d \varepsilon_t^d$$

where ε_t^d is a demand disturbance with variance 1 and α and σ_d are parameters. The demand for borrowed reserves is given by

$$BR_t = \beta FF_t - \gamma NBR_t + f_{BR}(\text{policy information}) + \sigma_b \varepsilon_t^b$$

where ε_t^b is the borrowing function disturbance with variance 1 and β, γ and σ_b are parameters. Finally, the Fed policy rule for setting nonborrowed reserves is

$$NBR_t = f_{NBR}(\text{policy information}) + \phi^d \sigma_d \varepsilon_t^d + \phi^b \sigma_b \varepsilon_t^b + \sigma_s \varepsilon_t^s$$

The quantity ε_t^s represents the exogenous monetary policy shock, again with variance 1, and ϕ^d, ϕ^b, and σ_s are parameters. It is assumed that the Fed knows the disturbances to the demand for total reserves and borrowed reserves when it makes policy decisions. The policy information set consists of instantaneous and lagged values of the variables:

- gdp_t – log of gross domestic product (GDP);
- p_t – log of the implicit GDP deflator; and
- pcom_t – index of commodity prices.

The functions $f_*(\cdot)$ are all linear functions. Thus, we have a three-equation VAR model with unmodeled variables $x_t = (\text{gdp}_t, p_t, \text{pcom}_t)'$. For an impulse response analysis these variables have to be modeled as well. However, for the present purposes where we focus on comparing alternative identification schemes for monetary policy shocks, there is no need to model them. The reduced form model is

$$y_t = \nu + A_1 y_{t-1} + \cdots + A_p y_{t-p} + C_0 x_t + C_1 x_{t-1} + \cdots + C_p x_{t-p} + u_t \tag{16}$$

and the reduced form error term u_t is related to the structural shocks $\varepsilon_t = (\varepsilon_t^d, \varepsilon_t^b, \varepsilon_t^s)'$ as $u_t = B\varepsilon_t$, where

$$B = \begin{bmatrix} \sigma_d \dfrac{\beta - \phi^d \alpha\gamma + \phi^d \alpha}{\beta + \alpha} & -\alpha\sigma_s \dfrac{\gamma - 1}{\beta + \alpha} & -\alpha\sigma_b \dfrac{-1 + \phi^b \gamma - \phi^b}{\beta + \alpha} \\ \sigma_d \phi^d & \sigma_s & \sigma_b \phi^b \\ \sigma_d \dfrac{\phi^d \gamma - \phi^d + 1}{\beta + \alpha} & \sigma_s \dfrac{\gamma - 1}{\beta + \alpha} & \sigma_b \dfrac{-1 + \phi^b \gamma - \phi^b}{\beta + \alpha} \end{bmatrix} \tag{17}$$

The nine elements of B are determined by the eight free parameters $\psi = (\alpha, \beta, \gamma, \phi^d, \phi^b, \sigma_d^2, \sigma_b^2, \sigma_s^2)'$. Since six restrictions are obtained from the relations $\Sigma_u = BB'$, we need two more restrictions to uniquely determine the eight parameters and, hence, the structural shocks in this general setup. Christiano et al. (1999) list a number of studies that report specific identifying assumptions. In particular, we consider the following four identification schemes from the given references.

- *FF policy shock*: $\phi^d = 1/(1-\gamma)$ and $\phi^b = -\phi^d$ (Bernanke & Blinder, 1992). These restrictions imply that the Fed targets the federal funds rate, hence the abbreviation FF policy shock.

- *NBR policy shock*: $\phi^d = \phi^b = 0$ (Christiano & Eichenbaum, 1992). With these restrictions the policy shocks can be associated with the errors of the equation for nonborrowed reserves.
- *NBR/TR policy shock*: $\alpha = \phi^b = 0$ (Bernanke & Mihov, 1998; Strongin, 1995). These restrictions assume that shocks to total reserves are demand shocks that are accommodated by the Fed.
- *BR policy shock*: $\phi^d = 1$, $\phi^b = \alpha/\beta$ and $\gamma = 0$ (Cosimano & Sheehan, 1994). These restrictions assume that the Fed targets borrowed reserves.

The first three sets of restrictions are just-identifying and, hence, they cannot be checked against the data in a conventional setting. Therefore LL use the heteroskedasticity in the residuals as additional identifying information. Thereby the foregoing restrictions on B become overidentifying and can be checked against the data.

Empirical Analysis

LL use monthly US data from Bernanke and Mihov (1998) for the period 1965M1–1996M12. The monthly values for GDP and its deflator are constructed from quarterly data by using a set of monthly interpolator variables.

A crucial question in the present framework is, of course, whether there are changes in volatility during the sample period. Indeed changes are found by other authors as well. Of course, one would like the data to confirm corresponding changes in the volatility of the shocks. For that purpose, models with heteroskedastic residuals must be estimated. LL do this with a multistep iterative procedure. The model (16) is first estimated with equation-wise OLS. The different covariance matrices are estimated from these residuals and then a feasible GLS estimator is used.

Based on these GLS estimators heteroskedasticity is checked with LR-type tests. It is confirmed that there is strong evidence for a change in volatility in 1984M2 in line with other literature. It also coincides with a shift in monetary policy under Volcker. There is some evidence that another change has occurred in 1979M10. Because only one volatility change is needed for identification of the shocks, LL go on under the assumption of a break in 1984M2 and leave open the possibility of another volatility change in 1979M10, that is, they leave the covariance matrix for the period prior to 1979M10 unrestricted and also do not use it for identification. Estimation under this assumption gives the diagonal elements of Λ shown in Table 1.[2] Apparently they are quite different and one-standard error intervals around the estimates do not overlap. Although this is not a particularly strict criterion, it shows that there is some evidence for distinct λ_m values. Given that these are reduced form parameters, we could, in fact, perform formal tests of equality. In the fifth section we discuss how to do that.

Thus, all the identification schemes presented in the previous subsection can be tested against the data. The results of the corresponding LR tests are presented in Table 2, where the degrees of freedom used for evaluating the p-values are equal to the number of restrictions specified under H_0. They are the asymptotically correct ones if the shocks are uniquely identified via heteroskedasticity, that is, under the assumption of distinct λ_m. If not all but only some of the λ_m are distinct, a smaller number of degrees of freedom may apply. However, even with the larger number assumed here, the p-values of all the tests are very small and at a 5% significance level all null hypotheses and, hence, all identification schemes are rejected. It is worth recalling, however, that LR tests in the presence of many parameters may be biased and in particular may reject too often. Thus, in the present case, one may want to work with very small significance levels. Even if a 1% level were chosen, only the FF scheme would survive. Thus, in the present framework only the Bernanke and Blinder (1992) model, based on the assumption that the Fed targets the federal funds rate, is not clearly rejected by the data. Our results agree to a large extent with those of Normandin and Phaneuf (2004) who also find that the data are not supportive of the standard identification procedures for monetary policy shocks

Table 1. Estimates of Structural Parameters with Standard Errors in Parentheses from LL.

Parameter	Estimate
λ_1	1.1463 (0.2583)
λ_2	0.0144 (0.0032)
λ_3	0.5520 (0.1246)

Table 2. LR Tests of Overidentifying Restrictions from LL.

Identification Scheme	H_0	LR statistic	p-value
FF	$\phi^d = 1/(1-\gamma)$ and $\phi^b = -\phi^d$	8.021	0.018
NBR	$\phi^d = \phi^b = 0$	55.880	7.342e-13
NBR/TR	$\alpha = \phi^b = 0$	66.119	2.890e-14
BR	$\phi^d = 1, \phi^b = \alpha/\beta$, and $\gamma = 0$	55.947	7.101e-13

used in much of the previous literature. Note, however, that these authors use a slightly different sample period, 1982M11–1998M12, and a different model for changes in volatility.

What the example shows is that it may pay to use the additional information coming from the heteroskedasticity of the data to discriminate between competing models that cannot be tested against the data in a conventional setting. Of course, rejecting all but one model is a very special case. Our framework could have ended up not discriminating between the alternative identification schemes. A further example is presented in the next section.

A MODEL FOR THE MARKET OF CRUDE OIL

In this subsection a model from Kilian (2009) for the crude oil market is reconsidered. It consists of three variables:

- Δprod_t – change in oil production;
- q_t – real activity; and
- p_t – real price of oil.

Thus, $y_t = (\Delta\text{prod}_t, q_t, p_t)'$. Monthly data for the period 1973M2–2006M12 from Kilian and Park (2009) are used. The change in oil production is measured as log differences of world crude oil production in millions of barrels pumped per day (averaged by month), as obtained from the US Department of Energy. For q_t Kilian constructs a measure based on dry cargo single voyage ocean freight rates which is meant to capture world-wide real economic activity. Finally, the real price of oil, p_t, is based on a series from the US Department of Energy for refiner acquisition cost of imported crude oil. Thus, all three variables refer to the world market.

Kilian (2009) considers three shocks:

- $\varepsilon_t^{\text{oil}-\text{s}}$ – oil supply shocks;
- $\varepsilon_t^{\text{aggr-d}}$ – aggregate demand shocks; and
- $\varepsilon_t^{\text{oil-d}}$ – oil market specific demand shocks that are innovations to the real price of oil that cannot be explained by oil supply shocks or aggregate demand shocks.

He identifies the shocks by imposing exclusion restrictions on the impact effects. More precisely, he assumes that oil-market specific demand shocks ($\varepsilon_t^{\text{oil-d}}$) do not have an instantaneous impact on oil production and real activity and aggregate demand shocks ($\varepsilon_t^{\text{aggr-d}}$) do not affect oil production instantaneously. Thereby B is a lower-triangular matrix and the reduced form and structural residuals are related as follows:

$$\begin{bmatrix} u_t^{\Delta\text{prod}} \\ u_t^q \\ u_t^p \end{bmatrix} = \begin{bmatrix} b_{11} & 0 & 0 \\ b_{21} & b_{22} & 0 \\ b_{31} & b_{32} & b_{33} \end{bmatrix} \begin{bmatrix} \varepsilon_t^{\text{oil-s}} \\ \varepsilon_t^{\text{aggr-d}} \\ \varepsilon_t^{\text{oil-d}} \end{bmatrix} \tag{18}$$

Of course, one may question these identification restrictions and, in fact, doubts are expressed by Kilian and Murphy (2012) and Baumeister and Peersman (2013) who consider sign restrictions instead. We will return to their studies later but will report results from Lütkepohl and Netšunajev (2013) (henceforth LN) first, who use heteroskedasticity in the residuals to check the identification assumptions. Thus, the following empirical study draws heavily on their results.

Empirical Analysis

LN consider a VAR(3) model with intercept because in a standard order selection exercise not accounting for changes in volatility, AIC favors that model. In contrast, Kilian (2009) uses a VAR(24). Such a large model would be difficult to estimate by ML with an MS structure for the residual variance, however. In any case, a finite order VAR should only be viewed as an approximation to the DGP. Below, a VAR(3), which preserves features of special interest in the present context, is used for expositional purposes. In particular, the reduced form residuals display clear changes in volatility. They are plotted in Fig. 1. The volatility of the residuals of the oil production equation are more volatile in the earlier part of the sample while the volatility in the price of oil residuals is larger in the later part of the sample. One may also spot some change in volatility in the equation for real activity although that change is not quite so obvious. Of course, visual inspection can usually not be a substitute for a formal statistical analysis that will be reported shortly. Before turning to that, it is worth pointing out that the kind of volatility changes observed by visual inspection are precisely what is needed for our analysis. Recall that the change in volatility has to be sufficiently heterogeneous in order to get identification through heteroskedasticity.

A formal statistical analysis of volatility changes is reported in Table 3 where AIC and SC values are presented for models with different numbers

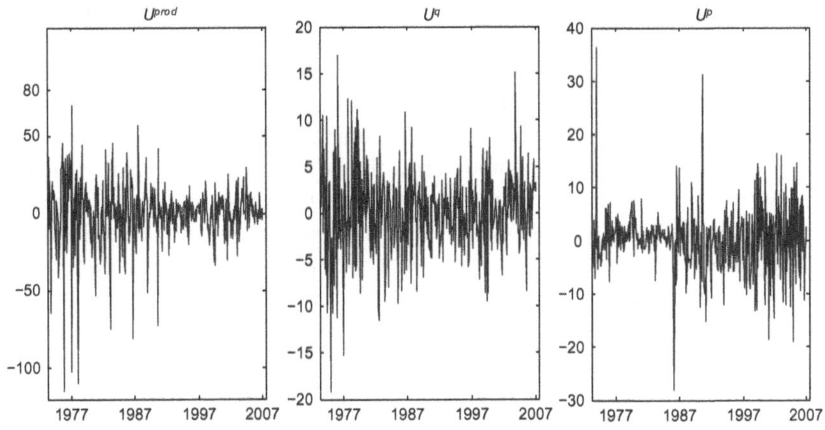

Fig. 1. Residuals of VAR(3) model from LN.

Table 3. Comparison of MS-VAR(3) Models for $y_t = (\Delta \text{prod}_t, q_t, p_t)'$ (results from LN, Table 2).

Model	AIC	SC
VAR(3) (without MS)	8615	8759
MS(2)-VAR(3)	8302	8478
MS(3)-VAR(3)	8206	8422
MS(4)-VAR(3)	8159	8423

Note: L_T – likelihood function, AIC $= -2\log L_T + 2 \times$ no of free parameters, SC $= -2\log L_T + \log T \times$ no of free parameters.

of volatility states. Notice that a VAR(p) model with m volatility states is denoted as MS(m)-VAR(p). From the results in Table 3 it is clear that both AIC and SC favor models that allow for volatility changes over a standard VAR model without MS. They do not agree on the number of states, however. AIC favors the largest model with four different volatility states while SC prefers a model with only three states. Since an MS(4) model is the largest one in our choice set, one may wonder whether AIC may even prefer more than four states. Admittedly, a maximum of four states was chosen for computational reasons. Even an MS(4)-VAR(3) model is already difficult to estimate from the given dataset.

Since the model selection criteria are not very reliable in choosing the number of states, we plot the estimated smoothed probabilities associated with the states in Fig. 2. The first two states of the MS(3) model are quite similar to those of the MS(2) model and the third state picks out some special events in the crude oil market (see, e.g., Barsky & Kilian, 2004 for a discussion of such events). Thus, a three-state model makes good sense from a subject matter point of view. In contrast, the MS(4) model has one state (the third state) that comes up only in a very few periods. While it may well be plausible to single out one state that covers only very special periods, such a situation is problematic from a computational and estimation point of view. Clearly, such a model may require that some parameters are basically estimated from very few observations and, hence, the estimates may be unreliable. If the estimation algorithm actually associates a separate state with a single observation, the associated covariance matrix estimate becomes singular and the algorithm runs into a problem because it cannot invert the matrix in evaluating the log-likelihood function. Given that a three-state model produces quite plausible states, we favor overall this model, like LN, and continue the analysis with an MS(3)-VAR(3) model.

Having decided on an MS(3)-VAR(3) model, the next important question in the context of the present analysis is whether the three covariance matrices can be decomposed as in expression (3). For answering that question we may use an LR test with $\frac{1}{2}MK(K+1) - K^2 - (M-1)K = 3$ degrees of freedom (see Section "Exogenous Changes in Volatility"). Such a test results in a p-value of 0.63 and, hence, there is little basis for rejecting the more restricted model. In other words, the covariance matrix decomposition in expression (3) is in line with the data and we can work with a model with state-invariant matrix of initial effects of the shocks.

The next question then concerns the uniqueness of the B transformation matrix and, hence, the uniqueness of the shocks. That uniqueness depends on the diagonal elements of the Λ_m matrices. The estimated quantities for the MS(3)-VAR(3) model are given in Table 4 together with estimated standard errors. Taking into account the estimated λ_{mi} values and the estimation uncertainty reflected in the standard errors, there is clearly hope for having enough heterogeneity for identification. To check this formally, the corresponding tests are presented in Table 5. Recall from Section "Markov Switching in Residual Volatility" that for each pair $i, j \in \{1, \ldots, K\}$, $i \neq j$, there has to be an $m \in \{1, \ldots, M\}$ such that $\lambda_{mi} \neq \lambda_{mj}$. Thus, identification is ensured if all three null hypotheses listed in Table 5 are rejected. Wald and LR tests are given in the table. The Wald tests are easier to compute because they only require estimates of the unrestricted model. The associated p-values are all below 10% and indicate that the identification

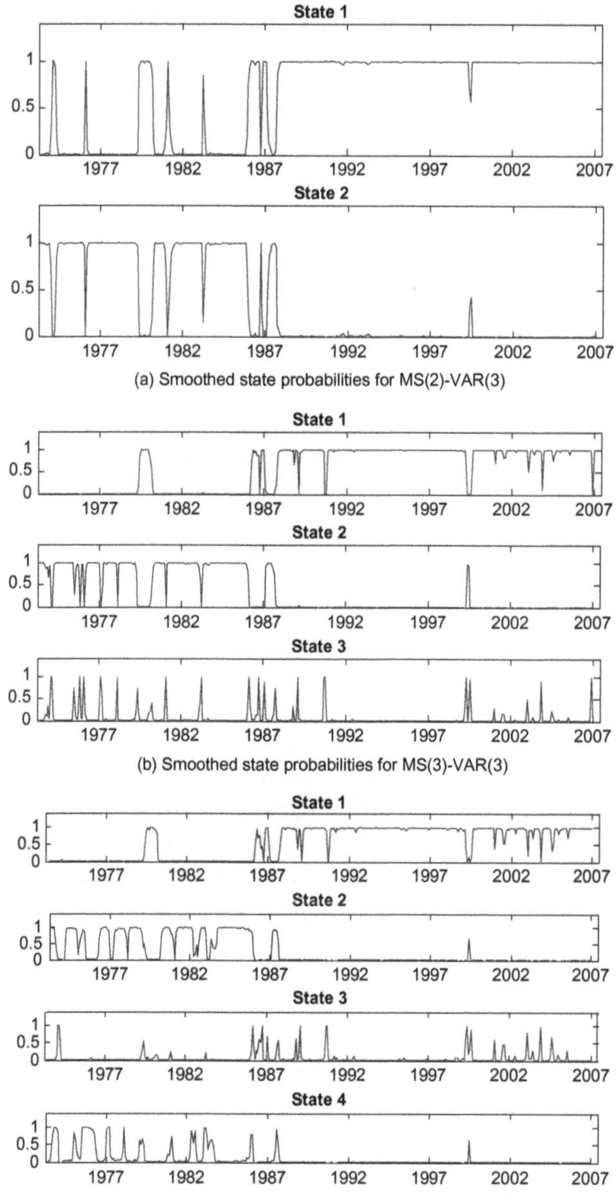

(a) Smoothed state probabilities for MS(2)-VAR(3)

(b) Smoothed state probabilities for MS(3)-VAR(3)

(c) Smoothed state probabilities for MS(4)-VAR(3)

Fig. 2. Smoothed state probabilities of MS(m)-VAR(3) models for $m = 2, 3, 4$ from Fig. 2 of LN (state-invariant B matrix imposed for MS(3) and MS(4)).

Table 4. Estimates of Structural Parameters of MS(3)-VAR(3) Models for $y_t = (\Delta\text{prod}_t, q_t, p_t)'$ with State-invariant B, adapted from LN.

Parameter	Estimate	SD
λ_{21}	5.384	0.933
λ_{22}	3.210	0.485
λ_{23}	0.056	0.009
λ_{31}	25.235	6.938
λ_{32}	3.387	1.242
λ_{33}	4.441	1.018

Note: Standard errors are obtained from the inverse of the outer product of numerical first order derivatives.

Table 5. Tests for Equality of λ_{mj}s for MS(3)-VAR(3) Model with State-invariant B (from LN).

H_0	Wald statistic	p-value	LR statistic	p-value
$\lambda_{21} = \lambda_{22}, \lambda_{31} = \lambda_{32}$	7.99	0.02	20.39	3.7×10^{-5}
$\lambda_{21} = \lambda_{23}, \lambda_{31} = \lambda_{33}$	7.87	0.02	21.04	2.7×10^{-5}
$\lambda_{22} = \lambda_{23}, \lambda_{32} = \lambda_{33}$	5.16	0.07	27.15	1.3×10^{-6}

conditions are satisfied. As Wald tests may not be very reliable in the present situation, we also report LR tests. They require two very nonlinear optimizations of the log-likelihood. Given that this function has many local optima and, hence, there is some danger of ending up in a local optimum and comparing inappropriate likelihood values in the LR tests, these tests are also problematic. Taking their values seriously, they do provide strong evidence for the identification conditions to be satisfied. Notice that the tests are performed for models where the λ_{2i} are ordered from largest to smallest. The ordering has to be fixed to ensure uniqueness of the decomposition of the covariance matrices.

Given the evidence in favor of the identification conditions, we may proceed with testing Kilian's identification conditions presented in Eq. (18). The corresponding LR test has a p-value of 0.84. Thus, we cannot reject the restrictions at conventional significance levels. In other words, in our framework the data do not object to Kilian's identification conditions. It is still worthwhile considering the case where these restrictions are not imposed. The corresponding impulse responses will be discussed next.

Impulse Response Analysis

Since we have identified shocks by assuming the MS structure for the volatility, we can go on and use these shocks for an impulse response analysis. We present the resulting responses to structural impulses obtained as $\varepsilon_t = B^{-1}u_t$ and 68% bootstrap confidence intervals in Fig. 3. These impulse responses are arbitrary in several respects. First, the scaling of the responses is determined by normalizing the shocks such that they have identity covariance matrix in the first state. Since the numbering of the states is arbitrary, this normalization is also arbitrary. The scaling does not affect the shape of the impulse responses, of course. Another element of arbitrariness results from the signs of the shocks. In Fig. 3 they are chosen such that an interpretation becomes possible, as we will see shortly. Finally, the ordering of the shocks is arbitrary. As mentioned in Section "Markov Switching in Residual Volatility," B is unique only up to column permutations. Because the columns of B correspond to the impact effects of the shocks, permuting the columns comes down to changing the order of the shocks.

Because the shocks are uniquely determined in our framework, they must correspond to the three shocks of interest in the system under investigation, namely an oil supply shock, an aggregate demand shock, and an oil-market specific demand shock, provided that these shocks are really the ones that drive the system. In our framework it is unclear a priori which of the three shocks deserves which label. However, by looking at the impulse responses we are able to assign labels and we have done so in Fig. 3.

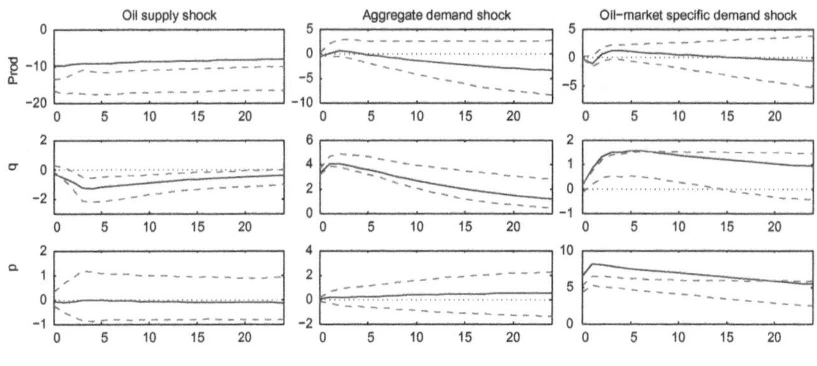

Fig. 3. Impulse responses with 68% confidence bounds of the MS(3)-VAR(3) model with state-invariant B (Fig. 4 from LN).

The first shock is called an oil supply shock because a surprise reduction in oil production goes together with a reduction in output and a possible increase in the price of oil. Notice that the reaction of the oil price is not clear in our system. However, taking into account the uncertainty in the response of the oil price, a price increase due to a surprise reduction in oil production is quite possible.

The second shock is labeled aggregate demand shock because it increases oil production, output, and the price of oil simultaneously, at least initially. Finally, the third shock is the only one that leads to a clear increase in the price of oil on impact. Therefore the shock is the only one that qualifies as an oil-market specific demand shock. The effects in Fig. 3 are in line with the sign restrictions imposed by Kilian and Murphy (2012) for the same system of variables. They characterize the shocks by the signs of their impact effects and in a sense we are doing the same thing by labeling the shocks obtained in our framework accordingly. Notice, however, that in the present analysis the shocks are point-identified and only the labeling is done by considering the effect or impulse responses whereas using sign restrictions results in set-identified shocks only that may be more difficult to interpret because their effects may be more diffuse.

Obviously, some of the effects on our models are not all that clear, given the substantial sampling uncertainty reflected in the confidence intervals associated with the impulse responses. In general there is no guarantee that the shocks of interest are actually in line with the model under study. Even if in an underlying theoretical model the desired shocks are the most important driving forces, they may not be found in an associated empirical model. The empirical model usually differs from the theoretical one in a number of dimensions. For example, variables that are not of interest in the theoretical model may be quite important in the empirical model, that is, there may be an omitted variables problem. Moreover, the empirical variables may not be exactly the same as the corresponding quantities in the theoretical model. In other words, there may be measurement errors and model specification errors. For example, the theoretical model may not account for trends and seasonality that may be present in practice. Hence, there are a number of reasons why the empirical shocks may not be those of interest from a theoretical point of view. LN use this point as an argument for proposing their framework for checking the validity of sign restrictions.

Regarding the confidence intervals in Fig. 3, it is remarkable that the original impulse responses in the upper left-hand and lower right-hand panels fall largely outside the intervals. Such an outcome is not impossible,

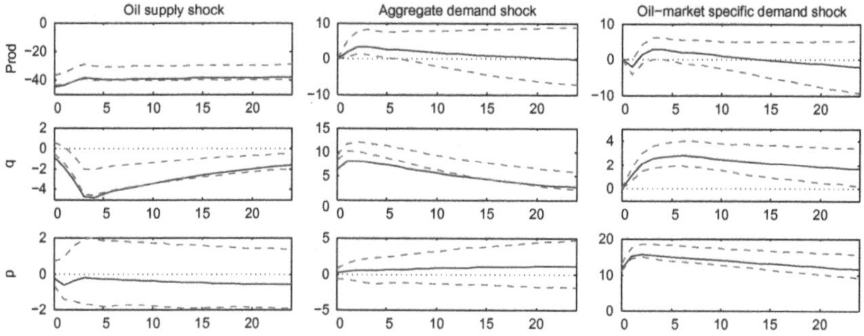

Fig. 4. Impulse responses with 68% confidence bounds of the MS(3)-VAR(3) model with state-invariant, lower-triangular B (Fig. 5 from LN).

of course, but it is still a bit unusual and may indicate that there is room for improvement in constructing suitable confidence intervals for the impulse responses in our model. The intervals are standard percentile intervals and it may have been worth trying other intervals for impulse responses (e.g., Benkwitz, Lütkepohl, & Wolters, 2001).

Given that Kilian's restrictions for the impact effects are not rejected in our analysis, it is also plausible to consider the corresponding impulse responses. They are presented in Fig. 4. Because the triangularity of B is very much in line with the data, it is not surprising that the impulse responses in Fig. 4 look quite similar to those in Fig. 3, except, of course, that now some of the impact effects are restricted to zero. Thus, in contrast to the example in the fourth section, in the present example using heteroskedasticity for the identification of the structural shocks supports the more stringent restrictions that were used in a previous analysis. The important point to note is that accounting for changes in volatility allows us to check the restrictions that had to be assumed in a conventional analysis.

CONCLUSIONS

In the present article it is argued that taking into account changes in the volatility of the residuals can provide useful additional identifying information for structural shocks. The basic assumptions are that changes in volatility occur during the sample period, that they just affect the volatility of the shocks but not the effects the shocks have on the variables and that the changes in volatility are sufficiently heterogeneous across variables. It is important to note that the requirements regarding the existence and heterogeneity of the volatility changes can be investigated by statistical procedures. Hence, the data are informative about some of the crucial conditions for identification. If full identification of the structural shocks is obtained in this framework, conventional identifying restrictions become overidentifying and, hence, can be checked with statistical tests. Thus, competing conventional identifying assumptions can be tested against the data, as illustrated in the fourth section. Moreover, the framework allows us to check that the empirical model is in line with the underlying theoretical model more generally and we have argued that differences between the empirical and theoretical models may stem from omitted variables or measurement error problems, for example.

It must be emphasized, however, that the shocks obtained in the present framework are unique due to the specific statistical model setup. Hence, they are statistically identified or normalized. There is no guarantee that these shocks necessarily correspond to economically interesting shocks. Therefore using this framework without additional economic reasoning why the shocks allow us to understand the economic relations in a system of variables is problematic. It is a useful tool for checking that an empirical model is compatible with the data, however.

NOTES

1. Milunovich and Yang (2013) discuss conditions for joint identification of the structural VAR and GARCH parameters.
2. Note that we report the corrected values from LL.

ACKNOWLEDGMENT

The author has benefitted from comments by Lutz Kilian, participants of a seminar at Instituto Universitário de Lisboa in September 2012, participants of the 12th Annual Advances in Econometrics Conference in Dallas, Texas, November 2012 and a seminar at the University of Sydney in February 2013.

REFERENCES

Amisano, G., & Giannini, C. (1997). *Topics in structural VAR econometrics* (2nd ed.), Berlin: Springer.

Barsky, R. B., & Kilian, L. (2004). Oil and the macroeconomy since the 1970s. *Journal of Economic Perspectives*, *18*, 115–134.

Baumeister, C., & Peersman, G. (2013). The role of time-varying price elasticities in accounting for volatility changes in the crude oil market. *Journal of Applied Econometrics*. doi:10.1002/jae.2283

Benkwitz, A., Lütkepohl, H., & Wolters, J. (2001). Comparison of bootstrap confidence intervals for impulse responses of German monetary systems. *Macroeconomic Dynamics*, *5*, 81–100.

Bernanke, B. S., & Blinder, A. (1992). The federal funds rate and the channels of monetary transmission. *American Economic Review*, *82*, 901–921.

Bernanke, B. S., & Mihov, I. (1998). Measuring monetary policy. *Quarterly Journal of Economics*, *113*, 869–902.

Blanchard, O., & Quah, D. (1989). The dynamic effects of aggregate demand and supply disturbances. *American Economic Review*, *79*, 655–673.

Bouakez, H., & Normandin, M. (2010). Fluctuations in the foreign exchange market: How important are monetary policy shocks? *Journal of International Economics*, *81*, 139–153.

Canova, F., & De Nicoló, G. (2002). Monetary disturbances matter for business fluctuations in the G-7. *Journal of Monetary Economics*, *49*, 1131–1159.

Christiano, L. J., & Eichenbaum, M. (1992). Identification and the liquidity effect of a monetary policy shock. In A. Cukierman, Z. Hercowitz, & L. Leiderman (Eds.), *Political economy, growth, and business cycles* (pp. 335–370). Cambridge, MA: MIT Press.

Christiano, L. J., Eichenbaum, M., & Evans, C. (1999). Monetary policy shocks: What have we learned and to what end? In J. B. Taylor & M. Woodford (Eds.), *Handbook of Macroeconomics* (Vol. 1A, pp. 65–148). Amsterdam: Elsevier.

Cosimano, T., & Sheehan, R. (1994). The federal reserve operating procedure, 1984–1990: An empirical analysis. *Journal of Macroeconomics*, *16*, 573–588.

Ehrmann, M., Fratzscher, M., & Rigobon, R. (2011). Stocks, bonds, money markets and exchange rates: Measuring international financial transmission. *Journal of Applied Econometrics*, *26*, 948–974.

Faust, J. (1998). The robustness of identified VAR conclusions about money. *Carnegie-Rochester Conference Series in Public Policy*, *49*, 207–244.

Garcia, R. (1998). Asymptotic null distribution of the likelihood ratio test in Markov switching models. *International Economic Review*, *39*, 763–788.

Goncalves, S., & Kilian, L. (2004). Bootstrapping autoregressions with conditional heteroskedasticity of unknown form. *Journal of Econometrics*, *123*, 89–120.

Hansen, B. E. (1992). The likelihood ratio test under nonstandard conditions: Testing the Markov switching model of GNP. *Journal of Applied Econometrics*, *7*, 61–82. (erratum: 11, 195–198).

Herwartz, H., & Lütkepohl, H. (2011). *Structural vector autoregressions with Markov switching: Combining conventional with statistical identification of shocks*. Working Paper Series, EUI, Florence.

Kilian, L. (2009). Not all oil price shocks are alike: Disentangling demand and supply shocks in the crude oil market. *American Economic Review*, *99*, 1053–1069.

Kilian, L., & Murphy, D. (2012). Why agnostic sign restrictions are not enough: Understanding the dynamics of oil market VAR models. *Journal of the European Economic Association, 10*, 1166–1188.

Kilian, L., & Park, C. (2009). The impact of oil price shocks on the U.S. stock market. *International Economic Review, 50*, 1267–1287.

King, R. G., Plosser, C. I., Stock, J. H., & Watson, M. W. (1991). Stochastic trends and economic fluctuations. *American Economic Review, 81*, 819–840.

Krolzig, H.-M. (1997). *Markov-switching vector autoregressions: Modelling, statistical inference, and application to business cycle analysis*. Berlin: Springer-Verlag.

Lanne, M., & Lütkepohl, H. (2008). Identifying monetary policy shocks via changes in volatility. *Journal of Money, Credit and Banking, 40*, 1131–1149.

Lanne, M., Lütkepohl, H., & Maciejowska, K. (2010). Structural vector autoregressions with Markov switching. *Journal of Economic Dynamics and Control, 34*, 121–131.

Lütkepohl, H. (2005). *New introduction to multiple time series analysis*. Berlin: Springer-Verlag.

Lütkepohl, H., & Netšunajev, A. (2013). Disentangling demand and supply shocks in the crude oil market: How to check sign restrictions in structural VARs. *Journal of Applied Econometrics*. doi:10.1002/jae.2330

Milunovich, G., & Yang, M. (2013). On identifying structural VAR models via ARCH effects. *Journal of Time Series Econometrics, 5*, 117–131.

Netšunajev, A. (2013). Reaction to technology shocks in Markov-switching structural VARs: Identification via heteroskedasticity. *Journal of Macroeconomics, 36*, 51–62.

Normandin, M., & Phaneuf, L. (2004). Monetary policy shocks: Testing identification conditions under time-varying conditional volatility. *Journal of Monetary Economics, 51*, 1217–1243.

Peersman, G. (2005). What caused the early millennium slowdown? Evidence based on vector autoregressions. *Journal of Applied Econometrics, 20*, 185–207.

Primiceri, G. E. (2005). Time varying structural vector autoregressions and monetary policy. *Review of Economic Studies, 72*, 821–852.

Psaradakis, Z., & Spagnolo, N. (2003). On the determination of the number of regimes in Markov-switching autoregressive models. *Journal of Time Series Analysis, 24*, 237–252.

Psaradakis, Z., & Spagnolo, N. (2006). Joint determination of the state dimension and autoregressive order for models with Markov regime switching. *Journal of Time Series Analysis, 27*, 753–766.

Rigobon, R. (2003). Identification through heteroskedasticity. *Review of Economics and Statistics, 85*, 777–792.

Rigobon, R., & Sack, B. (2003). Measuring the reaction of monetary policy to the stock market. *Quarterly Journal of Economics, 118*, 639–669.

Rigobon, R., & Sack, B. (2004). The impact of monetary policy on asset prices. *Journal of Monetary Economics, 51*, 1553–1575.

Rubio-Ramirez, J. F., Waggoner, D., & Zha, T. (2005). *Markov-switching structural vector autoregressions: Theory and applications*. Discussion Paper, Federal Reserve Bank of Atlanta.

Sentana, E., & Fiorentini, G. (2001). Identification, estimation and testing of conditionally heteroskedastic factor models. *Journal of Econometrics, 102*, 143–164.

Sims, C. A. (1980). Macroeconomics and reality. *Econometrica, 48*, 1–48.

Sims, C. A., Waggoner, D. F., & Zha, T. (2008). Methods for inference in large multiple-equation Markov-switching models. *Journal of Econometrics, 146*, 255–274.

Sims, C. A., & Zha, T. (1999). Error bands for impulse responses. *Econometrica, 67*, 1113–1155.

Sims, C. A., & Zha, T. (2006). Were there regime switches in U.S. monetary policy? *American Economic Review*, *96*, 54–81.

Strongin, S. (1995). The identification of monetary policy disturbances: Explaining the liquidity puzzle. *Journal of Monetary Economics*, *35*, 463–498.

Uhlig, H. (2005). What are the effects of monetary policy on output? Results from an agnostic identification procedure. *Journal of Monetary Economics*, *52*, 381–419.

van der Weide, R. (2002). GO-GARCH. A multivariate generalized orthogonal GARCH model. *Journal of Applied Econometrics*, *17*, 549–564.

Vrontos, I. D., Dellaportas, P., & Politis, D. N. (2003). A full-factor multivariate GARCH model. *Econometrics Journal*, *6*, 312–334.

PANEL VECTOR AUTOREGRESSIVE MODELS: A SURVEY☆

Fabio Canova and Matteo Ciccarelli

ABSTRACT

This article provides an overview of the panel vector autoregressive models (VAR) used in macroeconomics and finance to study the dynamic relationships between heterogeneous assets, households, firms, sectors, and countries. We discuss what their distinctive features are, what they are used for, and how they can be derived from economic theory. We also describe how they are estimated and how shock identification is performed. We compare panel VAR models to other approaches used in the literature to estimate dynamic models involving heterogeneous units. Finally, we show how structural time variation can be dealt with.

Keywords: Panel VAR; estimation; identification; inference

JEL classifications: C11; C30; C53

☆The views expressed in this article are those of the authors and do not necessarily reflect those of the ECB or the Eurosystem.

VAR Models in Macroeconomics – New Developments and Applications: Essays in Honor of Christopher A. Sims
Advances in Econometrics, Volume 32, 205–246

ISSN: 0731-9053/doi:10.1108/S0731-9053(2013)0000031006

INTRODUCTION

Macroeconomic analyses and policy evaluations increasingly require taking into account the interdependencies across sectors, markets, and countries. Even national economic issues, while often idiosyncratic, need now to be tackled from a global perspective. Thus, a number of additional channels of transmission need to be considered and additional spillovers accounted for when formulating policies, even for large and developed economies.

Domestic interdependencies have been known to produce domestic business cycle fluctuations from idiosyncratic sectoral shocks at least since Long and Plosser (1983). Spillovers from the financial sector to the real economy are key to understanding the recent global crisis (e.g., Ciccarelli, Ortega, & Valderrama, 2012a; Stock & Watson, 2012). Many authors have also argued that the rapidly rising degree of trade and financial market integration has induced closer international interdependencies within the developed world and between developing and developed world (Canova & Ciccarelli, 2012; Canova, Ciccarelli, & Ortega, 2007; Kose, Otrok, & Whiteman, 2003; Kose & Prasad, 2010; Pesaran, Schuerman, & Wiener, 2004). Thus, a multilateral perspective is crucial, and the failure to recognize this aspect of reality may induce distortions in the evaluation of economic outcomes and erroneous policy decisions.

There are two ways of examining economic issues in interdependent economies. One is to build multi-sector, multi-market, multicountry dynamic stochastic general equilibrium (DSGE) models, where agents are optimizers, and where preferences, technologies, and constraints are fully specified. Structures like these are now extensively used in the policy arena (see, e.g., the SIGMA model at the Federal Reserve Board; the global projection model at the IMF; or the EAGLE model at the ECB). Tightly parameterized DSGE models are useful because they offer clear answers to policy questions and provide easy-to-understand welfare prescriptions. However, by construction, they impose a lot of restrictions that are not always consistent with the statistical properties of the data. Thus, the policy prescriptions they provide are hardwired in the assumptions of the model, and must be considered more as a benchmark than as a realistic assessment of the options and constraints faced by policymakers in real-world situations.

An alternative approach to dealing with interdependent economies is to build panel VAR models. These models eschew most of the explicit microstructure present in DSGE models and, as their VAR counterparts, attempt to capture the dynamic interdependencies present in the data using a minimal set of restrictions. Shock identification can then transform these reduced-form models into structural ones, allowing typical exercises, such as impulse response analyses or policy counterfactuals, to be constructed in a relatively straightforward way. Structural panel VAR models are potentially liable to standard criticisms of SVAR models (see, e.g., Canova & Pina, 2005; Chari, Kehoe, & McGrattan, 2008; Cooley & Dweyer, 1998; Cooley & LeRoy, 1985; Faust & Leeper, 1997), and thus need to be implemented with care. The information they produce can effectively complement analyses conducted with DSGE models, help to point to the dimensions where these models fail, and provide stylized facts and predictions which can improve the realism of structural models.

The goal of this article is to describe what panel VAR models are and what their use is in applied work; how they can capture the heterogeneities present in interdependent economies and how the restricted specifications typically employed in the literature are nested in the general panel VAR framework we consider. We also examine how panel VAR models can be estimated, how shock identification is performed, and how one can conduct inference with such models. We highlight how the evolving nature of the cross unit interdependencies can be accounted for and how alternative frameworks such as factor models, global VAR (GVAR), collections of bilateral VARs, large-scale Bayesian VARs, or spatial VARs compare to them. Finally, we discuss the open challenges that researchers face when dealing with dynamic heterogeneous and interdependent panels (of countries, industries, or markets) in applied work.

The rest of the article is organized as follows. The next section discusses what the distinctive features of panel VAR models are and how they link to DSGE models. The third sectiondescribes how reduced form panel VARs are estimated. The fourth section adds time variation in the coefficients. The fifth section deals with shock identification and describes strategies to perform structural analyses. The sixth section compares panel VARs to other approaches that have been used to deal with dynamic models involving interdependent heterogeneous units. The conclusions and some additional considerations are given in the seventh section.

WHAT ARE PANEL VARS?

VAR models are now well-established in applied macroeconomics. In VAR models all variables are treated as endogenous and interdependent, both in a dynamic and in a static sense, although in some relevant cases, exogenous variables could be included (see, e.g., the dummy approach pioneered by Ramey & Shapiro, 1998). Let Y_t be a $G \times 1$ vector of endogenous variables. The VAR for Y_t is

$$Y_t = A_0(t) + A(\ell)Y_{t-1} + u_t \quad u_t \sim iid(0,\Sigma_u) \tag{1}$$

where $A(\ell)$ is a polynomial in the lag operator and *iid* means identically and independently distributed. Restrictions are typically imposed on the coefficient matrices A_j to make the variance of Y_t bounded and to make sure that $A(\ell)^{-1}$ exists – for example, one can impose that no roots of $A(e^{-\omega})^{-1}$ are on or inside the unit circle. Sometimes Eq. (1) is decomposed into its short-run and its long-run components, following the work of Beveridge and Nelson (1981) or Blanchard and Quah (1989), but for our purposes this distinction is not critical since the available time series dimension will be, at best, of medium length, rendering the long-run properties of the model very imprecisely pinned down. For the sake of notation, $A_0(t)$ is defined to include all deterministic components of the data. Thus, it should be understood that the representation (1) may include constants, seasonal dummies, and deterministic polynomials in time.

A typical variation of Eq. (1), used primarily in small open economy analyses, allows the G variables in Y_t to be linear function of W_t, a set of predetermined or exogenous variables. In the latter case, the VAR becomes

$$Y_t = A_0(t) + A(\ell)Y_{1t-1} + F(\ell)W_t + u_t$$

Such VARX structures have been described in Ocampo and Rodriguez (2011) and used, for example, by Cushman and Zha (1997) in their analysis of the effect of monetary policy shocks in Canada, or in exercises measuring how variables determined in world markets (such as commodity prices or world productivity) affect domestic economies.

Finite order, fixed coefficient VARs like Eq. (1) can be derived in many ways. The standard one is to use the Wold theorem (see, e.g., Canova, 2007) and assume linearity, time invariance, and invertibility of the resulting moving average representation. Under these assumptions, there exists an (infinite lag) VAR representation for any Y_t. To truncate this infinite dimension VAR and use a VAR(p), p finite and small, in empirical analyses we need to assume that the contribution of Y_{t-j} to Y_t is small when j is large.

Panel VARs have the same structure as VAR models, in the sense that all variables are assumed to be endogenous and interdependent, but a cross-sectional dimension is added to the representation. Thus, think of Y_t as the stacked version of y_{it}, the vector of G variables for each unit $i=1,\ldots,N$, that is, $Y_t=(y'_{1t}, y'_{2t}, \ldots y'_{Nt})'$. The index i is generic and could indicate countries, sectors, markets, or combinations of them. Then a panel VAR is

$$y_{it} = A_{0i}(t) + A_i(\ell)Y_{t-1} + u_{it} \qquad i=1,\ldots,N; \;\; t=1,\ldots,T \tag{2}$$

where u_{it} is a $G\times 1$ vector of random disturbances and, as the notation makes it clear, $A_{0i}(t)$ and A_i may depend on the cross-sectional unit.

When a panel VARX is considered, the representation is

$$y_{it} = A_{0i}(t) + A_i(\ell)Y_{1t-1} + F_i(\ell)W_t + u_{it} \tag{3}$$

where $u_t=[u_{1t}, u_{2t}, \ldots, u_{Nt}]' \sim iid(0,\Sigma)$, $F_{i,j}$ are $G\times M$ matrices for each lag $j=1,\ldots,q$, and W_t is a $M\times 1$ vector of exogenous variables, common to all units i.

Simple inspection of Eq. (2) or (3) suggests that a panel VAR has three characteristic features. First, lags of all endogenous variables of all units enter the model for unit i: we call this feature "dynamic interdependencies." Second, u_{it} are generally correlated across i: we call this feature "static interdependencies." In addition, since the same variables are present in each unit, there are restrictions on the covariance matrix of the the shocks. Third, the intercept, the slope, and the variance of the shocks u_{1it} may be unit-specific: we call this feature "cross-sectional heterogeneity." These features distinguish a panel VAR typically used in macroeconomic and finance from the panel VAR used in micro studies, such as Holtz-Eakin, Newey, & Rosen (1988) or more recently Vidangos (2009), where interdependencies are typically disregarded and sectoral homogeneity (allowing for certain time-invariant individual characteristics) is typically assumed. These features also distinguish the panel VAR setup from others used in the macroeconomic literature, where either cross sectional homogeneity is assumed and/or dynamic interdependencies are a priori excluded (see, e.g., Beetsma & Giuliadori, 2011; Lane & Benetrix, 2010). In a way, a panel VAR is similar to large-scale VARs where dynamic and static interdependencies are allowed for. It differs because cross-sectional heterogeneity imposes a structure on the covariance matrix of the error terms. A detailed comparison with other approaches designed to handle multiunit dynamics is provided in the sixth section.

An Example

For expository purposes, it is useful to consider a simple example. Suppose that $G=5$ variables, $N=3$ countries, and that there are $M=2$ weakly exogenous variables. Then, omitting deterministic terms, the panel VARX model is

$$y_{1t}=A_{11}(\ell)y_{1t-1}+A_{12}(\ell)y_{2t-1}+A_{13}(\ell)y_{3t-1}+F_1(\ell)W_t+u_{1t} \tag{4}$$

$$y_{2t}=A_{21}(\ell)y_{1t-1}+A_{22}(\ell)y_{2t-1}+A_{23}(\ell)y_{3t-1}+F_2(\ell)W_t+u_{2t} \tag{5}$$

$$y_{3t}=A_{31}(\ell)y_{1t-1}+A_{32}(\ell)y_{2t-1}+A_{33}(\ell)y_{3t-1}+F_3(\ell)W_t+u_{3t} \tag{6}$$

$$W_t=M(\ell)W_{t-1}+w_t \tag{7}$$

where $A_{ih,j}, i,h=1,2,3$ are 5×5 matrices for each j and $F_{i,j}, i=1,2,3$ are 5×2 matrices for each j. Furthermore, $E(u_tu_t')\equiv\Sigma_u=\begin{bmatrix}\sigma_{11} & \sigma_{12} & \sigma_{13}\\ \sigma_{21} & \sigma_{22} & \sigma_{23}\\ \sigma_{31} & \sigma_{32} & \sigma_{33}\end{bmatrix}$ is a full matrix and there is an additional structure on the 5×5 matrices σ_{ij} $i,j=1,2,3$, since the G variables are the same for each unit. In this setup, there are dynamic interdependencies ($A_{ik,j}=0, k\neq i$ for some j), there are static interdependencies ($\sigma_{i,k}\neq0, k\neq i$), and there are cross-sectional heterogeneities ($A_{i,k}\neq A_{i+1,k}, k\neq i, i+1$).

Clearly, not all three distinguishing features of panel VARs need to be used in all applications. For example, when analyzing the transmission of shocks across the financial markets of different countries, static interdependencies are probably sufficient if the time period of the analysis is a month or a quarter. Similarly, when analyzing countries in a monetary union, it may be more important to allow for slope heterogeneities (e.g., different countries may respond differently to a fiscal shock) than for variance heterogeneities (the shocks hitting different countries have different magnitude). Finally, dynamic cross-sectional differences are likely to be important when the panel includes developed and developing countries, or when it lumps together markets with different trading volumes, different transaction costs, etc.

Several interesting submodels are nested in the specification allowing us to test some of the model restrictions. For example, one might wish to know whether a model without dynamic interdependencies is sufficient to characterize the available data. This is the typical setup employed when all units are small and do not exercise dynamic effects on the other units, but shocks in different units have a common component. It is also the setup used in certain macro studies which treat units as isolated islands (see e.g., De Graeve & Karas, 2012; Rebucci, 2010a; Sa, Towbin, & Wieladek, 2012).

Another restricted specification nested in the general framework and often used in the literature is one where all interdependencies are eschewed and cross-sectional slope homogeneity is assumed. This is the typical setup used in micro studies, but it is potentially problematic in macroeconomic analyses dealing with countries or regions. Even within this restricted setup, micro and macro panel approaches differ in an important respect: the cross-sectional dimension is typically large in micro studies and small or moderate for macro panels. Vice versa, micro panels typically feature a very short time series dimension while macro panels have a moderate time series dimension. These differences have important implications for the identification of the dynamic effects of interest.

What have Panel VARs been used for?

Panel VARs have been used to address a variety of issues of interest to applied macroeconomists and policymakers. Within the realm of the business cycle literature, Canova et al. (2007) have employed a panel VAR to study the similarities and convergences among G7 cycles, while Canova and Ciccarelli (2012) employ them to examine the cross-sectional dynamics of Mediterranean business cycles. They can also be used to construct coincident or leading indicators of economic activity (see Canova & Ciccarelli, 2009) or to forecast out-of-sample, for example, output and inflation, taking into account potential cross unit spillovers effects. As we will see, cyclical indicators of both coincident and leading nature can be easily constructed from a panel VAR and (density) forecasts can be constructed with straightforward Monte Carlo methods.

Panel VARs are particularly suited to analyze the transmission of idiosyncratic shocks across units and time. For example, Canova, Ciccarelli, and Ortega (2012) have studied how shocks to U.S. interest rates are propagated to ten European economies, seven in the Euro area, and three outside of it, and how German shocks, defined as shocks which simultaneously increase output, employment, consumption, and investment in

Germany, are transmitted to the remaining nine economies. Ciccarelli et al. (2012a) investigate the heterogeneity of macro-financial linkages in a number of developed economies and compare the transmission of real and financial shocks during the 2008–2009 recession. Caivano (2006) examines how disturbances generated in the Euro area are transmitted to the United States and vice versa, when these two geographic units are included into a world economy. Beetsma and Giuliadori (2011) and Lane and Benetrix (2011) look at the transmission of government spending shocks, and Boubtane, Coulibaly, and Rault (2010) examine how immigration shocks are transmitted in a variety of countries. Finally, Love and Zicchino (2006) measure the effect of shocks to "financial factors" on a cross section of U.S. firms.

Panel VARs have also been frequently used to construct average effects – possibly across heterogeneous groups of units – and to characterize unit-specific differences relative to the average. For example, one may want to know if government expenditure is more countercyclical, on average, in countries or states which have fiscal restrictions included in the constitution, or whether the instantaneous fiscal rule depends on the type of fiscal restrictions that are in place (see Canova & Pappa, 2004). One may also be interested in knowing whether inflation dynamics in a monetary union may depend on geographical, political, institutional, or cultural characteristics, or on fiscal and monetary interactions (see Canova & Pappa, 2007). Alternatively, one may want to examine whether shocks generated outside a country (or an area) dominate the variability of domestic variables (see Canova, 2005; Rebucci, 2010a) what channels of transmission may make responses to international shocks different across countries. Finally, one may be interested in knowing how financial fragility may induce a different transmission mechanism of monetary policy shocks in different groups of countries (Ciccarelli, Maddaloni, & Peydró, 2012b).

Another potential use of panel VARs is in analyzing the importance of interdependencies, and in checking whether feedbacks are generalized or only involve certain pairs of units. Thus, one may want to use a panel VARs to test the small open economy assumption or to evaluate certain exogeneity assumptions, often made in the international economics literature. Finally, panel VARs may be used to examine the extent of dynamic heterogeneity and of convergence clubs (see Canova, 2004), to endogenously group units or to characterize their differences. For instance, De Graeve and Karas (2012) show that the dynamics of deposits and interest rates of "good" and "bad" banks differs in response to bank run shocks. They also show that differences in the health of their balance sheet are of second-order importance and what truly matters is whether banks are insured or not by regulators.

Are Panel VAR Models Consistent with Economic Theory?

As with standard VARs, one may wonder whether panel VARs can be used to "test" theories or to inform researchers about the relative validity of different economic paradigms. Panel VARs can be easily generated from standard intertemporal optimization problems under constraints, as long as the decision rules are log-linearized around the steady state. For example, Canova (2007, Chapter 8) shows that if all variables are observed, a small open economy version of the Solow Growth Model generates a heterogeneous panel VAR with static but without dynamic interdependencies. More importantly, he shows that the panel VAR that the theory generates either has both fixed effects and dynamic heterogeneity or none of them – that is, either the steady states and the dynamics are heterogeneous or both are homogeneous. Thus, it is difficult to justify the common practice of specifying panel VARs which allow only for intercept heterogeneity but impose dynamic homogeneity.

Panel VARs with dynamic interdependencies can be obtained if the small open economy assumption is dropped and at least one asset is traded in financial markets or if intermediate factors of production are exchanged in open markets (see, e.g., Canova & Marrinan, 1998). In this case, market clearing implies that the excess demands present in a unit is compensated by excess supply in other units, and these spillovers, together with adjustments in the relative prices of goods and/or assets, imply generalized feedbacks from one unit to all the others. Also here, whenever there are heterogeneities in the steady states, there will be heterogeneities in the dynamic responses to domestic and international shocks.

It is well known that if the VAR omits relevant states of the optimization problem, there is no insurance that the innovations obtained in a structural VAR (SVAR) will display the same characteristics as the innovations appearing in theoretical models (see, e.g., Fernandez Villaverde, Rubio Ramirez, Sargent, & Watson, 2007). Since the states of the problem are often not observed, this mismatch has caused several researchers to question the use of SVARs. However, it is also well known that this mismatch does not have an either/or consequence and that there are situations where the structural innovations the VAR recovers may have characteristics that are different from the innovations of the theoretical model but the dynamics they induce are similar (see, e.g., Sims, 2012). Panel VARs are not different in this respect: if important states are omitted from the list of variables for each unit, standard non-fundamentalness issues may arise. Thus, care must be exercised in choosing the variables for each unit. The presence of a cross section does not help in general to reduce the non-fundamentalness problem, unless it happens that cross-sectional data reveals information about the missing states that the data of single unit is not able to provide – an event which is, probabilistically, quite remote.

REDUCED-FORM ESTIMATION

Depending on the exact specification, different approaches can be used to obtain estimates of the unknowns of the model. Because of the added complexity, we first discuss the case of panel VARs without dynamic interdependencies and then analyze what happens when these dynamic interdependencies are allowed for. Both classical and Bayesian estimators are presented even though, for this particular problem, a Bayesian perspective is preferable, as it gives important insights into the estimation problem.

Panel VARs without Dynamic Interdependencies

Suppose there is a domestic VAR for each unit but the reduced-form shocks may be correlated across units. For later reference, we call this setup *a collection of unit-specific VARs*. Let the cross-sectional size N be large. If we are willing to assume that the data-generating process features dynamic homogeneity, and to condition on initial values of the endogenous variables, pooled estimation with fixed effects – potentially capturing idiosyncratic but constant heterogeneities across variables and/or units – is the standard classical approach to estimating the parameters of the model. However, when T is fixed, the pooled estimator is biased and one may want to employ the generalized method of moments (GMM) approach of Arellano and Bonds (1991), which is consistent even when T is small. A GMM approach, however, requires differencing the specification, throws away sample information, and may make inference less accurate when the information being ignored is important for the parameters of interest. Rather than differencing, one may want to impose a priori restrictions which ensure consistency in such an environment. Sims (2000) emphasizes that, in a model of this type, lagged dependent variables are not independent of the unit-specific intercepts and describes how this information can be used to recover parameter consistency, for both stationary and nonstationary models. Generally speaking, the inconsistency problem arises when the conditional pdf is used as the likelihood, because the number of parameters grows with the cross-sectional size. If the unconditional pdf is instead employed, where the density of the initial observations is a function of the unit-specific intercept and of the common slope parameters, consistency may be obtained. To be specific, let the model be

$$y_{it} = A_{0i} + A(\ell)y_{it-1} + u_{it} \tag{8}$$

where $u_{it}|y_{it-1} \sim N(0,\Sigma_i)$ and, for the sake of illustration, the roots of $A(e^{-iw})$ are all outside the unit circle. Then, the unconditional distribution of the initial conditions is $N((I - A(\ell))A_{0i}, \Omega_i)$, where Ω_i is implicitly defined by $\Omega_i = \Sigma_i + \sum_j A_j \Omega_i A'_j$. If this density is used together with the standard conditional density to build the likelihood, the maximum likelihood estimator of the parameters will be consistent, even with T short.

If T is large enough, one could also consider estimating the VAR for different units separately and averaging the results across units. Such a mean group estimator is inefficient relative to the pooled estimator under dynamic homogeneity, but gives consistent estimates of the average dynamic effect of shocks if dynamic heterogeneity is present, whereas the pooled estimator does not (see, e.g., Pesaran & Smith, 1995, and Rebucci, 2010b). The pooled estimator is inconsistent under dynamic heterogeneity because the regressors are correlated with the error term. If the data-generating process features dynamic heterogeneity, both a within and a between estimator will also give inconsistent estimates of the parameters, even when N and T are large, since the error term is also likely to be correlated with the regressors. With dynamic heterogeneity, a GMM strategy may be difficult to employ since it is hard to find instruments which are simultaneously correlated with the regressors and uncorrelated with the error term.

When T and N are moderate and dynamic heterogeneity suspected, some form of "partial pooling" may help to improve the quality of the coefficients' estimate. One standard format leading to partial pooling is a random coefficients model. The setup is the following. The model (without dynamic interdependencies) is $y_{it} = A_{0i} + A_i(\ell)y_{it-1} + u_{it}$, where now the slope parameters are potentially unit-specific. If we impose that

$$\alpha_i = \overline{\alpha} + v_i \tag{9}$$

where $\alpha_i = [\text{vec}(A_i(\ell)), \text{vec}(A_{0i})]'$ and $v_i \sim N(0, \Sigma_v)$, an estimator which partially pools the information present in different units can be constructed. Note that Eq. (9) implies that the coefficients of the VAR in different units are different, but are drawn from a distribution whose mean and variance are constant across i.

Given this structure, several estimators are available in the literature (see, e.g., Canova, 2007, Chapter 8). The two most popular ones are a classical estimator and a Bayesian one. In the classical estimator proposed by Swamy (1970), Eq. (9) is substituted into the model and generalized least square (GLS) is applied. GLS is required because of the particular error structure that substitution of Eq. (9) in the model generates. Importantly, this setup does not allow the estimation of the individual unit coefficients: only the mean $\overline{\alpha}$ is estimated. An estimate of the degree of cross-sectional heterogeneity is also difficult to obtain since Σ_v enters in a complicated way in the variance of the composite error term of the model.

The Bayesian alternative treats Eq. (9) as an exchangeable prior. This prior is then combined with the likelihood of the data to obtain the posterior distribution of the individual α_i, and of the average value, $\overline{\alpha}$, if that is of interest. Thus, a Bayesian perspective allows us to quantify the heterogeneity present in the dynamics, while a classical approach does not. If e_i and u_i are normally distributed, and $\overline{\alpha}$ and Σ_v known, conditional on the initial observations, the posterior of α_i is normal with mean

$$\tilde{\alpha}_i = (X_i'\Sigma_{i,\text{ols}}^{-1}X_i + \Sigma_v^{-1})^{-1}(X_i'\Sigma_{i,\text{ols}}^{-1}X_i\alpha_{i,\text{ols}} + \Sigma_v^{-1}\overline{\alpha}) \tag{10}$$

where $\alpha_{i,\text{ols}}$ is the ordinary least square (OLS) estimator of α_i, $\Sigma_{i,\text{ols}}^{-1}$ is the OLS estimate of Σ_i and X_i is the matrix containing the right-hand side variables, and variance

$$(X_i'\Sigma_{i,\text{ols}}^{-1}X_i + \Sigma_v^{-1})^{-1} \tag{11}$$

The moments of the posterior distribution of α_i have the usual convenient format: the posterior mean is a linear combination of sample and prior information, with weights given by the relative precision of the two types of information and the posterior variance is a weighted average of the prior and of the sample variance. Note that, if one wants to use the information present in the initial conditions for estimation (which could be very important when both N and T are short), a specification like the one suggested by Sims (2000), where a joint distribution for (α_i, y_{i0}) is a priori specified, could be used.

The above posterior distribution does not take into account the fact that the shocks in each unit VAR may be correlated. Thus, efficiency can be improved by stacking the sample information of all the units. Since the resulting model has a SUR structure, the convenient weighted average property of posterior moments is maintained.

Three important points need to be made regarding the posterior moment in Eqs. (10) and (11). First, the formulas are valid under the assumption that $\bar{\alpha}$ and Σ_v are known. If they are not, one can specify a prior distribution for these unknown and use the Gibbs sampler to construct draws for the marginal of α_i. We will describe how the Gibbs sampler can be used in a more complicated version of this hierarchical model inthe fourth section. Shortcuts designed to decrease the complexity of the computations are available. For example, modal estimates α_i^* can be easily computed plugging in

$$\bar{\alpha}^* = \frac{1}{n}\sum_{i=1}^{n} \alpha_i^* \tag{12}$$

$$(\sigma_i^*)^2 = \frac{1}{T+2}\left[(y_i - X_i\alpha_i^*)'(y_i - X_i\alpha_i^*)\right] \tag{13}$$

$$\Sigma_v^* = \frac{1}{n-\dim(\alpha)-1}\left[\sum_i (\alpha_i^* - \bar{\alpha}^*)(\alpha_i^* - \bar{\alpha}^*) + \Phi\right. \tag{14}$$

where "*" indicates modal estimates and $\Phi = \text{diag}[0.001]$ in the above formulas. In Eq. (14) an arbitrary diagonal matrix is added since Σ_v^* may not be positive definite. Alternatively, one could estimate $\bar{\alpha}$ and Σ_v from a training sample, if one is available, or from information contained in units left out from the cross section. In this situations, the formulas in Eqs. (10) and (11) are still valid with plug-in estimates in place of true values of $\bar{\alpha}$ and Σ_v. Clearly, with plug-in estimates, the uncertainty present in the posterior of α_i will be underestimated – the estimation error present in $\bar{\alpha}$ and Σ_v is disregarded, and one may want to use simple corrections to take these errors into account (see Canova, 2004).

Second, the mean of the posterior $\tilde{\alpha}_i$ collapses to a standard OLS estimator constructed from unit-specific information if heterogeneity is large, that is, $\Sigma_v \rightarrow \infty$, and it collapses to the prior mean, if the sample is uninformative. Thus, if one of the two types of information is highly imprecise, it is disregarded in the construction of the posterior. Third, the classical GLS estimator satisfies $\alpha_{\text{GLS}} = \frac{1}{n}\sum_{i=1}^{n} \tilde{\alpha}_i$. Thus, the classical estimator for the mean effect is the arithmetic average of the posterior means of the individual units.

Canova (2005), Canova and Pappa (2007), and more recently Calza, Monacelli, and Stracca (2013) apply such a Bayesian random coefficient approach to estimate the dynamic responses to shocks of a potentially heterogeneous collection of unit-specific VARs. These papers, however, rather than modeling cross-sectional heterogeneities in VAR coefficients, as described in Eq. (9), specify directly the nature of the heterogeneities present in the MA representation of the data for each unit. Thus, the model is

$$\tilde{y}_{it} = B_i(\ell)u_{it} \qquad i = 1, \ldots, N \tag{15}$$

where $\tilde{y}_{it}$ represents the original vector of series in country i in deviation from the deterministic components. Let $\beta_i = \text{vec}(B_i(\ell))$. Then, one can assume that the vector of MA coefficient β_i is random around a mean

$$\beta_i = \overline{\beta} + v_i \tag{16}$$

where $v_i \sim N(0, \Sigma_v)$. A Bayesian estimator of β_i still maintains a weighted average structure with weights given by the relative precision of the two types of information. The advantage of Eq. (16) is that an economically reasonable prior for the dynamic responses to shocks may be much easier to formulate than a prior for the VAR coefficients.

In many applications, an estimator of the average effect $\overline{\alpha}$ $(\overline{\beta})$ is of interest. If a Bayesian approach is followed, one may obtain the average effect by averaging the posterior mean over i. Thus, the classical and the Bayesian estimator of the mean effect coincide. An alternative is available if, in addition to treating $\alpha_i(\beta_i)$ as random around a mean, $\overline{\alpha}$, $(\overline{\beta})$, and Σ_v are also treated as random. In this case, the posterior distribution for, for example, $\overline{\beta}$ can be obtained by integrating out β_i and Σ_v from the joint posterior of $(\beta_i, \overline{\beta}, \Sigma_v)$ using standard hierarchical methods (see, e.g., Canova, 2007; Jarocinski, 2010).

Improved estimates of β_i can also be obtained in other ways. For example, Zellner and Hong (1989) suggested a prior specification for the α_i that results in a posterior distribution for the VAR coefficients combining unit-specific and average sample information. In particular, when Eq. (9) holds and v_i has a normal distribution with mean zero and variance $\Sigma_v = \phi^{-1}\sigma_u^2 I_k$ (k being the dimension of α_i), and $\Sigma_i = \Sigma = \sigma_u^2 I_G$, the (conditional) posterior distribution for $\alpha = (\alpha_1, \ldots, \alpha_n)'$ will be normal with mean $\tilde{\alpha}$ given by

$$\tilde{\alpha} = (Z'Z + \phi I_{nk})^{-1}(Z'Z\alpha_{\text{ols}} + \phi J\overline{\alpha}) \quad (17)$$

where $\alpha_{\text{ols}} = [\alpha_{1,\text{ols}}, \ldots, \alpha_{n,\text{ols}}]$ is the OLS estimator of α, unit by unit, Z is a block diagonal matrix containing in each diagonal block the regressors of each unit, and $J' = (I_k, \ldots, I_k)$., $\mathrm{I_k} = \mathrm{k} \times \mathrm{k}$ matrix The variance of the posterior is $(Z'Z + \phi I_{nk})^{-1}$. Zellner and Hong (1989) replace $\overline{\alpha}$ with the mean group estimator. Alternatively, an additional prior can be added for $\overline{\alpha}$ and its posterior distribution derived in a fully fledged hierarchical setup.

Improved classical estimators, which combine unit-specific and average information, also exist. For example, a James–Stein estimator for the above model is

$$\alpha_i = \alpha_{\mathrm{p}} + \left(1 - \frac{\kappa}{F}\right)(\alpha_{i,\text{ols}} - \alpha_p) \quad i = 1, \ldots, n \quad (18)$$

where $\alpha_{i,\text{ols}}$ is the OLS estimator with unit i data, α_p is the pooled estimator, F is the statistic for the null hypothesis $\alpha_i = \alpha, \forall i$, and $\kappa = [(\mathrm{NG} - 1)\dim(\alpha) - 2]/[\mathrm{NGT} - \dim(\alpha) + 2]$. Thus, the shrinkage factor κ depends on the dimension of α relative to T. When $\dim(\alpha) >> T$, $(1 - \frac{\kappa}{F})$ is smaller, therefore pulling α_i closer to α_p. It is rare to see estimators of this type in the macroeconomic literature primarily because the pretesting required to construct Eq. (18) is not typically performed.

Panel VARs with Dynamic Interdependencies

The estimation problem becomes more complicated if dynamic interdependencies are allowed for. The problem is related to the curse of dimensionality: excluding deterministic components, since there are $k = NGp$ coefficients in each equation, the total number of parameters (NGk) to be estimated in the model easily exceeds the sample one has available. One way to solve this problem is to selectively model the dynamic links across units while imposing zero-restrictions on others. Thus, for example, one can assume that only the variables of unit j enter the equations of unit i, as in the spatial VAR model discussed in Section "Spatial VARs." It is unclear how to do this in a way that avoids data mining. An alternative is to group cross-sectional units into clubs and assume random coefficients within each group but no relationship across groups. Such an approach has been used, for example, in Canova (2004) to examine regional convergence rates, but it naturally applies to a situation where dynamic interdependencies are excluded. Thus, some ingenuity is required to extend it to a framework where interdependencies are allowed for.

Canova and Ciccarelli (2004, 2009) have suggested different cross-sectional shrinkage approaches which can deal with the curse of dimensionality and thus allow the estimation of models with dynamic interdependencies. Del Negro and Schorfheide (2010) provide an overview of the approach.

To see what the procedure involves, rewrite Eq. (2) in a simultaneous equations format:

$$Y_t = Z_t\alpha + U_t \tag{19}$$

where $Z_t = I_{NG} \otimes X_t'$; $X_t' = (I, Y_{t-1}', Y_{t-2}', \ldots, Y_{t-p}')$, $\alpha = (\alpha_1', \ldots, \alpha_N')'$ and α_i are $Gk \times 1$ vectors containing, stacked, the G rows of the matrices $(A_{oi}(t), A_i(\ell))$, while Y_t and U_t are $NG \times 1$ vectors. Since α varies with cross-sectional units, its sheer dimensionality prevents any meaningful unconstrained estimation. Thus assume that α depends on a much lower dimension vector θ and posit the following linear structure:

$$\alpha = \Xi_1\theta_1 + \Xi_2\theta_2 + \Xi_3\theta_3 + \Xi_4\theta_4 + \ldots + e_t \tag{20}$$

where Ξ_1, Ξ_2, Ξ_3, Ξ_4 are matrices of dimensions $NGk \times q$, $NGk \times N$, $NGk \times G$, $NGk \times 1$ respectively and θ_i, $i = 1, 2 \ldots$, are factors, capturing the determinants of α. For example, θ_1 could capture components in the coefficient vector which are common across units and variables (or groups of them) – its dimension is, say, q; θ_2 could capture components which are common within units, thus its dimension equals N; θ_3 could capture components which are variable specific, thus its dimension is equal to G; θ_4 could capture components in the lagged coefficients and its dimension is equal to $p_1 < p$, and so on. Finally, e_t captures all the unmodeled features of the coefficient vector, which may have to do with time specific or other idiosyncratic effects.

Factoring α, as in Eq. (20), is advantageous in many respects. Computationally, it reduces the problem of estimating NGk coefficients into the one of estimating $q+N+G+p_1$ factors characterizing them. Practically, the factorization in Eq. (31) transforms an overparametrized panel VAR into a parsimonious SUR model, where the regressors are averages of certain right-hand side VAR variables. In fact, using Eq. (20) in Eq. (19) we have

$$Y_t = \sum_{j=1}^{r} \mathcal{Z}_{jt}\theta_j + \gamma_t \tag{21}$$

where $\mathcal{Z}_{jt} = Z_t \Xi_j$ capture respectively, common, unit-specific, variable-specific, lag-specific information present in the lagged dependent variables, and $\gamma_t = U_t + Z_t e_t$. Since, by construction, Z_{it} has a moving average structure, the regressors in Eq. (21) will capture low-frequency movements present in the VAR and this feature is valuable in medium-term out-of-sample forecasting exercises. Economically, the decomposition in Eq. (21) conveniently allows us to measure, for example, the relative importance of common and unit-specific influences for fluctuations in Y_t. In fact, $WLI_t = \mathcal{Z}_{1t}\theta_1$ plays the role of a (vector) of common indicators, while $CLI_t = \mathcal{Z}_{2t}\theta_2$ plays the role of a vector of unit-specific indicators. In general, WLI_t and CLI_t are correlated – a portion of the variables in $\mathcal{Z}_{1t}$ also enter in $\mathcal{Z}_{2t}$ – but the correlation tends to zero as N increases.

To illustrate the structure of $\mathcal{Z}_{jt}$, suppose there are $G=2$ variables, $N=2$ countries, $q=1$ common component, $p=1$ lags, and omit deterministic components, for convenience. Then:

$$\begin{bmatrix} y_t^1 \\ x_t^1 \\ y_t^2 \\ x_t^2 \end{bmatrix} = \begin{bmatrix} A_{1,1}^{1,y} & A_{2,1}^{1,y} & A_{1,2}^{1,y} & A_{2,2}^{1,y} \\ A_{1,1}^{1,x} & A_{2,1}^{1,x} & A_{1,2}^{1,x} & A_{2,2}^{1,x} \\ A_{1,1}^{2,y} & A_{2,1}^{2,y} & A_{1,2}^{2,y} & A_{2,2}^{2,y} \\ A_{1,1}^{2,x} & A_{2,1}^{2,x} & A_{1,2}^{2,x} & A_{2,2}^{2,x} \end{bmatrix} \begin{bmatrix} y_{t-1}^1 \\ x_{t-1}^1 \\ y_{t-1}^2 \\ x_{t-1}^2 \end{bmatrix} + U_t \tag{22}$$

$\alpha = [A_{1,1}^{1,y}, A_{2,1}^{1,y}, A_{1,2}^{1,y}, A_{2,2}^{1,y}, A_{1,1}^{1,x}, A_{2,1}^{1,x}, A_{1,2}^{1,x}, A_{2,2}^{1,x}, A_{1,1}^{2,y}, A_{2,1}^{2,y}, A_{1,2}^{2,y}, A_{2,2}^{2,y}, A_{1,1}^{2,x}, A_{2,1}^{2,x}, A_{1,2}^{2,x}, A_{2,2}^{2,x}]'$ is a 16×1 vector and the typical element of α, $\alpha_{l,h}^{i,j}$, is indexed by the unit i, the variable j, the variable in an equation l (independent of the unit), and the unit in an equation h (independent of variable). If we are not interested in modeling all these aspects, one possible factorization of α is

$$\alpha = \Xi_1\theta_1 + \Xi_2\theta_2 + \Xi_3\theta_3 + e_t \tag{23}$$

where e_t captures unaccounted features, and for each t, θ_1 is a scalar, θ_2 is a 2×1 vector, θ_3 is a 2×1 vector, Ξ_1 is a 16×1 vector of ones,

$$\underset{(16\times 2)}{\Xi_2} = \begin{bmatrix} \imath_1 & 0 \\ \imath_1 & 0 \\ 0 & \imath_2 \\ 0 & \imath_2 \end{bmatrix} \qquad \underset{(16\times 2)}{\Xi_3} = \begin{bmatrix} \imath_3 & 0 \\ 0 & \imath_4 \\ \imath_3 & 0 \\ 0 & \imath_4 \end{bmatrix}$$

with $\imath_1 = (1 \quad 1 \quad 0 \quad 0)'$, $\imath_2 = (0 \quad 0 \quad 1 \quad 1)'$, $\imath_3 = (1 \quad 0 \quad 1 \quad 0)'$, and $\imath_4 = (0 \quad 1 \quad 0 \quad 1)'$. Substituting Eq. (23) into the model, the panel VAR can be rewritten as

$$\begin{bmatrix} y_t^1 \\ x_t^1 \\ y_t^2 \\ x_t^2 \end{bmatrix} = \begin{bmatrix} \mathcal{Z}_{1t} \\ \mathcal{Z}_{1t} \\ \mathcal{Z}_{1t} \\ \mathcal{Z}_{1t} \end{bmatrix} \theta_1 + \begin{bmatrix} \mathcal{Z}_{2,1,t} & 0 \\ \mathcal{Z}_{2,1,t} & 0 \\ 0 & \mathcal{Z}_{2,2,t} \\ 0 & \mathcal{Z}_{2,2,t} \end{bmatrix} \theta_2 + \begin{bmatrix} \mathcal{Z}_{3,1,t} & 0 \\ 0 & \mathcal{Z}_{3,2,t} \\ \mathcal{Z}_{3,1,t} & 0 \\ 0 & \mathcal{Z}_{3,2,t} \end{bmatrix} \theta_3 + \gamma_t \tag{24}$$

where $\mathcal{Z}_{1t} = y_{t-1}^1 + x_{t-1}^1 + y_{t-1}^2 + x_{t-1}^2 + 1$, $\mathcal{Z}_{2,1,t} = y_{t-1}^1 + x_{t-1}^1$, $\mathcal{Z}_{2,2,t} = y_{t-1}^2 + x_{t-1}^2$, $\mathcal{Z}_{3,1,t} = y_{t-1}^1 + y_{t-1}^2$, $\mathcal{Z}_{3,2,t} = x_{t-1}^1 + x_{t-1}^2$.

The specification in Eq. (21) is preferable to a collection of VARs or bilateral VARs for two reasons. First, the parsimonious use of cross-sectional information helps to get more accurate estimates of the parameters and to reduce standard errors. Second, if the momentum that the shocks induce across countries is the result of a complicated structure of lagged interdependencies, the specification will be able to capture it. Such a structure would instead emerge as "common shocks" in the two alternative frameworks.

It is easy to estimate a model like Eq. (21). Stack the t observations in a vector so that

$$Y = \sum_{j=1}^{r} \mathcal{Z}_j\theta_j + \gamma \tag{25}$$

Thus, the reparametrized model is simply a multivariate regression model. If the factorization in Eq. (20) is exact, the error term is uncorrelated with the regressors and classical OLS can be used to estimate the vector θ and thus the vector α. Consistency is ensured as T grows. When the factorization in Eq. (20) allows for an error, γ_t has a particular heteroskedastic covariance matrix which needs to be taken into account. If a Bayesian framework is preferred, the posterior for the unknowns is easy to construct. Let $e_t \sim N(0, \Sigma_u \otimes V)$ and further restrict $V = \sigma^2 I$ as in Kadiyala and Karlsson (1997). Then $v_t \sim N(0, \sigma_t \Sigma_u)$ where $\sigma_t = (I + \sigma^2 X_t' X_t)$. Thus, if the prior for $(\theta, \Sigma_u, \sigma^2)$ is, for example, of the semi-conjugate type: $\theta \sim N(\theta_0, \Omega_0)$, $\Sigma_u^{-1} \sim W(z_0, Q_0)$, $\sigma^{-2} \sim G(0.5a_0, 0.5a_0 s^2)$, where $(\theta_0, \Omega_0, z_0, Q_0, a_0, s^2)$ are known quantities, W stands for Wishart distribution, and G for Gamma distribution, one can use the Gibbs sampler to construct sequences for $(\theta, \Sigma_u, \sigma^2)$ from their joint posterior distribution – see next section for details.

ADDING TIME VARIATION IN THE COEFFICIENTS

The modern macroeconomic literature is taking seriously the idea that the coefficients of a VAR and the variance of the shocks may be varying over time. For example, Cogley and Sargent (2005) and Primiceri (2005), pioneered a specification where the VAR coefficients evolve over time like random walks; Sims and Zha (2006) assume that VAR coefficients evolve over time according to the Markov switching process, while Auerbach and Gorodnichenko (2011) specify a smooth transition VAR model, where contemporaneous and lagged coefficients are a function of a prespecified variable indicator.

Specifications of this type can also be used in a panel VAR framework if time variation in the parameters is suspected. In general, the presence of time variation in the coefficients adds to the curse of dimensionality and some ingenuity is required if one is to obtain meaningful estimates of the parameters and of the responses to the shocks of interest. The approach employed in Canova and Ciccarelli (2004; 2009), and Canova et al. (2007; 2012), which extends the shrinkage structure previously described to the case of time-varying parameter models, can go a long way in that direction.

Let the time-varying panel VAR model be

$$y_{it} = A_{0i}(t) + A_{it}(\ell) Y_{t-1} + F_{it}(\ell) W_t + u_{it} \tag{26}$$

where $A_{it}(\ell)$ are the coefficients on the lag endogenous variables Y_{t-1}, W_t is a $M \times 1$ vector of weakly exogenous variables common to all units and time-varying coefficients are allowed in both $A_{it}(\ell)$ and $F_{it}(\ell)$. Such a specification could be employed, for example, to study time-varying business cycle features of a vector of countries, evolutionary patterns in the transmission of structural shocks across variables or units, or the effect of changes in the variance of the shocks on relevant endogenous variables. Time variation in the coefficients add realism to the specification but is costly, since there are $k=(NGp+Mq)$ parameters in each equation and there is only one time period per unit to estimate them.

Two approaches have been proposed to estimate such model. To see what they involve, rewrite Eq. (26) in a simultaneous equations format:

$$Y_t = Z_t \alpha_t + U_t \tag{27}$$

where $\alpha_t = (\alpha'_{1t}, \ldots, \alpha'_{Nt})'$ and α_{it} are $Gk \times 1$ vectors containing, stacked, the G rows of the matrices A_{0i}, A_{it}, and F_{it}.

A Panel-Type Hierarchical Prior

Here α_t is assumed to be the sum of two independent components: one which is unit-specific and constant over time; the other common across units but time-varying, that is:

$$\alpha_{it} = \delta_i + \lambda_t$$

An exchangeable prior is assumed for δ_i

$$\delta_i = \bar{\delta} + v_i, \quad v_i \sim N(0, \Sigma_v) \tag{28}$$

and, a further layer of dimensionality reducing hierarchy can be specified by setting $\bar{\delta} \sim N(\mu, \Psi)$. On the other hand, λ_t is assumed to follow an autoregressive process:

$$\lambda_t = \rho \lambda_{t-1} + (1-\rho)\lambda_0 + e_t \tag{29}$$

Additional assumptions on Σ_v, ρ, λ_0, and e_t complete the specification of the prior.

This setup is convenient and found to be useful in forecasting and turning point analysis (Canova & Ciccarelli, 2004). The fact that the time-varying parameter vector is common across units does not prevent unit-specific structural movements, since α_{it} can be rewritten as

$$\alpha_{it} = (1-\rho)(\delta_i + \lambda_{i0}) + (1-\rho)\alpha_{it-1} + e_t$$

where persistent movements in α_{it} are driven by the common coefficient ρ. Note also that the setup provides a general mechanism to account for structural shifts without explicitly modeling their sources.

The assumptions made allow us to recover the posterior of δ_i and of $\overline{\delta}$. Thus, one can distinguish between individual α_{it} and mean effects $\overline{\alpha}_t = \overline{\delta} + \lambda_t$, as in Lindley and Smith (1972). The difference is relevant in a forecasting context, since one may be concerned in predictions with the posterior of the average $\overline{\alpha}_t$ or with the posterior of the distribution of unit-specific effects. As in the case of a collection of VARs, the exchangeability assumption on δ_i allows for some degree of cross-sectional pooling and, again, this may be useful when there are similarities in the characteristics of the vector of variables across units.

The structure of the model can be summarized with the following hierarchical scheme:

$$
\begin{aligned}
Y_t &= Z_t\delta + Z_t S\lambda_t + U_t \qquad U_t \sim N(0,\Sigma_u)\\
\delta &= S\overline{\delta} + \zeta \qquad \zeta \sim N(0,\Delta)\\
\overline{\delta} &= \mu + \omega \qquad \omega \sim N(0,\Psi)\\
\lambda_t &= \rho\lambda_{t-1} + (1-\rho)\lambda_0 + e_t \qquad e_t \sim N(0,\Sigma_e)
\end{aligned} \tag{30}
$$

where $S = \mathbf{e} \otimes I$; $\mathbf{e} = vec(1, 1, ..., 1)$; and $\Delta = I \otimes \Sigma_v$. Canova and Ciccarelli (2004) describe how to construct the posterior distributions for (functions of) the parameters of interest under several prior assumptions on the variance covariance matrices Σ_u, Δ, Ψ, and Σ_e, the mean vector μ and the initial λ_0, using the Gibbs sampler.

A Factor Structure for the Coefficient Vector

Another possibility is to allow for time variation in the factorization in Eq. (20). Thus, let

$$
\alpha_t = \sum_j \Xi_j \theta_{jt} + e_t \tag{31}
$$

where Ξ_j are matrices with ones and zeros and θ_{jt} are factors. While in the setup of Eq. (20) θ's were fixed hyperparameters, now they are stochastic processes and thus a specification of their law of motion is needed to complete the model. Canova and Ciccarelli (2009) study different alternatives for this law of motion. A simple representation, nested in their specification, which illustrates the point is

$$\theta_t = \theta_{t-1} + \eta_t \qquad \eta_t \sim N(0,\Omega_t) \tag{32}$$

where $\theta_t = [\theta_{1t}, \theta_{2t} \ldots]'$, Ω_t is block diagonal and U_t, e_t, and η_t be mutually independent.

In Eq. (32) the factors driving the coefficients of the panel VAR evolve over time as random walks. This specification is similar to the one employed in the time-varying coefficient VAR literature, but it is parsimonious since it concerns θ_t, which is of much smaller dimension than the α_t vector, and allows us to focus on coefficient changes which are permanent. The variance of η_t is, in principle, allowed to be time-varying. Such a specification implies ARCH-M type effects in the representation for Y_t and it is a way of modeling time-varying conditional second moments that provides an alternative to the stochastic volatility specification used, for example, in Cogley and Sargent (2005) and many others. The main difference is that here volatility changes will be related to coefficient changes. Note that the computational costs involved in using this specification are moderate since the dimension of θ_t is considerably smaller than the dimensionality of Y_t. The block diagonality of Ω_t, on the other hand, guarantees the identifiability of the factors.

To make the specification composed of Eqs. (27), (31), and (32) estimable, we need assumptions on the error terms of Eq. (27) and of Eq. (31). If we let $U_t \sim N(0,\Sigma_u)$ and $e_t \sim N(0,\Sigma_u \otimes V)$, where $V = \sigma^2 I_k$ is a $k \times k$ matrix, the reparametrized model has the state space structure:

$$Y_t = (Z_t \Xi)\theta_t + \gamma_t \qquad \gamma_t = U_t + Z_t e_t \sim N(0,\sigma_t \Sigma_u)$$

$$\theta_t = \theta_{t-1} + \eta_t \qquad \eta_t \sim N(0,\Omega_t) \tag{33}$$

where $\sigma_t = (I + \sigma^2 Z_t' Z_t)$. Bayesian estimation requires prior distributions for $\Sigma_u, \Omega_t, \sigma^2$, and θ_0. Canova and Ciccarelli (2009) show how these joint prior densities can be specified so that the posterior distribution for the quantities of interest can be computed numerically with MCMC methods. Once these distributions are found, location and dispersion measures for any interesting continuous functions of the parameters can be obtained.

Similarly, the marginal likelihood and the predictive distributions needed for model checking and model comparisons are easy to construct.

If classical methods are preferred, notice that Eq. (33) is a linear state space system, where θ_{it} represents unobservable states. Thus, variations of the Kalman filter algorithm can be used to construct the likelihood function which then can be maximized with respect to the relevant parameters (see, e.g., Ljung & Soderstrom, 1984).

Implementation in Small Samples

The approaches described in the previous two subsections condition on the initial p observations. When T is large, the difference between the conditional and the unconditional likelihood is likely to be small. When T is small, the information present in the initial conditions may contain important information for the quantities of interest. Intuitively, with $T=20$ observations throwing away, say $p=4$, initial conditions effectively reduces the information by 20 percent, making the likelihood flatter and leaving to the prior the burden of producing enough curvature in the posterior. This problem is relevant for many applications since comparable macro time series across a number of countries or sectors exist only for the last 15–20 years, at best. Thus, one may want to take all the existing information into account when constructing the posterior of the quantities of interest.

In the case of Eq. (33), conditional on the first p observations, the likelihood is

$$L(\theta|y, \mathcal{Z}) \propto (\prod_t \sigma_t)|\Sigma_u|^{-0.5T} exp[-0.5 \sum_t (Y_t - \mathcal{Z}_t\theta_t)'(\sigma_t\Sigma_u)^{-1}(Y_t - \mathcal{Z}_t\theta_t)] \quad (34)$$

Since $\mathcal{Z}_t = Z_t\Xi_j$, the likelihood of $\mathcal{Z}_{jt}$ is proportional to the likelihood of Z_t. The likelihood of the initial conditions can be written as

$$L(\mathcal{Z}|\psi) \propto exp[-0.5 \sum_i (Z_t - \overline{Z})'(\Sigma_Z)^{-1}(Z_t - \overline{Z})] \quad (35)$$

where $\overline{Z}$ is a vector of mean parameters. The full likelihood of the sample is then simply the product of Eqs. (34) and (35) and can be combined with the prior to yield a posterior kernel or a conditional posterior for the unknowns which can be used in the Gibbs sampler.

IMPULSE RESPONSES AND SHOCK IDENTIFICATION

Shock identification can be performed with standard methods. To decrease the number of identification restrictions, it is typical to assume that Σ_u is block diagonal, with blocks corresponding to each unit, employ symmetric identification restrictions across units (these could be zero, long run, sign restrictions or a combination of the them) and require the structural shocks to be orthogonal. Block diagonality implies differences in the responses within and across units: within a unit, variables are allowed to move instantaneously; across units, variables may react but only with a lag. Symmetric identification restrictions imply that the nature of the disturbances (i.e., being demand or supply shocks) is independent of the unit.

Restrictions of this type have been used to obtain the (cross-sectional) average responses or the average responses of particular groups of units which are homogeneous in their dynamics. Ciccarelli and Rebucci (2006), for instance, study the transmission of a German monetary policy shock, identified as an innovation to the reaction function of the Bundesbank, in the four largest EMU countries using pre-EMU data. Jarocinski (2010) compares responses to monetary policy shocks in the Euro area countries before the EMU to those in the new member states from Eastern Central Europe. A monetary policy shock is identified with the same restrictions in each group of countries. Rebucci (2010a) is interested in assessing the role of external and policy shocks for domestic growth variability. External and policy shocks are identified by imposing the same Choleski decomposition in all countries. Finally, Ciccarelli et al. (2012b) analyze whether financial fragility has altered the transmission mechanism of monetary policy in the Euro area. A panel VAR is estimated for core countries and countries under financial stress, allowing the slopes and the contemporaneous impact matrix to be group dependent, but restricting them to be common within groups. A monetary policy shock is identified using the same restrictions in the two groups and average responses for each group are constructed.

Shock identification is somewhat more complicated when static interdependencies are allowed for and cross unit symmetry in shock identification cannot be assumed. A convenient tool to be used in this situation is described in Canova and Ciccarelli (2009). Researchers using panel VARs with static and dynamic interdependencies and, possibly, time variation in the coefficients may be interested in computing the responses of the endogenous variables to shocks in the variables or to shocks to the coefficients (via shocks to the common λ_t or shocks to the factors θ_t) and in describing their evolution over time. In this situation, responses can be obtained as the difference between two conditional forecasts: one where a particular variable (coefficient) is shocked and one where the disturbance is set to zero.

Formally, let y^t be the history for y_t, θ^t the trajectory for the coefficients up to t, Ω^t the trajectory for the variance of the coefficients up to t; $y_{t+1}^{t+\tau} = [y'_{t+1}, \dots y'_{t+\tau}]'$ be a collection of future observations and $\theta_{t+1}^{t+\tau} = [\theta'_{t+1}, \dots \theta'_{t+\tau}]'$ a collection of future trajectories for θ_t. Let $\mathcal{W} = (\Sigma_u, \sigma^2)$; set $\xi'_t = [u'_{1t}, u'_{2t}, \eta'_t]$, where u_{1t} are shocks to the endogenous variables and u_{2t} shocks to the exogenous variables (if there are any). Let $\xi_{j,t}^{\delta}$ be a realization of $\xi_{j,t}$ of size δ, and $\mathcal{F}_t^1 = \{y^t, \theta^t, \Omega^t, \mathcal{W}, J_t, \xi_{j,t}^{\delta}, \xi_{-j,t}, \xi_{t+1}^{t+\tau}\}$ and $\mathcal{F}_t^2 = \{y^t, \theta^t, \Omega^t, \mathcal{W}, J_t, \xi_t, \xi_{t+1}^{t+\tau}\}$ two conditioning sets, where $\xi_{-j,t}$ indicates all shocks, excluding the one in the j-th component, and $J_t J'_t = \Sigma_u$. Then, responses at horizon τ to a δ impulse in $\xi_{j,t}$, $j = 1, \dots$ are

$$IR_y^j(t, \tau) = E(y_{t+\tau} | \mathcal{F}_t^1) - E(y_{t+\tau} | \mathcal{F}_t^2) \qquad \tau = 1, 2, \dots \tag{36}$$

Notice that in Eq. (36) the history of the coefficient and of their variance is taken as given at each t and that the size of the impulse can be positive or negative, but it is also taken as given. This is because one may be interested in comparing responses over time for a given trajectory of the coefficients and their variance (rather than their average values) and because the relevant size of the impulse is generally determined by policy considerations. When the coefficients are constant, $\xi'_t = [u'_{1t}, u'_{2t}]$ and Eq. (36) produces the traditional impulse response function to structural shocks.

A proper identification strategy, that is, the selection of the (large) matrix J_t and its time evolution, is an open area for research since its sheer dimensionality makes it hard to find enough constraints to achieve identification of all shocks. Shortcuts, such as a block structure, may not be very appealing – one can envision situations where shocks are transmitted across unit within a time period. Alternative shortcuts, such as the one imposed in Canova et al. (2012), where shocks occurring in one unit (Germany) are allowed to feed contemporaneously on all other units (European countries) but not viceversa, may be acceptable if there are good economic reasons to justify them. In both cases, dynamics interdependencies are left unrestricted.

Which kind of restrictions should be used for identification is an open question. Zero restrictions and Choleski ordering for $J_t = J$, all t, are the most common ones and just identification is typically sought, even though overidentification requires only a simple extension of the tools described in Canova and Perez Forero (2012) for standard VARs. As in single-unit VARs, one could also employ external information to identify shocks in different units. For example, one could measure the elasticity of tax revenues and government expenditure to output shocks in each cross-sectional unit and use this information to identify government spending and tax revenue shocks in all units, adding the restriction that domestic government expenditure and domestic revenues do not instantaneously respond to shocks generated in other units. Long-run restrictions a la Blanchard and Quah (1989) as well as heteroskedasticity restrictions a la Lanne and Lütkepohl (2008) are also possible. In both cases, one has to clearly state what happens to the variables of other units and often the restrictions needed to achieve identification are economically difficult to justify. For example, when long-run restrictions are used, one has to impose that foreign supply shocks have no long-run effect on domestic real variables or that they have the same effect as domestic supply shocks, both of which are not very palatable.

Sign restrictions can also be used (see, e.g., Calza et al., 2013; Sa et al., 2012). In this case, it is typical to use the same type of restrictions on each cross-sectional unit. Recently, De Graeve and Karas (2012) have suggested using cross-sectional heterogeneity to identify certain structural shocks. Their approach involves imposing sign and inequality restrictions on $\partial Y^{\Lambda}_{m,t+s}/\partial u_{k,t}$, the response of variable Y_m at horizon $t+s$, $s=0,1,2,\ldots$, to shock k at time t, $u_{k,t}$, for a subset of the units $\Lambda=\{1,\ldots,M\}$.

The approach is best understood with an example. Suppose the cross-sectional dimension of the panel can be stratified according to an observable indicator, for example, in a sample of banks, whether bank deposits are insured or not. Suppose the endogenous variables are deposits D_t and the average interest rate they earn R_t. Then, a bank run is associated with a fall in deposits and an increase in the interest rate offered by uninsured banks. The insured banks may also respond, because of contagion effects, but the responses will be smaller because deposit insurance makes them less liable to the run. Thus, together with the sign restrictions, De Graeve and Karas impose that $|D_{t,u}|>|D_{t,I}|$ and $|R_{t,u}|>|R_{t,I}|$, where u stands for uninsured and I for insured banks. Hence, following a bank run, the deposit outflow in insured banks cannot be larger than in uninsured banks, and the corresponding increase in deposit rates must be smaller for insured than for uninsured banks.

Cross-sectional identification restrictions of this type seem useful if one can sharply stratify the data with some exogenous indicator. For example, one could impose such restrictions to identify shocks originating in less developed countries (LDC), when the sample includes LDC and developed countries, once it is recognized that shocks originating in LDCs are unlikely to generate the same amount of volatility in the two groups. Alternatively, one could identify shocks primarily affecting small open economies integrated in the world economy. If the impact of particular shocks is strong in open economies but weak in relatively more closed economy or if more closed economies respond to the shocks only via second round effects, sign and differential magnitude restrictions can help to isolate them. Finally, an approach that combines sign and inequality restrictions can be used also to distinguish shocks taking place in units with slow versus fast adjustments or in markets affected differently by the presence of frictions.

The combination of sign and relative magnitude restrictions appears to be a very powerful identification device if the stratification employed is relevant. De Graeve and Karas show that in their sample of banks these restrictions allow them to identify a bank run shock which has characteristics that are similar to those obtained with narrative approaches and with ex-post insight. Note that the set of constraints one can impose is quite large, making the combination of sign and relative magnitude restrictions potentially usable in many situations. For example, one could impose relative magnitude restrictions on a particular variables across subsets Λ_1, Λ_2 of the units $\partial Y^{\Lambda_1}_{m,t+s}/\partial u_{k,t} \geq \partial Y^{\Lambda_2}_{m,t+s}/\partial u_{k,t}$, across variables within a particular subset of units, $\partial Y^{\Lambda}_{m_1,t+s}/\partial u_{k,t} \geq \partial Y^{\Lambda}_{m_2,t+s}/\partial u_{k,t}$, or across variables and across subset of units $\partial Y^{\Lambda_1}_{m_1,t+s}/\partial u_{k,t} \geq \partial Y^{\Lambda_2}_{m_1,t+s}/\partial u_{k,t}$. Clearly, which one is used depends on the question and the available data. Theory-driven restrictions are clearly preferable, but restrictions obtained from reliable stylized facts characterizing different groups of units can also be used.

While the identification restrictions of De Graeve and Karas involve only the contemporaneous effect of shocks, one could also consider dynamic restrictions involving relative shape and/or relative magnitudes to identify shocks in a panel VAR. Future investigations need to clarify what kind of dynamic restrictions are consistent with economic theory and could be meaningfully employed to recover interesting shocks.

A COMPARISON WITH ALTERNATIVE APPROACHES

As we have emphasized, panel VARs are unique in their ability to model dynamic interdependencies, cross-sectional heterogeneities and, at the same time, account for evolving pattern of transmission. However, to estimate them, restrictions need to be imposed. Thus, it is natural to ask how panel VAR models compare to other models, which still allow us to study interdependencies and the transmission of shocks across units but impose alternative restrictions on the nature of the interdependencies present in the data.

This section sketches the main features of large-scale Bayesian VARs (e.g., Banbura, Giannone, & Reichlin, 2010), spatial econometric models (see Anselin, 2010), factor models (see, e.g., Stock & Watson, 1989, 2003), GVARs (see Dees, Di Mauro, Pesaran, & Smith, 2007; Pesaran et al., 2004) and bilateral VARs (see, e.g., Kim, 2001; Peersman, 2004), and highlights their differences with panel VARs.

Large-Scale VARs

A close cousin of panel Bayesian VARs is the large-scale Bayesian VAR model suggested by Mol, Giannone, and Reichlin (2008) and recently employed by Banbura et al. (2010). As in panel VARs, both static and dynamic interdependencies are allowed for, but the researcher gives no consideration to the existence of a panel dimension in the data. Thus, all variables are treated symmetrically, regardless of whether they belong to a unit or not, and of whether they measure the same quantity in different units or not. Given the large scale of the model, classical estimation methods are infeasible, especially if time-varying features are allowed for. The lack of a panel perspective is reflected in the type of priors imposed in Bayesian estimation, which are typically of Litterman–Minnesota type (see Doan, Litterman, & Sims, 1984), and do not exploit any cross-sectional information present in the data.

Failure to recognize that there is a cross-sectional dimension to the available data set may not be too damaging in terms of forecasting, since it is well known that dimensionality shrinkage is more important than the exact details on how it is implemented – the Litterman–Minnesota prior is indeed a shrinkage prior. However, the choice of priors may limit the type of analyses one can perform, since the covariance matrix of the error has a particular structure which is generally disregarded in the policy exercises.

As in panel VARs, time variation in the coefficients of a large-scale VARs are relatively easy to allow for (see, e.g., Koop & Korobilis, 2012). However, ingenuity needs to be used since unrestricted time variation on all coefficients is impossible to estimate. Thus a factor structure, like the one described in Eq. (31), may be necessary to make the estimation problem manageable, and simple processes for time variation need to be imposed for computational ease.

Spatial VARs

Largescale VARs are also popular in classical econometric frameworks. Here, dimensionality restrictions are directly imposed to make estimation feasible. Restrictions of this type are commonly employed in the spatial econometric literature (see Anselin, 2010). Assuming, for simplicity only one lag of the dependent variables and no deterministic components, a spatial VAR has the form:

$$Y_t = \rho S_1 Y_{t-1} + u_t \tag{37}$$

$$u_t = S_2 e_t \tag{38}$$

where S_1 and S_2 are fixed matrix of weights. For example, a typical structure for S_1 is

$$S_1 = \begin{bmatrix} s_{11} & s_{12} & 0 & 0 & \ldots & 0 & 0 & \\ s_{21} & s_{22} & s_{23} & 0 & \ldots & 0 & 0 & \\ 0 & s_{32} & \sigma_{33} & s_{34} & \ldots & 0 & 0 & \\ \ldots & \ldots & \ldots & \ldots & \ldots & \ldots & \ldots & \\ 0 & 0 & 0 & 0 & \ldots & s_{N-1,N-2} & s_{N-1,N-1} & s_{N-1,N} \\ 0 & 0 & 0 & 0 & \ldots & 0 & s_{N,N-1} & s_{N,N} \end{bmatrix}$$

Here only the neighbors will have dynamic repercussion on unit i within one period while the rest is assumed to have a negligible effects. This structure implies that a shock originating in unit i can be transmitted after one period to unit j if j is a neighbor of i. However, if j is not a neighbor of i, delayed effects are longer and will depends on how many neighbors are between unit j and unit i. The idea of restricting dynamic effects to neighbors has been implemented using, for example, regions which share a border or stores which are located in the same city. There are three main disadvantages to this procedure. First, the weights have to be chosen prior to the estimation and, depending on the focus of the analysis, different weights may be used by different researchers on the same data set. Second, the setup is difficult to entertain in analyzing, for example, countries in a monetary union since generalized feedbacks are possible, or situations where borders (national, regional, etc.) do not reflect the economic separation present across units. Third, while the approach is relatively easy to implement when y_{it} is a scalar, it is much more complicated when y_{it} is vector since different elements of y_{it} may have different relationships across units. For example, when y_{it} includes output and consumption and the units are countries, the neighbor scheme may be appropriate for output, if the units are grouped using their natural resources, but may be highly inappropriate for consumption if migrations to all units are important.

Recently, Chudik and Pesaran (2011) have extended this framework to allow S_1 and S_2 to be full matrices and have derived classical estimators under the assumption that there is weak or strong cross-sectional dependence across units. The basic idea of their approach is still to distinguish between neighbors and non-neighbors, where the former have non-negligible static and dynamic effect on unit i and the latter have negligible effects. However, even if the non-neighbor effects may be individually small, the sum of their absolute values may not be small, making aggregate feedback effects potentially large. To account for this possibility, the authors assume that the dynamic feedbacks produced by neighbors are important and independent of N, while the effects of the non-neighbors depend on N. As N increases, if there is weak cross correlation between a unit and the non-neighbors, the model approaches a spatial VAR where the non-neighbor effects are neither interesting nor estimable while the neighboring effects can be consistently estimated simply ignoring the non-neighbor feedbacks. On the other hand, when there is strong cross correlation between a unit and some non-neighbor, the structure approaches a factor model where, controlling for neighboring effects, one unit drives fluctuations in all the others. Here ignoring the feedbacks due to the factor may lead to inconsistent estimates of the parameters.

The main value of this setup is to provide a link between classical parameter shrinkage, as implied by spatial VAR models, and classical data shrinkage, as implied by factor models, both of which attempt to mitigate the curse of dimensionality present in large-scale VARs. The approach also provides a justification for using the GVAR approach described later. The main disadvantages of the procedure are similar to those of spatial VARs, namely that (i) neighbors and non-neighbors need to be chosen a priori; (ii) the approach is difficult to implement if y_{it} is a vector; and (iii) it is hard a-priori to know whether weak or strong cross-sectional dependence characterizes the units under consideration.

Dynamic Factor Models

Unobservable factor models are popular in the applied macroeconometric literature because they capture the idea that the comovements present in a large set of series may be driven by a small number of latent variables, and because they are relatively easy to specify and estimate (Forni et al., 2000, 2005; Stock & Watson, 2002a).

Factor models have another appealing feature: their format is consistent with economic theory. To see why, consider a prototypical factor model:

$$Y_t = \chi f_t + u_t$$

$$A(\ell) f_t = e_t$$

$$U(\ell) u_t = v_t \tag{39}$$

where Y_t is a $nG \times 1$ vector, f_t is a $m < nG \times 1$ vector of factors. The log-linearized solution of a DSGE model is

$$X_{2t} = AX_{1t} + Bv_t$$

$$X_{1t} = CX_{1t-1} + Du_t \tag{40}$$

For example, in the case of an international real business cycle (RBC) model, $X_{1t} = [k_{it}, \zeta_{it}]$, k_{it} is the capital stock, ζ_{it} the technology shock in country i and X_{2t} a vector including consumption, investment, output, hours, etc. in each country. Here $u_t = [u_{1t}, \ldots u_{nt}]$ is the vector of innovations in the technology process, and $U(\ell) = I$. Simple inspection indicates that the decision rules in Eq. (40) have a factor structure and this similarity allows statistical and economic analyses to be better linked.

When nG is small, one can use an EM algorithm or the Gibbs sampler to estimate the factors and free parameters of the model, if v_t and e_t are normally distributed. If the nG instead is large, averaging will insure that the idiosyncratic component u_t will cancel out. In this situation, one can use $Y_t = \chi f_t + u_t \approx \chi \overline{Y}_t + u_t$, where $\overline{Y}_t$ is a $m \times 1$ vector estimated averaging the variables in Y_t (see, e.g., Forni & Reichlin, 1998).

Multiunit (large-scale) dynamic factor models and panel VARs differ in a number of dimensions. In terms of specification, the complex structure of dynamic interdependencies is not modeled in the former and is instead captured with a set of unobservable factors. Furthermore, the presence of cross-sectional information is generally ignored.

The interpretation of impulse responses is usually much easier in a panel VAR than in factor models. In fact, in factor models dynamic analyses are typically performed by shocking the factors and seeing how impulses in these factors are transmitted to the endogenous variables. Thus, the structure does not allow us to study, say, how a shock generated in one unit is propagated to other units or identify the interdependencies that make an effect large or small. On the other hand, by appropriately selecting a combination of shocks, a panel VAR can mimic the exercises typically performed in factor models.

Finally note, that once the parameter dimensionality reduction described in the third and fourth sections is performed, a panel VAR with interdependencies can be written in a factor format. Still, two important differences arise. First, the regressors in Eq. (21) are combinations of the lags of the right hand side variables of the panel VAR and thus observable. In factor models, factors are unobservable and typically estimated using averages of (subsets) of the current values of endogenous variables. Hence, they are likely to have different characteristics and span a different informational space. Whether lags or current values of the endogenous variables provide superior information for the states of a theoretical model is an open question which deserves further investigation. Second, the regressors of Eq. (21) equally weight the information present in the subset of the variables used to construct them. The equal weighting scheme comes directly from Eq. (20) and the fact that all variables are measured in the same units (all variables will be demeaned and standardized). In factor models, instead, estimates of the factors reflect the relative variability of the variables used to construct them.

Global VARs

GVARs are similar in spirit to factor models and Dees et al. (2007) showed how they can be obtained when the DGP is a factor model with observable and unobservable factors. They are appealing to users because they intuitively capture important features of the cross-sectional relationships while maintaining a simple structure that makes estimation easy.

For our purposes, a GVAR can be thought as a collection of unit-specific VARs to which one tags on an unobservable common factor. This idea is explicitly put forward in Gilhooly, Weale, and Wieladek (2012), who estimate with Bayesian techniques a collection of unit-specific VARs and assume a common factor structure in the error term. Consider the structure in Eq. (26) where now the coefficients are time invariant, only lags of the variables for that particular unit appear and add a new vector of unobservable variables x_t:

$$y_{it} = A_i(\ell)y_{it-1} + F_i(\ell)W_t + H_i(\ell)x_t + e_{it} \tag{41}$$

x_t is, potentially, a vector of autoregressive processes with finite variance and $H_i(\ell)$ is a polynomial in the lag operator. Set for simplicity $W_t = 0, \forall t$. Then Eq. (41) is a collection of unit-specific VARs linked together by the presence of the unobservable vector of factors. Their presence complicates estimation since Kalman filter techniques need to be used, and unless the cross-sectional dimension is small, computations may be demanding.

As in large-scale factor models, the basic idea of GVARs is that, if N is sufficiently large, one can proxy x_t with cross unit averages of y_{it} (and W_t, when they are present). Thus, rather than estimating Eq. (41) one estimates

$$y_{it} = A_i(\ell)y_{it-1} + H_i(\ell)y_{it}^* + e_{it} \tag{42}$$

where $y_{it}^* = \sum_{j=1}^{N} \pi_{ij}y_{jt}$ with $\pi_{ii} = 0$ and π_{ij} is a set of country specific weights which reflect the relative importance of the unit in the aggregate. For example, if the units are countries, one does not expect them to be equally important in the world economy and may want to weight country specific variables by their share in world trade. Alternatively, they could capture relative variability if, for example, the cross section contains units featuring cyclical fluctuations with different amplitudes.

A model like Eq. (42) can be estimated in two steps. First, country-specific VARs are estimated and all endogenous variables of the model collected. Second, the vector $Y_t = [Y_{1t}, \ldots, Y_{NT}]$, where each $Y_{it} = [y_{it}, y_{1t}^*]$ is simultaneously solved from the model. Pesaran et al. (2007) show that this is equivalent to specifying a large-scale VAR of the form

$$Y_t = D(\ell)Y_{t-1} + u_t \tag{43}$$

where $I - D(\ell)$ is $NG \times 1$ matrix whose typical i element is $1 - D_i(\ell)\pi_i$.

Hence, a GVAR is a restricted large-scale VAR, where variables of different units in an equation are weighted according to π_i. Since the weights are country specific and a priori determined by the investigator, a GVAR imposes a particular structure on the interdependencies present in the data. In particular, it selectively chooses what feedbacks may be a-priori important based, for example, on trade or financial considerations, and forces the same dynamics on the variables belonging to all units, apart from a scale factor. In this sense, it resembles an extreme version of a Minnesota prior in that variables of the units different from the one appearing on the left hand side of the relationship have weights which are smaller than their own.

To state the concept differently, a GVAR becomes estimable imposing the restriction that the dynamics produced by variables of different units on the variables of unit i are proportional to the weights. This effectively collapses the number of estimated coefficients to a more manageable number, comparable to those one would estimate using a collection of single unit VARs.

Bilateral VARs

It is common to run bilateral VARs with units representing countries, sectors, or disaggregated components of important macro variables, even if the DGP is suspected to be much more complicated (see, e.g., Kim, 2001, or Peersman, 2004). One reason for doing so is ease of interpretation; another is to reduce the size of the parameter vector to be estimated. However, such an approach may distort both the properties of the estimated structural shocks and the dynamics of their transmission.

To see how this can happen, consider the following three unit structural panel VAR(1):

$$y_{1t} = A_{11}y_{1t-1} + A_{12}y_{2t-1} + A_{13}y_{3t-1} + J_{11}u_{1t} \tag{44}$$

$$y_{2t} = A_{21}y_{1t-1} + A_{22}y_{2t-1} + A_{23}y_{3t-1} + J_{21}u_{1t} + J_{22}u_{2t} + J_{23}u_{3t} \tag{45}$$

$$y_{3t} = A_{31}y_{1t-1} + A_{32}y_{2t-1} + A_{33}y_{3t-1} + J_{31}u_{1t} + J_{32}u_{2t} + J_{33}u_{3t} \tag{46}$$

where y_{it} is of dimension $G \times 1$ and all the roots of the A matrix are outside the unit circle. In this system the reduced-form innovations in y_{1t} are proportional to u_{1t} while the reduced-form innovations in y_{2t} and y_{3t} are linear combinations of the three structural shocks (u_{1t}, u_{2t}, u_{3t}).

Suppose that a researcher decides to use data from units 1 and 2 only to form a bilateral VAR. Then the estimated model would be

$$y_{1t} = (A_{11} + A_{13}(I - A_{33}\ell)^{-1}A_{31})y_{1t-1} + (A_{12} + A_{13}(I - A_{33}\ell)^{-1}A_{32})y_{2t-1} + e_{1t} \tag{47}$$

$$y_{2t} = (A_{21} + A_{23}(I - A_{33}\ell)^{-1}A_{31})y_{1t-1} + (A_{22} + A_{23}(I - A_{33}\ell)^{-1}A_{32})y_{2t-1} + e_{2t} \tag{48}$$

where

$$e_{1t} = J_{11}u_{1t} + (I - A_{33}\ell)^{-1}(J_{31}u_{1t} + J_{32}u_{2t} + J_{33}u_{3t}) \tag{49}$$

$$e_{2t} = J_{21}u_{1t} + J_{22}u_{2t} + J_{23}u_{3t} + (I - A_{33}\ell)^{-1}(J_{31}u_{1t} + J_{32}u_{2t} + J_{33}u_{3t}) \tag{50}$$

The dynamic responses induced by reduced-form shocks will be different in the two systems. In particular, in the estimated system, the true dynamics will be contaminated by the responses of the variables of unit 3 to the structural shocks. For example, the reduced-form dynamics of unit 1 will be correctly captured only when A_{13} is zero – which requires unit 1 to be exogenous with respect to the system. However, even in this special case, structural dynamics will be incorrectly measured in the estimated system for three reasons. First, while in the true model the reduced-form innovations to unit 1 are a scaled version of the true structural shocks, in the estimated system e_{1t} mixes structural shocks from different units and this is true even when the original system has only lagged interdependencies but no static interdependencies, i.e. $J = I$. In addition, while in the original system unit 1 was predetermined, in the estimated one it is not. Hence, while a particular Choleski block ordering would be able to recover the innovations to the first unit in the original system, the imposition of such a structure would induce important identification errors in the estimated system.

Second, even if the reduced-form innovations in the original system were serially uncorrelated, serial correlation would appear in the estimated ones, since the marginalization due the elimination of unit 3 data creates moving average components in the reduced-form errors of the estimated system. Thus, either the lag length of the estimated panel VAR is appropriately increased or the reduced-form errors will be serially correlated. In other words, to approximate the original panel VAR(1) with a smaller number of units, we need either a panel VARMA or a VAR(p) with p generally large.

Third, idiosyncratic shocks in the original system may show up as common shocks in the estimated system. For example, when $J=I$ structural shocks to the variables of the third unit will show up in the estimated system as common shocks to units 1 and 2. Hence, one should be cautious in interpreting the empirical evidence in favour of common shocks in such systems.

It general, it seems a bad practice to circumvent the curse of dimensionality problem using bilateral systems when the data generating process may be more complex. Omitted variables create important distortions in the estimated structural shocks and hamper the ability of researchers to interpret the estimated dynamic responses.

Summary

All available approaches impose restrictions. The large-scale Bayesian VARs and the Bayesian panel VARs leave the model unrestricted but employ a shrinkage prior to effectively reduce the dimensionality of the coefficient vector. The spatial econometric model, the factor model, the GVARs, and the bilateral VARs on the other hand impose that the interdependencies can be captured either with a set of factors, or that there is only a very limited number of neighbor effects, or that the off-diagonal elements of the matrix $D(\ell)$ are proportional to the diagonal elements, or that the estimated system is of lower dimension than the DGP. All restrictions may be violated in practice and it is unclear which ones are preferable. In theory, prior restrictions are superior to dogmatic restrictions. An interesting question for future research is whether and in what way different sets of restrictions affect our ability to capture and interpret interdependencies economies with heterogenous features.

CONCLUSIONS

Over the last 15 years, there has been considerable improvement and unification in the standards of data collection and substantial efforts to create detailed and comparable data (on banks, firms, industries) in various countries and regions of the world. This means that while empirical analyses were previously limited to a handful of developed countries, interdependencies were hardly explored, and cross-country comparisons very scant, now an important panel dimension is added to the exercises and studies analyzing differences between, say, emerging markets and developed economies, or open and relatively closed economies are more frequent.

Together with improvements in the data collection, there has been a gradual but steady increase in the interdependencies among regions, countries, and sectors. This phenomenon is not only the object of academic studies. Terms like "global economies," "global interdependencies," and "global transmission" have become part of everyday discussions in the popular press. Economies, regions, or sectors can no longer be treated in isolation and spillovers are now prevalent. In this new global order, shocks are quickly propagated and contagion effects are important, but heterogeneities across sectors and countries remain. Asymmetries both in the pace and in the magnitude of the recovery from the 2008 recession and the increasing North–South divide caused by the prolonged European debt crisis are simple examples of these heterogeneities. Their origins is probably complex, but factors such as income, initial conditions, geography, trade and financial developments, institutions, and culture are good suspects to consider.

Since the growth path, the dynamic responses to shocks and the transmission across sectors, markets or countries may differ substantially, it is inadvisable from an economic and from an econometric point of view to treat all units symmetrically or to disregard country specific features even within the EU or the Euro zone. The presence of dynamic heterogeneities suggests that there is ample room to study how shocks are transmitted across units; to characterize not only average effects but also cross-sectional differences that help understand the potential sources of heterogeneities; to analyze how past tendencies have created the current status quo and how one should expect the current situation to evolve in the future; and to provide policymakers with evidence that can help to build alternative scenarios and formulate policy decisions.

Panel VARs seem particularly suited to addressing issues that are currently at the center stage of policy discussions as they are able to (i) capture both static and dynamic interdependencies, (ii) treat the links across units in an unrestricted fashion, (iii) easily incorporate time variation in the coefficients and in the variance of the shocks, and (iv) account for cross-sectional dynamic heterogeneities. The recent boom in empirical analyses using panel VARs in macroeconomics, banking and finance, and international economics attests to this fact.

Panel VARs are built on the same logic as standard VARs but add a cross-sectional dimension. The purpose of this article was to point out their distinctive features and their potential applications, describe how they are estimated and how shocks are typically identified, and how one deals with structural time variation; what differences are there between the panel

VAR models used in microeconomic and macroeconomic studies; how panel VARs and a collection of VARs compare and how panel VARs relate to other popular alternatives such as large-scale VARs, factor models or GVARs.

The large dimension of panel VARs typically makes the curse of dimensionality an issue especially when researchers are interested in examining the input–output links of a region, such as Latin America, or an area, such as the Euro area, where the time series dimension of the panel is short. We present a shrinkage approach which goes a long way toward dealing with dimensionality issues without compromising too much on the structure and on the ability to address interesting economic questions.

Many challenges remain and future work can be expected to improve on existing approaches, both in terms of estimation and inference. For example, Koop and Korobilis (2012) have suggested fast algorithms to estimate large-scale time-varying coefficients VARs but it is unclear whether these will work also in time-varying coefficients panel VAR, where cross-sectional shrinkage becomes important.

Similarly, it is still unclear how to expand the Markov switching methods of Sims and Zha (2006) to a panel framework, especially when transition probabilities differ across heterogeneous units. The properties of estimators used have not been evaluated in relevant economic situations and it is unclear whether tests for model selection or model validation are powerful or not.

When it comes to identification, except for De Graeve and Karas (2012), the techniques are the traditional ones used in VARs and no effort has been made to exploit the richness of the cross-sectional information. Nor have there been efforts to directly link panel VARs to global DSGE models and to study whether they can be used as testing ground for different theories of transmission.

All in all, panel VARs have the potential to become as important as VARs to answer relevant economic questions that do not require the detailed specification of the structure of the economy.

ACKNOWLEDGMENT

We thank Lutz Kilian, Chris Sims, and the participants to the 12th Annual workshop on Avances in Econometrics, Dallas for comments and suggestions.

REFERENCES

Anselin, L. (2010). Thirty years of spatial econometrics. *Papers in Regional Science*, *89*, 3–25.

Arellano, M., & Bond, S. (1991). Some tests of specification for panel data: Monte Carlo evidence and an application to employment equations. *Review of Economic Studies*, *58*, 277–297.

Auerbach, A., & Gorodnichenko, Y. (2011). Measuring output responses to fiscal shocks. *American Economic Journal: Public Policy*, *4*, 1–27.

Banbura, M., Giannone, D., & Reichlin, L. (2010). Large scale Bayesian VARs. *Journal of Applied Econometrics*, *25*, 71–92.

Beetsma, R., & Giuliadori, M. (2011). The effects of government purchase shocks: Review and estimates for the EU. *Economic Journal*, *121*, F4–F32.

Beveridge, S., & Nelson, C. (1981). A new approach to decomposition of economic time series into permanent and transitory components with particular attention to measurement of the 'business cycle'. *Journal of Monetary Economics*, *7*, 151–174.

Blanchard, O. J., & Quah, D. (1989). The dynamic effect of aggregate demand and supply disturbances. *American Economic Review*, *79*, 655–673.

Boubtane, E., Coulibaly, D., & Rault, C. (2010). *Immigration, growth and unemployment: Panel VAR evidence from OECD countries*. Manuscript.

Caivano, M. (2006). *The transmission of shocks between the U.S. and the Euro area, Bank of Italy*. Manuscript.

Calza, A., Monacelli, T., & Stracca, L. (2013). Housing finance and monetary policy. *Journal of the European Economic Association*, *11*(1), 101–122.

Canova, F. (2004). Testing for convergence clubs in income per capita: A predictive density approach. *International Economic Review*, *45*, 49–77.

Canova, F. (2005). The transmission of U.S. shocks to Latin America. *Journal of Applied Econometrics*, *20*, 229–251.

Canova, F. (2007). *Methods for applied macroeconomic research*. Princeton, NJ: Princeton University Press.

Canova, F., & Ciccarelli, M. (2004). Forecasting and turning point predictions in a bayesian panel VAR Model. *Journal of Econometrics*, *120*(2004), 327–359.

Canova, F., & Ciccarelli, M. (2009). Estimating multi-country VAR models. *International Economic Review*, *50*, 929–961.

Canova, F., & Ciccarelli, M. (2012). ClubMed? Cyclical fluctuations in the Mediterranean basin. *Journal of International Economics*, *88*, 162–175.

Canova, F., Ciccarelli, M., & Ortega, E. (2007). Similarities and convergence in G-7 cycles. *Journal of Monetary Economics*, *54*, 850–878.

Canova, F., Ciccarelli, M., & Ortega, E. (2012). Do institutional changes affect business cycles? *Journal of Economic Dynamics and Control*, *36*, 1520–1533.

Canova, F., & Marrinan, J. (1998). Sources and propagation of international output cycles: Common shocks or transmission? *Journal of International Economics*, *46*, 133–166.

Canova, F., & Pappa, P. (2004). Does it cost to be virtuous? The macroeconomic effects of fiscal constraints. In R. Clarida, J. Frankel, F. Giavazzi, & K. West (Eds.), *NBER international seminar on macroeconomics* (pp. 89–124). National Bureau of Economic Research. Retrieved from http:link.springer.com/book/10.1007%2F3-540-28556-3

Canova, F., & Pappa, P. (2007). Price differentials in monetary unions: The role of fiscal shocks (with E. Pappa). *Economic Journal*, *117*, 713–737.
Canova, F., & Perez Forero, F. (2012). *Estimating overidentified, non-recursive, time varying coefficients, structural VARs*. UPF manuscript.
Canova, F., & Pina, J. (2005). Monetary policy misspecification in VAR models. In C. Diebolt & C. Kyrtsou (Eds.), *New trends in macroeconomic*. Springer Verlag.
Chari, V. V., Kehoe, P., & McGrattan, E. (2008). Are structural VAR with long run restrictions useful for developing business cycle theory. *Journal of Monetary Economics*, *55*, 1337–1352.
Chudik, A., & Pesaran, H. (2011). Infinite-dimensional VARs and factor models. *Journal of Econometrics*, *163*, 4–22.
Ciccarelli, M., Maddaloni, A., & Peydró, J. L. (2012b). Heterogeneous transmission mechanism: Monetary policy and financial fragility in the Euro area. Paper prepared for the 56th Economic Policy Panel.
Ciccarelli, M., Ortega, E., & Valderrama, M. T. (2012a). *Heterogeneity and cross-country spillovers in macroeconomic-financial linkages*. ECB Working Paper No. 1241.
Ciccarelli, M., & Rebucci, A. (2006). Has the transmission mechanism of European monetary policy changed in the run-up to EMU? *European Economic Review*, *50*, 737–776.
Cogley, T., & Sargent, T. (2005). Drifts and volatilities: Monetary policy and outcomes in the post WWII U.S. *Review of Economic Dynamics*, *8*, 262–302.
Cooley, T., & Dwyer, M. (1998). Business cycle analysis without much theory: A look at structural VARs. *Journal of Econometrics*, *83*, 57–88.
Cooley, T., & LeRoy, S. (1985). Atheoretical macroeconomics: A critique. *Journal of Monetary Economics*, *16*, 283–308.
Cushman, D., & Zha, T. (1997). Identifying monetary policy in a small open economy under flexible exchange rates. *Journal of Monetary Economics*, *39*, 433–448.
De Graeve, F., & Karas, A. (2012). *Identifying VARs through heterogeneity: An application to bank runs*. Riskbank manuscript.
Dees, S., Di Mauro, F., Pesaran, H., & Smith, V. (2007). Exploring the International linkages of the euro area: A global VAR analysis. *Journal of Applied Econometrics*, *22*, 1–38.
Del Negro, M., & Schorfheide, F. (2010). Bayesian macroeconometrics. In J. Geweke, G. Koop, & H. Van Dijk (Eds.), *The oxford handbook of bayesian econometrics* (pp. 293–389). Retrieve from http://www.amazon.com/Oxford-Handbook-Bayesian-Econometrics-Handbooks/dp/0199559082
Doan, T., Litterman, R., & Sims, C. (1984). Forecasting and conditional projections using realist prior distributions. *Econometric Review*, *3*, 1–100.
Faust, J., & Leeper, E. (1997). Do Long Run Restrictions Really Identify Anything? *Journal of Business and Economic Statistics*, *15*, 345–353.
Fernandez Villaverde, J., Rubio Ramirez, J., Sargent, T., & Watson, M. (2007). The A,B,C and D for understanding VARs. *American Economic Review*, *97*, 1021–1026.
Forni, M., Hallin, M., Lippi, M., & Reichlin, L. (2000). The generalized factor model: Identification and estimation. *The Review of Economics and Statistics*, *82*, 540–554.
Forni, M., Hallin, M., Lippi, L., & Reichlin, L. (2005). The generalized dynamic factor model: One-sided estimation and forecasting. *Journal of the American Statistical Association*, *100*, 830–840.
Forni, M., & Reichlin, L. (1998). Let's get real: A factor analytic approach to disaggregated business cycle dynamics. *Review of Economic Studies*, *65*, 453–473.
Gertler, M., & Comin, D. (2006). Medium term cycles. *American Economic Review*, *96*, 523–551.

Gilhooly, R., Weale M., & Wieladek T. (2012). *Estimation of short dynamic panels in the presence of cross-sectional dependence and dynamic heterogeneity*. Discussion Paper No. 38. Bank of England, External MPC Unit.

Holtz-Eakin, D., Newey, W., & Rosen, H. (1998). Estimating vector autoregressions with panel data. *Econometrica, 56*, 1371–1395.

Jarocinski, M. (2010). Responses to monetary policy shocks in the east and the west of Europe: A comparison. *Journal of Applied Econometrics, 25*, 833–868.

Kadiyala, K. R., & Karlsson, S. (1997). Numerical methods for estimation and inference in Bayesian-VAR models. *Journal of Applied Econometrics, 12*, 99–132.

Kilian, L. (2012). Structural Vector Autoregressions. Forthcoming. In N. Haschimzade & M. Thornton (Eds.), *Handbook or research methods and applications in empirical macroeconomics*. Camberley, UK: Edward Elgar.

Kim, S. (2001). International transmission of US monetary policy shocks. Evidence from VARs. *Journal of Monetary Economics, 48*, 339–372.

Koop, G., & Korobilis, D. (2012). *Large scale time varying VARs*. University of Strathclyde, manuscript.

Kose, A., Otrok, C., & Whiteman, C. (2003). International business cycles: World, region, and country-specific factors. *American Economic Review, American Economic Association, 93*, 1216–1239.

Kose, A., & Prasad, E. (2010). *Emerging markets: Resilience and growth amid global turmoil*. Washington, DC: Brooking Institution Press.

Lane, P., & Benetrix, A. (2010). *International differences in fiscal policy during the global crisis*. CEPR Working Paper No. 8009, London, UK.

Lanne, M., & Lütkepohl, H. (2008). Identifying monetary policy shocks via changes in volatility. *Journal of Money, Credit and Banking, 40*, 1131–1149.

Ljung, L., & Soderstrom, T. (1984). *Theory and practice of recursive identification*. Cambridge, Mass: MIT Press.

Lindlay, D. V., & Smith, A. F. M. (1972). Bays Estimates of the linear model. *Journal of the Royal Statistical Association, Ser B, 34*, 1–18.

Long, J., & Plosser, C. (1983). Real Business cycles. *Journal of Political Economy, 91*, 39–69.

Love, I., & Zicchino, L. (2006). Financial development and dynamic investment behavior: Evidence from a Panel VAR. *The Quarterly Review of Economic and Finance, 46*, 190–210.

Mol, C. D., Giannone, D., & Reichlin, L. (2008). Forecasting using a large number of predictors: Is Bayesian shrinkage a valid alternative to principal components? *Journal of Econometrics,* 146, 318–328.

Ocampo, S., & Rodriguez, N. (2011). *An introductory review of a structural VARX estimation and applications*. Borradores de Economia 686. Banco de la Republica, Colombia.

Peersman, G. (2004). The transmission of monetary policy shocks in the Euro area. Are the effects different across countries? *Oxford Bulletin of Economics and Statistics, 66*, 285–308.

Pesaran, H., Schuerman, T., & Wiener, S. (2004). Modelling regional interdependencies using a global error-correcting macroeconomic model. *Journal of Business and Economic Statistics, 22*, 129–162.

Pesaran, H., & Smith, R. (1995). Estimating long run relationship from dynamic heterogeneous panels. *Journal of Econometrics, 68*, 79–113.

Primiceri, G. (2005). Time varying structural vector autoregressions and monetary policy. *The Review of Economic Studies, 72*, 821–852.

Ramey, V., & Shapiro, M. (1998). Costly capital reallocation and the effects of government spending. *Carnegie-Rochester Conference Series on Public Policy*, *48*, 145–194.

Rebucci, A. (2010a). Is growth exogenous? Evidence from the 1970s and the 1980s. *Applied Economics*, *42*, 535–543.

Rebucci, A. (2010b). On the heterogeneity bias of pooled estimators in stationary panel VARs. *Economic Modeling*, *27*, 1183–1198.

Sa, F., Towbin, P., & Wieladek, T. (2012). Capital inflows, financial structure and housing booms. *Journal of the European Economic Association*.

Sims, C. (2000). Using a likelihood perspective to sharpen econometrics discourse: Three examples. *Journal of Econometrics*, *95*, 443–462.

Sims, C., & Zha, T. (2006). Were There Regime Switches in U.S. Monetary Policy? *American Economic Review*, *96*, 54–81.

Sims, E. (2012). News, non-invertibility, and structural VARs. *Advanced in Econometrics*, *28*, 81–135.

Stock, J., & Watson, M. (2003). Understanding changes in International business cycle dynamics. *Journal of European Economic Association*, *3*, 968–1006.

Stock, J., Watson, M. (2012). *Disentangling the channels of the 2007–2009 recession*. NBER Working Paper No. 18094, Boston, MA.

Stock, J. H., & Watson, M. W. (2002a). Forecasting using principal components from a large number of predictors. *Journal of the American Statistical Association*, *97*, 1167–1179.

Stock, J. H., & Watson, M. W. (1989). New indices of coincident and leading indicators. In O. Blanchard & S. Fisher (Eds.), *NBER macroeconomic annual* (p. 4), Boston, MA.

Swamy, V. (1970). Efficient inference in random coefficient regression models. *Econometrica*, *38*, 311–323.

Vidangos, I. (2009). *Fluctuations in individual labor income: A panel VAR analysis*. Finance and Economics Discussion Series. Divisions of Research & Statistics and Monetary Affairs, Federal Reserve Board, Washington, DC.

Zellner, A., & Hong, C. (1989). Forecasting international growth rates using Bayesian Shrinkage and other procedures. *Journal of Econometrics*, *40*, 183–202.

MIXED-FREQUENCY VECTOR AUTOREGRESSIVE MODELS☆

Claudia Foroni, Eric Ghysels
and Massimiliano Marcellino

ABSTRACT

The development of models for variables sampled at different frequencies has attracted substantial interest in the recent literature. In this article, we discuss classical and Bayesian methods of estimating mixed-frequency VARs, and use them for forecasting and structural analysis. We also compare mixed-frequency VARs with other approaches to handling mixed-frequency data.

Keywords: Mixed-frequency data; mixed-frequency VAR; MIDAS; nowcasting; forecasting

JEL classifications: E37; C53

☆This views expressed herein are solely those of the authors and do not necessarily reflect the views of the Norges Bank. The usual disclaimers apply.

VAR Models in Macroeconomics – New Developments and Applications: Essays in Honor of Christopher A. Sims
Advances in Econometrics, Volume 32, 247–272

ISSN: 0731-9053/doi:10.1108/S0731-9053(2013)0000031007

INTRODUCTION

Since the seminal work of Sims (1980), vector autoregressive (VAR) models have acquired a prominent role among the tools for macroeconomic analysis. VAR models have been used to characterize comovements in the macroeconomic series and to summarize the information contained in the data. In macroeconomic applications, these models often involve variables such as real GDP or fiscal aggregates that are measured at quarterly frequency. However, when the objective is to forecast quarterly real GDP growth, for example, leading and coincident indices of the business cycle as well as numerous financial indicators that are available at monthly or even weekly or daily frequency may be used to improve the VAR model's forecast accuracy. High-frequency data can also be helpful to policy makers in particular, who need to assess the current state of the economy in real time and forecast its short-term developments.

To exploit the information available in series released at different frequencies and to jointly analyze them, there is a growing literature on mixed-frequency models, which include low- and high-frequency data. One of the early approaches, still commonly used in particular in policy institutions, relies on bridge equations, see, for example, Baffigi, Golinelli, and Parigi (2004). A more recent approach is the mixed-data sampling (MIDAS) method, originally proposed by Ghysels, Santa-Clara, and Valkanov (2004) and later further developed and applied in a variety of contexts. A third approach relies on the use of the Kalman filter to handle mixed-frequency data. An early application in the VAR context is Zadrozny (1988). More recently, mixed frequencies have been incorporated into factor models, see, for example, Mariano and Murasawa (2003) for small-scale applications and Giannone, Reichlin, and Small (2008) for large-scale models. Marcellino and Schumacher (2010) introduce the Factor MIDAS method, a new approach to nowcasting and forecasting low-frequency variables that exploits information in large sets of higher-frequency predictors. Bai, Ghysels, and Wright (2011) study the relationship between the Kalman filter and MIDAS approach.

In this article, we focus on mixed-frequency (MF-) VAR models. In the second section, we review the literature, including both frequentist and Bayesian approaches. In the third section, we compare the mixed-frequency VAR model with the earlier mixed-frequency approaches in the literature. In the fourth section, we provide an in-depth empirical illustration of the implementation and uses of MF-VAR models. The fifth section contains our conclusions.

MIXED-FREQUENCY VARS: A REVIEW OF THE LITERATURE

The standard approach of specifying a VAR model for data observed at lower and at higher frequencies inevitably involves latent shocks because not all shocks are observed at the higher frequency. The bulk of the literature on MF-VARs is based on this approach. Starting with Zadrozny (1988), there has been a tradition of formulating high-frequency VAR models with latent shocks and linking them to data observed as stocks or flows at a different frequency. This approach fits very well in a linear Gaussian state space framework. Other examples of this approach include Mariano and Murasawa (2010), Chiu, Eraker, Foerster, Kim, and Seoane (2011), Kuzin, Marcellino, and Schumacher (2011), Schorfheide and Song (2012), and Foroni and Marcellino (2012, 2013).

While most of the analysis so far has focused on the forecasting performance of MF-VAR models, there is growing interest in studying the effects of shocks occurring at different frequencies. The construction of such impulse response functions in MF-VAR models is complicated by the fact that the responses are driven by latent shocks. This problem has been addressed by Ghysels (2012) who proposes an alternative MF-VAR specification that allows the construction of impulse responses to observable shocks similar to traditional VAR models. Related models have also been suggested by Chen et al. (2012) and McCracken et al. (2013). The remainder of this section discusses these two main competing approaches from a classical and a Bayesian point of view.

Mixed-Frequency VAR Models in a Classical Framework

The first attempt at a model that allows for data generated at a higher frequency than the observed data was Zadrozny (1988). Taking into account of the stock-flow relations among variables, he derived the implied state-space model of the discrete-time data and showed how to use it to compute the Gaussian likelihood function with the help of the Kalman filter. Notwithstanding this early attempt, mixed-frequency VAR models have only recently received much attention.

For example, Mariano and Murasawa (2010) estimate a VAR model for latent monthly real GDP and other coincident indicators using data at different frequencies. The authors derive a state-space model for the series observable at different frequency, in which the quarterly series are treated as monthly series with missing observations. They introduce the use of an EM algorithm for maximum-likelihood estimation of a linear Gaussian state-space model, which allows for missing observations. The goal of their paper is to exploit mixed-frequency information to obtain an estimate of monthly real GDP, to be used as a new coincident index and to be compared with the Stock-Watson coincident index and the NBER Business Cycle reference dates.

Kuzin et al. (2011) discuss the forecasting ability of MF-VAR models, as presented in Mariano and Murasawa (2010), and compare them with the MIDAS approach. In particular, they assess the relative performance of MF-VAR and MIDAS for short-term forecasting GDP growth in the Euro Area, using about 20 monthly relevant business cycle indicators. Since a theoretical comparison shows that both types of models exhibit specific relative advantages and disadvantages, an empirical comparison is more suited to evaluating their usefulness. Kuzin et al. (2011) show that forecasts with both types of models have information content up to one quarter ahead. Relative forecast comparisons based on selected indicators show that MIDAS performs well at short horizons, whereas MF-VAR outperforms at longer horizons.

Similar evidence is found in Foroni and Marcellino (2013) who compare different methods proposed in the literature to deal with mixed-frequency and ragged-edge datasets. The authors discuss the performance of the different methods when nowcasting and forecasting the quarterly growth rate of the Euro Area GDP and its components, using a very large set of monthly indicators. As mentioned, they also find that MF-VAR outperforms the MIDAS approach only at longer horizons.

Whereas the studies discussed so far focused on the forecasting performance of mixed-frequency models, Ghysels (2012) explored the problem of impulse response analysis in MF-VAR models. Ghysels introduced a different MF-VAR representation, in which he constructs the MF-VAR process as stacked skip-sampled processes. He characterizes explicitly the mis-specification of a traditional low-frequency VAR and the consequent mis-specifications in the impulse response functions. Moreover, since the MF-VAR specified in this way can also characterize the timing of information releases, he shows how Choleski factorizations are a more natural tool for impulse response analysis because the elements in the vector represent a sequence of time events. The advantage of the setup in Ghysels (2012) is that standard tools of the VAR literature – that is impulse response analysis, Choleski factorizations, structural VAR models, estimation, etc. – can be readily applied to a mixed-frequency setting. Foroni and Marcellino (2012) and Ghysels (2012) show how to use MF-VARs for structural analysis based on impulse response functions. They find that responses from single (low-) frequency VARs can be substantially biased, due to the problems in the identification of the structural shocks using low frequency information only, and to the missing dynamic interactions across the variables in high frequency. Both problems can be at least alleviated by a MF-VAR.

Mixed-Frequency VAR Models in a Bayesian Framework

While Bayesian estimation of VAR models is standard, the estimation of MF-VARs with Bayesian techniques has been developed only recently. One example is Chiu et al. (2011), who adopt an approach similar to that of Zadrozny with a hidden high frequency VAR model involving latent shocks. Chui et al. use a Gibbs sampler that allows for mixed-frequency and irregularly spaced data. Their algorithm involves iterating over the draws from the missing data and from the unknown parameters in the model. Under the assumption of a normally distributed error term, the algorithm allows for draws from Gaussian conditional distributions for estimating the missing data, and for draws from Gaussian and inverse Wishart conditional posterior distributions for the parameters in the model. The drawback of this approach, compared to Ghysels (2012), is the fact that the VAR is specified in terms of latent shocks and a latent data filtration. Latent high-frequency innovations obtained from the Gibbs sampler are only plausible guesses of what shocks might be. It should also be noted that the main focus of the article is on parameter estimation. In an exercise with simulated data, the authors show that taking into account mixed-frequency data yields smaller root mean squared errors for all the parameter estimates regardless of the sample size and of the correlation between the variables of the system. These results are also consistent with the two empirical examples, conducted with data respectively at monthly and quarterly and weekly and monthly frequency, for which Chui et al. (2011) compare the posterior distributions of the parameters and the impulse response functions. Note that the impulse response functions in their setting involve "guesses" of hidden shocks based on the Gibbs sampler. Qian (2013) also considers a mixed-frequency VAR in a spirit similar to Zadrozny and Chui et al. Unlike the recursive solution offered by the Kalman filter, he proposes an explicit solution to the temporal aggregation problem and mixed frequencies. Namely, disaggregated series have a joint distribution driven by a VAR process and observed aggregated data impose linear constraints on the distribution of latent variables. Using conditional normal density projections Qian (2013) is able to simulate the latent processes given the joint distribution of observed high frequency and aggregate low-frequency data.

In contrast, Ghysels (2012) formulates a model with impulse response functions driven by shocks that can be estimated from observable data. Using the same data as Chui et al., he demonstrates that the resulting impulse response functions are very different from those in Chui et al. Notably, whereas the model of Chui et al. is time invariant, in Ghysels's model the shocks are contingent on the time they occur (specific month of the quarter) as they are generated by a stacked system of mixed-frequency observations.

In an effort to exploit the relationship between MF-VARs and MIDAS regressions in describing the impact of the high-frequency variables into the low-frequency ones, Ghysels (2012) also proposes an alternative Bayesian estimation approach. Following Ghysels and Owyang (2011), Ghysels utilizes a Metropolis-in-Gibbs step to sample the MIDAS hyperparameters, which are introduced as restrictions to keep the parameter space low dimensional. Once the draws for the MIDAS polynomial are given, the Bayesian analysis of the MF-VAR is fairly similar to the traditional VAR. Ghysels (2012) also suggests a U-MIDAS approach – suggested by Foroni, Marcellino, and Schumacher (2012) and discussed later – combined with his MF-VAR stacked system which allows one to use standard Bayesian approaches to VAR models, namely Minnesota priors and conjugate prior/posterior formulations, to estimate mixed-frequency VAR models.

Another study by Viefers (2011) reconsiders the estimation of a MF-VAR as in Mariano and Murasawa (2010). First, Viefers makes use of the Bayesian MCMC algorithm to simulate and estimate the model, and second he extends the MF-VAR to allow for regime switching. In his model, inference is based on the joint posterior density of all the unknowns. With the use of Gibbs sampling, it is possible to take into account the joint dependence of the various unknowns to be estimated. Viefers' simulation study suggests that inference on the latent series and the regime processes is fairly precise, although there is a more pronounced imprecision in the estimation of the VAR parameters. In the empirical exercise, Viefers considers monthly and quarterly data for the U.S. economy. The model shows a relatively good ability to identify the NBER recession dates except in recent years when the probability estimates tend to be erratic and much less accurate.

The Bayesian papers we have surveyed so far focus on parameter estimation and in-sample analysis, rather than forecasting. The use of mixed-frequency data, however, is particularly relevant in this second case, since it allows exploiting more recent and timely information, which can improve the forecasting performance of the models. Schorfheide and Song (2012) develop a different algorithm to deal with mixed-frequency data in the VAR framework, and conduct a forecasting exercise on U.S. data. They represent the MF-VAR as a state-space model, and use MCMC methods to conduct Bayesian inference for model parameters and unobserved monthly variables. A Minnesota prior that shrinks the VAR coefficients toward univariate random walk representations is introduced to cope with the issue of dimensionality. The estimated model is used to generate monthly forecasts, which are then time aggregated to evaluate them at a quarterly level. The goal of Schorfheide and Song is to study the extent to which the incorporation of monthly information improves the forecasts compared to models based on quarterly aggregated data. The analysis is conducted for 11 U.S. variables, of which three observed at quarterly level, in a real-time context. The authors find that the monthly series provide important information in the short run, with significant RMSE reductions obtained with the mixed-frequency model. Moreover, the more intra-quarter information is available, the higher the improvement.

A COMPARISON OF ALTERNATIVE MIXED-FREQUENCY APPROACHES

As discussed in the introduction, in the literature there are alternative methods available that exploit the information in mixed-frequency datasets and avoid pre-filtering and temporal aggregation. The MF-VAR is one of them. In what follows we briefly depict the main features of the other available approaches and discuss in what aspects they differ from the MF-VAR models.

One of the early approaches to deal with mixed-frequency data focuses on forecasting and relies on bridge equations. Baffigi, Golinelli, and Parigi (2004) among others estimate equations that link low-frequency variables and time-aggregated indicators. Forecasts of the high-frequency indicators are obtained with high-frequency time series models (typically ARMA models), and then the forecast values are temporally aggregated and plugged into the bridge equations to obtain the forecast of the low-frequency variable.

A more recent and alternative approach is the mixed-data sampling (MIDAS) method originally proposed by Ghysels et al. (2004), see also Ghysels et al. (2007). This method employs distributed lag polynomials to ensure parsimonious specifications and at the same time to handle series sampled at different frequencies. While MIDAS models were initially introduced in financial applications (see, e.g., Ghysels, Santa-Clara, & Valkanov, 2006), this method has also been employed to forecast macroeconomic series, where typically a quarterly series like GDP growth is forecast by monthly indicators (see, e.g., Clements & Galvao, 2008, 2009). Foroni, Marcellino, and Schumacher (2012) propose an unrestricted version of the MIDAS model (U-MIDAS) and show that when the mismatch of the frequency is low, like in macroeconomic applications that typically involve monthly and quarterly data only, this unrestricted version provides a better nowcasting performance, combined with an easier estimation method.[1]

Yet another commonly used model is the mixed-frequency factor model. In this model, which relies on a state-space representation similar to MF-VAR models, the low-frequency variable is considered a high-frequency variable with missing observations. These models may also be used to extract an unobserved state of the economy and create a new coincident indicator, or to forecast and nowcast GDP growth, see, e.g., Mariano and Murasawa (2003, 2010) and Nunes (2005) for small-scale applications and Giannone et al. (2008) for large-scale models. Marcellino and Schumacher (2010) introduce Factor-MIDAS, an approach for nowcasting and forecasting low-frequency variables exploiting information in large sets of higher-frequency indicators.

It is evident from the overview of the different methods just sketched here, that in general there is an agreement on the fact that exploiting data at different frequencies matters. We try now to compare their most important features, and their advantages and disadvantages with respect to the MF-VAR approach.

Compared to non-Kalman filter based univariate approaches such as MIDAS models or bridge equations, the MF-VAR has the advantage of jointly specifying the dynamics of the indicators and of the variable to be explained, without imposing any a-priori restriction. Moreover, since the low-frequency series is seen as a high-frequency series with missing values, the use of the Kalman filter permits the estimation of these missing data.[2] Furthermore, as shown by Bai, Ghysels, and Wright (2011), the Kalman filter results in the optimal filter in population, when ignoring parameter estimation error and assuming that the model is correctly specified. Therefore, under these ideal circumstances, the state-space approach cannot be beaten by any other method. On the other side, there are some drawbacks to the use of this approach: first of all, in most cases it is computationally demanding, and the computational costs increase with the number of variables involved, so that most of the time only small-scale models can be estimated. Moreover, the state space approach requires the correct specification of the model at high frequency, which is even more difficult than usual given the missing observations in the dependent variable. Finally, finding a proper specification for some of the high-frequency variables can be difficult, and specification errors in one of the equations can spread to the entire system.

The alternative MIDAS approach, is less efficient than the optimal Kalman filter, when the process is correctly specified. However, it can be more robust in the presence of mis-specification. Moreover, MIDAS can be more easily estimated, by NLS or even by OLS in the unrestricted U-MIDAS version of Foroni et al. (2012). However, it is only possible to obtain a high-frequency update of the expected low-frequency realization, not an estimate of the missing values of the low-frequency variable.

Both the state-space and the MIDAS approaches can be combined with a factor specification, in order to use the information in a large dataset, possibly with a ragged edge. Whether factor methods provide more precise estimates and forecasts than VARs or single equation methods is a matter for empirical investigation, since there is a trade-off between model complexity and expanding information set.

FORECAST ACCURACY COMPARISON OF MIXED-FREQUENCY VARS AND MIDAS REGRESSIONS

As an empirical illustration, in this section we examine the nowcasting and short-term forecasting ability of MF-VARs, from a frequentist point of view. We follow the specification proposed in Mariano and Murasawa (2010), described in detail in the section "MF-VAR Specification," and applied for nowcasting and forecasting the U.S. GDP growth using additional monthly indicators in the section "Data and Results."[3]

The notation is as follows. We denote real GDP growth as y_{t_q}, where $t_q = 1, 2, 3, \dots T_q^y$ is a quarterly time index and T_q^y is the final quarter for which GDP is available. GDP growth can be expressed also as a monthly variable with missing values, by setting $t_m = t_q, \forall t_m = 3t_q$, where t_m is the monthly time index. GDP growth is observable only in $t_m = 3, 6, 9, \ldots T_m^y$, where $T_m^y = 3T_q^y$. Therefore, what we want to obtain is the nowcast or forecast of the economic activity h_q quarters ahead or, equivalently, $h_m = 3h_q$ months ahead. We exploit monthly stationary indicators x_{t_m}, with $t_m = 1, 2, 3, \ldots, T_m^x$, where T_m^x is the final month for which the indicator is available. Usually monthly indicators are available earlier during the quarter than the GDP release, so generally we condition the forecast on the information available up to month T_m^x, which includes GDP information up to T_q^y and indicator observations up to T_m^x with $T_m^x \geq T_m^y = 3T_q^y$. The GDP growth forecast is indicated as $y_{T_m^y + h_m | T_m^x}$.

MF-VAR Specification

As in Mariano and Murasawa (2010), we consider the state-space representation of a VAR model, treating quarterly series as monthly series with missing observations. The disaggregation of the quarterly GDP growth, y_{t_m}, into the unobserved month-on-month GDP growth, $y^*_{t_m}$, is based on the following aggregation equation:

$$
\begin{aligned}
y_{t_m} &= \frac{1}{3}\left(y^*_{t_m} + y^*_{t_m-1} + y^*_{t_m-2}\right) + \frac{1}{3}\left(y^*_{t_m-1} + y^*_{t_m-2} + y^*_{t_m-3}\right) \\
&\quad + \frac{1}{3}\left(y^*_{t_m-2} + y^*_{t_m-3} + y^*_{t_m-4}\right) \\
&= \frac{1}{3}y^*_{t_m} + \frac{2}{3}y^*_{t_m-1} + y^*_{t_m-2} + \frac{2}{3}y^*_{t_m-3} + \frac{1}{3}y^*_{t_m-4}
\end{aligned} \tag{1}
$$

where $t_m = 3, 6, 9, \ldots, T_m$, since GDP growth, y_{t_m}, is observed the last month of each quarter, while $y^*_{t_m}$ is never observed.

This aggregation equation comes from the assumption that the quarterly GDP series (in log levels), Y_{t_m}, is the geometric mean of the latent monthly random sequence $Y^*_{t_m}, Y^*_{t_m-1}, Y^*_{t_m-2}$. Taking the three-period differences and defining $y_{t_m} = \Delta_3 Y_{t_m}$ and $y^*_{t_m} = \Delta Y^*_{t_m}$, we obtain Eq. (1).

For all t_m, we let the latent month-on-month GDP growth $y^*_{t_m}$ and the corresponding monthly indicator x_{t_m} follow a bivariate VAR(p) process:

$$
\phi(L_m)\begin{pmatrix} y^*_{t_m} & -\mu^*_y \\ x_{t_m} & -\mu_x \end{pmatrix} = u_{t_m} \tag{2}
$$

where $u_{t_m} \sim N(0, \Sigma)$.[4]

For estimation purposes, we cast the VAR(p) process in Eq. (2) together with the aggregation Eq. (1) in state-space form. Assuming that $p \leq 4$, and defining s_{t_m} and z_{t_m} as

$$S_{t_m} = \begin{pmatrix} z_{t_m} \\ \vdots \\ z_{t_m - 4} \end{pmatrix}, \quad z_{t_m} = \begin{pmatrix} y^*_{t_m} & -\mu^*_y \\ x_{t_m} & -\mu_x \end{pmatrix},$$

the state-space representation of the MF-VAR is

$$S_{t_m} = F s_{t_m - 1} + G v_{t_m} \tag{3}$$

$$\begin{pmatrix} y_{t_m} & -\mu_y \\ x_{t_m} & -\mu_x \end{pmatrix} = H s_{t_m} \tag{4}$$

with $\mu_y = 3\mu^*_y$, $v_{t_m} \sim N(0, I_2)$, and

$$F = \begin{bmatrix} F_1 \\ F_2 \end{bmatrix}; \ F_1 = \begin{bmatrix} \phi_1 & \dots & \phi_p & 0_{2\times 2(5-p)} \end{bmatrix}; \ F_2 = [I_8 \ 0_{8\times 2}] \tag{5}$$

$$G = \begin{bmatrix} \Sigma^{1/2} \\ 0_{8\times 2} \end{bmatrix}; \ H = [H_0 \dots H_4] \tag{6}$$

where H contains the lag polynomial

$$H(L_m) = \begin{bmatrix} 1/3 & 0 \\ 0 & 1 \end{bmatrix} + \begin{bmatrix} 2/3 & 0 \\ 0 & 0 \end{bmatrix} L_m + \begin{bmatrix} 1 & 0 \\ 0 & 0 \end{bmatrix} L^2_m + \begin{bmatrix} 2/3 & 0 \\ 0 & 0 \end{bmatrix} L^3_m + \begin{bmatrix} 1/3 & 0 \\ 0 & 0 \end{bmatrix} L^4_m \tag{7}$$

In our exercise, we determine the number of lags, p, according to the Bayesian Information Criterion (BIC), with a maximum lag order of $p = 4$ months.[5]

The state-space representation of the mixed-frequency VAR model, described by Eqs. (3) and (4), can be estimated by maximum-likelihood even in the presence of missing observations due to publication lags and the low-frequency nature of GDP. However, when the number of parameters is large, the ML method can fail to converge.

Therefore, we follow the procedure suggested by Mariano and Murasawa (2010) and implement the EM algorithm modified to allow for missing observations. We therefore consider the missing values as realizations of some i.i.d. standard normal random variables, that is,

$$y^{+}_{t_m} = \begin{cases} y_{t_m} & \text{if } y_{t_m} \text{ is observable} \\ \zeta_{t_m} & \text{otherwise} \end{cases} \tag{8}$$

where ζ_{t_m} is a draw from a standard normal distribution independent of the model parameters.[6]

The measurement equation is modified accordingly in the first two months of each quarter, for which the upper row of H is set to zero and a standard normal error term is added, so that the Kalman filter skips the random numbers. Since the realizations of the random numbers do not matter in practice, we replace the missing values with zeros.

We use the Kalman smoother to obtain forecasts of economic activity. Although GDP growth for a particular month is not available, the smoother utilizes the monthly indicators available for the current quarter, to construct nowcasts. For the months in which no observations are available also for the monthly indicators, the Kalman smoother acts exactly as the Kalman filter. What we obtain are iterative multistep forecasts and an estimate of the expected value of GDP growth in each month.

Alternative Approaches

In our nowcasting exercise, we compare the performance of the MF-VAR as described in the fourth section to a quarterly AR benchmark whose lag length is recursively selected according to the BIC criterion, and to the MIDAS model as proposed by Ghysels et al. (2004) and extended by Clements and Galvao (2008), and its unrestricted (U-MIDAS) version as proposed by Foroni et al. (2012). Note that the extended AR-MIDAS regression suggested by Clements and Galvao (2008) is not the most general and in fact involves a common factor restriction which may not be valid and therefore may affect the forecast performance. The ADL-MIDAS model of Andreou et al. (2011, 2013) avoids this possible source of misspecification and therefore may yield more accurate forecasts. However, for the purpose of conciseness, we cover only the AR-MIDAS model.[7] Specifically, the alternative mixed-frequency models are the following.

- MIDAS with an autoregressive term as used in Clements and Galvao (2008, 2009): This yields different models for different forecasting horizons. The forecast model for horizon $h_q = h_m/3$ is

$$y_{t_q+h_q} = y_{t_m+h_m} = \beta_0 + \lambda y_{t_m} + \beta_1 b(L_m, \theta)\left(1 - \lambda L_m^3\right) x_{t_m+w}^{(3)} + \varepsilon_{t_m+h_m} \quad (9)$$

where y_{t_m} and x_{t_m} are respectively the GDP growth and the monthly indicator, $x_{t_m}^{(3)}$ is the corresponding skip sampled monthly indicator, $w = T_m^x - T_m^y$ and $b(L_m, \theta)$ is the exponential Almon lag,

$$b(L_m, \theta) = \sum_{k=0}^{K} c(k, \theta) L_m^k, \; c(k, \theta) = \frac{\exp\left(\theta_1 k + \theta_2 k^2\right)}{\sum_{k=0}^{K} \exp\left(\theta_1 k + \theta_2 k^2\right)} \quad (10)$$

The model is estimated using NLS, with the additional coefficient λ and the common factor structure imposed. As starting values, we compute the residual sum of squares in each replication for a wide range of parameter sets and select the pair (θ_1, θ_2) which minimizes it.
- U-MIDAS without considering the Almon lag polynomial, leaving the lag polynomial of the indicator HF variable x_t unrestricted. Furthermore, we do not impose the common factor restriction as in Clements and Galvao (2008) for the autoregressive term. Thus, the model becomes

$$y_{t_q+h_q} = y_{t_m+h_m} = \mu_0 + \mu_1 y_{t_m} + \psi(L) x_{t_m+w}^{(3)} + \varepsilon_{t_m+h_m} \quad (11)$$

with lag polynomial $\psi(L) = \sum_{j=0}^{K} \psi_j L^j = \psi_0 + \psi_1 L + \ldots + \psi_K L^K$. The coefficients μ_0, μ_1, and $\psi(L)$ are estimated by OLS. To specify the lag order, we use the BIC with a maximum lag order of K to choose the lags.

Both models are h–dependent, so we need to reestimate them for forecasts at longer horizons. They also need to be reestimated, as new information about the predictors becomes available.

Data and Results

We focus on nowcasting U.S. quarterly GDP growth in a pseudo real-time context, using a selection of common monthly indicators, as in Carriero,

Clark, and Marcellino (2013). In particular, we consider the following monthly variables: payroll employment, industrial production, real retail sales (nominal deflated by the CPI), housing starts, the ISM index (overall) for manufacturing, the ISM index for supplier delivery times, the ISM index for orders, stock prices as measured by the S&P 500 index, the 10-year Treasury bond yield, and the 3-month Treasury bill rate. The data are taken either from the Federal Reserve Bank of Philadelphia's Real-Time Data Set for Macroeconomists (RTDSM) or from the Federal Reserve Board's FAME database.[8]

The starting point of the estimation sample is 1970:Q1 for all models, the earliest possible starting date given data availability. The evaluation sample spans 1995-Q1 to 2011-Q3. We consider three nowcast horizons: $h_m = 1, 2, 3$, corresponding to nowcasting the current quarter using information available, respectively, at the beginning of the third, second and first month of the quarter. Furthermore, we consider forecasts up to 4 quarters ahead, or equivalently up to $h_m = 15$.[9]

We use the Mean Squared Prediction Error (MSPE) to compare the accuracy of the different models. We recursively estimate every individual model and compute its MSPE relative to the quarterly AR benchmark. Hence, a relative MSPE smaller than 1 indicates that the mixed-frequency model outperforms the AR benchmark, and therefore that exploiting the monthly information helps in improving the nowcasts and forecasts.

We first present the results on the accuracy of nowcasts from models using individual indicators, relative to our benchmark AR model. Looking at Table 1, we observe that in most of the cases the use of mixed-frequency data reduces the MSPE of the nowcasts relative to the quarterly AR model. Improvements in the forecast accuracy can be obtained also for one-quarter-ahead forecasts, while mixed-frequency data do not appear to improve the forecasting performance of the models at longer horizons, at least not systematically.

There is no clear winner among the mixed-frequency approaches. In Table 1, the numbers in bold show the best method at each horizon across the indicators, while the numbers in italics represent the best method for each monthly indicator (in the cases in which they beat the benchmark). As we can see, the best performance is obtained with the AR-MIDAS (and secondly AR U-MIDAS) approach for most of the nowcasting and short-term forecasting horizons. However, looking at the performance of the different approaches when using the same indicator, the MF-VAR model tends to be quite good. As the numbers in italics show, with the ISM, ISM for orders and the industrial production as monthly indicators,

Table 1. Relative MSPE Performance of Mixed-Frequency Models for Different Monthly Indicators against AR Benchmark.

		Horizon (h_m)														
		Nowcasts			1q-ahead forecasts			2q-ahead forecasts			3q-ahead forecasts			4q-ahead forecasts		
		1	2	3	4	5	6	7	8	9	10	11	12	13	14	15
ISM	AR MIDAS	*0.83*	*0.88*	*0.83*	*0.90*	*0.94*	*0.96*	1.02	1.00	1.00	1.00	1.00	1.02	1.03	1.03	1.02
	AR U-MIDAS	1.00	0.99	0.86	0.92	0.98	1.02	1.00	1.00	1.01	*0.99*	1.00	1.03	1.02	1.04	1.03
	MF-VAR	*0.83*	0.91	*0.81*	*0.90*	0.97	0.97	1.01	1.02	1.01	1.03	1.04	1.04	1.03	1.03	1.03
ISM supply	AR MIDAS	0.97	0.98	*0.93*	*0.92*	*0.93*	*0.98*	*0.98*	0.97	*0.97*	*0.97*	*0.96*	*0.98*	1.00	1.00	1.01
delivery	AR U-MIDAS	*0.94*	*0.93*	1.07	1.04	1.06	0.99	0.99	*0.96*	*0.97*	0.98	*0.96*	1.00	1.00	1.01	1.02
times	MF-VAR	1.06	1.10	1.03	1.02	1.04	1.02	1.02	1.01	1.00	1.00	1.00	1.00	1.00	1.00	1.00
ISM orders	AR MIDAS	0.89	0.97	0.79	*0.87*	0.97	*0.94*	*0.95*	*0.96*	1.00	1.01	0.98	1.01	1.02	1.03	1.02
	AR U-MIDAS	0.90	0.99	0.84	0.92	1.10	1.03	0.96	0.98	0.99	1.00	1.00	1.07	1.14	1.15	1.10
	MF-VAR	*0.86*	*0.94*	*0.80*	0.88	*0.96*	0.96	0.97	0.97	*0.98*	1.03	1.02	1.03	1.05	1.03	1.02
S&P 500	AR MIDAS	1.02	*0.81*	*0.72*	0.80	0.83	***0.84***	***0.87***	***0.88***	***0.93***	***0.92***	***0.94***	*0.99*	***0.93***	1.01	1.01
	AR U-MIDAS	*0.99*	0.90	0.94	0.79	***0.81***	0.91	0.97	0.91	0.96	0.95	0.95	1.02	0.96	1.01	1.05
	MF-VAR	1.06	1.02	0.93	1.02	0.99	1.01	1.00	1.00	1.00	1.00	1.00	*0.99*	0.99	***0.99***	1.00
10y T-bond	AR MIDAS	1.00	1.06	1.05	1.05	1.07	1.09	1.12	1.07	1.05	1.04	1.03	1.02	1.01	1.04	1.02
	AR U-MIDAS	*0.98*	1.04	1.02	1.02	1.03	1.11	1.14	1.14	1.06	1.05	1.07	1.05	1.05	1.03	1.05
	MF-VAR	1.00	1.14	1.10	1.13	1.14	1.10	1.11	1.10	1.11	1.11	1.10	1.10	1.10	1.09	1.10
3m T-bill	AR MIDAS	1.25	1.27	1.29	1.29	1.31	1.40	1.39	1.26	1.22	1.22	1.13	0.99	1.10	1.10	1.12
	AR U-MIDAS	1.15	1.14	1.16	1.22	1.27	1.42	1.36	1.21	1.25	1.27	1.28	1.20	1.22	1.44	1.40
	MF-VAR	1.22	1.29	1.25	1.23	1.25	1.33	1.26	1.20	1.22	1.24	1.23	1.22	1.17	1.19	1.19
Employment	AR MIDAS	0.78	*0.86*	0.72	***0.77***	*0.85*	0.93	0.92	0.97	1.00	1.03	1.05	1.09	1.07	1.11	1.11
	AR U-MIDAS	*0.76*	0.88	0.75	0.79	0.87	0.92	0.94	0.98	***0.98***	1.04	1.01	1.00	1.13	1.19	1.09
	MF-VAR	0.80	0.90	*0.69*	0.80	0.91	*0.89*	*0.89*	*0.93*	*0.96*	*0.97*	*0.98*	***0.98***	*0.99*	*0.99*	***0.98***
Industrial	AR MIDAS	*0.77*	0.87	0.67	0.96	*0.94*	1.02	*0.96*	1.04	1.04	1.00	1.01	1.04	*0.98*	1.07	1.10
production	AR U-MIDAS	0.80	0.97	***0.66***	*0.95*	0.97	1.00	0.97	1.00	1.05	1.05	1.13	1.07	1.09	1.02	1.12
	MF-VAR	*0.77*	*0.83*	0.70	0.96	0.96	*0.93*	0.98	*0.98*	*0.99*	1.00	1.00	*0.99*	0.99	*0.99*	1.00
Housing starts	AR MIDAS	1.12	1.08	1.08	1.04	1.01	1.04	1.01	1.00	1.02	1.09	1.22	1.20	0.97	1.25	1.21
	AR U-MIDAS	*0.96*	*0.97*	*0.82*	1.03	1.02	0.99	1.02	1.01	1.01	1.02	1.04	1.02	1.04	1.04	1.05
	MF-VAR	1.03	1.08	1.07	1.18	1.18	1.17	1.21	1.18	1.21	1.25	1.20	1.23	1.26	1.20	1.20

Table 1. *(Continued)*

		Horizon (h_m)														
		Nowcasts			1q-ahead forecasts			2q-ahead forecasts			3q-ahead forecasts			4q-ahead forecasts		
		1	2	3	4	5	6	7	8	9	10	11	12	13	14	15
Retail Sales	AR MIDAS	***0.64***	***0.76***	*0.73*	*0.87*	*0.92*	*0.87*	*0.90*	*0.93*	*0.95*	0.98	1.05	1.01	1.01	1.02	1.03
	AR U-MIDAS	0.67	0.82	0.87	0.93	0.93	0.96	0.99	0.97	0.96	*0.97*	1.00	1.03	1.01	1.01	1.05
	MF-VAR	0.82	0.91	0.95	1.05	1.07	1.00	1.00	1.00	1.00	1.00	1.00	0.99	*0.99*	*0.99*	1.00
MSE benchmark (absolute value)		0.398	0.398	0.486	0.486	0.486	0.515	0.515	0.515	0.518	0.518	0.518	0.520	0.520	0.520	0.520

Notes: The entries in the table are obtained by recursively estimating every individual model and then computing the relative MSPE with respect to the benchmark. The benchmark is the recursive estimate of an AR model with lag length specified accordingly to the BIC criterion with a maximum lag order of 4. The evaluation sample is 1995Q1–2011Q3. The numbers in italics display the best approach at each horizon for each indicator (when the benchmark is beaten), the number in bold the best model across indicators. At the bottom of the table, we report the absolute value of the MSPE of the benchmark.

the MF-VAR model is the most accurate. Furthermore, the MF-VAR is the mixed-frequency approach which seems to work best at longer horizons.

Among the monthly predictors, for nowcasting (horizons 1–3), retail sales are quite accurate, with gains of about 25–35% with respect to the benchmark AR. The S&P500 performs well for one quarter ahead forecasts (horizons 4–6), with gains in the range 10–20%. This indicator works well also two quarters ahead (horizons 7–9), but the MSPE reductions shrink to about 10%. Several other indicators are useful for nowcasting and one quarter ahead forecasting U.S. GDP growth, with the interesting exception of interest rates, while much fewer indicators can beat the AR benchmark at longer horizons.

Finally, we look at the average performance across all indicators for the different classes of mixed-frequency models. In Table 2 we report, for each model class, the average relative MSPE across all the indicators displayed in Table 1. The table highlights a general improvement when using monthly information for real GDP growth nowcasting, with worse results for forecasting, related to the poor performance of some of the indicators, as reported in 1. On average across the indicators, the MF-VAR does not perform well compared with the MIDAS and U-MIDAS approaches, with the former slightly better than the latter.

Combining Different Indicators

While in the section "Data and Results" we looked at the improvements obtained from using one monthly indicator at a time, in this section we focus on combining all monthly indicators. We look at the improvements both in combining forecasts with different schemes of forecast pooling, and in combining information ex-ante by extracting a common factor from the monthly indicators and then modeling it together with GDP growth with a mixed-frequency specification.

Starting with forecast combinations, we consider basic combination schemes: the mean, median, and a weighted mean (where the weights of the different models are inversely proportional to the MSPE of the models in the previous four quarters). We proceed as follows: first, forecasts from single indicator models are computed. Then, the mean, the median, and the weighted mean of all the forecasts within a single approach are computed. Finally, the MSPE of this forecast combination is calculated and divided by the MSPE of the benchmark. The results, reported in Table 3, are

Table 2. Average Relative MSPE Performance of Different Classes of Mixed-Frequency Models against AR Benchmark.

	Horizon (h_m)														
	Nowcasts			1q-ahead forecasts			2q-ahead forecasts			3q-ahead forecasts			4q-ahead forecasts		
	1	2	3	4	5	6	7	8	9	10	11	12	13	14	15
AR MIDAS	0.93	0.95	0.88	0.95	0.98	1.01	1.01	1.01	1.02	1.03	1.04	1.03	1.01	1.06	1.07
AR U-MIDAS	0.92	0.96	0.90	0.96	1.00	1.03	1.03	1.02	1.02	1.03	1.04	1.05	1.07	1.09	1.10
MF-VAR	0.95	1.01	0.93	1.02	1.05	1.04	1.04	1.04	1.05	1.06	1.06	1.06	1.06	1.05	1.05

Notes: The entries in the table are obtained as follows: first, estimate recursively every individual model and compute the relative MSPE with respect to the benchmark; then, take the average all across the indicators of the relative MSPE within a model class. The benchmark is the recursive estimate of an AR model with lag length specified accordingly to the BIC criterion with a maximum lag order of 4. The evaluation sample is 1995Q1–2011Q3.

Table 3. Relative MSPE Performance of Model Pooling within a Given Class of Models against AR Benchmark.

		Horizon (h_m)														
		Nowcasts			1q-ahead forecasts			2q-ahead forecasts			3q-ahead forecasts			4q-ahead forecasts		
		1	2	3	4	5	6	7	8	9	10	11	12	13	14	15
AR MIDAS	Mean	0.69	0.76	0.72	0.84	0.88	0.93	0.96	0.97	0.98	0.99	1.00	1.00	0.98	1.01	1.02
	Weighted mean	0.61	0.73	0.70	**0.81**	0.87	**0.91**	**0.94**	**0.95**	**0.97**	**0.97**	**0.99**	**0.99**	**0.97**	**1.00**	**1.00**
	Median	0.71	0.74	**0.69**	0.83	0.88	0.93	**0.94**	0.96	0.99	1.00	1.00	**0.99**	0.99	1.01	1.01
AR U-MIDAS	Mean	0.65	0.75	0.74	0.85	0.89	0.95	0.99	0.99	1.00	1.01	1.00	1.02	1.02	1.05	1.06
	Weighted Mean	**0.59**	**0.71**	0.71	0.83	**0.86**	0.93	0.98	0.98	1.00	1.00	1.00	1.02	1.02	1.04	1.04
	Median	0.65	0.73	0.76	0.84	0.92	0.96	0.98	0.98	0.99	1.00	1.00	1.01	1.02	1.01	1.04
MF-VAR	Mean	0.67	0.78	0.79	0.91	0.96	0.95	0.97	0.97	0.98	1.00	1.00	1.01	1.01	1.01	1.01
	Weighted mean	0.60	0.72	0.74	0.87	0.93	0.93	**0.94**	0.96	**0.97**	0.99	0.99	1.00	1.01	1.01	1.01
	Median	0.71	0.80	0.84	0.94	0.95	0.97	0.99	1.00	1.00	1.00	1.01	1.00	1.00	**1.00**	**1.00**

Notes: The entries in the table are obtained as follows: first, forecasts from single indicator models are computed, and the mean, weighted mean, and median of all the forecasts within a single model are obtained; second, the MSPE of the forecast combination is calculated and divided by the MSPE of the benchmark. The estimation is conducted recursively. The benchmark is the recursive estimate of an AR model with lag length specified accordingly to the BIC criterion with a maximum lag order of 4. The evaluation sample is 1995Q1–2011Q3. The numbers in bold show the best relative MSPE performance for each horizon.

very encouraging, since forecast combination allows us to beat the AR benchmark up to two quarters ahead with all the combination schemes and with all classes of mixed-frequency models. Furthermore, some improvements can be obtained even at longer horizons.

Although MIDAS methods display a slightly better performance, the pooling within the MF-VAR class of models performs very well, allowing for substantial gains relative to the benchmark. Overall, the gains are in the range of 20–30% for nowcasting (horizons 1–3), with the weighted mean of UMIDAS working particularly well, and 10–20% for one quarter ahead forecasts (horizons 4–6), though slightly lower for the MF-VAR pooled forecasts. The MSPE reductions shrink to less than 5% for two-quarter ahead forecasts, and basically disappear for longer horizons.

We now focus our attention on combining the information in the monthly series via the use of factor models. We estimate the same models as when using one indicator at a time, but we replace the single indicators with a monthly factor extracted from the 10 monthly indicators.

Since the dataset has a ragged-edge structure, we cannot extract directly the principal components, but we need to follow the procedure outlined by Stock and Watson (2002), which combines the EM algorithm with principal component analysis (PCA). To handle the fact that not all observations are available, due to publication lags, we write the relation between observed and not fully observed variables as

$$X_i^{obs} = A_i X_i \tag{12}$$

where X_i^{obs} contains the observations available for variable i, as a subset of X_i, and A_i is the matrix that tackles missing values. Taking this relation into account, the EM algorithm provides an estimate of the missing values (for more details, see Stock & Watson, 2002). Once we obtain the factor $\hat{f}_t$, we estimate Eqs. (2), (9), and (11), respectively replacing x_t with our $\hat{f}_t$.

Results are provided in Table 4, in which we report the MSPE of the different factor-based mixed-frequency models relative to the AR benchmark. For completeness, we report also the results for forecasting real GDP growth using a standard quarterly factor model as proposed by Stock and Watson (2002) applied to our dataset, with one factor and the number of lags chosen according to the BIC criterion with an upper bound of four lags.

The findings confirm the usefulness of mixed-frequency models, with a better performance of the mixed-frequency models relative not only to the

Table 4. Relative MSPE Performance of Different Classes of Factor Models against AR Benchmark.

	Horizon (h_m)														
	Nowcasts			1q-ahead forecasts			2q-ahead forecasts			3q-ahead forecasts			4q-ahead forecasts		
	1	2	3	4	5	6	7	8	9	10	11	12	13	14	15
Factor AR-MIDAS	0.73	**0.84**	0.76	**0.82**	**0.92**	0.94	0.96	0.98	1.00	1.01	1.02	**1.02**	**1.02**	**1.03**	1.04
Factor AR U-MIDAS	0.74	0.95	0.92	1.07	1.12	1.25	0.98	0.99	**0.99**	**1.00**	**0.99**	1.12	1.14	1.20	1.13
Factor MF-VAR	**0.70**	0.91	**0.74**	0.86	0.96	**0.93**	**0.93**	**0.97**	**0.99**	1.01	1.03	1.03	**1.02**	**1.03**	**1.01**
Quarterly factor model	0.90	0.90	0.92	0.92	0.92	1.01	1.01	1.01	1.02	1.02	1.02	1.03	1.03	**1.03**	1.03

Notes: The entries in the table are obtained as follows: first, estimate recursively every factor model, then compute the relative MSPE with respect to the benchmark. The factors are estimated with the EM algorithm together with PCA as in Stock and Watson (2002). The benchmark is the recursive estimate of an AR model with lag length specified accordingly to the BIC criterion with a maximum lag order of 4. The evaluation sample is 1995Q1–2011Q3. The numbers in bold show the best relative MSPE performance for each horizon.

AR benchmark but also to the quarterly factor model. The gains are sizable for nowcasting (about 20–30% with respect to the AR and 10–20% with respect to the quarterly factor model) and one-quarter ahead forecasting (about 10–20% with respect to the AR and 10% with respect to the quarterly factor model, except for the factor UMIDAS). Furthermore, inserting the factor in the MF-VAR approach seems to perform particularly well. The resulting model is the best performer at most forecasting horizons.

Summing up, the results in this section suggest that summarizing information via forecast pooling or the use of factors is a useful way to improve the forecasting performance of the mixed-frequency models. Overall, pooling seems slightly better than factor modeling in our empirical application. However, both methods emphasize that using monthly information plays a big role in improving the forecasts of the quarterly GDP growth.

CONCLUSIONS

The use of VAR models is a standard tool in policy analysis as well as forecasting and nowcasting. VARs are typically estimated with variables sampled at the same frequency. However, often higher frequency information is available for some of the variables under evaluation, and not using that information may involve a loss in forecast accuracy.

The development of mixed-frequency VAR models addresses this problem. The estimation and formulation of such models, is at an early stage. In this article, we surveyed the burgeoning literature in the field, and provided an empirical comparison of alternative mixed-frequency models for nowcasting and short-term forecasting U.S. GDP growth.

The main message is that incorporating additional monthly predictors into quarterly VAR models of real GDP growth can be quite useful. Mixed-frequency VAR models often produce more accurate forecasts, especially when the monthly predictors are summarized as a monthly factor and modeled jointly with quarterly real GDP growth.

NOTES

1. Andreou, Ghysels, and Kourtellos (2011, 2013) investigate the forecasting of GDP using daily financial series. They focus on MIDAS regressions and forecast combinations. Obviously, in their setting U-MIDAS is not feasible given the difference of daily/quarterly data.

2. Although this may fit better for stock variables, the same reasoning works for flow variables. As an example, a quarterly real GDP growth can be seen as an aggregation of monthly unobserved real GDP growth, which is then observed only every third period. See Mariano and Murasawa (2010) for a more comprehensive explanation.

3. Another state of the art methodology in the MF-VAR literature is the one in Schorfheide and Song (2011), described in the section "Mixed-frequency VAR models in a Bayesian framework." In their paper, Schorfheide and Song estimate a 11-variable MF-VAR and evaluate its forecasting performance against a quarterly benchmark. In our example, we want to compare the forecasting performance of MF-VAR models to other mixed-frequency approaches, such as MIDAS models. Since MIDAS models are estimated considering one single monthly indicator at the time, we prefer to use the Mariano and Murasawa methodology, which is tailored for small-scale models. Moreover, we choose to work only in a frequentist framework, and therefore we do not consider the Schorfheide and Song approach in our empirical exercise.

4. The assumption of normal errors is common in the literature on Kalman filter based mixed frequency models, though it might be questionable since high frequency variables often present non normal features. In principle, other distributions could be used in combination with a general filtering procedure, but we leave this as an interesting topic for future research.

5. We consider a maximum lag order of $p=4$ since with an higher lag order the state-space representation of the model changes slightly. For the sake of conciseness, we do not report here the representation for $p>4$. Details for this case can be found in Mariano and Murasawa (2010).

6. As pointed out by Mariano and Murasawa (2003), the distribution of ζ_{t_m} can be anything which does not depend on the model parameters. Typically normality is assumed for convenience.

7. See Andreou et al. (2011) for a further discussion of the comparison of ADL-MIDAS and AR-MIDAS models.

8. Carriero et al. (2013) conduct their analysis using real-time data, namely, different vintages as available in each month of the evaluation period. Here, for the sake of simplicity, we work with final vintage data only. There are some studies comparing mixed-frequency data models for real-time and ex post revised data. According to Foroni and Marcellino (2013), in the case of nowcasting GDP growth for the euro area the use of real time rather than final vintage data does not affect substantially the overall results.

9. A forecast horizon $h_m=15$ corresponds to a 4-quarters-ahead forecast, made on the first month of the quarter, e.g., a forecast of Q1:2010 made in January 2009.

ACKNOWLEDGMENT

We are grateful to Lutz Kilian for helpful comments on a previous draft. The second author likes to thank the financial support of a Fernand Braudel Senior Fellowship of the European University Institute, Florence.

REFERENCES

Andreou, E., Ghysels, E., & Kourtellos, A. (2011). Forecasting with mixed-frequency data. In M. Clements & D. Hendry (Eds.), *Oxford handbook of economic forecasting* (pp. 225–245).

Andreou, E., Ghysels, E., & Kourtellos, A. (2013). Should macroeconomic forecasters look at daily financial data? *Journal of Business and Economic Statistics*, *31*, 240–251.

Baffigi, A., Golinelli, R., & Parigi, G. (2004). Bridge models to forecast the Euro area GDP. *International Journal of Forecasting*, *20*(3), 447–460.

Bai, J., Ghysels, E., & Wright, J. (2011). State space models and MIDAS regressions. *Econometric Reviews*, *32*, 779–813.

Carriero, A., Clark, T., & Marcellino, M. (2013). *Real-time nowcasting with a Bayesian mixed frequency model with stochastic volatility*. CEPR WP 9312.

Chen, W., Anderson, B., & Deistler, M. (2012). Properties of blocked linear systems. *Automatica*, *48*, 2520–2525.

Chiu, C., Eraker, B., Foerster, A., Kim, T., & Seoane, H. (2011). *Estimating VAR's sampled at mixed or irregular spaced frequencies: A Bayesian approach*, mimeo. University of Wisconsin.

Clements, M. P., & Galvao, A. B. (2008). Macroeconomic forecasting with mixed-frequency data: Forecasting US output growth. *Journal of Business and Economic Statistics*, *26*, 546–554.

Clements, M. P., & Galvao, A. B. (2009). Forecasting US output growth using leading indicators: An appraisal using MIDAS models. *Journal of Applied Econometrics*, *24*(7), 1057–1217.

Foroni, C., & Marcellino, M. (2012). *Mixed-frequency structural models: Identification, estimation, and policy analysis*, mimeo, EUI.

Foroni, C., & Marcellino, M. (2013). A comparison of mixed frequency approaches for nowcasting euro area macroeconomic aggregates. *International Journal of Forecasting* (forthcoming).

Foroni, C., Marcellino, M., & Schumacher, C. (2012). *U-MIDAS: MIDAS regressions with unrestricted lag polynomials*. CEPR Discussion Papers, 8828.

Ghysels, E. (2012). *Macroeconomics and the reality of mixed frequency data*. Discussion Paper. University of North Carolina, US.

Ghysels, E., & Owyang, M. (2011). *Taylor monetary policy rules with financial market expectations*, Discussion Paper, Federal Reserve Bank of Saint-Louis and UNC.

Ghysels, E., Santa-Clara, P., & Valkanov, R. (2004). *The MIDAS touch: MIxed DAta Sampling regression models*, mimeo, Chapel Hill, N.C.

Ghysels, E., Santa-Clara, P., & Valkanov, R. (2006). Predicting volatility: Getting the most out of return data sampled at different frequencies. *Journal of Econometrics*, *131*(1–2), 59–95.

Ghysels, E., Sinko, A., & Valkanov, R. (2007). MIDAS regressions: Further results and new directions. *Econometric Reviews*, *26*(1), 53–90.

Giannone, D., Reichlin, L., & Small, D. (2008). Nowcasting GDP and inflation: The real-time informational content of macroeconomic data releases. *Journal of Monetary Economics*, *55*, 665–676.

Kuzin, V., Marcellino, M., & Schumacher, C. (2011). MIDAS vs. mixed-frequency VAR: Nowcasting GDP in the Euro Area. *International Journal of Forecasting*, *27*, 529–542.

Marcellino, M., & Schumacher, C. (2010). Factor-MIDAS for now- and forecasting with ragged-edge data: A model comparison for German GDP. *Oxford Bulletin of Economics and Statistics*, *72*, 518–550.

Mariano, R., & Murasawa, Y. (2003). A new coincident index of business cycles based on monthly and quarterly series. *Journal of Applied Econometrics*, *18*(4), 427–443.

Mariano, R. S., & Murasawa, Y. (2010). A coincident index, common factors, and monthly real GDP. *Oxford Bulletin of Economics and Statistics*, *72*(1), 27–46.

McCracken, M., Owyang, M., & Sekhposyan, T. (2013). *Real-time forecasting with a large Bayesian block model.* Discussion Paper, Federal Reserve Bank of St. Louis and Bank of Canada.

Nunes, L. C. (2005). Nowcasting quarterly GDP growth in a monthly coincident indicator model. *Journal of Forecasting*, *24*, 575–592.

Qian, H. (2013). *Vector autoregression with varied frequency data.* Discussion Paper, Iowa State University.

Schorfheide, F., & Song, D. (2012). *Real-time forecasting with a mixed frequency VAR.* FRB Minneapolis Working Papers, 701.

Sims, C. (1980). Marcoeconomics and reality. *Econometrica*, *48*(1), 1–48.

Stock, J., & Watson, M. (2002). Macroeconomic forecasting using diffusion indexes. *Journal of Business and Economic Statistics*, *20*, 147–162.

Viefers, P. (2011). *Bayesian inference for the mixed-frequency VAR model.* DIW Discussion Papers, 1172.

Zadrozny, P. A. (1988). Gaussian-likelihood of countinuous-time ARMAX models when data are stocks and flows at different frequencies. *Econometric Theory*, *4*(1), 108–124.

APPENDIX: DATA DESCRIPTION

Data Description.

Monthly Indicators	Description
ISM	ISM index for manufacturing – overall
ISM – supply delivery times	ISM index for supply delivery times
ISM – orders	ISM index for orders
S&P 500	Stock prices as measured by the S&P 500 index
10y T-bond	10-year Treasury bond yield
3m T-bill	3-month Treasury bill yield
Employment	Payroll employment
Industrial production	Industrial production
Housing starts	Housing starts
Retail sales	Real retail sales

Note: ISM, Institute for Supply Management.

THRESHOLDS AND SMOOTH TRANSITIONS IN VECTOR AUTOREGRESSIVE MODELS☆

Kirstin Hubrich and Timo Teräsvirta

ABSTRACT

This survey focuses on two families of nonlinear vector time series models, the family of vector threshold regression (VTR) models and that of vector smooth transition regression (VSTR) models. These two model classes contain incomplete models in the sense that strongly exogeneous variables are allowed in the equations. The emphasis is on stationary models, but the considerations also include nonstationary VTR and VSTR models with cointegrated variables. Model specification, estimation and evaluation is considered, and the use of the models illustrated by macroeconomic examples from the literature.

Keywords: Common nonlinearity; impulse response analysis; linearity testing; multivariate nonlinear model; nonlinear cointegration; threshold estimation

JEL classifications: C32; C51; C52; C53

☆The views expressed in this article are those of the authors and should not be interpreted as reflecting the views of the European Central Bank.

VAR Models in Macroeconomics – New Developments and Applications: Essays in Honor of Christopher A. Sims
Advances in Econometrics, Volume 32, 273–326

ISSN: 0731-9053/doi:10.1108/S0731-9053(2013)0000031008

INTRODUCTION

Since its appearance as a macroeconomic modelling tool more than 30 years ago (Sims, 1980), the basic linear vector autoregressive (VAR) model has been applied in various economic modelling situations and developed further in many ways. For a concise overview of the linear VAR model covering the basic statistical inference, interpretation of the estimated model through the impulse response analysis, and forecasting (see Lütkepohl, 2006).

Although linear VAR models have served economists well, there are situations where a nonlinear model would be appropriate. Economic theory may suggest employing a nonlinear model. Alternatively, observed time series may indicate that the variables of interest could be nonlinearly related. For example, the recent financial crisis and the Great Recession suggest that describing quantitative relationships between the financial sector and the macroeconomy would require a nonlinear model. Nonlinear models may also be useful in analyzing monetary policy. Positive and negative monetary policy shocks may have asymmetric effects, and the effect may depend nonlinearly on the size of the shock as opposed to linear models in which this effect is proportional to the size of the shock. Nonlinear models may also have a role to play in studies of effectiveness of fiscal policy. The effects of certain fiscal policy measures may depend on the phase of the business cycle, and investigating fiscal multiplier effects may require nonlinear tools.

There are different ways of modelling nonlinear relationships in economics. The leading examples of parametric nonlinear VAR models are the vector threshold autoregressive (VTAR), the vector smooth transition autoregressive (VSTAR) and the vector Markov-switching autoregressive (VMSAR) model. The first two models differ from the third one in that in the former all variables of the model are observable, whereas in the VMSAR model an unobservable, 'latent', variable plays a central role. In this survey the focus will be on the VTAR and VSTAR models. Incomplete models that contain strongly exogenous variables are considered as well: these are called the vector switching regression (VSR) and vector smooth transition regression (VSTR) model, respectively. Statistical theory behind these models is discussed and their use illustrated by empirical examples selected from the literature. The focus is on stationary models, but statistical inference in nonstationary models with cointegrated variables also receives attention. The important class of VMSAR models and their applications deserve a survey of their own and are not included in this review.

The choice between VTAR and VSTAR models depends on the economic problem to be investigated. The VTAR model is designed to describe situations in which the dynamic behavior of a set of random variables can be modelled by defining a limited number of linear states or regimes that the process can visit. VSTAR models can be viewed as models with a continuum of states, that is, the dynamic behavior of the variables changes smoothly between a number (often two) extreme states or regimes. An observable switch or transition variable determines the regime that generates the next observation.

The plan of this article is as follows. Single-equation switching regression and smooth transition regression models are introduced in the second section. The VSR and VSTR models, including some special cases, are defined in the third section. Parameter estimation by least squares is considered in the fourth section. The fifth section is devoted to building adequate VSR and VSTR models: a modelling cycle consisting of specification, estimation and evaluation of stationary models of this kind is presented. Approaches to inference in nonstationary models with cointegrated variables are considered as well. The sixth section contains a discussion of the possibility that nonlinearity in these models can be eliminated by a linear transformation, that is, nonlinearity in the individual equations of the model is a common feature. Generalized impulse response functions are considered in the seventh section, and forecasting with VSR and VSTR models in the eighth section. The purpose of the ninth section is to illustrate the use of the models by presenting applications to macroeconomic time series. Final remarks and suggestions for further reading can be found in the tenth section.

BACKGROUND: SINGLE-EQUATION MODELS

Switching Regression Model

The switching regression (SR) model was introduced by Quandt (1958), who also discussed parameter estimation. Subsequently, Quandt (1960) considered linearity testing but did not yet solve the problem of the asymptotic null distribution of his test statistic. A dynamic SR model is defined as follows:

$$y_t = \sum_{i=1}^{q}(\mu_i + \sum_{j=1}^{p} \varphi_{ij} y_{t-j} + \boldsymbol{\gamma}_i' \mathbf{x}_t + \varepsilon_{it}) I(c_{i-1} < s_t \leq c_i) \qquad (1)$$

where y_t and s_t are stationary random variables, $\mathbf{x}_t$ is a vector of strongly exogenous variables, $\{\varepsilon_{it}\} \sim \text{iid}(0,\sigma_i^2)$, μ_i and φ_{ij} are parameters, $i=1,\ldots,q$, and $j=1,\ldots,p$, $\boldsymbol{\gamma}_i$, $i=1,\ldots,q$, are $n\times 1$ parameter vectors, $c_0=-\infty$, and $c_q=\infty$. $I(A)$ is an indicator function: $I(A)=1$ when A occurs, zero otherwise.

Later, Howell Tong introduced an autoregressive version of Eq. (1), $\boldsymbol{\gamma}_i=\mathbf{0}$ for all i, and called it the self-exciting threshold autoregressive (SETAR) model. In the SETAR model, $s_t=y_{t-d}$, $d>0$, hence the term 'self-exciting'. It quickly became the most popular nonlinear univariate time series model in many areas of application. Chan (1993) worked out the asymptotic theory for least squares estimators of the SETAR model with a single-threshold, and Qian (1998) did the same for maximum likelihood estimators. In particular, they showed that under certain conditions, the least squares estimator of c_1 is super consistent. For an excellent review of SETAR models, see Tong (1990). For further developments of asymptotic inference, see Hansen (1997, 1999, 2000).

As an economic example, Pfann, Schotman, and Tschernig (1996) used a single-threshold SETAR model to describe the dynamic behavior of the three-month US T-bill interest rate. The time series extends from 1962(1) to 1990(6). The results show that when the interest rate is sufficiently low it behaves like a random walk. At high values, the rate is mean-reverting. The switch extends to the variance as in Eq. (11). The estimated variance is clearly higher in the high rate than in the low rate regime.

One obtains another form of a SR or threshold model by assuming that switching in Eq. (1) is controlled by a latent variable instead of an observable one. This is done by defining a latent discrete random variable s_t with q possible values and replacing the indicator in Eq. (1) by $I(s_t=k)$. In economic applications it is often assumed that s_t has memory. The most common case is the one where $\{s_t\}$ is a (first-order) Markov chain; see Lindgren (1978) for this type of model.

Smooth Transition Regression Model

The origins of smooth transition models can be traced back to Bacon and Watts (1971) who suggested that instead of fitting two linear regression lines to a bivariate point cloud, one could have a smooth transition from one line to the other as a function of the explanatory variable. The smooth transition regression (STR) model has the form

$$y_t = \varphi_0 + \sum_{j=1}^{p} \varphi_j y_{t-j} + \boldsymbol{\gamma}'\mathbf{x}_t + (\psi_0 + \sum_{j=1}^{p} \psi_j y_{t-j} + \boldsymbol{\eta}'\mathbf{x}_t)G(\gamma, \mathbf{c}; s_t) + \varepsilon_t \quad (2)$$

where φ_j and ψ_j, $j=0,1,\ldots,p$, are parameters, $\boldsymbol{\gamma}$ and $\boldsymbol{\eta}$ are $n \times 1$ parameter vectors, and $G(\gamma, \mathbf{c}; s_t)$ is a bounded continuous function of s_t, called the transition function. The error sequence $\{\varepsilon_t\} \sim \text{iid}(0, \sigma^2)$. The transition function of Bacon and Watts (1971) was a hyperbolic tangent function, and the parameters of the ensuing nonlinear model were estimated using Bayesian methods. Maddala (1977, p. 396) suggested a logistic function for the purpose, and that has become a standard choice. For $\mathbf{c}=c$ (scalar),

$$G(\gamma, c; s_t) = (1 + \exp\{-\gamma(s_t - c)\})^{-1}, \quad \gamma > 0 \quad (3)$$

where $\gamma > 0$ is an identifying restriction. Eqs. (2) and (3) define a single-transition logistic STR model. A pure time series smooth transition autoregressive (STAR) model, $\boldsymbol{\gamma} = \boldsymbol{\eta} = \mathbf{0}$ and $s_t = y_{t-d}$, $d > 0$, in Eq. (2), was introduced by Chan and Tong (1986). An alternative to the logistic function (3) is the exponential transition function:

$$G(\gamma, c; s_t) = 1 - \exp\{-\gamma(s_t - c)^2\} \quad (4)$$

and (2) with (4) define an exponential STR (ESTR) model. Haggan and Ozaki (1981) already considered a univariate version of this model in the special case $c=0$ in Eq. (4). Teräsvirta (1994) constructed a model-building strategy for STAR models that Eitrheim and Teräsvirta (1996) completed.

Smooth transition type models can describe aggregate dynamics in situations where individual agents respond differently to a stimulus. LSTAR models are often an alternative to single-threshold TAR models. ESTAR models have been used for modelling real exchange rates in cases where the mean reversion is symmetric around an equilibrium and the strength of the attraction depends nonlinearly on the size of the deviation from this equilibrium. See Taylor, Peel, and Sarno (2001) for discussion and examples. More discussion on all these models and modelling time series with them can be found in Teräsvirta, Tjøstheim, and Granger (2010). For forecasting with STAR models, see also Lundbergh and Teräsvirta (2002), Teräsvirta (2006), and Kock and Teräsvirta (2011).

Switching Mean Autoregressive Model

Lanne and Saikkonen (2002) introduced a univariate threshold autoregressive (TAR) model in which only the intercept is switching, whereas the autoregressive structure remains unchanged. Their model can be written as follows:

$$y_t = \sum_{j=1}^{r} \varphi_{0j} I(c_{j-1} < y_{t-d} \leq c_j) + \sum_{j=1}^{p} \varphi_j y_{t-j} + \varepsilon_t \quad (5)$$

where $\varepsilon_t \sim \text{iid}(0, \sigma^2)$. This specification is used for describing 'near unit root' behavior. That means modelling series which may 'look' nonstationary but are stationary and fluctuate within bounds. Interest rate or unemployment rate series serve as examples. The switching intercept in Eq. (5) causes level shifts in realizations, although the sequence $\{y_t\}$ itself is stationary and ergodic when the roots of the lag polynomial $1 - \sum_{j=1}^{p} \varphi_j z^j$ lie outside the unit circle.

One of the two examples in Lanne and Saikkonen (2002) consists of fitting the intercept-switching TAR model with one and two thresholds to the monthly UK Treasury bill series, 1964(1) to 1997(9). The results show that the persistence (the sum of the autoregressive coefficients) considerably decreases when one moves from a pure linear autoregressive model, for which the unit root hypothesis is not rejected, to the intercept-switching TAR model with two thresholds. When the TAR model is tested against a linear nonstationary autoregressive model containing a unit root, the null hypothesis is not rejected. For the test and the asymptotic theory behind it, see Lanne and Saikkonen (2002).

The shifting-mean autoregressive (SM-AR) model, see González and Teräsvirta (2008) or González, Hubrich, and Teräsvirta (2009), is a variant of Eq. (5). In the SM-AR model the switching intercept has been replaced by an explicitly nonstationary component. The model has the following form:

$$y_t = \delta(t) + \sum_{j=1}^{p} \varphi_j y_{t-j} + \varepsilon_t \quad (6)$$

where $\varepsilon_t \sim \text{iid}(0, \sigma^2)$. The deterministic time-varying intercept is defined as

$$\delta(t) = \delta_0 + \sum_{j=1}^{q} \delta_j G(\gamma_j, c_j, t/T)$$

where δ_j, $j=0,1,\ldots,q$, are parameters and $G(\gamma_j, c_j, t/T)$ logistic functions. The roots of $1-\sum_{j=1}^{p}\varphi_j z^j$ are assumed to lie outside the unit circle, so $\{y_t-\delta(t)\}$ is a stationary sequence. The unconditional (switching) mean is

$$\mathsf{E}_t y_t = (1-\sum_{j=1}^{p}\varphi_j \mathsf{L}^j)^{-1}\delta(t)$$

where E_t indicates that the expectation is time-varying, and L is the lag operator.

It is also possible to reparameterise (6) as follows:

$$y_t = \delta(t) + \sum_{j=1}^{p}\varphi_j\{y_{t-j}-\delta(t-j)\} + \varepsilon_t \qquad (7)$$

In that case, $\mathsf{E}_t y_t = \delta(t)$. This parameterisation is analogous to that in the Markov-switching model Hamilton (1989) introduced. In that model the deterministic $\delta(t)$ in Eq. (7) is replaced by a latent random variable $\mu(s_t)$, where $\mu(s_t)$ represents the local (switching) mean of y_t. The latent variable s_t is assumed to have a (first-order) Markov structure, which implies that the switching-mean obtains q different values.

VECTOR MODELS

General Nonlinear Dynamic Vector Model

A general nonlinear dynamic vector model can be written as follows:

$$\mathbf{y}_t = \mathbf{f}(\mathbf{y}_{t-1},\ldots,\mathbf{y}_{t-p};\mathbf{x}_t;\boldsymbol{\theta}) + \boldsymbol{\varepsilon}_t \qquad (8)$$

where $\mathbf{y}_t$ is a stochastic $m\times 1$ vector, $\mathbf{x}_t$ is an $n\times 1$ vector of exogenous variables, $\boldsymbol{\varepsilon}_t$ is an $m\times 1$ error vector with $\mathsf{E}\boldsymbol{\varepsilon}_t=\mathbf{0}$ and $\mathrm{cov}(\boldsymbol{\varepsilon}_t)=\boldsymbol{\Sigma}>0$, and $\boldsymbol{\theta}$ is a parameter vector. The $m\times 1$ vector $\mathbf{f}(\mathbf{y}_{t-1},\ldots,\mathbf{y}_{t-p};\mathbf{x}_t;\boldsymbol{\theta})$ has the form

$$\mathbf{f}(\mathbf{y}_{t-1},\ldots,\mathbf{y}_{t-p};\mathbf{x}_t;\boldsymbol{\theta}) = (f_1(\mathbf{y}_{t-1},\ldots,\mathbf{y}_{t-p};\mathbf{x}_t;\boldsymbol{\theta}),\ldots,f_m(\mathbf{y}_{t-1},\ldots,\mathbf{y}_{t-p};\mathbf{x}_t;\boldsymbol{\theta}))' \qquad (9)$$

The n-vector $\mathbf{x}_t$ may contain an intercept, other deterministic components and strongly exogenous variables. In some cases the whole vector, excluding the deterministic components, may be lagged. All variables in Eq. (8) are observable and all parameters non-random. The sequence $\{\mathbf{y}_t\}$ is often assumed weakly stationary, but sometimes it is only assumed that the sequence of first differences $\{\Delta\mathbf{y}_t\}$ is weakly stationary.

This survey concentrates on special cases of Eq. (8). A useful simplification is the one in which nonlinearity is additive. This means that the function $\mathbf{f}$ defined in Eq. (9) can be decomposed as follows:

$$\mathbf{f}(\mathbf{y}_{t-1},\ldots,\mathbf{y}_{t-p};\mathbf{x}_t;\boldsymbol{\theta}) = \boldsymbol{\mu} + \sum_{j=1}^{p} \boldsymbol{\Phi}_j \mathbf{y}_{t-j} + \boldsymbol{\Gamma}\mathbf{x}_t + \mathbf{f}_1(\mathbf{y}_{t-1},\ldots,\mathbf{y}_{t-p};\mathbf{x}_t;\boldsymbol{\theta}_1) \quad (10)$$

where $\boldsymbol{\mu}$ is an $m \times 1$ intercept vector (other deterministic components are assumed away for notational simplicity), $\boldsymbol{\Phi}_j$ is an $m \times m$ parameter matrix, $j = 1,\ldots,p$, and $\boldsymbol{\Gamma}$ is an $m \times n$ parameter matrix. The models to be considered below have this or a similar structure. The function $\mathbf{f}_1$ can sometimes be linear in parameters, if not in variables.

Vector Threshold Autoregressive Model

Definition

The VSR or VTAR model is a straightforward generalization of the corresponding single-equation model. The former term applies to incomplete models that contain strongly exogenous regressors, whereas the VTAR model generalizes the linear VAR model. The stationary VSR model is defined as follows:

$$\mathbf{y}_t = \sum_{i=1}^{q} (\boldsymbol{\mu}_i + \sum_{j=1}^{p} \boldsymbol{\Phi}_{ij} \mathbf{y}_{t-j} + \boldsymbol{\Gamma}_i \mathbf{x}_t + \boldsymbol{\varepsilon}_{it}) I(c_{i-1} < s_t \leq c_i) \quad (11)$$

where $\mathbf{y}_t$ and $\boldsymbol{\varepsilon}_{it}$, $j = 1,\ldots,q$, are stochastic $m \times 1$ vectors, and $\boldsymbol{\mu}_i$ is an $m \times 1$ vector of intercepts, $i = 1,\ldots,q$. Furthermore, $\boldsymbol{\Phi}_{ij}$ are $m \times m$ coefficient matrices, $i = 1,\ldots,q$, and $j = 1,\ldots,p$, $\boldsymbol{\Gamma}_i$ are $m \times n$ coefficient matrices, $i = 1,\ldots,q$, $c_0 = -\infty$, and $c_q = \infty$. When $\boldsymbol{\Gamma}_i = \mathbf{0}$ for all i, (11) becomes a VTAR model. The errors $\boldsymbol{\varepsilon}_{it}$ are serially uncorrelated with mean $\mathbf{0}$ and positive definite covariance matrix $\boldsymbol{\Sigma}_i$, $i = 1,\ldots,q$. The vector $\mathbf{x}_t$ contains n stationary variables that are strongly exogenous for the parameters of the model. If the stationary variable s_t is replaced by t/T, where T is the number of observations, (11) becomes a linear VR model with $q-1$ breaks.

In Eq. (11), a single stationary and continuous switch-variable s_t determines the regime of the whole system. This assumption has been relaxed by Tena and Tremayne (2009) such that each equation may have a separate switch-variable and some equations may even be linear. The authors argue that there are good economic reasons for that, but in economic applications other than theirs, see Section "Applications of Vector TAR Models," this has not been the case. A systematic modelling strategy for VSR or VTAR models with applications can be found in Tsay (1998) and is discussed in Section "Building Stationary Vector Switching Regression Models."

Finally, it may be noted that substituting a latent indicator for the observable s_t in Eq. (11) yields the following vector regression model:

$$\mathbf{y}_t = \sum_{i=1}^{q} (\boldsymbol{\mu}_i + \sum_{j=1}^{p} \boldsymbol{\Phi}_{ij}\mathbf{y}_{t-j} + \boldsymbol{\Gamma}_i\mathbf{x}_t + \boldsymbol{\varepsilon}_{it}) I(s_t = \nu_i) \tag{12}$$

in which the indicator variable s_t is defined similarly to the single-equation model of Lindgren (1978). Typically, the switches in all equations are controlled by the same indicator function, as is the case in Eq. (12). This assumption can be relaxed, however, see Sims, Waggoner, and Zha (2008). Recent economic applications include Sims and Zha (2006) and Hubrich and Tetlow (2012). Their approach is Bayesian; for more details and discussion in the classical framework, see Krolzig (1997) and Jacobson, Lindh, and Warne (2002).

Nested Vector Threshold Autoregressive Model

The VTAR model (11) may be generalized in various ways. It is possible to add thresholds and threshold variables such that the switches are controlled by more than one variable. One of these models is a multivariate generalization of the univariate Nested TAR (NTAR) model by Astatkie, Watts, and Watt (1997). It is a four-regime model with the following form:

$$\begin{aligned}\mathbf{y}_t &= \{(\boldsymbol{\mu}_{11} + \sum_{j=1}^{p} \boldsymbol{\Phi}_{11j}\mathbf{y}_{t-j}) I(c_1 < s_{1t}) + (\boldsymbol{\mu}_{12} + \sum_{j=1}^{p} \boldsymbol{\Phi}_{12j}\mathbf{y}_{t-j}) I(c_1 \geq s_{1t})\} \\ &\quad \times I(c_2 < s_{2t}) + \{(\boldsymbol{\mu}_{21} + \sum_{j=1}^{p} \boldsymbol{\Phi}_{21j}\mathbf{y}_{t-j}) I(c_1 < s_{1t}) \\ &\quad + (\boldsymbol{\mu}_{22} + \sum_{j=1}^{p} \boldsymbol{\Phi}_{22j}\mathbf{y}_{t-j}) I(c_1 \geq s_{1t})\} I(c_2 \geq s_{2t}) + \boldsymbol{\varepsilon}_t\end{aligned} \tag{13}$$

where the error process $\{\boldsymbol{\varepsilon}_t\} \sim \text{iid}(0, \boldsymbol{\Sigma})$ and thus does not switch. In this model, the first threshold variable s_{1t} controls switching as in Eq. (11). But then, the two states between which the process switches according to s_{1t} can change due to the other threshold variable s_{2t}. The standard VTAR model is thus nested in the more general Vector NTAR (VNTAR) model.

There may not be any applications of the VNTAR model (13) with stochastic stationary threshold variables s_{1t} and s_{2t} to economic time series. A special case, however, in which $s_{2t} = t$, has been considered by Galvão (2006). In this model, the two threshold regimes determined by s_{1t} change once during the observation period: the second threshold indicates a structural break in the VTAR process. The author calls this model the Structural Break Threshold Vector Autoregressive (SBTVAR) model. She discusses specification and estimation issues and applies the model to predicting recessions, the 2001 recession in the United States in particular.

Threshold Vector Error Correction

Following the way linear cointegrated VAR models are parameterised, there is an extension of the VTAR model applicable to nonstationary time series. Balke and Fomby (1997) introduced this extension to allow discontinuous changes in the attraction towards the equilibrium characterised by the cointegrating relationships. To define the model, assume for simplicity that $\boldsymbol{\Gamma}_i = \mathbf{0}$ for all i and that $q = 3$. The model (11) is then rewritten as follows:

$$\Delta \mathbf{y}_t = \sum_{j=1}^{3} (\boldsymbol{\mu}_j + \boldsymbol{\Pi}_j \mathbf{y}_{t-j} + \sum_{k=1}^{p-1} \boldsymbol{\Psi}_{jk} \Delta \mathbf{y}_{t-k} + \boldsymbol{\varepsilon}_{jt}) I(c_{j-1} < s_t \leq c_j) \qquad (14)$$

for $p \geq 2$. When $p = 1$, the weighted sum of lagged first differences vanishes. Assume that in each regime, $\mathbf{y}_t$ is integrated of order one ($\Delta \mathbf{y}_t$ stationary in the mean) and the $m \times m$ matrix $\boldsymbol{\Pi}_j = \mathbf{A}_j \mathbf{B}'$, where $\mathbf{A}_j$, $j = 1, \ldots, q$, and $\mathbf{B}$ are $m \times r$ matrices with full column rank. Furthermore, s_t is continuous and stationary as before. If $m = 2$, say, then $\mathbf{A}_j = (\alpha_{1j}, \alpha_{2j})'$ and $\mathbf{B} = (1, \beta_2)'$, and the variables y_{1t} and y_{2t} are cointegrated with the cointegrating vector $\mathbf{B}$. This model is called the threshold vector error correction (TVEC) model. The difference between the linear cointegrated VAR model and (14) with $q = 3$ is that strength of the attraction varies in the three regimes according to $\mathbf{A}_j$. The cointegrating relationships defined by the matrix $\mathbf{B}$ remain linear.

A special case with economic applications has $q = 3$ and $\mathbf{A}_2 = \mathbf{0}$. This allows modelling relationships in which error correction operates outside an interval, but is incomplete in the sense that there is a band within which no adjustment takes place. The intercept is restricted in the cointegrating relationship, so that

$$\Delta \mathbf{y}_t = \sum_{j=1}^{3} \{\mathbf{A}_j(\mathbf{B}'\mathbf{y}_{t-1} - \boldsymbol{\mu}_j) + \sum_{k=1}^{p-1} \boldsymbol{\Psi}_{jk}\Delta \mathbf{y}_{t-k} + \boldsymbol{\varepsilon}_{jt}\} I(c_{j-1} < s_t \leq c_j) \tag{15}$$

where $\boldsymbol{\mu}_j$ is the regime-specific mean vector of $\mathbf{B}'\mathbf{y}_{t-1}$. Model (15) is called the band-TVEC model; see Lo and Zivot (2001) for discussion. The corresponding univariate model is called the band-TAR model.

In applications of Eq. (15), the interval $(c_1, c_2]$ often contains zero. For example, when considering trade with identical commodities, transportation costs help sustain a difference in the price of this commodity in two regions. Price convergence can only occur when the absolute difference between the prices is sufficiently large, so profitable export or import is possible. In this example, the transition variable s_t in Eq. (15) is the difference between the prices.

Stationarity in Threshold Vector Error Correction Models

Stationarity and ergodicity of the equilibrium component $\mathbf{B}'\mathbf{y}_{t-1}$ and the differences $\Delta \mathbf{y}_t$ defined by the TVEC process (14) are considered in Bec and Rahbek (2004). Let $\mathbf{z}_t = (\mathbf{y}_t'\mathbf{B}, \Delta \mathbf{y}_{t-1}', \ldots, \Delta \mathbf{y}_{t-p+2}')'$. The focus is on the case where the threshold variable $s_t = \|\mathbf{z}_{t-1}\|$ and the indicator has the form $I(\|\mathbf{z}_{t-1}\| \geq c)$ in Eq. (14). This implies that the resulting three-regime model is 'symmetric': the mid-regime is different from the identical outer regimes, $\mathbf{A}_1 = \mathbf{A}_3 \neq \mathbf{A}_2$, and the two thresholds separating them are $c_1 = -c$ and $c_2 = c > 0$. Under the conditions given by Bec and Rahbek (2004), see their Theorem 2, $\mathbf{B}'\mathbf{y}_t$ and $\Delta \mathbf{y}_t$ are geometrically ergodic, and assuming the process starts at $t = 0$, they can be given initial distributions such that they are stationary. Stationarity is seen to depend only on the parameters of the outer regimes. For stationarity, the characteristic polynomial matrix for those regimes or regime

$$C_j(z) = \mathbf{I}_m(1-z) - \mathbf{A}_j\mathbf{B}'z - \sum_{k=1}^{p-1} \boldsymbol{\Psi}_{jk}(1-z)z^k, \quad j = 1 \text{ or } 3$$

must have $m - r$ roots equal to unity, while the remaining roots $|z| > 1$ (Johansen, 1995, Theorem 4.2). Bec and Rahbek (2004) also look at the case in which $s_t = \|\mathbf{B}'\mathbf{y}_{t-1}\|$ but then they assume that $\mathbf{\Psi}_{jk} = \mathbf{\Psi}_k$, $k = 1, \ldots, p$, that is, the short-run coefficients do not switch.

Saikkonen (2005) studies a rather general class of nonlinear error correction models with the error correction component $\mathbf{B}'\mathbf{y}_t$ as before. Conditional heteroskedasticity in the errors is also allowed. A three-regime VTAR model with $\mathbf{A}_1 = \mathbf{A}_3 \neq \mathbf{A}_2$ is a member of this class. The threshold variable is a function of $\mathbf{z}_{t-1}$ defined as before. Under the conditions given in the article and using a transformation of the general model, Saikkonen (2005) is able to show that given a proper choice of initial distributions, $\mathbf{B}'\mathbf{y}_t$ and $\Delta\mathbf{y}_t$ are both strictly and weakly stationary.

It should be mentioned, however, that the results in Bec and Rahbek (2004) are not restricted to mere VTAR models either but apply to a more general class containing, among other things, the case in which the threshold is not completely observed but contains an unobserved random element. The results in Saikkonen (2005) cover this extension as well. See also Section "Stationarity in Smooth Transition Vector Error Correction Models," where extensions to smooth transition VEC models are discussed.

Vector Smooth Transition Autoregressive Model

Definition

When the stationary smooth transition regression model is generalized to the vector case, one obtains a logistic vector STR (LVSTR) model of order p. This model first appeared in Anderson and Vahid (1998) who were interested in testing common nonlinearities in vector models. Their approach to this testing problem is discussed in Section "Testing Common Nonlinearity in VSTR Models." For notational simplicity, a single-transition (two extreme regimes) version of the LVSTR model is considered here. For more general representations, see Teräsvirta and Yang (2013). The model may be defined as follows:

$$\begin{aligned}\mathbf{y}_t &= \boldsymbol{\mu}_0 + \sum_{j=1}^{p}\mathbf{\Phi}_j\mathbf{y}_{t-j} + \mathbf{\Gamma}\mathbf{x}_t + \mathbf{G}(\boldsymbol{\gamma}, \mathbf{c}; \mathbf{s}_t) \\ &\quad \times \{\boldsymbol{\mu}_1 + \sum_{j=1}^{p}(\mathbf{\Psi}_j\mathbf{y}_{t-j} + \mathbf{\Xi}\mathbf{x}_t)\} + \boldsymbol{\varepsilon}_t \\ &= \boldsymbol{\mu}_0 + \mathbf{G}(\boldsymbol{\gamma}, \mathbf{c}; \mathbf{s}_t)\boldsymbol{\mu}_1 + \sum_{j=1}^{p}\{\mathbf{\Phi}_j + \mathbf{G}(\boldsymbol{\gamma}, \mathbf{c}; \mathbf{s}_t)\mathbf{\Psi}_j\}\mathbf{y}_{t-j} \\ &\quad + (\mathbf{\Gamma} + \mathbf{G}(\boldsymbol{\gamma}, \mathbf{c}; \mathbf{s}_t)\mathbf{\Xi})\mathbf{x}_t + \boldsymbol{\varepsilon}_t\end{aligned} \tag{16}$$

where $\mathbf{y}_t$ is an $m \times 1$ vector, $\boldsymbol{\mu}_0$ and $\boldsymbol{\mu}_1$ are $m \times 1$ intercept vectors, $\boldsymbol{\Phi}_j$ and $\boldsymbol{\Psi}_j$, $j = 1, \ldots, p$, are $m \times m$ parameter matrices, $\boldsymbol{\Gamma}$ and $\boldsymbol{\Xi}$ are $m \times n$ parameter matrices, and

$$\mathbf{G}(\boldsymbol{\gamma}, \mathbf{c}; \mathbf{s}_t) = \mathrm{diag}\{G_1(\gamma_1, \mathbf{c}_1, s_{1t}), \ldots, G_m(\gamma_m, \mathbf{c}_m, s_{mt})\} \tag{17}$$

is an $m \times m$ diagonal matrix of bounded logistic transition functions. It is usually assumed that $\{\mathbf{y}_t\}$ defined by (16) and (17) is a stationary and ergodic sequence.

Eq. (16) defines quite a general LVSTR model. One way of simplifying it is to assume that the transition variables as well as the parameters of the transition functions are the same for all equations. As already noted, a similar assumption is made in VTAR models. In that case, $\mathbf{G}(\boldsymbol{\gamma}, \mathbf{c}; \mathbf{s}_t) = G(\gamma, \mathbf{c}; s_t)\mathbf{I}_m$. The LVSTAR model suggested by Weise (1999) contains this assumption. When $\gamma \to \infty$, this model becomes a two-regime VTAR model. Setting $s_{jt} = t, j = 1, \ldots, m$, leads to an LVSTAR model that defines the alternative model in testing parameter constancy of linear VAR models by LM-type tests; see He, Teräsvirta, and González (2009).

Another simplification is to assume that the LVSTAR model is intercept-switching, that is, $\boldsymbol{\Psi}_j = \mathbf{0}$, $j = 1, \ldots, p$, and $\boldsymbol{\Xi} = \mathbf{0}$ in Eq. (16). The intercept-switching model is already fairly flexible specification. Besides, as the intercept $\boldsymbol{\mu}_0 + \mathbf{G}(\boldsymbol{\gamma}, \mathbf{c}; \mathbf{s}_t)\boldsymbol{\mu}_1$ is bounded for all values of $\mathbf{s}_t$, the model has the same stability condition as the linear VAR model. The expectation

$$\mathsf{E}\mathbf{y}_t = \left(\mathbf{I}_m - \sum_{j=1}^{p} \boldsymbol{\Phi}_j \mathsf{L}^j\right)^{-1} \{\boldsymbol{\mu}_0 + \mathsf{E}\mathbf{G}(\boldsymbol{\gamma}, \mathbf{c}; \mathbf{s}_t)\boldsymbol{\mu}_1\}$$

can only be computed numerically. An empirical example of the intercept-switching LVSTAR model can be found in Section "Applications of Vector STAR and STR Models." It would also be possible to generalize the intercept-switching AR model of Lanne and Saikkonen (2002) to the vector case, but it seems that this has not been done in the literature on vector TAR models.

It would be possible to define smooth transition counterparts to the VNTAR model (13) and its special case, the SBTVAR model. The latter in fact exists as a univariate model and is called the Time-Varying STAR (TV-STAR) model, see Lundbergh, Teräsvirta, and van Dijk (2003).

Smooth Transition Vector Error Correction
A Smooth Transition Vector Error Correction (STVEC) model can be defined along the same lines as the TVEC model (14). As an example, a single-transition STVEC model (without exogenous variables) can be written as follows:

$$\Delta \mathbf{y}_t = \boldsymbol{\mu}_1 + \mathbf{A}_1\mathbf{B}'\mathbf{y}_{t-1} + \sum_{k=1}^{p-1} \boldsymbol{\Phi}_k \Delta \mathbf{y}_{t-k} + \mathbf{G}(\boldsymbol{\gamma}, \mathbf{c}; \mathbf{s}_t)(\boldsymbol{\mu}_2 + \mathbf{A}_2\mathbf{B}'\mathbf{y}_{t-1} + \sum_{k=1}^{p-1} \boldsymbol{\Psi}_k \Delta \mathbf{y}_{t-k}) + \boldsymbol{\varepsilon}_t \tag{18}$$

where $\Delta \mathbf{y}_t$ is I(0), $\mathbf{A}_j\mathbf{B}'$, $j=1,2$, are reduced rank $m \times m$ matrices as in Eq. (14) and $\mathbf{G}(\boldsymbol{\gamma}, \mathbf{c}; \mathbf{s}_t)$ is a diagonal matrix defined in Eq. (17). Rothman, van Dijk, and Franses (2001) defined the STVEC model assuming $\mathbf{G}(\boldsymbol{\gamma}, \mathbf{c}; \mathbf{s}_t) = G(\gamma, \mathbf{c}; s_t)\mathbf{I}_m$. Camacho (2004) constructed a two-dimensional ($m=2$) STVEC model without this assumption. His application is discussed in Section "Applications of Vector TAR Models." Goodwin, Holt and Prestemon (2012) also specified a two-dimensional STVEC model without assuming identical transition functions and applied it to modelling the exchange rate pass-through in timber products crossing the border between Canada and the United States. Their model could be viewed as being of type (16) in the sense that it contains strongly exogenous variables. The authors, however, closed the system by constructing separate linear equations for these variables.

Stationarity in Smooth Transition Vector Error Correction Models
Conditions for stationarity and ergodicity of nonstationary STVEC models can be found as special cases of the general nonstationary model introduced in Saikkonen (2008). The general model in that article is defined by

$$\begin{aligned} \Delta \mathbf{y}_t = & \sum_{j=1}^{q} h_j(\mathbf{S}'\mathbf{z}_{t-1}, \eta_t)(\boldsymbol{\mu}_j^* + \mathbf{A}_j^*\mathbf{B}'\mathbf{y}_{t-1} + \sum_{k=1}^{p-1} \boldsymbol{\Phi}_{jk}^* \Delta \mathbf{y}_{t-j}) \\ & + (\sum_{j=1}^{q} h_j(\mathbf{S}'\mathbf{z}_{t-1}, \eta_t)\boldsymbol{\Sigma}_j)^{1/2}\boldsymbol{\varepsilon}_t \end{aligned} \tag{19}$$

where $\mathbf{B}$ and $\mathbf{A}_j^*$, $j=1,\ldots,q$, are $m \times r$ matrices with full column rank, $\boldsymbol{\varepsilon}_t \sim$ iid$(\mathbf{0}, \boldsymbol{\Sigma})$, and $\mathbf{S}$ is typically a selection matrix or vector, selecting the threshold or transition variable or a linear combination of variables from the elements of $\mathbf{z}_{t-1}$ defined as before. For example, setting $\mathbf{S}' = [\mathbf{I}_r\ \mathbf{0}]$ gives $\mathbf{S}'\mathbf{z}_{t-1} = \mathbf{B}'\mathbf{y}_{t-1}$, and one can for example choose $h_j(\mathbf{S}'\mathbf{z}_{t-1}, \eta_t) = I(\|\mathbf{B}'\mathbf{y}_{t-1}\| > c)$ as in Bec and Rahbek (2004). Furthermore, η_t is a scalar random variable making the threshold or the location parameter in the logistic transition function unobservable: here we set $\eta_t = 0$. It is assumed that $\sum_{j=1}^{q} h_j(\mathbf{S}'\mathbf{z}_{t-1}, \eta_t) = 1$. Since we are interested in STVEC models, we assume that $h_j(\mathbf{s}'\mathbf{z}_{t-1}), j = 1, \ldots, q$, where $\mathbf{s}$ is a selection vector, are functions of logistic functions. For example, for $q = 2$ we have

$$h_1(\mathbf{s}'\mathbf{z}_{t-1}) = G(\gamma, c, \mathbf{s}'\mathbf{z}_{t-1}) = (1 + \exp\{-\gamma(\mathbf{s}'\mathbf{z}_{t-1} - c)\})^{-1}, \quad \gamma > 0$$

so $h_2(\mathbf{s}'\mathbf{z}_{t-1}) = 1 - G(\gamma, c, \mathbf{s}'\mathbf{z}_{t-1})$. This yields a reparameterised version of Eq. (18) with a common transition function for all equations. Selecting $q = 3$, one obtains

$$h_j(\mathbf{s}'\mathbf{z}_{t-1}) = \begin{cases} 1 - G(\gamma, c_1; \mathbf{s}'\mathbf{z}_{t-1}) & \text{for } j = 1 \\ G(\gamma, c_1; \mathbf{s}'\mathbf{z}_{t-1}) - G(\gamma, c_2; \mathbf{s}'\mathbf{z}_{t-1}) & \text{for } j = 2 \\ G(\gamma, c_2; \mathbf{s}'\mathbf{z}_{t-1}) & \text{for } j = 3 \end{cases} \tag{20}$$

where $c_1 < c_2$. The positive slope parameter γ is the same for both logistic functions. The argument can also be a function of $\mathbf{s}'\mathbf{z}_{t-1}$ such as $\|\mathbf{s}'\mathbf{z}_{t-1}\|$. When $\gamma \to \infty$, assuming $\mathbf{A}_1 = \mathbf{A}_3$ and that the transition variable equals $\|\mathbf{B}'\mathbf{y}_{t-1}\|$ the stationarity results for $\mathbf{B}'\mathbf{y}_t$ and $\Delta\mathbf{y}_t$ in models with conditional heteroskedasticity are obtained from Saikkonen (2005). When $\mathbf{A}_1 \neq \mathbf{A}_3$, this is no longer the case, and the corresponding results are worked out by Saikkonen (2008).

In proving stationarity and ergodicity of the long-run equilibrium relations and the differences is that the relevant matrix polynomials

$$C_j(z) = \mathbf{I}_m(1 - z) - \mathbf{A}_j\mathbf{B}'z - \sum_{k=1}^{p-1} \mathbf{\Phi}_{jk}^*(1 - z)z^k \tag{21}$$

have $m - r$ unit roots, whereas $|z| > 1$ for the rest. In our example with the logistic transition function and $q = 2$ this concerns both $C_1(z)$ and $C_2(z)$. When $h_j(\mathbf{s}'\mathbf{z}_{t-1}), j = 1, 2, 3$, are defined by Eq. (20), the relevant polynomials are $C_1(z)$ and $C_3(z)$. The roots of $C_2(z)$ are not restricted.

The proof is based on a model transformation similar to the one in Saikkonen (2005) but requires additional conditions and is technically more complex than its predecessor. For details, see Saikkonen (2008). These results apply to the case in which all equations of the STVEC model have the same transition function. As yet, no results seem to be available for the more general situation characterised by (16) and (17).

Vector Shifting Mean Autoregressive Model

Holt and Teräsvirta (2012) introduced a multivariate version of the shifting-mean autoregressive model. It is a straightforward generalization of the univariate model in Section "Switching Mean Autoregressive Model." Accordingly, it can be defined in two ways, either

$$\mathbf{y}_t = \boldsymbol{\delta}(t) + \sum_{j=1}^{p} \boldsymbol{\Phi}_j \mathbf{y}_{t-j} + \boldsymbol{\varepsilon}_t \tag{22}$$

where $\boldsymbol{\delta}(t) = (\delta_1(t), \ldots, \delta_m(t))'$ with $\delta_j(t) = \delta_{j0} + \sum_{i=1}^{q_j} \delta_{ji} G(\gamma_{ji}, c_{ji}, t/T), j = 1, \ldots, m$, $\boldsymbol{\varepsilon}_t : \text{iid}(0, \boldsymbol{\Sigma})$, $\boldsymbol{\Sigma} > 0$, or

$$\mathbf{y}_t = \boldsymbol{\delta}(t) + \sum_{j=1}^{p} \boldsymbol{\Phi}_j \{\mathbf{y}_{t-j} - \boldsymbol{\delta}(t-j)\} + \boldsymbol{\varepsilon}_t \tag{23}$$

In both cases, $|\mathbf{I} - \sum_{j=1}^{p} \boldsymbol{\Phi}_j z^j| \neq 0$ for $z \leq 1$, that is, $\mathbf{y}_t - \boldsymbol{\delta}(t)$ is a weakly stationary sequence. The transition functions are logistic:

$$G(\gamma_{ji}, c_{ji}, t/T) = (1 + \exp\{-\gamma_{ji}(t/T - c_{ji})\})^{-1}, \quad \gamma_{ji} > 0$$

for all i and j. In Eq. (22), the shifting-mean equals

$$\mathsf{E}_t \mathbf{y}_t = (\mathbf{I} - \sum_{j=1}^{p} \boldsymbol{\Phi}_j \mathsf{L}^j)^{-1} \boldsymbol{\delta}(t) \tag{24}$$

whereas for (23), $\mathrm{E}_t \mathbf{y}_t = \boldsymbol{\delta}(t)$. As in single-equation SM-AR models, the intercept can be a very flexible function of time, and in some situations the VSM-AR model can be viewed as a reasonable alternative to models with unit roots and stochastic trends. Since $\boldsymbol{\delta}(t)$ is bounded, the expectation (24), even if time-varying, is also bounded.

Vector Contemporaneous-Threshold Model

Dueker, Psaradakis, Sola, and Spagnolo (2011) introduced the multivariate (or vector) contemporaneous-threshold (C-MSTAR) model which they write belongs to the class of vector STAR models. It is actually an m-dimensional mixture model consisting of q^2 basic linear equations that are weighted together by time-varying probabilities. The first-order (for simplicity) model can be written as follows:

$$\mathbf{y}_t = \sum_{i=1}^{q^2} G_i(\mathbf{z}_{t-1})(\boldsymbol{\mu}_i + \mathbf{A}_{i1}\mathbf{y}_{t-1} + \boldsymbol{\Sigma}_i^{1/2}\boldsymbol{\varepsilon}_t)$$

where $\boldsymbol{\varepsilon}_t \sim \text{iid}\mathcal{N}(\mathbf{0}, \mathbf{I}_m)$, $\mathbf{z}_{t-1}$ is a linear function of $\mathbf{y}_{t-1}$,

$$G_i(\mathbf{z}_{t-1}) = \frac{\Phi(\boldsymbol{\Sigma}_i^{-1/2}\{\mathbf{c}_i - \boldsymbol{\mu}_i - \mathbf{A}_{i1}\mathbf{y}_{t-1}\})}{\sum_{j=1}^{q^2} \Phi(\boldsymbol{\Sigma}_i^{-1/2}\{\mathbf{c}_i - \boldsymbol{\mu}_i - \mathbf{A}_{i1}\mathbf{y}_{t-1}\})}$$

$i = 1, \ldots, q^2$, is bounded between zero and one, $\Phi(\cdot)$ is the cumulative distribution function of the standard normal distribution, and $\mathbf{c}_i = (c_{i1}, c_{i2})'$, $i = 1, \ldots, q^2$, are the location vectors. For example, if $q = 2$, then $\mathbf{c}_1 = (c_1, c_2)'$, $\mathbf{c}_2 = (c_1, -c_2)'$, $\mathbf{c}_3 = (-c_1, c_2)'$, and $\mathbf{c}_4 = (-c_1, -c_2)'$. The authors discuss linearity testing ($q = 1$), stability, and maximum likelihood estimation of the parameters, and apply their model to consider the relationship between stock prices and interest rates using a long annual time series from 1900 to 2000. Since the number of parameters increases rapidly as a function of q and the lag length, the applications of the C-MSTAR model are likely to be low-dimensional.

ESTIMATION

Estimation of Vector Switching Regression Models

Parameters of the VSR model (11) can be estimated using least squares. This includes estimating the thresholds c_i, $i = 1, \ldots, r-1$. Consider the case in which $q = 2$, so there is one threshold, c_1. First set the threshold at $c^{(i)}$ where $c^{(i)}$ is the ith value of the order statistic of the transition variable s_t.

Then form two sets of observations: $\mathcal{T}_1$ containing the n_1 observations for which $s_t \leq c^{(i)}$ and $\mathcal{T}_2$ comprising the remaining $T - n_1$ ones, and the sum of squares function

$$Q(c^{(i)}) = \text{tr} \sum_{t \in \mathcal{T}_1} \boldsymbol{\varepsilon}_t'(c^{(i)})\boldsymbol{\varepsilon}_t(c^{(i)}) + \text{tr} \sum_{t \in \mathcal{T}_2} \boldsymbol{\varepsilon}_t'(c^{(i)})\boldsymbol{\varepsilon}_t(c^{(i)})$$

where $\boldsymbol{\varepsilon}_t(c^{(i)}) = \mathbf{y}_t - (\boldsymbol{\mu}_i + \sum_{j=1}^{p} \boldsymbol{\Phi}_{ij}\mathbf{y}_{t-j} + \boldsymbol{\Gamma}_i\mathbf{x}_t)$ for $t \in \mathcal{T}_i$, $i = 1, 2$. The parameters $\boldsymbol{\mu}_i, \boldsymbol{\Phi}_{ij}$, $j = 1, \ldots, p$, $\boldsymbol{\Gamma}_i$, $i = 1, 2$, are estimated by minimizing $Q(c^{(i)})$ with respect to these parameters. The final estimates, including that of c_1, are obtained by repeating this for all $c^{(i)} \in \mathrm{C} = \{c^{(i)} : c(q) \leq c^{(i)} \leq c(1-q)\}$ where $c(q)$ is the qth quantile of the order statistic of s_t and selecting the parameter estimates including c_1 that minimize $Q(c_1)$

$$Q(\hat{c}_1) = \arg \min_{c^{(i)} \in \mathcal{C}} Q(c^{(i)})$$

If T is very large, $\mathcal{C}$ can be replaced by a grid to save computations. Tsay (1998) showed that under regularity conditions including the existence of fourth moments for all variables and the error process, the least squares estimators, including that of c_1, are consistent. As already mentioned, in the univariate TAR case Chan (1993) and Qian (1998) showed that $\hat{c}_1$ is super consistent and that the estimators other than $\hat{c}_1$ are asymptotically normal. These results hold when the conditional mean $\mathsf{E}\{\mathbf{y}_t|\mathcal{F}_{t-1}\}$ is discontinuous at the threshold, see Tsay (1998) for discussion. Chan and Qian also showed that if the transition variable is selected from a set $\mathcal{S} = \{s_t^{(1)}, \ldots, s_t^{(D)}\}$ of stationary switch-variables, then minimizing $Q(c_1, \mathcal{S})$ even leads to consistent selection of s_t, provided $\mathcal{S}$ contains the true switch-variable. The covariance matrix is then estimated as

$$\hat{\boldsymbol{\Sigma}} = n_1^{-1} \sum_{t \in \mathcal{T}_1} \boldsymbol{\varepsilon}_t(\hat{c}_1)\boldsymbol{\varepsilon}_t(\hat{c}_1)' + (T - n_1)^{-1} \sum_{t \in \mathcal{T}_2} \boldsymbol{\varepsilon}_t(\hat{c}_1)\boldsymbol{\varepsilon}_t(\hat{c}_1)'$$

Seo (2011b) considers estimation of nonlinear error correction models. One of his objects of interest is the TVEC model with one cointegrating relationship such that the cointegrating vector is the transition variable denoted by s_t in Eq. (14). This was the case in the original TVEC model by Balke and Fomby (1997). The model contains a single threshold, but as Seo (2011b) points out, replacing s_t by $|s_t|$, a symmetric three-regime model (the band-TVEC model (15) with $c_1 = -c_2$) is also included in his considerations. The interesting result is that under assumptions given in the article, the least squares estimator of the cointegrating vector is $T^{3/2}$-consistent (the convergence is extremely rapid), and the estimator of the threshold parameter (there is one threshold if the transition variable is an absolute-valued) is super consistent. Seo (2011b) suggests that estimation of these parameters be carried out using a grid search (see also Hansen & Seo, 2002).

VTAR models may also be estimated using heuristic (derivative-free) methods such as simulated annealing and genetic algorithms. These methods can be useful in estimating TVEC models, in which the threshold and cointegration parameters have to be estimated jointly for efficiency. This is true in particular if the number of cointegrating vectors exceeds one. Hansen and Seo (2002) have solved this estimation problem by a grid search, because their cointegrating vector only contained one parameter and the model had just a single threshold. In higher-dimensional models, the grid search could be very slow. Yang, Tian, and Yuan (2007) suggested a hybrid genetic-simulated annealing algorithm to this problem. El-Shagi (2011) compared various genetic algorithms applied to estimating TVEC models. For a good overview of heuristic methods, including genetic algorithms and simulated annealing, see Winker & Gilli (2004).

Estimation of Vector Smooth Transition Regression Models

The single-transition VLSTAR model (16) can be estimated by nonlinear least squares. If the transition functions are not assumed to be the same, estimation can proceed equation by equation as in single-equation STR models (see, e.g., Teräsvirta et al., 2010, Section 16.3). In the present section the focus is on the case where all equations have the same transition function. The model contains the parameters $\theta = \{\mathbf{\Phi}_1, \ldots, \mathbf{\Phi}_p; \mathbf{\Psi}_1, \ldots, \mathbf{\Psi}_p; \mathbf{\Gamma}, \mathbf{\Xi}; \gamma, c\}$. Set

$$\mathbf{B} = (\boldsymbol{\mu}_0, \mathbf{\Phi}_1, \ldots, \mathbf{\Phi}_p, \boldsymbol{\mu}_1, \mathbf{\Psi}_1, \ldots, \mathbf{\Psi}_p, \mathbf{\Gamma}, \mathbf{\Xi})$$

and

$$\begin{aligned}\mathbf{z}_t(\gamma, c) = (&\mathbf{1}', \mathbf{y}'_{t-1}, \ldots, \mathbf{y}'_{t-p}, G(\gamma, c; s_t)\mathbf{1}', G(\gamma, c; s_t)\mathbf{y}'_{t-1}, \ldots, \\ &G(\gamma, c; s_t)\mathbf{y}'_{t-p}, \mathbf{x}'_t, G(\gamma, c; s_t)\mathbf{x}'_t)'\end{aligned}$$

The NLS estimator of θ is obtained by solving the optimization problem

$$\hat{\theta}_{NLS} = \arg\min_{\theta} \sum_{t=1}^{T} \left(\mathbf{y}_t - \mathbf{B}\mathbf{z}_t(\gamma, c)\right)' \mathbf{\Sigma}^{-1} \left(\mathbf{y}_t - \mathbf{B}\mathbf{z}_t(\gamma, c)\right) \tag{25}$$

assuming $\mathbf{\Sigma} = \mathbf{I}$. When γ and c are known the solution for $\mathbf{B}$ is analytic. This offers a key to simplifying the nonlinear optimization problem. In general, finding the optimum may be numerically demanding in the sense that convergence may be slow and the algorithm may converge to some local minimum. This is due to the shape of the sum of squares function in Eq. (25): the function can be rather 'flat' in some directions and possess many local optima.

For this reason, finding a suitable starting-value of θ for the algorithm is crucial. The search to find one can be carried out by means of a grid search similar to the one in the previous section. The basic idea is to construct a discrete grid in the parameter space of γ and c and estimate the parameters in $\mathbf{B}$ conditionally on each pair (γ, c) in the grid. Choosing the pair (γ, c) which produces the smallest residuals sum of squares yields a starting-value for the nonlinear optimization. More specifically, the corresponding average residual sum of squares function is concentrated with respect to $\mathbf{B}$. Therefore the optimization is only performed with respect to γ and c.

The same procedure is employed in solving the nonlinear optimization problem (25). After estimating $\mathbf{B}$ conditionally on γ and c, one retains the estimated $\mathbf{B}$ and solves the two-dimensional nonlinear estimation problem to obtain new values for γ and c. The matrix $\mathbf{B}$ is then re-estimated conditionally on these new values of γ and c, and this is continued to convergence. Dividing each iteration into these two components reduces the dimension of the nonlinear estimation problem and thus saves computation time. If the grid is dense, the initial step-length of any derivative-based optimization algorithm must be sufficiently short so that optimization with a high probability leads to the local maximum closest to the value found using the grid approach and not to a lower local maximum further away.

The covariance matrix $\mathbf{\Sigma}$ can be estimated from the residual matrices

$$\hat{\mathbf{E}}'\hat{\mathbf{E}} = \sum_{t=p+1}^{T} (\mathbf{y}_t - \hat{\mathbf{B}}\mathbf{z}_t(\hat{\gamma}, \hat{c}))(\mathbf{y}_t - \hat{\mathbf{B}}\mathbf{z}_t(\hat{\gamma}, \hat{c}))'$$

as $\hat{\mathbf{\Sigma}} = (T-p)^{-1}\hat{\mathbf{E}}'\hat{\mathbf{E}}$. If a more efficient estimator is desired, one has to iterate the solution (25) by plugging in $\hat{\mathbf{\Sigma}}^{-1}$ and applying generalized least squares to solve the optimization problem. In practice, the estimates of $\mathbf{B}$, γ and c are not likely to change much by these iterations, however.

It is also possible to apply heuristic methods such as simulated annealing or genetic algorithms mentioned in Section "Estimation of Vector Switching Regression Models" to solving the optimization problem (25). These methods may also be used for obtaining starting-values for an appropriate derivative-based algorithm. There is not yet much empirical evidence available of how well they perform in LVSTAR models. Auerbach and Gorodnichenko (2012) estimated a three-dimensional LVSTAR model with a single-transition function with $c=0$ by the Monte Carlo Markov Chain method. A complication is that the error covariance matrix is assumed time-varying and controlled by the same transition function as the conditional mean. In our notation

$$\mathbf{\Sigma}_t = \mathbf{\Sigma}_0 + G(\gamma, c; s_t)\mathbf{\Sigma}_1$$

where $\mathbf{\Sigma}_0$ and $\mathbf{\Sigma}_0 + \mathbf{\Sigma}_1 > 0$. The authors simply assume this form for the covariance matrix and do not test the hypothesis $\mathbf{\Sigma}_t \equiv \mathbf{\Sigma}$.

BUILDING NONLINEAR VECTOR MODELS

General Considerations

In a few modelling situations, economic theory may offer guidance concerning the form of the relationship the researcher wants to use for characterising the dynamic behavior of a number of economic variables. For example, given the model classes considered in this article, an appropriate economic or other theory may suggest thresholds rather than smooth transitions, or the other way round. It may also be clear already at the outset what the switch or transition variable should be.

Nevertheless, there will almost always be choices to be made before settling for a final specification. The choice between linearity and nonlinearity is a very important one. It is possible that the linear model adequately describes the relationship considered, in which case a nonlinear model is not needed. In fact, the first step of a proper modelling strategy is to test linearity. This and other specification issues will be discussed below.

After the model has been specified, it has to be estimated, but that is not enough. Before an estimated model can be put to use, it has to be evaluated. The purpose of the evaluation is to find out whether the assumptions made when estimating the model appear satisfied, given the estimates. A battery of misspecification tests are available for this purpose, and some of them are discussed in Section "Evaluation." A complete modelling cycle thus consists of specifying the model, estimating its parameters and evaluating the estimated model. An early example is the cycle Box and Jenkins (1970) constructed for building linear ARMA or ARIMA models. Tsay (1989) considered specification and estimation of univariate TAR models. A modelling technique for STAR models was designed in Teräsvirta (1994) and Eitrheim and Teräsvirta (1996), and an analogous cycle for STR models discussed in Teräsvirta (1998).

Building Stationary Vector Switching Regression Models

Testing Linearity

It is not advisable to fit a VSR model to any set of stationary time series without first testing linearity. An obvious practical reason is that the linear VR model is easier to work with than a VSR model, so linearity should be checked first. There is also an important statistical reason: the VSR model is not identified when the data-generating process is linear. If the model contains only one regime, the time series are not informative about the switch-point which therefore cannot be estimated consistently. This lack of identification invalidates the standard testing approach: the asymptotic null distribution of the likelihood ratio statistic is unknown and does not have an analytic expression. This difficulty and how to deal with it was first discussed by Davies (1977). For a recent overview, see Teräsvirta et al. (2010, Chap. 5).

In VSR models, this identification problem can be handled in different ways. Following Tsay (1998), let

$$\mathbf{y}_t = \Theta \mathbf{z}_t + \boldsymbol{\varepsilon}_t \tag{26}$$

where $\Theta = (\boldsymbol{\mu}\ \boldsymbol{\Phi}_1\ \ldots\ \boldsymbol{\Phi}_p\ \boldsymbol{\Gamma})$ and $\mathbf{z}_t = (\mathbf{1}', \mathbf{y}'_{t-1}, \ldots, \mathbf{y}'_{t-p}, \mathbf{x}'_t)'$ be a linear VR model (only one regime in Eq. (11)) with independent errors. In order to test this VR model against a VSR model with one threshold and the threshold or switch-variable s_t, arrange the observations in the ascending order according to the values of s_t. Since the error vectors $\boldsymbol{\varepsilon}_t$ are independent, this can be done without affecting the dynamic properties of the model. Let the arranged model be

$$\mathbf{y}_{t(i)} = \Theta \mathbf{z}_{t(i)} + \boldsymbol{\varepsilon}_{t(i)}, \quad i = 1, \ldots, T \tag{27}$$

Next, estimate Θ from (27) using the first K arranged observations, call the least squares estimator $\tilde{\Theta}^{(K)}$, and compute the residuals

$$\tilde{\boldsymbol{\varepsilon}}_{t(i)} = \mathbf{y}_{t(i)} - \tilde{\Theta}^{(K)} \mathbf{z}_{t(i)}, \quad i = 1, \ldots, K$$

and the predictive residuals

$$\tilde{\boldsymbol{\eta}}_{t(i)} = \tilde{\boldsymbol{\varepsilon}}_{t(i)} (1 + \mathbf{z}'_{t(i)} (\mathbf{V}^{(K)})^{-1} \mathbf{z}_{t(i)})^{-1}, \quad i = 1, \ldots, K$$

where $\mathbf{V}^{(K)} = (1/K) \sum_{i=1}^{K} \mathbf{z}_{t(i)} \mathbf{z}'_{t(i)}$. Do this for $K = T_0 + 1, \ldots, T$, where $T_0 \geq dim(\mathbf{z}_t)$ and form the regression

$$\tilde{\boldsymbol{\eta}}_{t(i)} = \boldsymbol{\Psi} \mathbf{z}_{t(i)} + \boldsymbol{\upsilon}_{t(i)}, \quad i = T_0 + 1, \ldots, T$$

If (26) is the true model, the predictive residuals are asymptotically vector white noise and uncorrelated with $\mathbf{z}_{t(i)}$. Testing H_0: $\boldsymbol{\Psi} = \mathbf{0}$ against H_1: $\boldsymbol{\Psi} \neq \mathbf{0}$ thus reveals misspecification in Eq. (26). Under regularity conditions, the test statistic, based on the residuals $\tilde{\boldsymbol{\eta}}_{t(i)}$ and $\tilde{\boldsymbol{\upsilon}}_{t(i)}$ and given in Tsay (1998), has an asymptotic χ^2-distribution under H_0.

This test does not explicitly have the VSR model as the alternative hypothesis. Another possibility would be to generalize the approach of Hansen (1996) to VSR models. The test would be carried out as follows:

1. Estimate the VR (or VAR) model (26) and a VSR (or a VTAR) model with a single threshold (the threshold variable is assumed known). Test the null hypothesis that Eq. (26) is the true model. For example, this can be done using a Wald statistic for testing the hypothesis that the coefficients in the two regimes are identical. The problem is that the resulting statistic is a function of the unknown nuisance parameter c (the threshold) and thus not operational. To solve the problem, carry out the test for a set of values of $c \in C$. Typically, C contains the values between the β and $1 - \beta$ quantiles of the empirical distribution of the assumed threshold variable. Often $\beta = 0.15$. Calculate the maximum or the average value of these statistics over $c \in C$. This yields two nuisance parameter

free statistics, call them sup $W(c)$ and ave $W(c)$. The null distribution of these statistics is obtained numerically as follows:

2. Generate T observations from the estimated VR model either by simulation (drawing the error vectors from a distribution) or by a bootstrap using the estimated residuals, estimate a VR model and a VSR model from these observations and calculate the value of sup $W(c)$ or ave $W(c)$ as in the previous step.
3. Repeat the second step N times. Arrange the N values of the test statistic obtained this way to form its empirical cumulative distribution function.
4. Compare the original value of the statistic with the distribution. If it exceeds the $(1-\alpha)$th quantile of the distribution (large values are critical), reject the null hypothesis at the significance level α.

Balke (2000) and Atanasova (2003) may be mentioned as examples of the use of this method. They applied the Wald test but the likelihood ratio test would also be a possibility.

A third, and so far untested, possibility would be to generalize the approach of Strikholm and Teräsvirta (2006) to vector models. This would require testing the linear VR model against a vector STR model by a test described in Section "Specification and Estimation." Since the alternative hypothesis only contains the VSR model as a special case, this method must have lower power than that of Hansen (1996). Its main advantage would be computational simplicity.

Specification of lag lengths, exogenous variables and the switch or threshold variable requires estimation. This has already been discussed in Section "Estimation of Vector Switching Regression Models" in the case where $q=2$. Generalizing this to a larger number of regimes is obvious. However, the single-threshold model seems to be the most popular one in applications to stationary time series.

Testing for Threshold Cointegration

Cointegration is a major issue in the analysis of vectors of nonstationary time series. Consequently, testing the null hypothesis of no cointegration is an important problem in building TVEC models presented in Section "Threshold Vector Error Correction." There exist tests testing linear against threshold cointegration and testing the null of no cointegration against threshold cointegration. The early tests assume a two-regime (single-threshold) model as the alternative. Hansen and Seo (2002) developed a supremum LM-test of testing a VAR with linear cointegration against a TVEC model in the case where both the null model and the alternative contain a single cointegrating vector and the cointegrating vector (the error correction term) is the threshold variable. The LM approach makes it possible to use the cointegrating vector estimated under the null hypothesis in the test. Seo (2006) constructed a supremum Wald test of no cointegration against a TVEC model in this framework. He also showed how to approximate the small-sample distribution of the test using a residual-based bootstrap in order to reduce the size distortion present in the asymptotic test.

Gonzalo and Pitarakis (2006) generalized the approach of Hansen and Seo (2002) for testing linear against threshold cointegration such that the cointegration rank is not restricted to one. Furthermore, the threshold variable can be other than the error correction term. It has to be strictly stationary and ergodic, however. An additional restriction is that the threshold variable is what the authors call ‘external’: it is independent of the iid error vector at all lags and leads. Under these and other assumptions Gonzalo and Pitarakis (2006) derived the limiting null distribution of their supremum Wald statistic that they tabulated for some values of m. Recently, Krishnakumar and Neto (2012) constructed a supremum Wald test of no cointegration against the band-TVEC model. In their case, the threshold variable is assumed stationary and ergodic with a continuous and symmetric distribution. A set of critical values for the test can be found in the article.

Finally, it may be mentioned that the tests are derived for alternatives in which the short-run coefficients of the model do not switch and the error covariance matrices remain constant. These models are therefore somewhat less general than the ones defined by Eqs. (14) and (15) in Section “Threshold Vector Error Correction.”

Evaluation

As already mentioned, the estimated VSR model has to be evaluated before putting it into use. This implies testing the model by appropriate misspecification tests. Tsay (1998) mentions testing for serial correlation and conditional heteroskedasticity, without providing any details. Testing is straightforward in cases where the objective function is discontinuous at the threshold, because the estimator of the threshold is super or T-consistent instead of being $\sqrt{T}$-consistent like the other estimators. This does not, however, concern tests of the switch parameter. The existing tests relate to single-equation models; see Hansen (1997, 2000) and Section “Testing Common Nonlinearity in VSR Models” for discussion. The faster convergence rate means that in testing the threshold may be assumed known, which in turn implies that the standard test of no error autocorrelation for linear VAR or VR models is available in this case as well.

Other available tests include tests of constancy of the error covariance matrix. The test by Eklund and Teräsvirta (2007) makes use of the decomposition of the conditional covariance matrix by Bollerslev (1990) and contains several parametric alternatives to the null hypothesis. There is also a recent test by Yang (2013) that relies on the spectral decomposition of the covariance matrix. Under the alternatives the eigenvalues of the covariance matrix vary over time. These tests, however, require the errors to be independent and normal.

As Strikholm and Teräsvirta (2006) pointed out, super consistency of the threshold parameter estimator also allows using the linearity test against STAR to select the number of regimes in the SR model. This generalizes to VSR models, although at this moment small-sample properties of this technique are not known. It may be expected, however, that the LM-type test discussed in Section "Specification and Estimation" is size-distorted in small samples and has to be modified to correct the bias.

One can also check stability of the estimated AR model. In the absence of analytical stability results, this may be done by simulation. One then switches off the noise and generates observations from the estimated model using different initial values or "histories." If the generated realizations converge to the same point, the conclusion is that the simulations do not contradict the stability assumption. They cannot confirm it either, but then a single realization converging to a different point suffices to invalidate this assumption.

Building Stationary Vector Smooth Transition Regression Models

Specification and Estimation

Camacho (2004) was the first to develop a modelling strategy for stationary vector STR models. He considered bivariate single-transition models. Teräsvirta and Yang (2013) generalized this to multitransition multivariate LVSTR models. As already discussed, the modelling strategy has three stages: specification, estimation, and evaluation. Specification involves testing linearity and, if the null hypothesis is rejected, finding the number of transitions. Furthermore, possibilities to reduce the size of the model, that is, imposing parameter restrictions, are considered, which already requires estimation.

Testing Linearity. As already mentioned, the first step of the specification stage is to test linearity of the model against VLSTAR. The arguments for doing so are the same as in the VSR case. The linear VR model is easier to handle than the VLSTAR one, and the latter is not identified when the data-generating process is linear.

The way linearity is tested depends on how the model is defined. If the model is (16), linearity can be tested equation by equation. Lagrange multiplier tests are preferred, because they only require estimating the model under the linearity hypothesis. A computationally easy way to perform the test is to approximate the transition function by a Taylor series expansion around the null hypothesis $\gamma_j = 0$ (the jth equation) and base the test on this approximation, as Luukkonen, Saikkonen, and Teräsvirta (1988) suggested. This LM-type test has an asymptotic χ^2-distribution under the null hypothesis. Alternatively, one may construct the empirical distribution of the test statistic as in Hansen (1996). Doing so is computationally more demanding than performing the Taylor series alternative, but the advantage is that the test has higher power than the corresponding LM-type test if the true alternative is indeed a STAR equation. See Teräsvirta et al. (2010, Chap. 5) for discussion.

If it is assumed that the transition function is the same for the whole system, one can carry out a systemwise test as discussed in Yang and Teräsvirta (2013). Consider the m-dimensional two-regime logistic VSTR model

$$\mathbf{y}_t = \mathbf{B}_1'\mathbf{z}_t + \mathbf{G}_t\mathbf{B}_2'\mathbf{z}_t + \boldsymbol{\varepsilon}_t \tag{28}$$

where $\mathbf{B}_1' = (\boldsymbol{\mu}_0\ \boldsymbol{\Phi}_1\ \ldots\ \boldsymbol{\Phi}_p\ \boldsymbol{\Gamma})$, $\mathbf{B}_2' = (\boldsymbol{\mu}_1\ \boldsymbol{\Psi}_1\ \ldots\ \boldsymbol{\Psi}_p\ \boldsymbol{\Xi})$, $\mathbf{z}_t = (\mathbf{1}', \mathbf{y}_{t-1}', \ldots, \mathbf{y}_{t-p}', \mathbf{x}_t')'$, the sequence $\{\boldsymbol{\varepsilon}_t\} \sim \text{iid}\,\mathcal{N}(\mathbf{0}, \boldsymbol{\Omega})$, and $\mathbf{G}(\boldsymbol{\gamma}, \mathbf{c}; \mathbf{s}_t) = G(\gamma, c; s_t)\mathbf{I}_m$. The null hypothesis of linearity can be written as H_0: $\gamma = 0$. When the null is true, $\mathbf{G}_t \equiv (1/2)\mathbf{I}_m$ and (28) becomes linear. Neither the location parameter c in the logistic function nor the parameters in the linear combination $\mathbf{B}_1 + (1/2)\mathbf{B}_2$ are identified. The alternative hypothesis H_1: $\gamma > 0$. Using a third-order Taylor expansion approximation of the logistic transition function $G(\gamma, c; s_t)$ around $\gamma = 0$ helps circumvent the identification problem and yields

$$\begin{aligned}\mathbf{G}_t &= \text{diag}\{a_{01} + a_{11}(\gamma)s_t + a_{21}(\gamma^2)s_t^2 + a_{31}(\gamma^3)s_t^3 + r_{1t}, \ldots, a_{0m} + a_{1m}(\gamma)s_t \\ &\quad + a_{2m}(\gamma^2)s_t^2 + a_{3m}(\gamma^3)s_t^3 + r_{mt}\} \\ &= \mathbf{A}_0 + \mathbf{A}_1 s_t + \mathbf{A}_2 s_t^2 + \mathbf{A}_3 s_t^3 + \mathbf{R}_t\end{aligned} \tag{29}$$

where $\mathbf{A}_0 = \text{diag}(a_{01}, \ldots, a_{0m})$, $\mathbf{A}_j = \text{diag}(a_{j1}(\gamma^j), \ldots, a_{jm}(\gamma^j))$, $j = 1, 2, 3$, are functions of γ such that $\gamma = 0$ implies $a_{ji}(\gamma^j) = 0$, and $\mathbf{R}_t = \text{diag}(r_{1t}, \ldots, r_{mt})$. Inserting (29) into (28) gives

$$\begin{aligned}
\mathbf{y}_t &= \mathbf{B}_1'\mathbf{z}_t + (\mathbf{A}_0 + \mathbf{A}_1 s_t + \mathbf{A}_2 s_t^2 + \mathbf{A}_3 s_t^3 + \mathbf{R}_t)\mathbf{B}_2'\mathbf{z}_t + \boldsymbol{\varepsilon}_t \\
&= (\mathbf{B}_1' + \mathbf{A}_0\mathbf{B}_2')\mathbf{z}_t + \mathbf{A}_1\mathbf{B}_2'\mathbf{z}_t s_t + \mathbf{A}_2\mathbf{B}_2'\mathbf{z}_t s_t^2 + \mathbf{A}_3\mathbf{B}_2'\mathbf{z}_t s_t^3 + \mathbf{R}_t\mathbf{B}_2'\mathbf{z}_t + \boldsymbol{\varepsilon}_t \\
&= \boldsymbol{\Theta}_0'\mathbf{z}_t + \boldsymbol{\Theta}_1'\mathbf{z}_t s_t + \boldsymbol{\Theta}_2'\mathbf{z}_t s_t^2 + \boldsymbol{\Theta}_3'\mathbf{z}_t s_t^3 + \boldsymbol{\varepsilon}_t^*
\end{aligned} \tag{30}$$

where $\boldsymbol{\Theta}_0 = \mathbf{B}_1 + \mathbf{B}_2\mathbf{A}_0$, $\boldsymbol{\Theta}_j = \mathbf{B}_2\mathbf{A}_j$, $j = 1, 2, 3$, and $\boldsymbol{\varepsilon}_t^* = \mathbf{R}_t\mathbf{B}_2'\mathbf{z}_t + \boldsymbol{\varepsilon}_t$. The null hypothesis implies $\mathbf{A}_0 = (1/2)\mathbf{I}_m$, $\mathbf{A}_j = \mathbf{0}$, $j = 1, 2, 3$, and $\mathbf{R}_t = \mathbf{0}$ in Eq. (30).

Thus, under the null hypothesis, the model (28) is linear, with $\boldsymbol{\Theta}_0 = \mathbf{B}_1 + (1/2)\mathbf{B}_2$ and $\boldsymbol{\Theta}_j = \mathbf{0}$, $j = 1, 2, 3$. The linearity test is therefore equivalent to the test of $\text{H}_0 : \boldsymbol{\Theta}_j = \mathbf{0}$, $j = 1, 2, 3$, in Eq. (30). Moreover, due to the fact that $\mathbf{R}_t = \mathbf{0}$ under the null hypothesis, the error term $\boldsymbol{\varepsilon}_t^* = \boldsymbol{\varepsilon}_t$ when H_0 holds. Since the Lagrange multiplier test only requires estimating the model under the null hypothesis, the remainder term does not affect the normality of the errors or the standard asymptotic inference.

The corresponding Lagrange multiplier test under the null is derived from the score

$$\frac{\partial \log L(\tilde{\theta})}{\partial \boldsymbol{\Theta}_1} = \sum_{t=1}^{T} (\mathbf{z}_t s_t \;\; \mathbf{z}_t s_t^2 \;\; \mathbf{z}_t s_t^3)\left(\mathbf{y}_t - \tilde{\mathbf{B}}_1'\mathbf{z}_t\right)'\tilde{\boldsymbol{\Omega}}^{-1} = \mathbf{Z}_3'(\mathbf{Y} - \mathbf{Z}_0\tilde{\mathbf{B}}_1)\tilde{\boldsymbol{\Omega}}^{-1} \tag{31}$$

where

$$\mathbf{Y} = \begin{bmatrix} \mathbf{y}_1' \\ \mathbf{y}_2' \\ \vdots \\ \mathbf{y}_T' \end{bmatrix}, \mathbf{Z}_0 = \begin{bmatrix} \mathbf{z}_1' \\ \mathbf{z}_2' \\ \vdots \\ \mathbf{z}_T' \end{bmatrix}, \mathbf{Z}_3 = \begin{bmatrix} \mathbf{z}_1' s_1 & \mathbf{z}_1' s_1^2 & \mathbf{z}_1' s_1^3 \\ \mathbf{z}_2' s_2 & \mathbf{z}_2' s_2^2 & \mathbf{z}_2' s_2^3 \\ \vdots & \vdots & \vdots \\ \mathbf{z}_T' s_T & \mathbf{z}_T' s_T^2 & \mathbf{z}_T' s_T^3 \end{bmatrix}$$

and $\tilde{\mathbf{B}}_1$ and $\tilde{\boldsymbol{\Omega}}$ are estimates from the restricted model. Under regularity conditions, the score converges to a matricvariate normal distribution with zero mean and variance $\mathbf{Z}_3'(\mathbf{I}_T - \mathbf{P}_Z)\mathbf{Z}_3 \otimes \boldsymbol{\Omega}^{-1}$, where $\mathbf{P}_Z \equiv \mathbf{Z}_0(\mathbf{Z}_0'\mathbf{Z}_0)^{-1}\mathbf{Z}_0'$ is the projection matrix.

Luukkonen et al. (1988) pointed out that the third-order Taylor expansion is strictly needed only when the only nonlinear term in the equation is the intercept, that is, $\mathbf{B}_2 = (\boldsymbol{\beta}_{21} \;\; \mathbf{0})$ and s_t equals one of the remaining stochastic elements in $\mathbf{z}_t$. In that case the test based on the first-order Taylor expansion has only trivial power against the alternative. The second-order Taylor expansion does not remedy this problem, as $a_{2j}(\gamma^2) = 0$ whenever $c_j = 0$.

The test can be performed in two vector regressions as follows:

1. Estimate the restricted model: regress $\mathbf{Y}$ on $\mathbf{Z}_0$. Collect the residuals $\tilde{\mathbf{E}} = (\mathbf{I}_T - \mathbf{P}_Z)\mathbf{Y}$, and compute the matrix residual sum of squares $\mathbf{RSS}_0 = \tilde{\mathbf{E}}'\tilde{\mathbf{E}}$.
2. Run an auxiliary regression of $\tilde{\mathbf{E}}$ on $(\mathbf{Z}_0, \mathbf{Z}_3)$. Compute the residuals $\hat{\mathbf{E}}$ and the matrix residual sum of squares $\mathbf{RSS}_1 = \hat{\mathbf{E}}'\hat{\mathbf{E}}$.
3. Compute the test statistic

$$LM = T\,\mathrm{tr}\{\mathbf{RSS}_0^{-1}(\mathbf{RSS}_0 - \mathbf{RSS}_1)\} = T(m - \mathrm{tr}\{\mathbf{RSS}_0^{-1}\mathbf{RSS}_1\}) \qquad (32)$$

The joint test statistic defined in Eqs. (31) and (32) collapses into the univariate LM-type linearity test statistic when $m = 1$. This joint test can also be applied to any subset of equations in the system, for instance, to check whether some equations in the system are nonlinear with a common transition variable. It is also possible to test subhypotheses. For example, it may be assumed that the exogenous variables $\mathbf{x}_t$ only enter Eq. (28) linearly, that is, $\mathbf{\Xi} = \mathbf{0}$.

In small samples, the empirical size of this LM-type test is sensitive to the dimension of the model. The size distortion increases with m. If the errors are assumed independent and identically distributed, Rao's F-test (Rao, 1973, Section 8c.5) effectively corrects the size. If the errors contain higher-order dependence, the problem can be solved by applying the wild bootstrap. These alternatives are discussed in Yang and Teräsvirta (2013).

It should be noted, however, that a rejection of the null hypothesis does not automatically imply that all m equations of the alternative VLSTAR model have the same transition function. This is because the same approximation (30) can also be obtained by assuming that all equations have the same transition variable but not necessarily the same transition function.

Harvill and Ray (1999) generalized the univariate linearity test of Tsay (1986) to vector series. The test is based on the following model

$$\mathbf{y}_t = \sum_{j=1}^{p} \mathbf{\Phi}_j \mathbf{y}_{t-j} + \mathbf{\Psi}\mathrm{vech}(\mathbf{z}_t\mathbf{z}_t') + \boldsymbol{\varepsilon}_t$$

where $\mathbf{z}_t = (\mathbf{y}'_{t-1}, \ldots, \mathbf{y}'_{t-p})'$ is an $mp \times 1$ vector, $\mathbf{\Psi}$ is an $m \times mp(mp+1)/2$ parameter matrix, and $\boldsymbol{\varepsilon}_t \sim \text{iid}\mathcal{N}(\mathbf{0}, \mathbf{\Sigma})$. The null hypothesis is $\mathbf{\Psi}=\mathbf{0}$. It is tested using Wilks's Λ:

$$\Lambda = \frac{|\mathbf{RSS}_0|}{|\mathbf{RSS}_0 + \mathbf{RSS}_1|} \tag{33}$$

where, as previously, $\mathbf{RSS}_0$ is the residual sum of squares matrix from estimating the null model, and $\mathbf{RSS}_1$ is obtained by regressing the residuals from that regression on $\mathbf{y}_{t-1}, \ldots, \mathbf{y}_{t-p}$ and $\text{vech}(\mathbf{z}_t\mathbf{z}'_t)$. Statistic (33) has an approximate F-distribution under H_0, see Harvill and Ray (1999) for details. One could also use Rao's F-statistic instead of Eq. (33). Simulations in Yang and Teräsvirta (2013) suggest that although Wilks's Λ is less size-distorted in small samples than most other statistics, Rao's F-test is the best performer.

Weise (1999) has yet another solution. Take a linear VAR model ($\mathbf{\Xi}=0$ in Eq. (16) but a generalization is possible) and test linearity equation by equation as in Teräsvirta (1998) using the Taylor expansion approach and the same transition variable. Let Let $\tilde{\boldsymbol{\varepsilon}}_t = (\tilde{\varepsilon}_{1t}, \ldots, \tilde{\varepsilon}_{mt})'$ be the residual vector of the estimated linear VAR model and $\tilde{\boldsymbol{\eta}}_t = (\tilde{\eta}_{1t}, \ldots, \tilde{\eta}_{mt})'$ the corresponding vector from the m auxiliary regressions. Form the covariance matrices $\tilde{\mathbf{\Sigma}}_\varepsilon = (T-p)^{-1} \sum_{j=p+1}^{T} \tilde{\boldsymbol{\varepsilon}}_t \tilde{\boldsymbol{\varepsilon}}'_t$ and $\tilde{\mathbf{\Sigma}}_\eta = (T-p)^{-1} \sum_{j=p+1}^{T} \tilde{\boldsymbol{\eta}}_t \tilde{\boldsymbol{\eta}}'_t$ and the likelihood ratio-type test statistic

$$LR = T(\ln|\tilde{\mathbf{\Sigma}}_\varepsilon| - \ln|\tilde{\mathbf{\Sigma}}_\eta|)$$

that has an asymptotic χ^2-distribution with pm^2 degrees of freedom when the linearity hypothesis holds. This approach provides results both for individual equations and for the whole system. Weise (1999) simulates the statistic and finds its small-sample behavior satisfactory. For an application, see Section "Applications of Vector STAR and STR Models."

Finding the Number of Transitions. When linearity is rejected, the VLSTAR model with a single-transition is estimated. The next step is to test one transition against two. This is either done equation by equation or for the whole system, depending on the situation. In the former case, the relevant single-equation test of no additive nonlinearity is described in Teräsvirta (1998). If it is assumed that one transition function controls the whole system, the same test is designed for the whole system (see Yang & Teräsvirta, 2013).

Specifying the Lag Structure. Linearity testing is, when the number of observations allows it, preferably carried out with a null model that is free of error autocorrelation, because it may distort the empirical size of the test. It is then desirable to find and impose restrictions on the parameters of the VLSTAR model. There are three types of intraequation restrictions. As an example, consider the parameter matrices $\mathbf{\Phi}_j = [\varphi_{ij}]$ and $\mathbf{\Psi}_j = [\psi_{ij}]$ in Eq. (16). First, the jth lag $y_{i,t-j}$ may only enter linearly: $\psi_{ij} = 0$. Second, this lag may be completely redundant: $\varphi_{ij} = \psi_{ij} = 0$. Third, the effect of $y_{i,t-j}$ is nonlinear and zero at low values of s_t: $\varphi_{ij} = 0$. Finally, the effect of $y_{i,t-j}$ is nonlinear and approaches zero at high values of s_t: $\varphi_{ij} = -\psi_{ij}$. Testing these restrictions and imposing them requires estimation of the VLSTAR model.

Evaluation

Misspecification Tests. Misspecification tests are the most important and useful tool for evaluating VLSTR models. Out-of-sample forecasting may also be used to select between different models. The available tests include the test of no serial correlation, no remaining nonlinearity, and the test of parameter constancy. For space reasons it is not possible to consider all these tests in detail. Suffice it so say that the estimated model can be tested equation by equation see Teräsvirta (1998), or as a system. As an example of the latter, the test of no error autocorrelation can be carried out by assuming that under the alternative, the error sequence $\{\boldsymbol{\varepsilon}_t\}$ in Eq. (16) has the following structure

$$\boldsymbol{\varepsilon}_t = \sum_{j=1}^{q} \mathbf{\Lambda}_j \boldsymbol{\varepsilon}_{t-j} + \boldsymbol{\eta}_t$$

where $\boldsymbol{\eta}_t \sim \text{iid}\mathcal{N}(\mathbf{0}, \mathbf{\Sigma}_\eta)$; see Camacho (2004) or Yang and Teräsvirta (2013). Under the null hypothesis, $\mathbf{\Lambda}_j = \mathbf{0}$, $j = 1, \ldots, q$. If the corresponding hypothesis is tested equation by equation, the contemporaneous cross-correlations between errors are ignored. The tests of no additive nonlinearity and parameter constancy can likewise be conducted either equation by equation or jointly for the whole system.

These tests have been discussed in the situation in which the equations contain a single transition, but they can be generalized to multitransition equations. As in the case of VTAR models, it is possible to test the constancy of the error covariance matrix against various parametric alternatives using the tests of Eklund and Teräsvirta (2007) and Yang (2013). The stability of a VLSTAR model may be checked numerically in the same way as that of the VTAR model.

Bayesian Approach to Building Vector Smooth Transition Autoregressive Models

It is possible to construct VSTAR models using Bayesian techniques, as shown by Gefang and Strachan (2010). The authors consider a bivariate ($m=2$) variant of the logistic (16) with $\mathbf{x}=\mathbf{0}$ and $\mathbf{G}(\boldsymbol{\gamma},\mathbf{c};\mathbf{s}_t)=G(\gamma,c;s_t)\mathbf{I}_2$. Determination of prior distributions contains one detail worth mentioning. As $\gamma=0$ in the logistic transition function implies linearity and an unidentified LVSTAR model, this value is excluded from the support of the prior distribution of γ. For a general discussion of this identification problem in the Bayesian framework, see Bauwens, Lubrano, and Richard (1999, pp. 241–242).

Specification of the model consists of estimating a large number of models with varying transition functions and lag lengths (these are not fixed in advance) and using Bayes factors to choose between models. Since linearity cannot be excluded a priori, the set of models compared also contains linear VAR models. Posterior distributions of the parameters are estimated using Gibbs sampling. The aim of the application in Gefang and Strachan (2010) is to study (pairwise) business cycle linkages between the United Kingdom, the United States, and Germany. They find that nonlinear VLSTAR models are superior to linear VARs in describing these relationships.

Inference in Nonstationary Vector Smooth Transition Autoregressive Models

The statistical inference discussed in the preceding section builds on the assumption that $\{\mathbf{y}_t\}$ is stationary. The situation becomes more complex when that is no longer true. Inference in STVEC models is studied in Kristensen and Rahbek (2010, 2013). In the former article the authors consider maximum likelihood estimation of an m-dimensional nonlinear vector error correction model in which the adjustment is symmetric and the lags of $\Delta\mathbf{y}_t$ enter linearly. The model is written as follows

$$\Delta\mathbf{y}_t=\mathbf{h}(\mathbf{B}'\mathbf{y}_{t-1},\boldsymbol{\theta})+\sum_{j=1}^{p-1}\boldsymbol{\Phi}_j\Delta\mathbf{y}_{t-j}+\boldsymbol{\varepsilon}_t \tag{34}$$

where $\{\boldsymbol{\varepsilon}_t\} \sim$ iid$(\mathbf{0}, \boldsymbol{\Sigma})$. The symmetry arises from the fact that

$$\mathbf{h}(\mathbf{B}'\mathbf{y}, \boldsymbol{\theta}) = \mathbf{A}_1\mathbf{B}'\mathbf{y} + o(\|\mathbf{B}'\mathbf{y}\|)$$

for some $m \times r$ matrix $\mathbf{A}_1$ as $\|\mathbf{B}'\mathbf{y}\| \to \infty$. The characteristic polynomial

$$C_1(z) = \mathbf{I}_m(1-z) - \mathbf{A}_1\mathbf{B}'z - \sum_{k=1}^{p-1} \boldsymbol{\Phi}_k(1-z)z^k$$

has $m-r$ unit roots, whereas $|z|>1$ for the remaining ones. The 'symmetric' STVEC model is obtained as a special case of Eq. (34) by defining

$$\mathbf{h}(\mathbf{B}'\mathbf{y}_{t-1}, \boldsymbol{\theta}) = \mathbf{A}_1\mathbf{B}' + \mathbf{A}_2\mathbf{B}'[1 + \exp\{(\mathbf{B}'\mathbf{y}_{t-1} - \boldsymbol{\omega})'\mathbf{C}(\mathbf{B}'\mathbf{y}_{t-1} - \boldsymbol{\omega})]^{-1} \tag{35}$$

where $\mathbf{C}$ is a positive definite matrix, and $\boldsymbol{\theta} = (\boldsymbol{\omega}', \text{vech}(\mathbf{C})', \text{vec}(\mathbf{A}_1)', \text{vec}(\mathbf{A}_2)')'$. A single transition function controls nonlinearity in all equations, and the transition variable is a function of the r error-correcting vectors. Under regularity conditions for $\mathbf{h}(\mathbf{B}'\mathbf{y}_{t-1}, \boldsymbol{\theta})$ that exclude VTAR models, and other conditions given in the article, Kristensen and Rahbek (2010) derive consistency and the asymptotic distribution of the (quasi) maximum likelihood estimators of the parameters of (the normalized version of) Eq. (34). The short-run parameter vector estimators $\hat{\boldsymbol{\theta}}$ and vec$(\hat{\boldsymbol{\Phi}}_k)$, $k=1,\ldots,p-1$, are $\sqrt{T}$-consistent, whereas the estimator of (normalized) $\mathbf{B}$ is super or T-consistent. The estimators have nonstandard distributions, so confidence intervals have to be constructed numerically.

Kristensen and Rahbek (2013) consider the same model as in Kristensen and Rahbek (2010) but allow what they call asymmetric adjustment: the transition function in Eq. (34) equals (Saikkonen, 2008)

$$G_t(\mathbf{y}_{t-1}, \boldsymbol{\theta}) = [1 + \exp\{\mathbf{s}'(\mathbf{B}'\mathbf{y}_{t-1} - \boldsymbol{\omega})\}]^{-1} \tag{36}$$

where $\mathbf{s}$ is an $r \times 1$ vector. In the former case (symmetric adjustment), $\|\mathbf{B}'\mathbf{y}_{t-1} - \boldsymbol{\omega}\| \to 0$ as $\|\mathbf{y}_{t-1}\| \to \infty$, whereas in the latter, the nonlinear component will contribute to $\Delta\mathbf{y}_t$ even when $\|\mathbf{y}_{t-1}\| \to \infty$. The authors derive tests for testing hypothesis about the short-run parameters $\boldsymbol{\theta} = (\boldsymbol{\omega}', \mathbf{s}', \text{vec}(\mathbf{A}_1)', \text{vec}(\mathbf{A}_2)')'$ and vec$(\boldsymbol{\Phi}_j)$, $j=1,\ldots,p$, and the long-run parameters vec$(\mathbf{B})$. Note that the hypothesis vec$(\mathbf{A}_2)=\mathbf{0}$ is a special case because it is a linearity hypothesis. As in the stationary case, testing it directly, that is, without approximating the alternative (see below), leads to nonstandard asymptotic inference because $\boldsymbol{\omega}$ and vech($\mathbf{C}$) in Eq. (35) or $\boldsymbol{\omega}$ and $\mathbf{s}$ in Eq. (36) are not identified when the null hypothesis holds. Testing linear against nonlinear cointegration will be discussed next.

Testing for Nonlinear Cointegration in Smooth Transition Error Correction Models

Kristensen and Rahbek (2013) discuss the standard solution to testing H_0: $\text{vec}(\mathbf{A}_2) = \mathbf{0}$; see for example Hansen (1996). First define a set Θ_1 of values of $\boldsymbol{\theta}_1 = (\boldsymbol{\omega}', \text{vech}(\mathbf{C})')'$ (symmetric model) or $\boldsymbol{\theta}_1 = (\boldsymbol{\omega}', \mathbf{s}')'$ (asymmetric model) and compute the test statistic using these values in turn. These statistics are not operational because they are functions of $\boldsymbol{\theta}_1$. The proposed test statistic is the supremum statistic, for example the LM statistic LM_{NL} is defined by

$$LM_{\text{NL}} = \sup_{\boldsymbol{\theta}_1 \in \Theta_1} LM(\boldsymbol{\theta}_1)$$

where $LM(\boldsymbol{\theta}_1)$ is an inoperational statistic. The empirical null distribution of LM_{NL} is computed by simulation as discussed in Section "Testing Linearity." The power of the test strongly depends on the number of values in Θ_1 and the selected values of $\boldsymbol{\theta}_1$.

Seo (2011a) also tests linear cointegration against nonlinear one. Furthermore, he presents a test of linearity of the short-run component. In testing linear against nonlinear cointegration, the model is

$$\Delta\mathbf{y}_t = \mathbf{A}\{\mathbf{B}'\mathbf{y}_{t-1} + \mathbf{g}(\mathbf{y}_{t-1})\} + \sum_{j=1}^{p-1} \boldsymbol{\Phi}_j \Delta\mathbf{y}_{t-j} + \mathbf{u}_t \tag{37}$$

where $\{\mathbf{u}_t\}$ is a martingale difference sequence with respect to $\mathbf{y}_{t-j}$, $j \geq 1$, and no specific parametric structure is assumed for the $r \times 1$ vector $\mathbf{g}(\mathbf{y}_{t-1})$. The null hypothesis is that $\mathbf{g}(\mathbf{y}_{t-1}) = \mathbf{0}$. When the null hypothesis concerns the short-run parameters, the alternative model is

$$\Delta\mathbf{y}_t = \mathbf{A}\mathbf{B}'\mathbf{y}_{t-1} + \sum_{j=1}^{p-1} \boldsymbol{\Phi}_j \Delta\mathbf{y}_{t-j} + \mathbf{h}(\mathbf{B}'\mathbf{y}_{t-1}, \Delta\bar{\mathbf{y}}_{t-1}) + \mathbf{u}_t \tag{38}$$

where $\Delta\bar{\mathbf{y}}_{t-1} = \text{vec}(\Delta\mathbf{y}_{t-1}, \ldots, \Delta\mathbf{y}_{t-p+1})$. The null hypothesis is that $\mathbf{h}(\mathbf{B}'\mathbf{y}_{t-1}, \Delta\bar{\mathbf{y}}_{t-1}) = \mathbf{0}$. The argument is that under the null hypothesis, $\mathbf{u}_t$ is neither dependent on $\mathbf{y}_{t-1}$ in Eq. (37), nor on $\mathbf{B}'\mathbf{y}_{t-1}$ or $\Delta\overline{\mathbf{y}}_{t-1}$ in Eq. (38). The idea of the test is to approximate the alternative, which is done by replacing the nonlinear component by a polynomial approximation. Thus in Eq. (37)

$$\mathbf{g}(\mathbf{y}_t) \approx \mathbf{Z}'_{1t}\boldsymbol{\delta}_1$$

where $\mathbf{Z}_{1t}$ is an $s_1 \times r$ matrix, whose elements are powers and cross-products of the elements of $\mathbf{y}_t$, and $\boldsymbol{\delta}$ is an $s_1 \times 1$ parameter vector. The null hypothesis is $\boldsymbol{\delta}_1 = \mathbf{0}$. Analogously, in Eq. (38), it is assumed that

$$\mathbf{h}(\mathbf{B}'\mathbf{y}_{t-1}, \Delta\overline{\mathbf{y}}_{t-1}) \approx \mathbf{Z}'_{2t}\boldsymbol{\delta}_2$$

where the $s_2 \times p$ matrix $\mathbf{Z}_{2t}$ consists of powers and cross-products of the elements of $\mathbf{B}'\mathbf{y}_{t-1}$ and $\Delta\overline{\mathbf{y}}_{t-1}$, and the null hypothesis is $\boldsymbol{\delta}_2 = \mathbf{0}$. The resulting tests have the advantage that assuming $\mathbf{u}_t \sim \text{iid}\mathcal{N}(\mathbf{0}, \boldsymbol{\Sigma})$ the test statistics have a χ^2-distribution under H_0; for details see Seo (2011a). The test of $\boldsymbol{\delta}_2 = \mathbf{0}$ can be viewed as a generalization of the univariate linearity test by Luukkonen et al. (1988) or the one in Teräsvirta, Lin, and Granger (1993), see also Section "Specification and Estimation."

Kapetanios, Shin, and Snell (2006) test the null hypothesis of no cointegration against smooth transition cointegration in VAR models assuming that under the alternative the model contains exactly one cointegrating relationship. This implies that the $\mathbf{A}_1, \mathbf{A}_2$, and $\mathbf{B}$ in Eq. (18) are $m \times 1$ vectors. Compared to Eq. (18), the model of Kapetanios et al. (2006) does not contain intercepts, and the lags of $\Delta\mathbf{w}_t$ enter the model linearly. Writing the (normalized) cointegrating relationship as

$$u_t = y_{1t} - \mathbf{B}'\mathbf{w}_t$$

where $\mathbf{w}_t = (y_{2t}, \ldots, y_{mt})'$, the alternative can be expressed as the following transformed system

$$\begin{aligned}\Delta y_{1t} &= \alpha_1 u_{t-1} + \sum_{j=1}^{p-1} \boldsymbol{\varphi}'_j \Delta\mathbf{y}_{t-j} + \boldsymbol{\varphi}'_0 \Delta\mathbf{w}_t \\ &\quad + \{\alpha_2 u_{t-1} \exp\{-\gamma(u_{t-1} - c)^2\} + \sum_{j=1}^{p-1} \boldsymbol{\psi}'_j \Delta\mathbf{y}_{t-j} + \varepsilon_{1t}, \ \gamma > 0 \qquad (39) \\ \Delta\mathbf{w}_t &= \sum_{j=1}^{p-1} \boldsymbol{\Phi}'_{2j} \Delta\mathbf{y}_{t-j} + \boldsymbol{\varepsilon}_{2t}.\end{aligned}$$

Several things are worth noticing in Eq. (39). First, $\Delta \mathbf{w}_t$ is assumed exogenous to the parameters of interest in the equation of Δy_{1t}. Second, the nonlinearity is ESTAR or "symmetric" type. Third, the transition variable is the lagged error correction term itself. Finally, nonlinearity is not extended to the short-run dynamic structure of the model.

Testing is carried out by first regressing y_{1t} on $\mathbf{w}_t$ assuming that $y_{2t}, \ldots, y_{mt}$ are weakly exogenous to the parameters of interest in this equation. The residuals are $\tilde{u}_t = y_{1t} - \tilde{\boldsymbol{\beta}}' \mathbf{w}_t$, $t = 1, \ldots, T$. Then, following Luukkonen et al. (1988), $\exp\{-\gamma(u_{t-1} - c)^2\}$ is approximated with its first-order Taylor expansion at $\gamma = 0$ and the auxiliary regression

$$\Delta y_{1t} = \delta_1 \tilde{u}_{t-1} + \delta_2 \tilde{u}_{t-1}^2 + \delta_3 \tilde{u}_{t-1}^3 + \sum_{j=1}^{p-1} \boldsymbol{\varphi}_j' \Delta \mathbf{y}_{t-j} + \boldsymbol{\varphi}_0' \Delta \mathbf{w}_t + \varepsilon_{1t}^*$$

formed to test the null hypothesis $\delta_1 = \delta_2 = \delta_3 = 0$. The resulting tests are LM-type tests.

Bayesian Approach to Building Smooth Transition Vector Error Correction Models

Building smooth transition vector error correction models using Bayesian techniques is discussed in Gefang (2012). The purpose of the article is to study Granger causality between money and output using monthly US time series from 1959(1) to 2006(12) using a four-dimensional STVEC model. The model can be written as follows

$$\begin{aligned}\Delta \mathbf{y}_t = \boldsymbol{\mu}_1 + \mathbf{A}_1 \mathbf{B}_1' \mathbf{y}_{t-1} + \sum_{j=1}^{p-1} \boldsymbol{\Phi}_j \Delta \mathbf{y}_{t-j} \\ + \{\boldsymbol{\mu}_2 + \mathbf{A}_2 \mathbf{B}_2' \mathbf{y}_{t-1} \sum_{j=1}^{p-1} \boldsymbol{\Psi}_j \Delta \mathbf{y}_{t-j}\} G(\gamma, c; s_t) + \boldsymbol{\varepsilon}_t, \quad \gamma > 0\end{aligned} \tag{40}$$

where $\mathbf{y}_t$ is a $m \times 1$ vector of nonstationary variables such that $\Delta \mathbf{y}_t$ is I(0), and $\boldsymbol{\varepsilon}_t \sim \text{iid}\mathcal{N}(\mathbf{0}, \boldsymbol{\Sigma})$. The matrices $\mathbf{A}_1$ and $\mathbf{B}_1$ are both $m \times r_1$, $r_1 < m$, whereas $\mathbf{A}_2$ and $\mathbf{B}_2$ are $m \times r_2$, $r_2 < m$, The ranks r_1 and r_2 are not fixed in advance, and they are not assumed equal. This means that the extreme regimes corresponding to $G(\gamma, c; s_t) = 0$ and $G(\gamma, c; s_t) = 1$, respectively, can contain a different number of cointegrating relationships. In the application, $m=4$, so the ranks can be between one and three, which implies nine possible combinations for r_1 and r_2. How to interpret the changing ranks (and the "fractional" ranks when $G(\gamma, c; s_t) \neq 0$) is not discussed. Models without cointegration in Eq. (40) are not considered separately, as the author argues that they are special cases of models with ranks between one and three, $\mathbf{A}_1 = \mathbf{0}$ and $\mathbf{A}_2 = \mathbf{0}$. In that case, the ranks are not identified, however. The transition variable s_t is not predetermined, and the set of possible candidates is defined. Some of the stochastic ones used in the study are nonstationary, which means that the local dynamic structure of the corresponding model mostly changes in one direction and does not fluctuate between the two extreme regimes.

As in the stationary case in Section "Bayesian Approach to Building Vector Smooth Transition Autoregressive Models," the posterior distributions are obtained using Gibbs sampling. Model selection is carried out analogously to the stationary case using Bayes factors. The most frequently selected models support the assumption of nonlinear Granger causality from money to output. They are in fact ones in which $r_1 \neq r_2$.

COMMON NONLINEARITIES

Definitions

In a system of nonlinear equations such as the VTAR or VLSTAR model in which nonlinearity is additive, common nonlinearities may occur. Anderson and Vahid (1998) were the first to consider this possibility. They discussed both the general case in which the nonlinear model (8) is additive as in Eq. (10), and the common nonlinearity in the VLSTAR and VTAR models. The concept of common nonlinearity is defined as follows. Consider the model (8) with Eq. (10) assuming that it contains m equations. Suppose now that there is an $m \times r$ matrix $\mathbf{M}$, $r < m$, such that

$$\mathbf{M}'\mathbf{f}_1(\mathbf{y}_{t-1}, \ldots, \mathbf{y}_{t-p}; \mathbf{x}_t; \boldsymbol{\theta}_1) = \mathbf{0}$$

Consequently, nonlinearity in the relationship in Eq. (8) with Eq. (10) can be eliminated by defining a linear combination $\mathbf{z}_t = \mathbf{M}'\mathbf{y}_t$ which removes the nonlinear component $\mathbf{f}_1(\mathbf{y}_{t-1}, \ldots, \mathbf{y}_{t-p}; \mathbf{x}_t; \boldsymbol{\theta}_1)$ from the system. Using the terminology of Engle and Kozicki (1993), nonlinearity is a feature, and if it is eliminated by left-multiplying the Eq. (10) by $\mathbf{M}'$, it is a common feature.

Apply now this concept to the VLSTAR model (16). A necessary but not sufficient condition for common nonlinearity is that the transition function matrix

$$\mathbf{G}(\boldsymbol{\gamma}, \mathbf{c}; \mathbf{s}_t) = \mathrm{diag}\{G_1(\gamma_1, \mathbf{c}_1, s_{1t}), \ldots, G_m(\gamma_m, \mathbf{c}_m, s_{mt})\} = G(\gamma, \mathbf{c}; s_t)\mathbf{I}_m \tag{41}$$

that is, that the transition is common to all equations in the model. In Eq. (16), define the $m \times ((p+1)m+n)$ coefficient matrix $\boldsymbol{\Pi} = (\boldsymbol{\mu}_1\ \boldsymbol{\Psi}_1 \ldots \boldsymbol{\Psi}_p\ \boldsymbol{\Xi})$. For common nonlinearity it is required that $\mathbf{M}'\boldsymbol{\Pi} = \mathbf{0}$, that is, the $m \times r$ matrix $\mathbf{M}$, $r < m$, $\mathrm{rk}(\mathbf{M}) = r$, eliminates the nonlinear component from Eq. (16). This is a very strong condition if m, n, and p are large, as it requires that $\boldsymbol{\Pi}$ has a reduced rank. For this reason it is useful to define partial common nonlinearity such that the orthogonality condition only applies to a subset of columns of $\boldsymbol{\Pi}$. For example, consider the following m-dimensional first-order VLSTAR model

$$\mathbf{y}_t = \boldsymbol{\mu}_0 + \boldsymbol{\Phi}_1\mathbf{y}_{t-1} + (\boldsymbol{\mu}_1 + \boldsymbol{\Psi}_1\mathbf{y}_{t-1})G(\gamma, \mathbf{c}; s_t) + \boldsymbol{\varepsilon}_t \tag{42}$$

where $\boldsymbol{\varepsilon}_t \sim \mathrm{iid}\ \mathcal{N}(0, \boldsymbol{\Sigma})$. The interest may now be focussed on finding out whether the model may be transformed into a model in which there are $r < m$ linear combinations of the elements of $\mathbf{y}_t$ defined by $\mathbf{M}$ such that the resulting r-dimensional model is intercept-switching, that is, $\mathbf{M}'\boldsymbol{\Psi}_1 = 0$ in Eq. (42).

The definition of common nonlinearity is similar to that of co-breaking. As defined in Hendry and Massmann (2007), the matrix $\mathbf{M}$ is co-breaking of order q if left-multiplying a linear vector model with breaks with $\mathbf{M}'$ eliminates the breaks. A similar definition can be developed for a vector TAR model with a common threshold. Consider the following two-regime generalization of the VTAR model (11)

$$\begin{aligned}\mathbf{y}_t = \boldsymbol{\mu}_1 + \sum_{j=1}^{p}\boldsymbol{\Phi}_{1j}\mathbf{y}_{t-j} + \boldsymbol{\Gamma}_1\mathbf{x}_t \\ + \mathbf{I}(\mathbf{s}_t > \mathbf{c}_1)(\boldsymbol{\mu}_2 + \sum_{j=1}^{p}\boldsymbol{\Phi}_{2j}\mathbf{y}_{t-j} + \boldsymbol{\Gamma}_2\mathbf{x}_t) + \boldsymbol{\varepsilon}_t\end{aligned} \tag{43}$$

where the $m \times m$ indicator matrix

$$\mathbf{I}(\mathbf{s}_t > \mathbf{c}_1) = \mathrm{diag}(I(s_{1t} > c_1), \ldots, I(s_{mt} > c_m)) \tag{44}$$

A necessary condition for a common threshold is

$$\mathbf{I}(\mathbf{s}_t > \mathbf{c}_1) = I(s_t > c_1)\mathbf{I}_m \tag{45}$$

that is, $s_{1t} = ... = s_{mt}$ and $c_{11} = ... = c_{1m}$ in Eq. (44). This may be completed by a condition analogous to the one for vector STAR or STR models. As already discussed, the condition (45) is satisfied in the original VTAR or VSR model (11) of Tsay (1998).

Co-Shifting

Common nonlinearity in shifting-mean VAR models is called co-shifting. In this case, the transition functions have the same transition variable, and so a necessary condition for strong co-shifting is

$$\mathbf{G}_j(\boldsymbol{\gamma}_j, \mathbf{c}_j; t/T) = G(\gamma_j, c_j; t/T)\mathbf{I}_m, \quad j = 1, \ldots, q \tag{46}$$

This requires that all equations of the model have the same number of shifts. If q, the number of shifts, is less than m, the number of equations, the condition (46) is also sufficient. If $q \geq m$, then the necessary condition has to be completed by the condition

$$\mathbf{M}'(\boldsymbol{\delta}_1, \boldsymbol{\delta}_2, \ldots, \boldsymbol{\delta}_q) = \mathbf{0}$$

where $\mathbf{M}$ is again an $m \times q$ matrix, $\text{rk}(\mathbf{M}) = q$. Partial co-shifting may be defined analogously to partial common nonlinearity. There may then be a subset of diagonal matrices $\mathbf{G}_j(\boldsymbol{\gamma}_j, \mathbf{c}_j; t/T)$ satisfying (46), and if their number $q_1 < m$, then this is sufficient for weak co-shifting. For an application of co-shifting see Holt and Teräsvirta (2012) who fit an VSM-AR model to a pair of annual hemispheric temperature series and discuss comovements in them.

Testing Common Nonlinearity in VSTR Models

Anderson and Vahid (1998) consider testing common nonlinearity in a nonlinear model of additive type (10). In their notation, $\mathbf{z}_t$ is the $n \times 1$

vector of nonlinear regressors. It is implicitly assumed that the nonlinear model is linear in parameters. As an example, suppose that one is testing linearity against STAR equation by equation assuming that the transition variable is s_t using the technique of Luukkonen et al. (1988). Consider the ith equation of Eq. (42). Then, if s_t is not an element in $\mathbf{z}_t$, the auxiliary regression has the following form

$$y_{it} = \boldsymbol{\theta}_i'\mathbf{z}_t + \boldsymbol{\theta}_i'\mathbf{z}_t s_t + \boldsymbol{\theta}_i'\mathbf{z}_t s_t^2 + \boldsymbol{\theta}_i'\mathbf{z}_t s_t^3 + \varepsilon_{it}^*$$

and $\mathbf{z}_t = (\mathbf{z}_t's_t, \mathbf{z}_t's_t^2, \mathbf{z}_t's_t^3)'$ is a $3(p+1+n)\times 1$ vector. Let $\mathbf{y}_t^\dagger$ and $\mathbf{z}_t^\dagger$ be the corresponding variables after the linear influence of these variables have been eliminated. Then, there is common nonlinearity of at least dimension one, if there exists an $m\times 1$ vector $\mathbf{m}_1$ such that

$$\mathsf{E}(\mathbf{m}_1'\mathbf{y}_t^\dagger \otimes \mathbf{z}_t^\dagger) = 0.$$

This vector can be estimated using the generalized method of moments and the normalization constraint $\mathbf{m}_1'\mathbf{Y}'\mathbf{Y}\mathbf{m}_1 = 1$, where $\mathbf{Y}=(\mathbf{y}_1^{\dagger\prime}, \ldots, \mathbf{y}_T^{\dagger\prime})'$. Anderson and Vahid (1998) show that the solution is the eigenvector of $\mathbf{y}_t^\dagger$ that corresponds to the smallest squared canonical correlation λ_1^2 between $\mathbf{y}_t^\dagger$ and $\mathbf{z}_t^\dagger$. Under regularity conditions, the test statistic $Q_T = T\lambda_1^2$ has a χ^2-distribution with $3(p+1+n)\times 1$ degrees of freedom when the null hypothesis of a single common nonlinearity vector is valid. If one tests the null hypothesis of $q<m$ common nonlinearities, the statistic is $Q_T = T \sum_{i=1}^{q} \lambda_i^2$, where λ_i^2, $i=1,\ldots,q$, are the q smallest squared canonical correlations between $\mathbf{y}_t^\dagger$ and $\mathbf{z}_t^\dagger$.

For a full information maximum likelihood approach to this testing problem using techniques originally developed for cointegration analysis (see Hungnes, 2012).

Testing Common Nonlinearity in VSR Models

Anderson and Vahid (1998) base a test of common nonlinearity on the arranged regression of Tsay (1998) considered in Section "Testing Linearity"; see their article for details. Testing the hypothesis (45) directly against the alternative that the equations have the same switch-variable but different switch-points or thresholds would be difficult, because there is no statistical theory for testing the hypothesis $c_{11} = \ldots = c_{1m}$ in the VTAR model (43). There is, however, asymptotic distribution theory for testing the null hypothesis $c_1 = c_1^0$ in a univariate two-regime TAR model, $q = 2$ in Eq. (1). The likelihood ratio test statistic is

$$LR_T(c_1^0) = T\frac{Q(c_1^0) - Q(\hat{c}_1)}{Q(\hat{c}_1)}$$

where $Q(c_1^0)$ is the sum of squared residuals from the model estimated under the null hypothesis, and $Q(\hat{c}_1)$ is the same from the estimated alternative, Under a set of assumptions including $\varepsilon_{it} = \varepsilon_t \sim \text{iid}\,\mathcal{N}(0, \sigma^2)$; see Hansen (1997, 2000), $LR_T(c_1^0)$ has an analytic (nonstandard) asymptotic null distribution, and Hansen tabulates critical values for customary significance levels.

Tena and Tremayne (2009) use this result to test Eq. (45). They estimate a VTAR model equation by equation and choose an estimated threshold from one of the equations to represent c_1^0 and test this null hypothesis in the other equations. This is repeated for all equations. They also suggest testing the hypothesis $c_{1i} = 0$ equation by equation for $i = 1, \ldots, m$ and comparing the results, because $c_1^0 = 0$ is a sensible value in the application the authors are working with. Doing so only provides indirect evidence against the null hypothesis, but in some cases the results may be sufficiently convincing for the researcher to reject the null hypothesis, albeit without a formal significance level to support the decision.

GENERALIZED IMPULSE RESPONSE FUNCTIONS

Interpreting the coefficient estimates of a VLSTAR model is not possible, except possibly for $\hat{\gamma}$ and $\hat{c}$ that indicate the shape and location of the transition function. A similar remark is valid for VTAR models. In order to understand the dynamic behavior of the model, impulse response functions are most helpful. As Koop, Pesaran, and Potter (1996) have pointed out, estimating impulse response functions for nonlinear models is not as easy as it may be for linear VAR models. The reason is that the impulse response is no longer proportional to the size of the shock, nor is it independent of the "history," that is, the starting-values. These authors thus defined the so-called generalized impulse response (GIR) function which is a random variable and a function of both the size of the shock and the history. It is defined as follows

$$GIRF(h, \boldsymbol{\varepsilon}_t, \Omega_{t-1}) = \mathsf{E}\{\mathbf{y}_{t+h} | \boldsymbol{\varepsilon}_t^{\delta}, \Omega_{t-1}\} - \mathsf{E}\{\mathbf{y}_{t+h} | \Omega_{t-1}\} \quad (47)$$

where $\boldsymbol{\varepsilon}_t^\delta$ is the vector of shocks, and $\Omega_{t-1} = \{\omega_{t-j} : j \geq 1\}$ is the set of possible histories but can be replaced by an appropriate subset. The conditioning variables in Eq. (47), $\boldsymbol{\varepsilon}_t^\delta$ and Ω_{t-1}, are assumed random. That Ω_{t-1} is random means that its values in computing densities of the random GIR function are drawn from the set $\{\omega_{t-j} : j \geq 1\}$ or a subset.

As is the case for standard impulse responses from VAR models, estimating the GIR function requires ordering the variables in the vector $\mathbf{y}_t$. This is equivalent to making weak exogeneity assumptions in the model. Once this has been done, the GIR function can be estimated by simulation or bootstrapping the residuals of the estimated model. How this is done is described in Tena and Tremayne (2009). First draw a history from the set of histories. Second, form the Cholesky decomposition of the residual covariance matrix: $\hat{\boldsymbol{\Sigma}} = \hat{\mathbf{C}}\hat{\mathbf{C}}'$, where $\hat{\mathbf{C}}$ is a lower triangular matrix defined by the appropriate weak exogeneity assumptions. Then orthogonalize the residuals: $\mathbf{e}_s = \hat{\mathbf{C}}^{-1}\hat{\boldsymbol{\varepsilon}}_s$, $s = p+1, \ldots, T$. Next, either sample from the distribution of $\mathbf{e}_s$ or draw the residuals $\{\mathbf{e}_t^*, \mathbf{e}_{t+1}^*, \ldots, \mathbf{e}_{t+h}^*\}$ from this set with replacement. Suppose one wants to shock the ith variable. Then form another set by replacing the ith element of $\mathbf{e}_t^*$ by a shock e_i^δ drawn from a set of shocks. For example, one may be interested in shocks that are at least as large as the 75% quantile of the (empirical) distribution of the transformed residual. This yields another sequence of residuals $\{\mathbf{e}_t^\delta, \mathbf{e}_{t+1}^*, \ldots, \mathbf{e}_{t+h}^*\}$, where $\mathbf{e}_t^\delta = (e_{1t}, \ldots, e_{i-1,t}, e_i^\delta, e_{i+1,t}, \ldots, e_{mt})'$. Transform back to the original residuals: $\hat{\boldsymbol{\varepsilon}}_{t+j}^* = \hat{\mathbf{C}}\mathbf{e}_{t+j}^*$, $j = 1, \ldots, h$, and $\hat{\boldsymbol{\varepsilon}}_t^\delta = \hat{\mathbf{C}}\mathbf{e}_t^\delta$. Generate two sequences of $\mathbf{y}_{t+j}$, $j = 0, 1, \ldots, h$, from the estimated models using the two sequences of residuals and compute the difference element by element. This gives one observation of Eq. (47) for periods $t, t+1, \ldots, t+h$, when the system is shocked at time t.

An estimate of the conditional distribution of the random GIR function given Ω_{t-1} is generated by repeating these steps by new draws for $\mathbf{e}_t^\delta$ and thus generating new observations of Eq. (47). Repeating everything up until this phase with histories drawn from the whole set yields an unconditional empirical distribution of Eq. (47). This distribution may also be conditional if the size of the shock and/or a subset of histories is restricted. For example, the interest may lie in finding out whether certain shocks affect the output differently when the economy is in a recession than when it is in expansion. One may then estimate a GIR function for both cases by selecting appropriate subsets of histories and shocks and compare the results.

If one computes GIR functions for several horizons, as is normally the case, the question is how to present them. The highest density regions (HDR) for characterizing distributions, see Hyndman (1996) or Teräsvirta et al. (2010, Chap. 15), can be recommended for the purpose. They may be conveniently graphed using fan charts, see, for example, Wallis (1999) or Lundbergh et al. (2002). Teräsvirta and Yang (2013) apply box and whiskers graphs. The latter have the drawback that they miss potential multimodality of the HDR, so the user of those has to make sure that the estimated HDR is unimodal.

FORECASTING WITH VECTOR THRESHOLD AUTOREGRESSIVE AND SMOOTH TRANSITION MODELS

The forecasts considered here are conditional means: the forecast of $\mathbf{y}_{T+h}$ given the conditioning information $\mathcal{F}_T$ at time T, denoted $\mathbf{y}_{T+h|T}$, in theory equals

$$\mathbf{y}_{T+h|T} = \mathsf{E}\{\mathbf{y}_{T+h}|\mathcal{F}_T\} \tag{48}$$

Supposing that there is a nonlinear model, such as the VTAR or VSTAR model, $\mathcal{F}_T = \Omega_T$, the history of $\{\mathbf{y}_t\}$ up until T. This means that forecasting $\mathbf{y}_{T+h}$ has computationally much in common with computing GIR functions. Forecasting is in fact simpler as history is not a random variable. Furthermore, the residuals need not be orthogonalized if the error sequences $\{\boldsymbol{\varepsilon}_{T+1}^{(i)}, \ldots, \boldsymbol{\varepsilon}_{T+h}^{(i)}\}$, $i = 1, \ldots, B$, are generated by a bootstrap, that is, by B independent draws from the set of residuals $\{\hat{\boldsymbol{\varepsilon}}_{p+1}, \ldots, \hat{\boldsymbol{\varepsilon}}_T\}$. Some researchers construct the forecast by assuming $\boldsymbol{\varepsilon}_{T+1} = \ldots = \boldsymbol{\varepsilon}_{T+h} = \mathbf{0}$, but this simplification leads to biased forecasts. For discussion, see for example Teräsvirta (2006), Kock and Teräsvirta (2011) or Teräsvirta et al. (2010, Chap. 14). The forecast, an unbiased approximation of the conditional mean in Eq. (48), becomes

$$\mathbf{y}_{T+h|T} = (1/B)\sum_{j=1}^{B} \mathbf{y}_{T+h}^{(j)}$$

where $\mathbf{y}_{T+h}^{(j)}$ is the jth realization of $\mathbf{y}_{T+h}$ obtained from the model, given the history Ω_T, and the error sequence $\{\boldsymbol{\varepsilon}_{T+1}^{(j)}, \ldots, \boldsymbol{\varepsilon}_{T+h}^{(j)}\}$. Note that if one makes an assumption of the multivariate distribution of $\boldsymbol{\varepsilon}_t$ it becomes possible to also draw the error sequence from this distribution using the estimated error covariance matrix $\hat{\boldsymbol{\Sigma}}$. This requires that the errors are homoskedastic, whereas the bootstrap allows unconditional heteroskedasticity.

It seems that most applications focus on describing macroeconomic relationships rather than forecasting. The article by Galvão (2006) mentioned in Section "Nested Vector Threshold Autoregressive Model" constitutes an exception. This may be a reason for the fact that as yet there are few studies (Galvão's is one) comparing forecasting performance of nonlinear VAR models to that of linear VARs. Consequently, usefulness of VTAR and VSTAR models in macroeconomic forecasting is still an open question.

APPLICATIONS

Vector TAR and STAR models have been fitted to macroeconomic and financial time series. These models have been used to describe asymmetries, for example in the response of real variables to changes in monetary policy. A relevant question for monetary policy decisions and their evaluation is whether or not the response of these variables to monetary policy shocks varies nonlinearly according to the business cycle. The recent financial crisis has made the dependence of the interaction between the financial sector and the macroeconomy on the financial conditions an interesting research topic. In the following we consider four macroeconomic applications of vector TAR and STAR models that illustrate how these models can be used to address a variety of relevant economic questions. In addition to the empirical results, we highlight several model-building issues and choices.

Applications of Vector TAR Models

Both stationary VTAR and nonstationary TVEC models have been considered in the empirical literature. In this section we discuss two examples to illustrate both economic issues that might be investigated with this model class and different modelling approaches. We begin by considering Balke (2000) who discusses the possibility that credit plays a role of a nonlinear propagator of shocks to US macroeconomic variables. The author constructs a four-dimensional TVAR model in which the variables appear in the following order: output growth, inflation, the Fed funds rate and an indicator of credit conditions (measured in a number of alternative ways).

A noteworthy detail is that the two-regime TVAR model is in a triangular form

$$\mathbf{y}_t = \mathbf{A}\mathbf{y}_t + \mathbf{B}(\mathsf{L})\mathbf{y}_{t-1} + (\mathbf{C}\mathbf{y}_t + \mathbf{D}(\mathsf{L})\mathbf{y}_{t-1})I(s_{t-d} < c) + \boldsymbol{\upsilon}_t \tag{49}$$

where **A** and **C** are 4×4 upper triangular matrices with zeroes on the main diagonal, $\mathbf{B}(\mathsf{L}) = \sum_{j=1}^{p_b} \mathbf{B}_j \mathsf{L}^j$, $\mathbf{D}(\mathsf{L}) = \sum_{j=1}^{p_d} \mathbf{D}_j \mathsf{L}^j$, $\boldsymbol{\upsilon}_t$ is the error vector with uncorrelated elements, and s_t is the regime indicator. The upper triangular matrices determine the contemporaneous 'structural' relationships between the four variables and thus identify the shocks.

Since one does not know which one of the potential credit condition variables, if any, would be the switch or threshold variable, the analysis begins by testing linearity against (49). This is done by applying the method of Hansen (1996) and generating the null distribution of the test statistic by simulation. Balke considers both the supremum and the average Wald tests, as discussed in Section "Testing Linearity." His tests reject the linearity hypothesis for all three credit condition variables: the spread of the 4–6-month commercial paper and the 6-month Treasury bill, the mix of bank loans and commercial paper in total firm external finance, and difference of growth rate in short-term debt in small and large manufacturing firms. Balke then estimates a triangular TVAR model using each of these indicators as the threshold variable. He notes that the three indicators split the sample into two regimes in more or less the same way.

Balke (2000) does not report any evaluation results for the estimated model(s) but concentrates on generalized impulse responses. The set of histories is divided into periods of "normal" and "tight" credit conditions, and the shock is plus/minus one or two standard deviations computed from the distribution of the corresponding residuals. Results for both regimes are reported separately for the four different shocks assuming that the shocked variable is any of the four included in $\mathbf{y}_t$. What is graphed is not the density of the estimated GIR function but rather its mean. The shocks are identified as the model is triangular with properly ordered equations, and the errors independent. It is found that the shocks generally have a larger (average) effect on output growth when the economy is in the tight credit regime than when the opposite is true. Evidence of asymmetry in the response to shocks is also found.

Tena and Tremayne (2009) apply the VTAR model to studying the transmission of monetary policy to industrial output in the United Kingdom. Their article contains both a general (aggregate output) and a sectoral analysis, of which only the former is considered here. The reason why we review this particular application is that in setting up the model the authors do not assume a common threshold a priori, and modelling proceeds equation by equation. The seasonally adjusted quarterly time series are the industrial production index, the retail price index, an interest rate, and the Pound sterling/US Dollar exchange rate. The models are built on first differences of these series, and the period of observation is 1972(1)–2007(1). The variables appear in the vector to be modelled in the aforementioned order that is based on relevant weak exogeneity assumptions.

Linearity is tested using the lagged industrial production growth rate as the threshold variable. This is due to the fact that the interest lies in the asymmetries in the transmission mechanism which depends on the state of the economy. It turns out that the inflation (growth rate of the retail price index) equation is linear. Parameters are estimated equation by equation following Hansen (1997). The hypothesis of a common threshold for the three threshold equations is tested as discussed in Section "Testing Common Nonlinearity in VSR Models" and rejected. Evaluation involves testing for another threshold (Hansen, 2000), but the null hypothesis of a single threshold is not rejected.

The sectoral analysis also contains estimating GIR functions. Due to the focus on monetary policy, the interest rate is shocked. The size of the shock comprises two values, and a separate GIR function is estimated for them. Two sets of histories are considered: expansionary and recessionary episodes (measured using the industrial output). The whole HDR is not reported, but 95% confidence intervals from the estimated densities are given for horizons extending from one to ten quarters. The main conclusion is that negative interest rate shocks have a bigger impact on output than positive ones.

Most applications of the VTAR or TVEC model have been to macroeconomic series. But then, validity of the Law of One Price has been investigated by a TVEC model using prices of individual commodities in different cities as in Lo and Zivot (2001). Further, an empirical example in Tsay (1998) concerns the high-frequency (one-minute) S&P 500 stock index and the corresponding index futures data. These time series have also been modelled using the TVEC model; see Martens, Kofman, and Vorst (1998). A comprehensive list of applications of TAR models, both vector and univariate ones, can be found in Hansen (2011).

Applications of Vector STAR and STR Models

Camacho (2004) makes good use of the modelling strategy described in Section "Building Stationary Vector Smooth Transition Regression

Models" when building a VSTAR model for describing the relationship between the quarterly US GDP (y_{1t}) and the Composite Leading Indicator (CLI) of the Conference Board (y_{2t}). The time period is 1959(1)–2002(1). The latter variable is monthly, and the last month of every quarter is taken to construct a quarterly series to match the quarterly GDP series. Choosing the STVEC model as the starting-point allows for the possibility that these two nonstationary variables be cointegrated. Cointegration is tested for using the nonparametric cointegration test of Bierens (1997), the result being that the two variables are not cointegrated. This leads to eliminating the cointegration terms from Eq. (18).

The next step is to test linearity, which is done equation by equation. This is done using the Taylor expansion approach LM-type tests in Luukkonen et al. (1988). Camacho (2004) also considers the Exponential VSTAR (EVSTAR) model, which is Eq. (16) with the exponential transition function

$$G_j(\gamma_j, \mathbf{c}_j; s_{jt}) = 1 - \exp\{-\gamma_j(s_{jt} - c_j)^2\}, \quad \gamma_j > 0$$

in Eq. (17). Even a third VSTAR model, the so-called LVSTAR-D model is considered but not discussed in detail here. The linearity tests reject the null hypothesis for most transition variables, and test sequences as in Teräsvirta (1994) are used to select the transition function. This leaves the model builder with three alternatives: an LVSTAR, an EVSTAR and an LVSTAR-D model. A final choice is left to the evaluation stage. In the LVSTAR model, $s_t = \Delta y_{1,t-2}$ and in the EVSTAR one $s_t = \Delta y_{2,t-2}$ (in both cases the same transition variable for both equations). The estimated equations only contain one lag of each variable, except for the transition variables. A noteworthy detail is that the estimated models are intercept-switching STAR models; see Section "Definition."

The results of the evaluation stage indicate that the LVSTAR model is superior to the other two VSTAR variants. It has the best fit and passes most misspecification tests that include testing the hypotheses of no error autocorrelation, no additive nonlinearity, and parameter constancy. Furthermore, one-quarter-ahead real-time forecasts for the GDP growth from this model are more accurate than the corresponding ones from the other two models.

Camacho (2004) also carries out tests of equal forecasting ability with forecasts from the LVSTAR model as the benchmark. The idea is to construct the one-period-ahead forecasts from 1978(1) to 2002(2) in such a way that they as accurately as possible reflect the forecasting situation, that is, the information available for the forecasters at the time of forecasting.

The details can be found in the article. The main outcome is that all tests reject the null hypothesis of equal forecastability in favor of the intercept-switching LVSTAR forecasts.

The goals of Weise (1999) are to investigate whether (a) positive and negative monetary shocks have asymmetric effects on output and prices, (b) whether the effects of monetary shocks vary over the business cycle, and (c) whether the effects of monetary shocks vary disproportionately with the size of the shock. After first developing a simple structural model incorporating asymmetric nominal rigidities and discussing its potential implications for the nonlinear econometric model, the author chooses an LVSTAR framework (16) to pursue his goals. The quarterly US time series of first differences of the logarithmic industrial production, the consumer price index, and M1 cover the period from 1960(2) to 1995(2). Before proceeding, the series are filtered by regressing them separately on a constant, seasonal dummies, two intercept dummies accounting for presumed structural breaks, a linear trend, and two breaking linear trends in which the breaks are synchronized with the intercept dummies. This makes the filtered series stationary.

A linear trivariate VAR model with four lags forms the basis of the analysis. Linearity is tested using the likelihood ratio test approach discussed in Section "Specification and Estimation." It is found that linearity is never rejected when the transition variable is money (M1). Lags of the other two variables produce rejections, some of which are very strong. Selection of final models based on the results of the tests yields three LVSTAR models with the first lag of industrial production, the second lag of the consumption prices and the first lag of their differences as transition variables.

The models are estimated by assuming a single transition function and fixing the location parameter so it is not estimated. Estimation is equation by equation as discussed in Section "Estimation of Vector Smooth Transition Regression Models." No evaluation of the estimated models is carried out, but the dynamic properties of the estimated models are investigated by impulse response analysis. The interest lies in the response of output and prices to monetary shocks. As discussed in the seventh section, the functions are generated by drawing from different sets of histories and shocks. Monetary shocks are identified in the customary way by appropriately ordering the equations and applying the Cholesky decomposition in generating the shocks. Randomness of the GIR functions is not taken fully into account, however, as only averages and not complete estimated densities are reported. Shocks to the money supply are found to have stronger output effects and weaker price effects when output growth is initially low than when it is high. Positive and negative monetary shocks have nearly symmetric effects to these variables. Only shocks of ± 1 and ± 2 standard deviations have been considered.

Vector STAR and STR models have been applied less frequently than their threshold counterparts. Rothman et al. (2001) specify and estimate an LVSTAR model to investigate the direction of nonlinear Granger causality in the relationship between money and output in the United States. Rahman and Serletis (2010) apply the LVSTAR model to studying asymmetries in the effects of oil price and monetary policy shocks to output growth in the United States. Also using US time series, Auerbach and Gorodnichenko (2012) investigate whether fiscal policy is more effective in recessions than in expansions. Differences across regimes are allowed in the variance-covariances of the shocks and in the transmission. The authors find that fiscal policy is more effective in recessions. They also investigate fiscal multipliers of disaggregate spending variables and find that military spending has the largest multiplier of the ones considered.

FINAL REMARKS

Nonlinear models have become a useful tool in modelling economic relationships given the experiences gained and conclusions drawn from the financial crisis and the Great Recession. The purpose of this article is to provide an overview of VTAR and smooth transition models. This includes modelling nonstationary cointegrated variables with models of this kind. Model building is emphasized, and various testing situations described. Space limitations have made it necessary to restrict the discussion of applications to a few interesting examples.

Readers interested in more details about modelling strategies and issues of VTAR and smooth transition models discussed in this survey are referred to Tsay (1998) for vector TAR models and to Camacho (2004), Teräsvirta and Yang (2013) and Yang and Teräsvirta (2013) for their smooth transition counterparts. Hansen (2011) provides a comprehensive review of both univariate and vector TAR models and contains a large number of references. Gonzalo and Pitarakis (2006) present a useful survey of testing for cointegration in the TVEC framework; for an update see Gonzalo and Pitarakis (2013). For a more general discussion on vector nonlinear models, see Teräsvirta et al. (2010, Chap. 11).

ACKNOWLEDGMENT

This research has been supported by the Danish National Research Foundation.

REFERENCES

Anderson, H. M., & Vahid, F. (1998). Testing multiple equation systems for common nonlinear components. *Journal of Econometrics*, *84*, 1–36.

Astatkie, T., Watts, D. G., & Watt, W. E. (1997). Nested threshold autoregressive (NeTAR) models. *International Journal of Forecasting*, *13*, 105–116.

Atanasova, C. (2003). Credit market imperfections and business cycle dynamics: A nonlinear approach. *Studies in Nonlinear Dynamics & Econometrics*, *7*(4), Article 5.

Auerbach, A. J., & Gorodnichenko, Y. (2012). Measuring the output responses to fiscal policy. *American Economic Review: Economic Policy*, *4*, 1–27.

Bacon, D. W., & Watts, D. G. (1971). Estimating the transition between two intersecting straight lines. *Biometrika*, *58*, 525–534.

Balke, N. (2000). Credit and economic activity: Credit regimes and nonlinear propagation of shocks. *Review of Economics and Statistics*, *82*, 344–349.

Balke, N. S., & Fomby, T. B. (1997). Threshold cointegration. *International Economic Review*, *38*, 627–645.

Bauwens, L., Lubrano, M., & Richard, J.-F. (1999). *Bayesian inference in dynamic econometric models*. Oxford: Oxford University Press.

Bec, F., & Rahbek, A. (2004). Vector equilibrium correction models with non-linear discontinuous adjustments. *Econometrics Journal*, *7*, 628–651.

Bierens, H. J. (1997). Nonparametric cointegration analysis. *Journal of Econometrics*, *77*, 379–404.

Bollerslev, T. (1990). Modelling the coherence in short-run nominal exchange rates: A multivariate generalized ARCH model. *Review of Economics and Statistics*, *72*, 498–505.

Box, G. E. P., & Jenkins, G. M. (1970). *Time series analysis, forecasting and control*. San Francisco, CA: Holden-Day.

Camacho, M. (2004). Vector smooth transition regression models for US GDP and the composite index of leading indicators. *Journal of Forecasting*, *23*, 173–196.

Chan, K. S. (1993). Consistency and limiting distribution of the least squares estimator of a threshold autoregressive model. *Annals of Statistics*, *21*, 521–533.

Chan, K. S., & Tong, H. (1986). On estimating thresholds in autoregressive models. *Journal of Time Series Analysis*, *7*, 178–190.

Davies, R. B. (1977). Hypothesis testing when a nuisance parameter is present only under the alternative. *Biometrika*, *64*, 247–254.

Dueker, M. J., Psaradakis, Z., Sola, M., & Spagnolo, F. (2011). Multivariate contemporaneous-threshold smooth transition autoregressive models. *Journal of Econometrics*, *160*, 311–325.

Eitrheim, Ø., & Teräsvirta, T. (1996). Testing the adequacy of smooth transition autoregressive models. *Journal of Econometrics*, *74*, 59–75.

Eklund, B., & Teräsvirta, T. (2007). Testing constancy of the error covariance matrix in vector models. *Journal of Econometrics*, *140*, 753–780.

El-Shagi, M. (2011). An evolutionary algorithm for the estimation of threshold vector error correction models. *International Economics and Economic Policy*, *8*, 341–362.

Engle, R. F., & Kozicki, S. (1993). Testing for common features. *Journal of Business and Economic Statistics*, *11*, 369–386.

Galvão, A. B. C. (2006). Structural break threshold VARs for predicting US recessions using the spread. *Journal of Applied Econometrics*, *21*, 463–487.

Gefang, D. (2012). Money-output causality revisited – A Bayesian logistic smooth transition VECM perspective. *Oxford Bulletin in Economics and Statistics*, *74*, 131–151.

Gefang, D., & Strachan, R. (2010). Nonlinear impacts of business cycles on the U.K. – A Bayesian smooth transition VAR approach. *Studies in Nonlinear Dynamics and Econometrics*, *14*(1), Article 2.

González, A., Hubrich, K., & Teräsvirta, T. (2009). *Forecasting inflation with gradual regime shifts and exogenous information*. CREATES Research Paper 2009-3, Aarhus University, Denmark.

González, A., & Teräsvirta, T. (2008). Modelling autoregressive processes with a shifting mean. *Studies in Nonlinear Dynamics and Econometrics*, *12*(1), Article 1.

Gonzalo, J., & Pitarakis, J.-Y. (2006). Threshold effects in multivariate error correction models. In T. C. Mills & K. Patterson (Eds.), *Palgrave Handbooks of Econometrics, Econometric Theory* (Vol. 1, 578–609). Basingstoke: Palgrave Macmillan.

Gonzalo, J. & Pitarakis, J.-Y. (2013). Estimation and inference in threshold type regime switching models. In N. Hashimzade & M. A. Thornton (Eds.), *Handbook of research methods and applications in empirical macroeconomics* (pp. 189–204). Cheltenham: Edward Elgar.

Goodwin, B. K., Holt, M. T., & Prestemon, J. P. (2012). *Exchange rate pass-through in timber products: The case of oriented strand board in Canada and the United States*. Technical report, University of Alabama at Tuscaloosa.

Haggan, V., & Ozaki, T. (1981). Modelling non-linear random vibrations using an amplitude-dependent autoregressive time series model. *Biometrika*, *68*, 189–196.

Hamilton, J. D. (1989). A new approach to the economic analysis of nonstationary time series and the business cycle. *Econometrica*, *57*, 357–384.

Hansen, B. E. (1996). Inference when a nuisance parameter is not identified under the null hypothesis. *Econometrica*, *64*, 413–430.

Hansen, B. E. (1997). Inference in TAR models. *Studies in Nonlinear Dynamics and Econometrics*, *2*, 1–14.

Hansen, B. E. (1999). Testing for linearity. *Journal of Economic Surveys*, *13*, 551–576.

Hansen, B. E. (2000). Sample splitting and threshold estimation. *Econometrica*, *68*, 575–603.

Hansen, B. E. (2011). Threshold autoregression in economics. *Statistics and Its Interface*, *4*, 123–127.

Hansen, B. E., & Seo, B. (2002). Testing for two-regime threshold cointegration in vector error-correction models. *Journal of Econometrics*, *110*, 293–318.

Harvill, J. L., & Ray, B. K. (1999). A note on tests for nonlinearity in a vector time series. *Biometrika*, *86*, 728–734.

He, C., Teräsvirta, T., & González, A. (2009). Testing parameter constancy in vector autoregressive models against continuous change. *Econometric Reviews*, *28*, 225–245.

Hendry, D. F., & Massmann, M. (2007). Co-breaking: Recent advances and a synopsis of the literature. *Journal of Business and Economic Statistics*, *25*, 33–51.

Holt, M. T., & Teräsvirta, T. (2012). *Global hemispheric temperature trends and co-trending: A shifting mean vector autoregressive analysis*. CREATES Research Paper 2012-54, Aarhus University, Denmark.

Hubrich, K., & Tetlow, R. J. (2012). *Financial stress and economic dynamics: The transmission of crises*. Finance and Economics Discussion Series 2012-82, The Federal Reserve Board.

Hungnes, H. (2012). *Testing for co-non-linearity*. Discussion Paper 699, Statistics Norway, Research Department.

Hyndman, R. J. (1996). Computing and graphing highest density regions. *American Statistician*, *50*, 120–126.

Jacobson, T., Lindh, T., & Warne, A. (2002). Growth, saving, financial markets and Markov switching regimes. *Studies in Nonlinear Dynamics and Econometrics*, *5*(4), 242–259. Article 1.

Johansen, S. (1995). *Likelihood-based inference in cointegrated vector autoregressive models*. Oxford: Oxford University Press.

Kapetanios, G., Shin, Y., & Snell, A. (2006). Testing for cointegration in nonlinear smooth transition error correction models. *Econometric Theory*, *22*, 279–303.

Kock, A. B., & Teräsvirta, T. (2011). Forecasting with nonlinear time series models. In M. P. Clements & D. F. Hendry (Eds.), *The Oxford Handbook of Economic Forecasting* (pp. 61–87). Oxford: Oxford University Press.

Koop, G., Pesaran, M. H., & Potter, S. M. (1996). Impulse response analysis in nonlinear multivariate models. *Journal of Econometrics*, *74*, 119–147.

Krishnakumar, J., & Neto, D. (2012). Testing uncovered interest rate parity and term structure using a three-regime threshold unit root VECM. An application to the Swiss 'isle' of interest rates, *Oxford Bulletin of Economics and Statistics*, *74*, 180–202.

Kristensen, D., & Rahbek, A. (2010). Nonlinear error-correction models. *Journal of Econometrics*, *158*, 78–94.

Kristensen, D., & Rahbek, A. (2013). Testing and inference in nonlinear cointegrating vector error correction models. *Econometric Theory*, *29*, 1238–1288.

Krolzig, H.-M. (1997). *Markov-Switching Vector Autoregressions: Modelling, Statistical Inference and Applications to Business Cycle Analysis*. Berlin: Springer.

Lanne, M., & Saikkonen, P. (2002). Threshold autoregressions for strongly autocorrelated time series. *Journal of Business and Economic Statistics*, *20*, 282–289.

Lindgren, G. (1978). Markov regime models for mixed distributions and switching regressions. *Scandinavian Journal of Statistics*, *5*, 81–91.

Lo, M. C., & Zivot, E. (2001). Threshold cointegration and nonlinear adjustment to the law of one price. *Macroeconomic Dynamics*, *5*, 533–576.

Lundbergh, S., & Teräsvirta, T. (2002). Forecasting with smooth transition autoregressive models. In M. P. Clements, & D. F. Hendry (Eds.), *A Companion to Economic Forecasting* (pp. 485–509). Oxford: Blackwell.

Lundbergh, S., Teräsvirta, T., & van Dijk, D. (2003). Time-varying smooth transition autoregressive models. *Journal of Business and Economic Statistics*, *21*, 104–121.

Lütkepohl, H. (2006). Vector autoregressive models. In T. C. Mills & K. Patterson (Eds.), Palgrave handbook of econometrics (Vol. 1, pp. 477–510). Economic Theory. Basingstoke: Palgrave Macmillan.

Luukkonen, R., Saikkonen, P., & Teräsvirta, T. (1988). Testing linearity against smooth transition autoregressive models. *Biometrika*, *75*, 491–499.

Maddala, D. S. (1977). *Econometrics*. New York, NY: McGraw-Hill.

Martens, M., Kofman, P., & Vorst, T. C. F. (1998). A threshold error-correction models for intraday futures and index returns. *Journal of Applied Econometrics*, *13*, 245–263.

Pfann, G. A., Schotman, P. C., & Tschernig, R. (1996). Nonlinear interest rate dynamics and implications for the term structure. *Journal of Econometrics*, *74*, 149–176.

Qian, L. (1998). On maximum likelihood estimators for a threshold autoregression. *Journal of Statistical Planning and Inference, 75*, 21–46.

Quandt, R. E. (1958). The estimation of parameters of a linear regression system obeying two separate regimes. *Journal of the American Statistical Association, 53*, 873–880.

Quandt, R. E. (1960). Tests of the hypothesis that a linear regression system obeys two separate regimes. *Journal of the American Statistical Association, 55*, 324–330.

Rahman, S., & Serletis, A. (2010). The asymmetric effects of oil price and monetary policy shocks: A nonlinear VAR approach. *Energy Economics, 32*, 1460–1466.

Rao, C. R. (1973). *Linear Statistical Inference and its Applications.* New York, NY: Wiley.

Rothman, P., van Dijk, D., & Franses, P. H. (2001). A multivariate STAR analysis of the relationship between money and output. *Macroeconomic Dynamics, 5*, 506–532.

Saikkonen, P. (2005). Stability results for nonlinear error correction models. *Journal of Econometrics, 127*, 69–81.

Saikkonen, P. (2008). Stability of regime switching error correction models under linear cointegration. *Econometric Theory, 24*, 294–318.

Seo, B. (2011a). Nonparametric testing for linearity in cointegrated error-correction models. *Studies in Nonlinear Dynamics and Econometrics, 15*(2), Article 6.

Seo, M. (2006). Bootstrap testing for the null of no cointegration in a threshold vector error correction model. *Journal of Econometrics, 134*, 129–150.

Seo, M. H. (2011b). Estimation of nonlinear error correction models. *Econometric Theory, 27*, 201–234.

Sims, C., & Zha, T. (2006). Were there regime shifts in U.S. monetary policy? *American Economic Review, 96*, 54–81.

Sims, C. A. (1980). Macroeconomics and reality. *Econometrica, 48*, 1–48.

Sims, C. A., Waggoner, D. F., & Zha, T. (2008). Methods for inference in large multple equation Markov-switching models. *Journal of Econometrics, 146*, 255–274.

Strikholm, B., & Teräsvirta, T. (2006). A sequential procedure for determining the number of regimes in a threshold autoregressive model. *Econometrics Journal, 9*, 472–491.

Taylor, M. P., Peel, D. A., & Sarno, L. (2001). Nonlinear mean-reversion in real exchange rates: Towards a solution to the Purchasing Power Parity puzzles. *International Economic Review, 42*, 1015–1042.

Tena, J. D., & Tremayne, A. R. (2009). Modelling monetary transmission in UK manufacturing industry. *Economic Modelling, 26*, 1053–1066.

Teräsvirta, T. (1994). Specification, estimation, and evaluation of smooth transition autoregressive models. *Journal of the American Statistical Association, 89*, 208–218.

Teräsvirta, T. (1998). Modeling economic relationships with smooth transition regressions. In A. Ullah and D. E. Giles (Eds.), *Handbook of Applied Economic Statistics* (pp. 507–552). New York, NY: Dekker.

Teräsvirta, T. (2006). Forecasting economic variables with nonlinear models. In G. Elliott, C. W. J. Granger, & A. Timmermann (Eds.), *Handbook of Economic Forecasting* (Vol. 1, pp. 413–457). Amsterdam: Elsevier.

Teräsvirta, T., Lin, C.-F., & Granger, C. W. J. (1993). Power of the neural network linearity test. *Journal of Time Series Analysis, 14*, 309–323.

Teräsvirta, T., Tjøstheim, D., & Granger, C. W. J. (2010). *Modelling Nonlinear Economic Time Series.* Oxford: Oxford University Press.

Teräsvirta, T., & Yang, Y. (2013). *Specification, estimation and evaluation of vector smooth transition autoregressive models with applications.* CREATES Research Paper, Aarhus University, Denmark.

Tong, H. (1990). *Non-Linear Time Series. A dynamical system approach*. Oxford: Oxford University Press.

Tsay, R. S. (1986). Nonlinearity tests for time series. *Biometrika*, *73*, 461–466.

Tsay, R. S. (1989). Testing and modeling threshold autoregressive processes. *Journal of the American Statistical Association*, *84*, 231–240.

Tsay, R. S. (1998). Testing and modeling multivariate threshold models. *Journal of the American Statistical Association*, *93*, 1188–1202.

Wallis, K. F. (1999). Asymmetric density forecasts of inflation and the Bank of England's fan chart. *National Institute Economic Review*, (January), 106–112.

Weise, C. L. (1999). The asymmetric effect of monetary policy: A nonlinear vector autoregression approach. *Journal of Money, Credit, and Banking*, *31*, 85–108.

Winker, P., & Gilli, M. (2004). Applications of optimization heuristics to estimation and modelling problems. *Computational Statistics and Data Analysis*, *47*, 211–223.

Yang, Y. (2013). *Testing constancy of the error covariance matrix in vector models against parametric alternatives using a spectral decomposition*. CREATES Research Paper, Aarhus University, Denmark.

Yang, Y., & Teräsvirta, T. (2013). *Linearity and misspecification tests for vector smooth transition regression models*. CREATES Research Paper, Aarhus University, Denmark.

Yang, Z., Tian, Z., & Yuan, Z. (2007). GSA-based maximum likelihood estimation for threshold vector error correction model. *Computational Statistics and Data Analysis*, *52*, 109–120.

NONPARAMETRIC VECTOR AUTOREGRESSIONS: SPECIFICATION, ESTIMATION, AND INFERENCE

Ivan Jeliazkov

ABSTRACT

For over three decades, vector autoregressions have played a central role in empirical macroeconomics. These models are general, can capture sophisticated dynamic behavior, and can be extended to include features such as structural instability, time-varying parameters, dynamic factors, threshold-crossing behavior, and discrete outcomes. Building upon growing evidence that the assumption of linearity may be undesirable in modeling certain macroeconomic relationships, this article seeks to add to recent advances in VAR modeling by proposing a nonparametric dynamic model for multivariate time series. In this model, the problems of modeling and estimation are approached from a hierarchical Bayesian perspective. The article considers the issues of identification, estimation, and model comparison, enabling nonparametric VAR (or NPVAR) models to be fit efficiently by Markov chain Monte Carlo (MCMC) algorithms and compared to parametric and semiparametric alternatives

VAR Models in Macroeconomics – New Developments and Applications: Essays in Honor of Christopher A. Sims
Advances in Econometrics, Volume 32, 327–359

ISSN: 0731-9053/doi:10.1108/S0731-9053(2013)0000031009

by marginal likelihoods and Bayes factors. Among other benefits, the methodology allows for a more careful study of structural instability while guarding against the possibility of unaccounted nonlinearity in otherwise stable economic relationships. Extensions of the proposed non-parametric model to settings with heteroskedasticity and other important modeling features are also considered. The techniques are employed to study the postwar U.S. economy, confirming the presence of distinct volatility regimes and supporting the contention that certain nonlinear relationships in the data can remain undetected by standard models.

Keywords: Additive model; VAR model; Bayesian model comparison; Markov chain Monte Carlo

JEL classifications: C11; C14; C15; C32; C52; E31; E32; E37; E43; E47

INTRODUCTION AND MOTIVATION

Following the seminal work of Sims (1980), the vector autoregressive (VAR) model has played a central role in empirical macroeconomics. The basic model postulates that a q-dimensional vector of time-series variables $\boldsymbol{y}_t = (y_{1t}, \ldots, y_{qt})'$ depends on its past realizations through the specification

$$\boldsymbol{y}_t = \boldsymbol{c} + \sum_{j=1}^{p} \boldsymbol{B}_j \boldsymbol{y}_{t-j} + \boldsymbol{\varepsilon}_t, \quad t = 1, \ldots, T \tag{1}$$

where $\mathbf{c}$ is a q-vector of intercepts, $\{\boldsymbol{B}_j\}_{j=1}^{p}$ are $q \times q$ matrices of parameters, $\{\boldsymbol{y}_{t-j}\}_{j=1}^{p}$ are lags of $\boldsymbol{y}_t$, and $\boldsymbol{\varepsilon}_t$ is an error term with mean zero and $q \times q$ covariance matrix $\boldsymbol{\Omega}$. VAR models are general and can capture sophisticated dynamic behavior, even when the lag length p is relatively small. Moreover, VAR models are very versatile – in the past few decades they have been adapted to incorporate structural instability, regime switching, time-varying parameters, dynamic factors, threshold-crossing behavior, and discrete data, among others. Consequently, VAR methodology has been an important instrument in policy analysis, forecasting, and academic discourse (for a recent review, see Koop & Korobilis, 2009).

While many extensions of the model in Eq. (1) are possible, it has been common practice to maintain a parametric, typically linear, functional form for the conditional mean of y_t given its lags. Important extensions of the basic setup are afforded by models in which the parameters are allowed to change over time as in regime switching, changepoint, time-varying parameter, and threshold models.[1] For instance, following Hamilton (1989), much work has been done on estimating models subject to regime shifts in the mean, variance, or dynamics (e.g., Chauvet, 1998; Chib, 1996; Hansen, 1992; Kim & Nelson, 1999; Kim, Morley, & Piger, 2005; Sims & Zha, 2006). Threshold regressions have been considered in Beaudry and Koop (1993), Potter (1995), and Pesaran and Potter (1997), while time-varying parameter applications have been examined in Canova (1993), Stock and Watson (1996), Cogley and Sargent (2001), Primiceri (2005), and Chan and Jeliazkov (2009), among others. Much more rare in time-series analysis has been the application of nonparametric methods in the modeling and estimation of the conditional mean of a time-series process, and this represents the basic econometric problem motivating this work.

The discussion in this article is primarily concerned with methods for allowing considerable flexibility in estimating the dependence on lags in VAR models, while maintaining simplicity, computational tractability, and accommodating other modeling features that may be present in the model. The VAR paradigm came to prominence as a methodological framework involving only minimal restrictions. Consequently, continuing in this tradition, this article discusses ways of estimating multivariate dynamic systems without assuming a priori knowledge of the functional form. Even though the nonparametric literature is vast and diverse, applications to time series have been limited despite their potential appeal and importance. In an application to exchange rates, Härdle, Tsybakov, and Yang (1998) use local polynomial methods based on kernel-weighted least squares, to estimate a nonparametric bivariate dynamic system in which both the conditional mean and variance are unknown functions of the past. The methodology in that paper relates to single-equation techniques and exchange rate applications considered in Härdle and Tsybakov (1997) and Yang, Härdle, and Nielsen (1999). Univariate nonparametric regressions have been used in Dahl and Gonzalez-Rivera (2003a) to study the evolution of U.S. GNP growth using the method of Hamilton (2001), who employed the techniques to address nonlinearity in the inflation-unemployment trade-off in an example involving the Phillips curve. Dahl and Gonzalez-Rivera (2003b) apply the methodology to study the evolution of industrial production for a subsample of OECD countries. Their results support the contention that much nonlinearity is neglected if standard linear models are applied in these settings.

These and other papers have provided a growing body of evidence that allowing for various kinds of nonlinearity in empirical macroeconomics can be very valuable in uncovering important features of time-series relationships. Building upon these advances, this article seeks to add to the literature by examining a nonparametric dynamic model for multivariate time series. Specifically, this article considers a dynamic system of q regression equations for data $\{\boldsymbol{y}_t\}_{t=1}^{T}$, where $\boldsymbol{y}_t = (y_{1t}, \ldots, y_{qt})'$, in which the ith equation $(i=1, \ldots, q)$ is modeled through the additive form (Hastie & Tibshirani, 1990)

$$y_{it} = \sum_{j=1}^{q} \sum_{k=1}^{p} g_{ijk}(y_{j,t-k}) + \varepsilon_{it}, \quad t = 1, \ldots, T \tag{2}$$

where $\varepsilon_t = (\varepsilon_{1t}, \ldots, \varepsilon_{qt})' \sim N(\mathbf{0}, \boldsymbol{\Omega})$ and the unknown functions $\{g_{ijk}(y_{j,t-k})\}$ will be modeled and estimated nonparametrically. For this reason, in the remainder of this article, the specification in Eq. (2) will be referred to as a nonparametric VAR (or NPVAR) model. The model in Eq. (2) provides a natural extension of the traditional linear VAR model in Eq. (1) – relative to its parametric counterpart, the NPVAR model maintains additivity but does not require that the estimated regression relationships lie in a particular class of functions. Functional flexibility is desirable because nonlinearity is common in both economic theory and practice. Moreover, nonparametric additive modeling has desirable practical and theoretical properties and can serve as a useful exploratory tool that is easily inserted in more complex models. Even though the estimation of unknown functions is a complex high-dimensional problem, the additive framework is well suited for dealing with the "curse of dimensionality" because the argument of each function is a single variable.

The specification of the NPVAR model will be approached from a hierarchical Bayesian perspective with special emphasis on the issues of identification, estimation, and model comparison, enabling NPVAR models to be fit efficiently by Markov chain Monte Carlo (MCMC) algorithms and compared to nested and non-nested parametric and semiparametric alternatives by marginal likelihoods and Bayes factors. The methodology is useful in its own right as an exploratory and modeling tool, but is also appealing because it enables a more careful study of other structural features while guarding against the possibility of unaccounted nonlinearity. Doing so is important for theoretical and practical reasons, and because the consequences of ignored nonlinearity can be severe.

The types of misspecification that arise from assuming an inappropriate functional form can be illustrated by considering two simple motivating examples. Imagine that data are generated from the model $y_t = g(x_t) + \varepsilon_t$, $\varepsilon_t \sim N(0, \sigma^2)$, and $g(\cdot)$ is the nonlinear function in panel (a) of Fig. 1. If estimation is by linear regression (the resulting regression line is also shown in the panel (a) of Fig. 1), it is easy to see that the regression residuals will be heteroskedastic, owing to the neglected nonlinearity in $g(\cdot)$. If the covariate x_t is a lag of y_t, the misspecification can also lead to erroneous findings of serial correlation in the errors. Furthermore, even though the original errors used to generate the data were Gaussian, in a linear regression they will appear non-Gaussian (see panel (b) of Fig. 1). Due to the omitted nonlinearity, the error distribution will be a location mixture of normals, and consequently one would conclude that the Gaussian assumption is inadequate when the real culprit is neglected nonlinearity. Note that these problems will not be resolved by using estimators that are robust to distributional misspecification.

For our second example, consider panel (c) of Fig. 1. In this case, imagine that the researcher is aware that the data generating process involves a nonlinear mean, but chooses to restrict attention to the class of piecewise linear polynomials. Although heteroskedasticity, autocorrelation, and non-Gaussianity may not be significant (or even discernable) problems if a bilinear model is fit to the particular data in panel (c) of the Fig. 1, one can erroneously conclude that there is evidence of structural instability. For instance, regime switching, changepoint, threshold, and time-varying parameter models may appreciably improve the fit relative to a linear model,

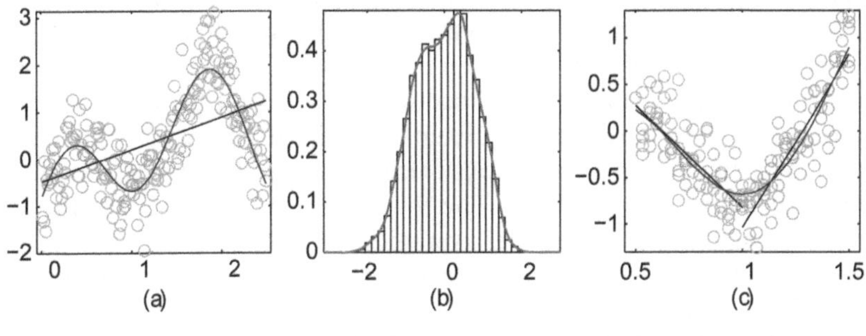

Fig. 1. Ignored nonlinearity can lead to erroneous conclusions about the presence of heteroskedasticity or autocorrelation, the adequacy of the distributional assumptions, the structural stability of the regression, or lead to conclusions that more profligate models are required.

although one should bear in mind that these would be spurious findings of instability or structural breaks since the underlying data generating process is a stable, although nonlinear, function of the covariates that is not properly accommodated in the regressions. Such spurious instability, unfortunately, is not the only pitfall that can be induced by this type of misspecification. The poor fit of low order linear dynamic systems may also lead researchers to explore more profligate models involving more lags. This would lead to loss of parsimony as additional lag components are incorporated but simply act as atheoretical fitting parameters in the model.

These examples provide strong motivation for studying the NPVAR model because the problems they identify can not be addressed satisfactorily without directly addressing the flexibility of the functional form. This, of course, is not to say that features such as heteroskedasticity, autocorrelation, non-Gaussianity, or structural instability can not be present in nonlinear models. On the contrary, they can be important integral parts of the NPVAR model, and many such extensions will be considered in the fifth section and the application in the sixth section. However, the examples do suggest that before jumping to conclusions about the presence of any of the aforementioned features, one must ensure that they are not spuriously induced by functional form misspecifcation. To enable this task to be carried out, this article provides methodology for the specification, estimation, and comparison of nonparametric models, which can be useful in this pursuit, as demonstrated in a study of U.S. macroeconomic data. The application reveals that the NPVAR model supports the existence of distinct volatility regimes in the data, and provides evidence that means remain stable but exhibit interesting nonlinearities.

The remainder of this article deals with the hierarchical structure of NPVAR models and their implementation in practice. Specifically, the second section presents the specification of the NPVAR model together with a computationally convenient identification restriction on the unknown additive functions. The third section presents an efficient fitting algorithm based on MCMC simulation techniques, which subsumes frequentist estimation by backfitting as a special case. The fourth section addresses the problem of model comparison and model averaging by discussing the computation of marginal likelihoods and Bayes factors. The fifth section outlines extensions to settings with heteroskedasticity, Student-*t* errors, structural instability, heteroskedasticity or stochastic volatility, dynamic factors, and discrete outcomes, and provides references to the relevant literature. The sixth section considers the application of the NPVAR model to data for the postwar U.S. economy, whereas the seventh section offers concluding remarks.

HIERARCHICAL MODEL SPECIFICATION

The NPVAR model will be specified through the following distributional hierarchy. The likelihood function $f(y|\{g_{ijk}(\cdot)\},\mathbf{\Omega})$ obtained from the additive model in Eq. (2) will be augmented with a prior distribution (or model) for the each unknown function $\pi(g_{ijk}(\cdot)|\tau^2_{ijk})$ that will, in turn, depend on a hyperparameter τ^2_{ijk}. The hierarchy will be completed by the priors on τ^2_{ijk} and $\mathbf{\Omega}$, denoted by $\pi(\tau^2_{ijk})$ and $\pi(\mathbf{\Omega})$, respectively. The prior $\pi(g_{ijk}(\cdot)|\tau^2_{ijk})$ is often referred to as a "smoothness prior" because it aims at penalizing rough functions but does not absolutely rule out any values that the function can take. The prior is very similar to the roughness penalty in frequentist penalized likelihood estimation. The parameters τ^2_{ijk} are often called smoothness parameters because they control the degree of smoothness of $g_{ijk}(\cdot)$ in $\pi(g_{ijk}(\cdot)|\tau^2_{ijk})$. Details will be provided in the remainder of this section, but it is important to note that the methodology leans on a vast literature in Bayesian nonparametric estimation in a variety of areas with continuous, discrete, and censored responses, including cross-sectional settings (Besag, Green, Higdon, & Mengersen, 1995; Fahrmeir & Lang, 2001; Hastie & Tibshirani, 2000; Koop & Poirier, 2004; Wood & Kohn, 1998; Wood, Kohn, Shively, & Jiang, 2002), multiple equation systems (Holmes, Denison, & Mallick, 2002; Koop, Poirier, & Tobias, 2005; Smith & Kohn, 2000), panel data (Chib & Jeliazkov, 2006), sample selection models (Chib & Greenberg, 2007; Chib, Greenberg, & Jeliazkov, 2009), and time-series applications (Hamilton, 2001). Extensions to Bayesian models with free-knot splines have been pursued in Denison, Mallick, and Smith (1998) and DiMatteo, Genovese, and Kass (2001), while Bayesian estimation techniques for multivariate functions have been provided in Shively, Kohn, and Wood (1999) and Wood et al. (2002). Useful reviews and introduction to many aspects of nonparametric modeling can be found in Hastie and Tibshirani (1990), Denison, Holmes, Mallick, and Smith (2002), Koop (2003), Ruppert, Wand, and Carroll (2003),Wasserman (2006), and Ahamada and Flachaire (2010).[2] Nonparametric functional modeling has appealing frequentist and Bayesian properties, and many of its advantages have been illustrated in the aforementioned works.

Some simplification in the notation can be obtained by denoting the $r=qp$ lagged variables on the right hand side of the ith equation ($i=1,\ldots,q$) in Eq. (2) by $\{s_{ijt}\}_{j=1}^{r}$ and writing that equation as

$$y_{it}=g_{i1}(s_{i1t})+\ldots+g_{ir}(s_{irt})+\varepsilon_{it}, \quad (t=1,\ldots,T) \tag{3}$$

Then, to motivate the hierarchical model for the functions, it is useful to stack the observations and write the model in matrix notation. Let $\boldsymbol{y}_i = (y_{i1}, \ldots, y_{iT})'$, $\varepsilon_i = (\varepsilon_{i1}, \ldots, \varepsilon_{iT})'$, and for each of the $j = 1, \ldots, r$ functions in Eq. (3), let the T observations in the covariate vectors $\boldsymbol{s}_{ij} = (s_{ij1}, \ldots, s_{ijT})'$, determine the corresponding $m_j \times 1$ *design point vectors* $\boldsymbol{v}_{ij} = (v_{ij1}, \ldots, v_{ijm_j})'$ with entries equal the *unique ordered* values of $\boldsymbol{s}_{ij}$, that is $v_{ij1} < \ldots < v_{ijm_j}$. Let the corresponding function evaluation vectors be denoted by $\boldsymbol{g}_{ij} = (g_{ij}(v_{ij1}), \ldots, g_{ij}(v_{ijm_j}))'$. Then, stacking over time, the ith equation of the system can be written in matrix notation as

$$\boldsymbol{y}_i = \boldsymbol{Q}_{i1}\boldsymbol{g}_{i1} + \boldsymbol{Q}_{i2}\boldsymbol{g}_{i2} + \ldots + \boldsymbol{Q}_{ir}\boldsymbol{g}_{ir} + \boldsymbol{\varepsilon}_i \tag{4}$$

where $\boldsymbol{Q}_{ij}$ are $T \times m_{ij}$ *incidence matrices* with entries $\boldsymbol{Q}_{ij}(h,k) = 1$ if $s_{ijh} = v_{ijk}$ and 0 otherwise, which establishes the correspondence between $\boldsymbol{s}_{ij}$ and $\boldsymbol{v}_{ij}$. Note that because there may be repeating values in $\boldsymbol{s}_{ij}$, we have that $m_j \leq T$ for $j = 1, \ldots, r$. Since all rows of $\boldsymbol{Q}_j$ contain a single 1, row t of the product $\boldsymbol{Q}_{ij}\boldsymbol{g}_{ij}$ is given by $g_{ij}(s_{ijt})$.

The idea behind nonparametric modeling is to view the function evaluations in each $\boldsymbol{g}_{ij}$ as the realization of a stochastic process which controls the degree of local variation between neighboring elements. Despite differing theoretical foundations and assumptions, the vectors of function evaluations $\boldsymbol{g}_{ij}$ can eventually be written, in a wide range of nonparametric modeling approaches, as random fields of the form

$$\boldsymbol{g}_{ij} | \tau_{ij}^2 \sim N\left(\boldsymbol{g}_{ij0}, \tau_{ij}^2 \boldsymbol{K}_{ij}^{-1}\right), \quad j = 1, \ldots, r \tag{5}$$

where τ_{ij}^2 is a smoothness parameter and $\boldsymbol{K}_{ij}$ is a matrix whose structure will be discussed shortly. From a Bayesian perspective, Eq. (5) can be viewed as a smoothness prior for $\boldsymbol{g}_{ij}$, where as from a frequentist perspective, it is often viewed as a roughness penalty term in penalized likelihood estimation (Wahba, 1978). In either case, the goal of the modeling is to introduce a penalty to local variation between successive elements of $\boldsymbol{g}_{ij}$, without absolutely ruling out any possible value that the elements of $\boldsymbol{g}_{ij}$ can take.

The focus in this article is on models involving banded precision matrices $\boldsymbol{K}_{ij}$, that is, matrices which have nonzero elements only in small bands around the main diagonal. Matrix bandedness is a feature that significantly reduces the computational costs and makes the analysis of high-dimensional problems feasible and inexpensive. This article will examine the construction of $\boldsymbol{g}_{ij0}$ and $\boldsymbol{K}_{ij}$ in Eq. (5) for a class of smoothness priors, which are conceptually simple and easily adaptable, can approximate unknown functions arbitrarily well, and have been widely used (see, e.g., Besag et al., 1995; Chib & Jeliazkov, 2006; Chib et al., 2009; Fahrmeir & Lang, 2001; Koop & Poirier, 2004; Koop et al., 2005; Poirier, 1973; Shiller, 1984). The roots of this method can be traced back to Whittaker (1923), and its relationship with state space models has been discussed in Chan and Jeliazkov (2009). It should be noted that despite the focus on a specific smoothness prior, the estimation methodology described in this article is generic and can be applied with various modeling approaches for the unknown functions, such as splines (Poirier, 1973; Shiller, 1984), B-splines (Silverman, 1985), wavelets (Denison et al., 2002), or the approach of Hamilton (2001). Although bandedness of the precision matrix is a useful characteristic of many of the preceding nonparametric approaches, it is not a feature of other popular modeling methods, for example, regression splines or integrated Wiener process priors, which may be more computationally intensive in high-dimensional cases.

Because the modeling follows identical steps for each of the functions, for the time being we can simplify notation by suppressing the *ij* subscripts that denote the equation and function numbers. With this convention, a Markov process prior views the elements of $\boldsymbol{g} = (g(v_1), \ldots, g(v_m))' \equiv (g_1, \ldots, g_m)'$ as a stochastic process observed at the unique and ordered values in $\boldsymbol{v}$. Specifically, letting $h_\ell \equiv v_\ell - v_{\ell-1}$, a first-order Markov process prior can be defined as

$$g_\ell = g_{\ell-1} + u_\ell \tag{6}$$

while a second-order Markov process prior is given by

$$g_\ell = \left(1 + \frac{h_\ell}{h_{\ell-1}}\right) g_{\ell-1} - \frac{h_\ell}{h_{\ell-1}} g_{\ell-2} + u_\ell \tag{7}$$

where $u_\ell \sim N(0, \tau^2 h_\ell)$ and τ^2 is a smoothness parameter, such that small values of τ^2 produce smoother functions, while larger values allow the function to be more flexible and interpolate the data more closely. The weights h_ℓ adjust the variance to account for possibly irregular spacing between consecutive points in each design vector; the one given here implies that the variance grows linearly with the distance h_ℓ, although other weights are also possible. A distribution for the initial states of the stochastic process is necessary in order to complete the specification of the smoothness prior. For example, for the first-order prior, the initial state can be modeled as

$$g_1 \sim N(g_{10}, \tau^2 G_{10}) \tag{8}$$

whereas in the second-order case, we have

$$\begin{pmatrix} g_1 \\ g_2 \end{pmatrix} | \tau^2 \sim N\left(\begin{pmatrix} g_{10} \\ g_{20} \end{pmatrix}, \tau^2 \boldsymbol{G}_0 \right) \tag{9}$$

where $\boldsymbol{G}_0$ is a 2×2 symmetric positive definite matrix. The prior on the initial conditions in Eqs. (8) and (9) is very important because it induces a *proper* prior on the remaining observations (see Chib & Jeliazkov, 2006). Specifically, Eq. (6), starting with the initial condition in Eq. (8), implies a penalty on abrupt jumps between successive function evaluations, whereas Eq. (7), starting with Eq. (9), induces a more general prior on linear functions of $\boldsymbol{v}_j$ that is conceptually similar to the usual priors placed on the intercept and slope parameters in linear regression. This can be seen more precisely by iterating (7) in expectation (to eliminate u_ℓ which is the source of the nonlinearity), starting with initial states in Eq. (9).

The interpretability of the directed Markovian structure of the priors specified by Eqs. (6)–(9) is a convenient aspect of this approach, however, it also leads to an equivalent undirected representation that is used in deriving the random field version of the smoothness prior in Eq. (5). This can be shown by recongnizing that upon defining

$$\boldsymbol{H} = \begin{pmatrix} 1 & & & \\ -1 & 1 & & \\ & \ddots & \ddots & \\ & & -1 & 1 \end{pmatrix}, \quad \boldsymbol{\Sigma} = \begin{pmatrix} G_{10} & & & \\ & h_2 & & \\ & & \ddots & \\ & & & h_m \end{pmatrix}$$

for the first-order case in Eqs. (6) and (8), and similarly letting

$$\boldsymbol{H} = \begin{pmatrix} 1 & & & & & \\ & 1 & & & & \\ \frac{h_3}{h_2} & -\left(1 + \frac{h_3}{h_2}\right) & 1 & & & \\ & \ddots & & \ddots & \ddots & \\ & & & \frac{h_m}{h_{m-1}} & -\left(1 + \frac{h_m}{h_{m-1}}\right) & 1 \end{pmatrix}, \quad \boldsymbol{\Sigma} = \begin{pmatrix} \boldsymbol{G}_0 & & & \\ & h_3 & & \\ & & \ddots & \\ & & & h_m \end{pmatrix}$$

for the second-order Markov process in Eqs. (7) and (9), one can write $\boldsymbol{Hg}=\boldsymbol{u}$, where $\boldsymbol{u}\sim N(\boldsymbol{u}_0,\boldsymbol{\Sigma})$ is used to denote the errors in the Markov process with $\boldsymbol{u}_0=(g_{10},0,\ldots,0)'$ and $\boldsymbol{u}_0=(g_{10},g_{10},0,\ldots,0)'$ in the first- and second-order cases, respectively. A simple change of variables technique leads to the distribution $\boldsymbol{g}|\tau^2\sim N(\boldsymbol{g}_0,\tau^2\boldsymbol{K}^{-1})$, where the penalty matrix $\boldsymbol{K}$ is given by $\boldsymbol{K}=\boldsymbol{H}'\boldsymbol{\Sigma}^{-1}\boldsymbol{H}$ and $\boldsymbol{g}_0=\boldsymbol{H}^{-1}\boldsymbol{u}_0$. This derivation leads to the distributions presented in Eq. (5), where the indices i and j are explicitly present. Note that $\boldsymbol{g}_0$ can alternatively be derived by taking recursive expectations of either Eq. (6) or Eq. (7) starting with the mean in Eqs. (8) or (9), respectively.

Two key features of the class of priors are that (*i*) they are proper, which allows for formal Bayesian model selection, and (*ii*) the $m\times m$ penalty matrices $\boldsymbol{K}$ are banded, which is of considerable convenience, as manipulations involving such matrices take $O(m)$ operations, rather than the usual $O(m^3)$ operations for inversions and determinant computations, or $O(m^2)$ operations for multiplication by a vector. Given that m may potentially be as large as the total number of observations T in the sample, this has important ramifications for the numerical efficiency.

Since the priors on $\{\boldsymbol{g}_{ij}\}$ are defined conditionally on the hyperparameters $\{\tau_{ij}^2\}$, the hierarchical structure of the model is completed by specifying the prior distributions $\tau_{ij}^2\sim IG(\nu_{ij0}/2,\delta_{ij0}/2)$. Similarly, the prior distribution on the covariance matrix $\boldsymbol{\Omega}$ is taken as $\boldsymbol{\Omega}^{-1}\sim W(r_0,R_0)$. In setting these priors it is generally very helpful to consider their mapping to the mean and variance of the inverse gamma and Wishart distributions (see Gelman, Carlin, Stern, & Rubin, 2003, Appendix A), as the choice of these parameters plays a role in determining the trade-off between smoothness and goodness of fit. An example of how different settings of the prior parameters can lead to over- or under-smoothing is presented in Chib and Jeliazkov (2006).

Before we can focus on estimating the model, we must address the likelihood identification problem that emerges due to the additive structure in Eq. (3). Because the likelihood will remain unchanged if we simultaneously let $g_j^*(\cdot)=g_j(\cdot)+\alpha$ and $g_k^*(\cdot)=g_k(\cdot)-\alpha$ for $k\neq j$, it is obvious that neither an intercept, nor the level of the individual functions is likelihood identified. This can also be seen by recognizing that all rows of every incidence matrix $\boldsymbol{Q}_j$ in Eq. (4) sum to 1, leading to perfect multicollinearity because model (4) can be thought of as a saturated dummy variable model. Therefore, the functions must be appropriately "anchored" in order to achieve likelihood identification.

It is well known that Bayesian models with proper priors do not suffer from identification problems even when the likelihood is not identified

(Lindley, 1971; Poirier, 1998). However, because the nonparametric components of additive models are correlated by construction (since they enter the mean function additively), likelihood identification is essential for providing a model with well-behaved conditional posterior distributions that will produce quickly-mixing MCMC algorithms for efficient posterior sampling. To achieve likelihood identification, I remove free constants in the likelihood by employing the identification restrictions proposed in Jeliazkov (2011). The approach formally identifies the model by centering the functions in the likelihood and integrating out – instead of holding fixed – any unidentified quantities that enter the specification.[3] Such quantities are marginalized out with respect to a proper prior that is of no relevance in the likelihood and does not affect marginal likelihood estimation, thus relating this approach to the idea of marginal data augmentation discussed in Meng and van Dyk (1999), van Dyk and Meng (2001), and Imai and van Dyk (2005).

One approach to identification is to remove the free constants in the likelihood by restricting $r-1$ of the functions $\{\boldsymbol{g}_{ij}\}$ in each equation to start at zero (e.g., Chib et al., 2009; Koop et al., 2005; Shively et al., 1999). While this approach is quite natural as it corresponds to creating a baseline category in dummy variable models, in the context of nonparametric regression it tends to produce funnel-shaped error bands for the function estimates due to the identification restriction. Consequently, the information content in the data can be confounded with the repercussions of the identification restriction, so that narrower bands need not correspond to regions with more data or better identification of the function.

Another possibility for anchoring the functions is to consider the following version of Eq. (4)

$$\boldsymbol{y}_i = \boldsymbol{Q}_{i1}\boldsymbol{g}_{i1} + \boldsymbol{Q}_{i2}\boldsymbol{M}_{02}\boldsymbol{g}_{i2} + \ldots + \boldsymbol{Q}_{ir}\boldsymbol{M}_{0r}\boldsymbol{g}_{ir} + \boldsymbol{\varepsilon}_i \tag{10}$$

where

$$\boldsymbol{M}_{0j} = \left(\boldsymbol{I}_{m_j} - \frac{\mathbf{1}_{m_j}\mathbf{1}'_{m_j}}{m_j}\right), \quad j = 1, \ldots, r$$

are $m_j \times m_j$ symmetric and idempotent mean-differencing matrices (Hastie & Tibshirani, 1990; Lin & Zhang, 1999). Unfortunately, this identification scheme does not lend itself to computationally efficient posterior simulation and, as pointed out by Gelfand (2000), it has been applied in ways that do not correspond to well-defined Bayesian models, with centering typically introduced "on the fly" merely as a step in the fitting algorithm.

For these reasons, this article employs the closely related, yet computationally very distinct, identification scheme presented in Jeliazkov (2011), where the additive functions are identified through

$$\boldsymbol{y}_i = \boldsymbol{Q}_{i1}\boldsymbol{g}_{i1} + \boldsymbol{M}_0\boldsymbol{Q}_{i2}\boldsymbol{g}_{i2} + \ldots + \boldsymbol{M}_0\boldsymbol{Q}_{ir}\boldsymbol{g}_{ir} + \boldsymbol{\varepsilon}_i \tag{11}$$

where the $T \times T$ symmetric and idempotent mean-differencing matrix

$$\boldsymbol{M}_0 = \left(\boldsymbol{I}_T - \frac{\mathbf{1}_T\mathbf{1}_T'}{T}\right)$$

now pre-multiplies the incidence matrices $\{\boldsymbol{Q}_{ij}\}$ and centers the expanded vector of functional evaluations. The benefits from this identification method are discussed next.

ESTIMATION

To motivate the general approach, this section begins by considering the important special case of a single equation univariate regression model. Given data $\{y_t, s_t\}_{t=1}^{T}$, the scalar responses y_t are assumed to depend on the (scalar) covariate s_t according to

$$y_t = g(s_t) + \varepsilon_t, \quad (t = 1, \ldots, T) \tag{12}$$

where $\varepsilon_t \sim N(0, \sigma^2)$, and $g(\cdot)$ is an unknown smooth function. The model in Eq. (12) can be written in stacked form as

$$\boldsymbol{y} = \boldsymbol{Q}\boldsymbol{g} + \varepsilon, \quad \varepsilon \sim N(\mathbf{0}, \sigma^2\boldsymbol{I}) \tag{13}$$

where $\boldsymbol{Q}$ is the incidence matrix defined after Eq. (4). Given the Gaussian likelihood implied by (13), and assuming the Gaussian smoothness prior in Eq. (5) for either a first- or second-order process together with inverse Gamma priors $\tau^2 \sim IG(\nu_0/2, \delta_0/2)$ and $\sigma^2 \sim IG(s_0/2, d_0/2)$, yields full-conditional distributions which are conjugate, that is, they are in the same family as the priors (see, e.g., Greenberg, 2008; Koop, 2003). Sequential sampling from those full-conditional distributions lays the foundations for the following algorithm.

Algorithm 1. Univariate Gaussian Nonparametric Model: MCMC Implementation

1. *Sample* $[\mathbf{g}|\mathbf{y}, \tau^2, \sigma^2] \sim N(\hat{\mathbf{g}}, \mathbf{G})$, *where* $\mathbf{G}$ *and* $\hat{\mathbf{g}}$ *are the usual Bayes updates for linear regression, namely* $\mathbf{G} = (\mathbf{K}/\tau^2 + \mathbf{Q}'\mathbf{Q}/\sigma^2)^{-1}$ *and* $\hat{\mathbf{g}} = \mathbf{G}(\mathbf{K}\mathbf{g}_0/\tau^2 + \mathbf{Q}'\mathbf{y}/\sigma^2)$. *Remark 1 presents important notes on the sampling in this step.*
2. *Sample* $[\tau^2|\mathbf{g}] \sim IG\left(\frac{\nu_0 + m}{2}, \frac{\delta_0 + (\mathbf{g} - \mathbf{g}_0)' \mathbf{K}(\mathbf{g} - \mathbf{g}c_0)}{2}\right)$, *where conditionally on* $\mathbf{g}$, τ^2 *is independent of the remaining parameters and the data.*
3. *Sample* $[\sigma^2|\mathbf{y}, \mathbf{g}] \sim IG\left(\frac{s_0 + n}{2}, \frac{d_0 + (\mathbf{y} - \mathbf{Q}\mathbf{g})'(\mathbf{y} - \mathbf{Q}\mathbf{g})}{2}\right)$.

While steps 2 and 3 of Algorithm 1 are fairly straightforward, step 1 requires careful consideration because the quantities involved there can be of dimension as high as the sample size n. For this reason, estimation is performed as follows (see Fahrmeir & Lang, 2001).

Remark 1. Sampling of g.

To sample $\mathbf{g}$, *note that* $\mathbf{Q}'\mathbf{Q}$ *is a diagonal matrix whose t-th diagonal entry equals the number of values in* $\mathbf{s}$ *corresponding to the design point* $\mathbf{v}$. *Since* $\mathbf{K}$ *and* $\mathbf{Q}'\mathbf{Q}$ *are banded,* $\mathbf{G}^{-1}$ *is banded as well. Thus sampling of* $\mathbf{g}$ *need not include an inversion to obtain* $\mathbf{G}$ *and* $\hat{\mathbf{g}}$. *The mean* $\hat{\mathbf{g}}$ *is found instead by solving* $\mathbf{G}^{-1}\hat{\mathbf{g}} = (\mathbf{K}\mathbf{g}_0/\tau^2 + \mathbf{Q}'\mathbf{y}/\sigma^2)$, *which is done in* $O(T)$ *operations by back substitution. Also, let* $\mathbf{P}'\mathbf{P} = \mathbf{G}^{-1}$, *where* $\mathbf{P}$ *is the Cholesky decomposition of* $\mathbf{G}^{-1}$ *and is also banded. To obtain a random draw from* $N(\hat{\mathbf{g}}, \mathbf{G})$ *efficiently, sample* $\mathbf{u} \sim N(\mathbf{0}, \mathbf{I})$, *and solve* $\mathbf{P}\mathbf{w} = \mathbf{u}$ *for* $\mathbf{w}$ *by back substitution. It follows that* $\mathbf{w} \sim N(\mathbf{0}, \mathbf{G})$. *Adding the mean* $\hat{\mathbf{g}}$ *to* $\mathbf{w}$, *one obtains a draw* $\mathbf{g} \sim N(\hat{\mathbf{g}}, \mathbf{G})$.

Turning attention to the additive case, let $\boldsymbol{\theta}$ denote the vector of all model parameters, that is, the elements of $\{\mathbf{g}_{ij}\}$, $\{\tau_{ij}^2\}$, and the unique entries of $\boldsymbol{\Omega}$. Then, based on the identifying restrictions in Eq. (11) and the priors discussed in the second section, MCMC estimation can proceed through iterative sampling of the following steps.

Algorithm 2. NPVAR Model: MCMC Implementation

1. *Sample* $[\mathbf{g}_{i1}|\mathbf{y}, \boldsymbol{\theta}\backslash\mathbf{g}_{i1}] \sim N(\hat{\mathbf{g}}_{i1}, \hat{\mathbf{G}}_{i1})$, *where*

$$\hat{G}_{i1} = \left(\frac{1}{\tau^2_{i1}} K_{i1} + \frac{1}{\sigma^2_{i\setminus i}} Q'_{i1} Q_{i1} \right)^{-1}$$

$$\hat{g}_{i1} = \hat{G}_{i1} \left(\frac{1}{\tau^2_{i1}} K_{i1} g_{i10} + \frac{1}{\sigma^2_{i\setminus i}} Q'_{i1} \left(y_i - \mu_{i\setminus i} - \sum_{j=2}^{r} M_0 Q_{ij} g_{ij} \right) \right)$$

with $\mu_{i\setminus i} = E(\varepsilon_i | \varepsilon_{\setminus i})$ *and* $\sigma^2_{i\setminus i} = Var(\varepsilon_i | \varepsilon_{\setminus i})$. *The sampling in this step is carried out efficiently in* $O(T)$ *operations as discussed in Remark 1.*

1. *Sample* $[\mathbf{g}_{ij} | \mathbf{y}, \boldsymbol{\theta} \setminus \mathbf{g}_{ij}] \sim N(\hat{\mathbf{g}}_j, \hat{\mathbf{G}}_j)$ *for* $j = 2, \ldots, r$ *and* $i = 1, \ldots, q$, *where*

$$\hat{G}_{ij} = \left(\frac{1}{\tau^2_{ij}} K_{ij} + \frac{1}{\sigma^2_{i\setminus i}} Q'_{ij} M_0 Q_{ij} \right)^{-1}$$

and

$$\hat{g}_{ij} = \hat{G}_{ij} \left(\frac{1}{\tau^2_{ij}} K_{ij} g_{ij0} + \frac{1}{\sigma^2_{i\setminus i}} Q'_{ij} M_0 \left(y - \mu_{i\setminus i} - Q_{i1} g_{i1} - \sum_{k \geq 2, k \neq j} M_0 Q_{ik} g_{ik} \right) \right)$$

Remark 2 below shows how the sampling in this step can be carried out efficiently in $O(T)$ *operations, even though* $\hat{\mathbf{G}}_j$ *is not banded.*

2. *Sample* $[\tau^2_{ij} | \mathbf{g}_{ij}] \sim IG\left([\nu_{ij0} + m_j]/2, [\delta_{ij0} + (\mathbf{g}_{ij} - \mathbf{g}_{ij0})' \mathbf{K}_{ij} (\mathbf{g}_{ij} - \mathbf{g}_{ij0})]/2 \right)$ *for* $i = 1, \ldots, q$, *and* $j = 1, \ldots, r$, *where, given* $\mathbf{g}_{ij}$, τ^2_{ij} *is independent of the other elements in* $\boldsymbol{\theta}$ *and the data* $\mathbf{y}$.
3. *Sample* $[\boldsymbol{\Omega}^{-1} | \mathbf{y}, \boldsymbol{\theta} \setminus \boldsymbol{\Omega}] \sim W\left(r_0 + T, [R_0^{-1} + \sum_{t=1}^{T} \mathbf{e}_t \mathbf{e}'_t]^{-1} \right)$, *where* $\mathbf{e}_t$ *denotes the* $q \times 1$ *vector of residuals in time period* t.

Algorithm 2 generalizes Algorithm 1 in a straightforward fashion by sampling each unknown function conditionally on the remaining ones, making simulation manageable. Importantly, however, step 2 of Algorithm 2 involves $r-1$ non-banded matrices in each equation, and at first glance it would appear that simulation will be very demanding. Fortunately, however, as shown in Jeliazkov (2011), an application of the Sherman-Morrison formula makes it possible to sample these functions efficiently. The approach is presented in greater detail in Remark 2 further below. The modularity and computational advantages of this estimation strategy can provide important benefits in a variety of settings because simulating the functions by brute force methods is not always practical owing to the algorithmic complexity of working with high-dimensional matrices. Moreover, because the frequentist backfitting approach to estimating the unknown functions can be viewed as a (non-stochastic) simplification of Gibbs sampling (see Hastie & Tibshirani, 2000), Algorithm 2 can also be useful in frequentist estimation.

Remark 2. Sampling of Centered Functions.

To draw $\mathbf{g}_{ij} \sim N(\hat{\mathbf{g}}_{ij}, \hat{\mathbf{G}}_{ij})$ *in step 2 of Algorithm 2, use the definition of* $\mathbf{M}_0$ *to write*

$$\hat{\mathbf{G}}_{ij} = \left(\frac{1}{\tau_{ij}^2}\mathbf{K}_{ij} + \frac{1}{\sigma_{i\backslash i}^2}\mathbf{Q}'_{ij}\mathbf{M}_0\mathbf{Q}_{ij}\right)^{-1}$$

$$= \left(\frac{1}{\tau_{ij}^2}\mathbf{K}_{ij} + \frac{1}{\sigma_{i\backslash i}^2}\mathbf{Q}'_{ij}\mathbf{Q}_{ij} - \frac{\mathbf{c}_{ij}\mathbf{c}'_{ij}}{\sigma_{i\backslash i}^2 T}\right)^{-1}$$

where $\mathbf{c}_{ij} = \mathbf{Q}'_{ij}\mathbf{1}$. *Letting* $\mathbf{A}_{ij} = \frac{1}{\tau_{ij}^2}\mathbf{K}_{ij} + \frac{1}{\sigma_{i\backslash i}^2}\mathbf{Q}'_{ij}\mathbf{Q}_{ij}$, $\mathbf{u}_{ij} = \frac{1}{\sqrt{\sigma_{i\backslash i}^2 T}}\mathbf{c}_{ij}$, *and* $\lambda_{ij} = \mathbf{u}'_{ij}\mathbf{A}_{ij}^{-1}\mathbf{u}_{ij}$, *one can write, by the Sherman-Morrison formula*

$$\hat{\mathbf{G}}_{ij} = \left(\mathbf{A}_{ij} - \mathbf{u}_{ij}\mathbf{u}'_{ij}\right)^{-1}$$

$$= \mathbf{A}_{ij}^{-1} + \frac{\mathbf{A}_{ij}^{-1}\mathbf{u}_{ij}\mathbf{u}'_{ij}\mathbf{A}_{ij}^{-1}}{1 - \lambda_{ij}} \tag{14}$$

Significant efficiency benefits can be derived from (14) because $\hat{\mathbf{g}}_j$ *in step 2 of Algorithm 2 can be obtained by working with* $\mathbf{A}_{ij}$ *without inverting to* $\mathbf{A}_{ij}^{-}1$ *as outlined in Remark 1. Furthermore, let*

$$\mathbf{B}_{ij} = \left(\mathbf{A}_{ij} + \frac{\mathbf{u}_{ij}\mathbf{u}'_{ij}}{1 - \lambda_{ij}}\right)$$

which implies that $\hat{\mathbf{G}}_{ij} = \mathbf{A}_{ij}^{-1}\mathbf{B}_{ij}\mathbf{A}_{ij}^{-1}$. *Thus, if* $\mathbf{x} \sim N(\mathbf{0}, \mathbf{B}_{ij})$, *then* $\mathbf{z} = \mathbf{A}_{ij}^{-1}\mathbf{x}$ *is distributed* $\mathbf{z} \sim N(\mathbf{0}, \hat{\mathbf{G}}_{ij})$, *and a draw* $\mathbf{g}_{ij} \sim N(\hat{\mathbf{g}}_{ij}, \hat{\mathbf{G}}_{ij})$ *is obtained as* $\mathbf{g}_{ij} = \hat{\mathbf{g}}_{ij} + \mathbf{z}$. *To generate* $\mathbf{x} \sim N(\mathbf{0}, \mathbf{B}_{ij})$, *draw* $\mathbf{w}_1 \sim N(\mathbf{0}, \mathbf{A}_{ij})$ *and* $w_2 \sim N(0, 1)$ *and let* $\mathbf{x} = \mathbf{w}_1 + w_2\mathbf{u}_{ij}/\sqrt{1 - \lambda_{ij}}$.

As a consequence of the shortcuts afforded by Remark 2, all operations are $O(T)$ rather than $O(T^3)$.

MODEL COMPARISON

Empirical studies must inevitably address uncertainty not only about the parameters of a given model, but also about the model specification itself. This makes model comparison a central issue in statistical analysis. Given a collection of models $\{\mathcal{M}_1, \ldots, \mathcal{M}_L\}$, the formal Bayesian approach to model comparison (or testing the validity of the alternative hypotheses captured by each model) is based on the posterior model probabilities and their ratios, the posterior odds. Specifically, for any two models $\mathcal{M}_i$ and $\mathcal{M}_j$, a simple application of Bayes' theorem suggests that the posterior odds can be represented as the product of the prior odds and the ratio of the marginal likelihoods (the Bayes factor) as follows

$$\frac{\Pr(\mathcal{M}_i|\mathbf{y})}{\Pr(\mathcal{M}_j|\mathbf{y})} = \frac{\Pr(\mathcal{M}_i)}{\Pr(\mathcal{M}_j)} \times \frac{m(\mathbf{y}|\mathcal{M}_i)}{m(\mathbf{y}|\mathcal{M}_j)}$$

In turn, for any model $\mathcal{M}_l$, $l = 1, \ldots, L$, the marginal likelihood is given by

$$m(\mathbf{y}|\mathcal{M}_l) = \int f(\mathbf{y}|\boldsymbol{\theta}_l, \mathcal{M}_l)\pi_l(\boldsymbol{\theta}_l|\mathcal{M}_l)d\boldsymbol{\theta}_l \tag{15}$$

which is the integral of the likelihood function $f(\mathbf{y}|\boldsymbol{\theta}_l, \mathcal{M}_l)$ with respect to the prior distribution on the model parameters $\pi(\boldsymbol{\theta}_l|\mathcal{M}_l)$. Because in the case of nonparametric additive models the dimension of $\boldsymbol{\theta}$ can be very large, it should be clear that direct analytical integration will generally be infeasible. However, this difficulty can be addressed by using the approach of Chib (1995), where after rearranging Bayes' theorem $m(\mathbf{y}|\mathcal{M}_l)$ can alternatively be expressed as

$$m(\mathbf{y}|\mathcal{M}_l) = \frac{f(\mathbf{y}|\boldsymbol{\theta}_l^*, \mathcal{M}_l)\pi(\boldsymbol{\theta}_l^*|\mathcal{M}_l)}{\pi(\boldsymbol{\theta}_l^*|\mathbf{y}, \mathcal{M}_l)} \tag{16}$$

so that the integral in Eq. (15) is reduced to the more tractable problem of evaluating the likelihood, prior, and posterior ordinates at a single point $\boldsymbol{\theta}_l^*$

(e.g., the posterior mean). Because the numerator terms in Eq. (16) are available by direct calculation, the marginal likelihood can be computed by finding an estimate of the posterior ordinate $\pi(\boldsymbol{\theta}^*|\mathbf{y})$.

In the current context, the hierarchical structure of NPVAR models allows application of (16) in two different ways. One approach relies on

$$m(\mathbf{y}) = \frac{f\left(\mathbf{y}|\left\{\tau_{ij}^{2*}\right\}, \boldsymbol{\Omega}^*, \left\{\mathbf{g}_{ij}^*\right\}\right)\pi\left(\left\{\tau_{ij}^{2*}\right\}, \boldsymbol{\Omega}^*, \left\{\mathbf{g}_{ij}^*\right\}\right)}{\pi\left(\left\{\tau_{ij}^{2*}\right\}, \boldsymbol{\Omega}^*, \left\{\mathbf{g}_{ij}^*\right\}|\mathbf{y}\right)}$$

where the nonparametric functions are explicitly included in the identity (see Chib & Jeliazkov, 2006; Chib et al., 2009). However, owing to the Gaussian structure of the model, the marginal likelihood can also be computed using

$$m(\mathbf{y}) = \frac{f\left(\mathbf{y}|\{\tau_{ij}^{2*}\}, \boldsymbol{\Omega}^*\right)\pi\left(\{\tau_{ij}^{2*}\}, \boldsymbol{\Omega}^*\right)}{\pi\left(\{\tau_j^{2*}\}, \boldsymbol{\Omega}^*|\mathbf{y}\right)}$$

where all quantities are marginalized over the high-dimensional blocks $\{\mathbf{g}_{ij}\}$. This marginalization is possible because conditionally on $\left(\left\{\tau_{ij}^{2*}\right\}, \boldsymbol{\Omega}^*\right)$, the density $f\left(\mathbf{y}|\{\tau_{ij}^{2*}\}, \boldsymbol{\Omega}^*\right)$, marginalized over $\{\mathbf{g}_{ij}\}$ with respect to the prior distributions in Eq. (5), is also normal (Koop & Poirier, 2004) and can be evaluated directly for the typical sample sizes T encountered in macroeconomic applications. Because of this analytical tractability, $m(\mathbf{y})$ can then be found after the main run where, using the conditional independence of the densities in steps 3 and 4 of Algorithm 2, one computes

$$\pi\left(\left\{\tau_{ij}^{2*}\right\}, \boldsymbol{\Omega}^*|\mathbf{y}\right) \approx T^{-1}\sum_{t=1}^{T}\left\{f_{IW}\left(\boldsymbol{\Omega}^*|\mathbf{y}, \left\{\mathbf{g}_j^{(t)}\right\}\right)\prod_{i=1}^{p} f_{IG}\left(\tau_j^{2*}|\mathbf{g}_j^{(t)}\right)\right\}$$

using draws $\{\mathbf{g}_j^{(t)}\}$ from the main MCMC run. In instances where T is large or there are other complications (e.g., discrete outcomes), the decomposition involving $\{\mathbf{g}_{ij}\}$ is more appropriate and readers are referred to Chib and Jeliazkov (2006) and Chib et al. (2009) for methods that use reduced runs or to Jeliazkov and Lee (2010) for a method that employs the Gibbs kernel and invariance of the Markov chain to estimate the posterior ordinate $\pi(\{\tau_{ij}^{2*}\}, \boldsymbol{\Omega}^*, \{\mathbf{g}_{ij}^*\}|\mathbf{y})$.

An important point mentioned earlier is that the marginal likelihood for the additive NPVAR model does not depend on the (likelihood unidentified) levels of the functions that are centered for identification. This can be seen by recognizing that if $f(\mathbf{y}|\boldsymbol{\theta}_1, \boldsymbol{\theta}_2) = f(\mathbf{y}|\boldsymbol{\theta}_1)$, that is, the likelihood depends only on $\boldsymbol{\theta}_1$ whereas $\boldsymbol{\theta}_2$ is unidentified, but we have a proper prior $\pi(\boldsymbol{\theta}_1, \boldsymbol{\theta}_2)$, then the marginal likelihood

$$
\begin{aligned}
m(\mathbf{y}) &= \int f(\mathbf{y}|\boldsymbol{\theta}_1, \boldsymbol{\theta}_2)\pi(\boldsymbol{\theta}_1, \boldsymbol{\theta}_2)d\boldsymbol{\theta}_1 d\boldsymbol{\theta}_2 \\
&= \int f(\mathbf{y}|\boldsymbol{\theta}_1)\pi(\boldsymbol{\theta}_1) \int \pi(\boldsymbol{\theta}_2|\boldsymbol{\theta}_1)d\boldsymbol{\theta}_2 d\boldsymbol{\theta}_1 \\
&= \int f(\mathbf{y}|\boldsymbol{\theta}_1)\pi(\boldsymbol{\theta}_1)d\boldsymbol{\theta}_1
\end{aligned}
$$

is not influenced by the prior on $\boldsymbol{\theta}_2$. In practice this is important for modeling, because it implies that researchers with different beliefs about unidentified parameters will nevertheless reach identical conclusions about the relative ranking of alternative models.

Finally, note that because estimation of the marginal likelihood does not require maximization, it is less computationally intensive in nonparametric additive models than evaluation of information criteria such as AIC and BIC. This point has been overlooked and not fully appreciated in the literature despite its importance for model selection and model averaging on the basis of $\{\Pr(\mathcal{M}_l|\mathbf{y})\}$.

MODEL EXTENSIONS

The estimation techniques presented in this article are fully modular and readily applicable in various other settings since estimation of the unknown functions $\{\mathbf{g}_{ij}\}$ can be done conditionally on modifications in other parts of the model. The goal of this discussion is to briefly review the relevant literature and provide references that could guide researchers interested in pursuing such extensions.

One should note that the methods in the second and third sections trivially generalize to cases where one or more exogenous covariates enter the regression as in seemingly unrelated regression models (e.g., Holmes et al., 2002; Koop et al., 2005; Smith & Kohn, 2000). Further extensions of the framework to Bayesian models with nonparametric endogeneity or sample selection can be pursued following Chib and Greenberg (2007) or Chib et al. (2009), respectively; the modeling would also be useful in guiding future research on structural NPVAR models and impulse response analysis. Many of the aforementioned papers also trivially subsume semiparametric and partially linear cases where some of the covariates enter the model linearly. Estimation of such models is a straightforward extension of Algorithms 1 and 2, and proceeds by using the partial residuals $\boldsymbol{y}_i - \boldsymbol{X}_i\boldsymbol{\beta}$ when simulating $\{\boldsymbol{g}_{ij}\}$, followed by simulating $\boldsymbol{\beta}$ conditionally upon the functions $\{\boldsymbol{g}_{ij}\}$.

The methods in this article, including the above extensions, can also be applied, using data augmentation techniques (Albert & Chib, 1993; Tanner & Wong, 1987), to the analysis of dynamic systems involving binary, polychotomous, censored, and other discrete outcomes such as the qualitative VAR model of Dueker (2005). The main advantage of this approach is that conditionally on the latent data, estimation of the parameters and the unknown functions closely mirrors the methods for continuous data. The methodology presented here is also applicable to the class of additive mixed models for continuous and discrete data (e.g., Lin & Zhang, 1999). For example, Chib and Jeliazkov (2006) discuss the specification and estimation of a semiparametric partially linear model for dynamic binary panel data with multivariate heterogeneity. The estimation algorithm in that paper can be easily modified to include an additive structure whose estimation can be carried out by the methods presented in the third section.

While the specification and estimation of NPVAR models was discussed in detail for homoskedastic Gaussian models, extensions to other distributions (e.g., Student's t, mixtures of normals, or Dirichlet process priors for nonparametric distributional modeling) and heteroskedasticity (e.g., regime switching models with different variance regimes, or models with stochastic volatility) may be very desirable in certain applications. One such example, relating to different variance regimes, will be studied in the sixth section. Fortunately, such extensions can be estimated using data augmentation techniques that could build upon the homoskedastic Gaussian specification discussed earlier. In particular, consider a heteroskedastic model in which the ith equation can be written as

$$\boldsymbol{y}_i = \boldsymbol{Q}_{i1}\boldsymbol{g}_{i1} + \boldsymbol{M}_0\boldsymbol{Q}_{i2}\boldsymbol{g}_{i2} + \ldots + \boldsymbol{M}_0\boldsymbol{Q}_{ir}\boldsymbol{g}_{ir} + \boldsymbol{\varepsilon}_i$$

with $\boldsymbol{\Sigma}_i \equiv \mathrm{Var}(\boldsymbol{\varepsilon}_i) = \mathrm{diag}(\sigma^2_{i1}, \ldots, \sigma^2_{iT})$. Due to the heteroskedasticity, the covariance matrix of $[\boldsymbol{g}_{ij}|\boldsymbol{y}, \boldsymbol{\theta}]$ is not of the form presented in Remark 2, and estimation can not be performed efficiently by relying on the Sherman–Morrison formula. This poses an important computational difficulty because estimation in large dimensional models would be very difficult and potentially infeasible. In this article, I propose a solution to this problem that employs data augmentation to reduce the heteroskedastic model to a homoskedastic one enabling application of the methods discussed in Algorithm 2 and Remark 2. In particular, following Chib and Jeliazkov (2006), we can write

$$\boldsymbol{y}_i = \boldsymbol{Q}_{i1}\boldsymbol{g}_{i1} + \boldsymbol{M}_0\boldsymbol{Q}_{i2}\boldsymbol{g}_{i2} + \ldots + \boldsymbol{M}_0\boldsymbol{Q}_{ir}\boldsymbol{g}_{ir} + \boldsymbol{\eta}_i + \boldsymbol{\nu}_i$$

where $\boldsymbol{\eta}_i \overset{iid}{\sim} N(\mathbf{0}, \boldsymbol{\Sigma}_i - \kappa_i \boldsymbol{I})$ and $\boldsymbol{\nu}_i \overset{iid}{\sim} N(\mathbf{0}, \kappa_i \boldsymbol{I})$ with $0 < \kappa_i \leq \min\{\sigma_{it}^2\}$. Consequently, given a draw of $\boldsymbol{\eta}_i$, which is simple and inexpensive to obtain, the model

$$\boldsymbol{y}_i - \boldsymbol{\eta}_i = \boldsymbol{Q}_{i1}\boldsymbol{g}_{i1} + \boldsymbol{M}_0\boldsymbol{Q}_{i2}\boldsymbol{g}_{i2} + \ldots + \boldsymbol{M}_0\boldsymbol{Q}_{ir}\boldsymbol{g}_{ir} + \boldsymbol{\nu}_i$$

is homoskedastic because $\mathrm{Var}(\boldsymbol{\nu}_i) = \kappa_i \boldsymbol{I}$. In our context, it would actually be optimal to set $\kappa_i = \min\{\sigma_{it}^2\}$ because this would imply that the corresponding elements of $\boldsymbol{\eta}_i$ would be identically 0 and will not need to be sampled. This leads to the following extension of Algorithm 2:

Algorithm 3. NPVAR Model: MCMC Estimation of Heteroskedastic Model.

1. *For* $i = 1, \ldots, q$:
   (a) *Sample* $[\boldsymbol{\eta}_i | \boldsymbol{y}, \boldsymbol{\theta}]$ *by drawing, for* $t = 1, \ldots, T$, $\eta_{it} \sim N(\hat{\eta}_{it}, \hat{H}_{it})$, *where* $\hat{H}_{it} = \kappa_i(\sigma_{it}^2 - \kappa_i)/\sigma_{it}^2$ *and* $\hat{\eta}_{it} = (\sigma_{it}^2 - \kappa_i)(y_{it} - \mu_{it|\backslash i,t} - m_{it})/\sigma_{it}^2$, *where* m_{it} *is the t-th row of* $\boldsymbol{Q}_{i1}\boldsymbol{g}_{i1} + \sum_{k \geq 2} \boldsymbol{M}_0 \boldsymbol{Q}_{ik}\boldsymbol{g}_{ik}$, $\mu_{it|\backslash i,t}$ *is the t-th row of* $\boldsymbol{\mu}_{i|\backslash i} = E(\varepsilon_i | \varepsilon_{\backslash i})$, *and* $\kappa_i = \min\{\sigma_{it}^2\}$, *where* $\sigma_{it}^2 = Var(\varepsilon_{it} | \varepsilon_{\backslash i,t})$. *Note that for cases where* $\kappa_i = \sigma_{it}^2$, *the corresponding entry in* $\boldsymbol{\eta}_i$ *is identically zero and need not be sampled.*
   (b) *Sample* $[\boldsymbol{g}_{i1} | \boldsymbol{y}, \boldsymbol{\eta}_i, \boldsymbol{\theta} \backslash \boldsymbol{g}_{i1}] \sim N\left(\hat{\boldsymbol{g}}_{i1}, \hat{\boldsymbol{G}}_{i1}\right)$, *where*

$$\hat{\boldsymbol{G}}_{i1} = \left(\frac{1}{\tau_{i1}^2}\boldsymbol{K}_{i1} + \frac{1}{\kappa_i}\boldsymbol{Q}_{i1}'\boldsymbol{Q}_{i1}\right)^{-1}$$

$$\hat{\boldsymbol{g}}_{i1} = \hat{\boldsymbol{G}}_{i1}\left(\frac{1}{\tau_{i1}^2}\boldsymbol{K}_{i1}\boldsymbol{g}_{i10} + \frac{1}{\kappa_i}\boldsymbol{Q}_{i1}'\left(\boldsymbol{y}_i - \boldsymbol{\eta}_i - \boldsymbol{\mu}_{i|\backslash i} - \sum_{j=2}^{r}\boldsymbol{M}_0\boldsymbol{Q}_{ij}\boldsymbol{g}_{ij}\right)\right)$$

with $\boldsymbol{\mu}_{i|\backslash i} = E(\varepsilon_i|\varepsilon_{\backslash i})$ *and* $\kappa_i = \min\{\sigma_{it}^2\}$, *where* $\sigma_{it}^2 = Var(\varepsilon_{it}|\varepsilon_{\backslash i,t})$. *The sampling in this step is carried out efficiently as in Remark 1.*

(c) *Sample* $[\boldsymbol{g}_{ij}|\boldsymbol{y}, \boldsymbol{\eta}_i, \boldsymbol{\theta}\backslash \boldsymbol{g}_{ij}] \sim N(\hat{\boldsymbol{g}}_j, \hat{\boldsymbol{G}}_j)$ *for* $j = 2, \ldots, r$ *and* $i = 1, \ldots, q$, *where*

$$\hat{\boldsymbol{G}}_{ij} = \left(\frac{1}{\tau_{ij}^2}\boldsymbol{K}_{ij} + \frac{1}{\kappa_i}\boldsymbol{Q}_{ij}'\boldsymbol{M}_0\boldsymbol{Q}_{ij}\right)^{-1}$$

and

$$\hat{\boldsymbol{g}}_{ij} = \hat{\boldsymbol{G}}_{ij}\left(\frac{1}{\tau_{ij}^2}\boldsymbol{K}_{ij}\boldsymbol{g}_{ij0} + \frac{1}{\kappa_i}\boldsymbol{Q}_{ij}'\boldsymbol{M}_0\left(\boldsymbol{y} - \boldsymbol{\eta}_i - \boldsymbol{\mu}_{i|\backslash i} - \boldsymbol{Q}_{i1}\boldsymbol{g}_{i1} - \sum_{k \geq 2, k \neq j} \boldsymbol{M}_0\boldsymbol{Q}_{ik}\boldsymbol{g}_{ik}\right)\right)$$

This is done as in Remark 2.

2. *Sample* $[\tau_{ij}^2|\boldsymbol{g}_{ij}]$ *for* $i = 1, \ldots, q$, *and* $j = 1, \ldots, r$, *as in Algorithm 2.*
3. *Sample* $[\boldsymbol{\Omega}_{it}^{-1}|\boldsymbol{y}, \boldsymbol{\theta}\backslash\boldsymbol{\Omega}]$ *according to the volatility process under consideration.*

The above machinery also applies to mixture-of-normals and scale mixture-of-normals models. In particular, a model with t errors with ν degrees of freedom can be represented as a conditionally Gaussian model, whose variance, given a set of a priori gamma latent variables $\lambda_t \sim G(\nu/2, \nu/2)$, $t = 1, \ldots, T$, is given by $\text{Var}(\varepsilon_t|\lambda_t) = \sigma^2/\lambda_t$ (Albert & Chib, 1993; Andrews & Mallows, 1974). Estimation of these models is straightforward because given $\{\lambda_i\}$, one can decompose $\boldsymbol{\varepsilon}_i$ into $\boldsymbol{\eta}_i$ and $\boldsymbol{\nu}_i$ and proceed as above. In this way, NPVAR models can be adapted to a variety of specifications for the error variance, including changepoint and regime switching models (Chib, 1996, 1998; Sims & Zha, 2006), time-varying parameter models (Chan & Jeliazkov, 2009; Primiceri, 2005), factor models (Belviso & Milani, 2006; Chan & Jeliazkov, 2009; Kose, Otrok, & Whiteman, 2003, 2008), and others.

APPLICATION TO U.S. MACROECONOMIC DATA

The data sample for this application contains postwar quarterly macroeconomic data for the United States from 1948:Q1 to 2005:Q1. The set of variables includes output growth g_t measured by log differences of real GDP between two consecutive quarters, average quarterly unemployment rate u_t, inflation π_t measured by the percentage change in the Consumer Price Index between consecutive quarters, and interest rates i_t measured by the average quarterly secondary market yield on the three-month Treasury bill. The first three of these variables are seasonally adjusted. These variables, summarized in Table 1, reflect the general state of the economy, and have been widely used in empirical macroeconomics.[4] From the table, we see that the average quarterly GDP growth over the sample period is 0.85 percent, which amounts to annual GDP growth of 3.4 percent. A similar computation shows an average annual inflation rate of approximately 3.7 percent. Unemployment and interest rates average at 5.63 and 4.81 percent, respectively.

These data are analyzed using the econometric techniques discussed earlier. The empirical strategy for studying the behavior of the dynamic system in Eq. (2) is to address both model and functional form uncertainty. The first area of model uncertainty in the macroeconomic system has to do with determination its dynamics – that is, the number of lags needed in Eq. (2). This one-lag model was compared to several more richly parameterized models in order to gauge whether restricting attention to an NPVAR(1) specification is a sensible empirical strategy. The baseline NPVAR(1) model, which contains a single lag of $\boldsymbol{y}_t$ with 16 unknown functions, was compared with an NPVAR(2) model (with 32 such functions). The baseline model overwhelmingly outperformed the longer lag specification – its log-marginal likelihood exceeded that of the larger model by over 40, implying a Bayes factor of over e^{40} in favor of the NPVAR(1) specification.[5]

Guided by earlier research findings suggesting the possibility of a structural break (at least in error volatility as in Stock & Watson, 2003 and Sims & Zha, 2006), I also used split-sample estimation to capture the possibility of structural breaks in the series. Specifically, an NPVAR(1) model was fit on data in the pre-Volcker era (prior to 1979:Q2), and a separate model was fit on the data thereafter (following 1979:Q3). The marginal likelihood for the pre-Volcker model was −539.5, and that for the second

Table 1. Descriptive Statistics for the Data Sample (in Percentage Points).

Variable	Mean	SD	Min	Max
Quarterly growth in real GDP	0.85	1.00	−2.76	4.02
Unemployment rate	5.63	1.52	2.60	10.70
Nominal interest rate	4.81	2.92	0.79	15.05
Quarterly inflation	0.92	0.85	−1.24	4.08

part of the data sample was −438.7. Compared to the log-marginal likelihood of −908.8 for the baseline NPVAR(1) model on the entire data sample, the split sample measure of fit, as captured by the marginal likelihood, was far worse (the sum of the log-marginal likelihoods for the two subsamples is −978.2, which is far below marginal likelihood for the overall NPVAR(1) model of −908.8). These results are interesting because (*i*) they suggest that the simpler and parsimonious NPVAR(1) fit on the entire specification appears preferable to the (twice as big) split sample model and (*ii*) they demonstrate the ability of the Bayesian model comparison framework to penalize overparameterized specifications.

Fig. 2 presents the estimated functions for the full-sample NPVAR(1) model. The figure shows that a linear model would be reasonable for many of the economic relationships – particularly in modeling the effects of lagged unemployment. To a lesser extent, the same is true in other instances (e.g., the dependence of interest on its past value), where the function

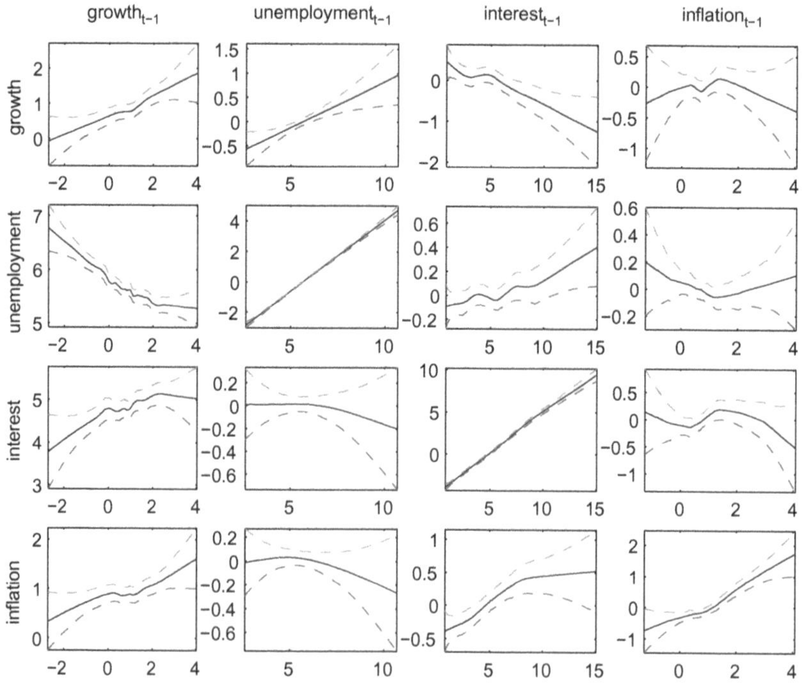

Fig. 2. Full sample estimates: the rows represent the functions in each equation, columns contain the functions of a given lagged variable across equations.

estimates do not reveal drastic departures from linearity. On the other hand, however, in many equations, the effects of lagged financial variables (interest and inflation), as well as the effects of lagged growth, appear to be quite nonlinear. This finding concurs with earlier studies that have found nonlinearity in growth behavior (Dahl & Gonzalez-Rivera, 2003a, 2003b) and financial markets (Härdle & Tsybakov, 1997; Härdle et al., 1998). The figure shows, for instance, that lagged inflation exhibits significant nonlinearities in every equation, whereas the function estimates for lagged interest exhibit nonlinearities in three of the four equations. For this reason, future analysis of financial variables might benefit considerably from employing nonparametric methods. Regarding the effects of lagged GDP growth, a review of the estimated functions reveals that although there is much nonlinearity, there are also large regions where the function estimates are approximately linear. This suggests that an interesting future research question would be to examine whether those types of nonlinearities can be adequately captured through threshold models.

For comparison purposes, Figs. 3 and 4 show function estimates for the pre- and post-Volcker periods, respectively. It is interesting to note that, although the function estimates differ in some respects, most point to the same types of nonlinearities as the estimates from the overall sample.[6] This is quite instructive, as it provides evidence that the econometric relationships may be stable but nonlinear, and therefore omitted nonlinearity may be a significant driver in findings of structural instability (cf. Hamilton, 2001). Resolving this issue should be an important item on the research agenda of studies focusing on structural (in)stability.

The apparent stability of the nonparametric function estimates across subsamples naturally leads to another important research question that has attracted much attention recently. Specifically, it would be of interest to consider whether an NPVAR model would exhibit evidence of a structural change in variances, which has been widely documented in contexts utilizing linear models. Such findings (e.g., Sims & Zha, 2006; Stock & Watson, 1996, 2003) have led to the conclusion that a reduction in error volatility has been a driving force in the "Great Moderation" of the 1980s and 1990s. To formally test the stability of the mean relationships while allowing for structural breaks in variances, I have estimated three additional NPVAR models. The first allows for a single structural break between 1979:Q2 and 1979:Q3 with the Volcker appointment. The second model employs a single structural break between 1982:Q4 and 1983:Q1 with the following the Fed's disinflation of the early 1980s. The third model allows for both of these break points. The models were estimated using Algorithm 3 of the fifth section, and the marginal likelihoods were estimated as discussed in

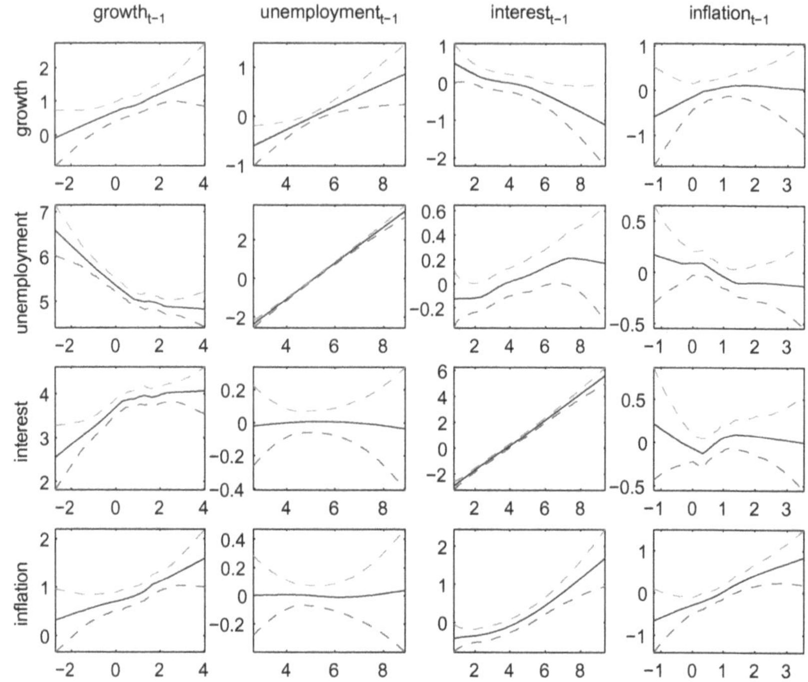

Fig. 3. Pre-Volcker estimates: the rows represent the functions in each equation, columns contain the functions of a given lagged variable across equations.

the fourth section. The marginal likelihood for the first model was estimated to be −891.4, whereas that of the second was estimated to be −880.1, showing that, conditionally on a single break, the data favor the 1982/83 breakpoint. However, a much more dramatic improvement is offered by the third model, the one which allows for both a 1979 and a 1982/83 breakpoints. The marginal likelihood for that model is −838.4, leading to the conclusion that these three periods in the U.S. sample are indeed dramatically different. This if further confirmed by examining the estimated covariance matrices for the three subperiods:

$$\mathbf{\Omega}_{48:79} = \begin{pmatrix} 1.135 & -0.254 & 0.085 & -0.060 \\ -0.254 & 0.144 & -0.033 & 0.002 \\ 0.085 & -0.033 & 0.255 & 0.044 \\ -0.060 & 0.002 & 0.044 & 0.463 \end{pmatrix}$$

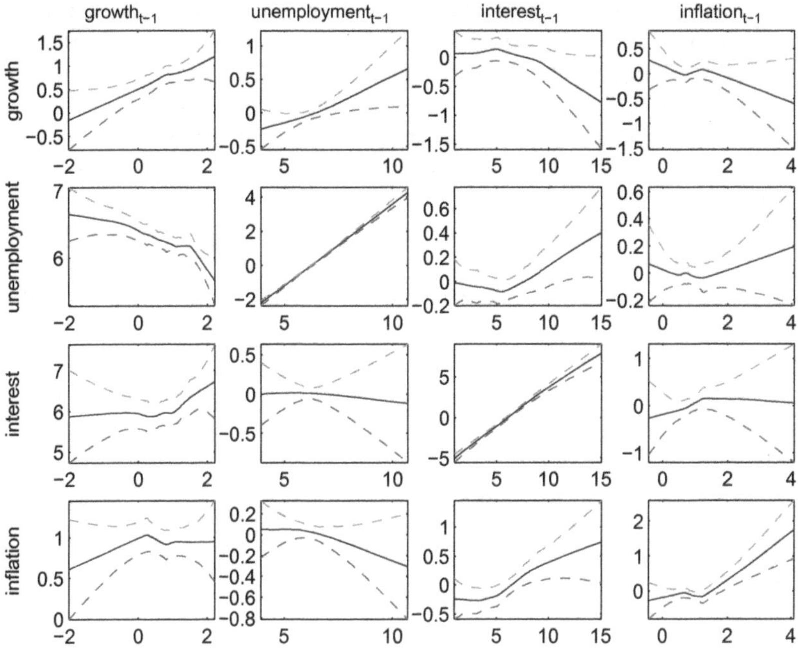

Fig. 4. Post-Volcker estimates: the rows represent the functions in each equation, columns contain the functions of a given lagged variable across equations.

$$\mathbf{\Omega}_{79:82} = \begin{pmatrix} 1.327 & -0.285 & 1.095 & 0.636 \\ -0.285 & 0.367 & -0.575 & -0.261 \\ 1.095 & -0.575 & 4.348 & 1.445 \\ 0.636 & -0.261 & 1.445 & 1.048 \end{pmatrix}$$

and

$$\mathbf{\Omega}_{83:05} = \begin{pmatrix} 0.234 & -0.032 & 0.051 & 0.008 \\ -0.032 & 0.066 & -0.031 & -0.010 \\ 0.051 & -0.031 & 0.202 & 0.044 \\ 0.008 & -0.010 & 0.044 & 0.216 \end{pmatrix}$$

These covariance matrices clearly demonstrate the dramatic peak in the error variances of all variables except growth (i.e., unemployment, interest rates, and inflation) during the disinflation period and the subsequent "moderation" of all 4 variables in 1983. A notable feature is the large jump, and subsequent decrease, in the estimated error variance in the interest rate equation during the period 1979–1982, which can be accounted for by the Fed's change of policy instrument from the federal funds rate to reserve targeting, as well as the unprecedented increase in interest rates during the disinflation period.

Fig. 5 presents the function estimates from the model with three variance regimes. The figure demonstrates that the same type of nonlinearities that were present in the homoskedastic models are still present here. Therefore, even though the heteroskedastic NPVAR model has confirmed earlier conclusions that changes were large due to breaks in variances, it also shows that there is much nonlinearity that would remain unexplored by linear

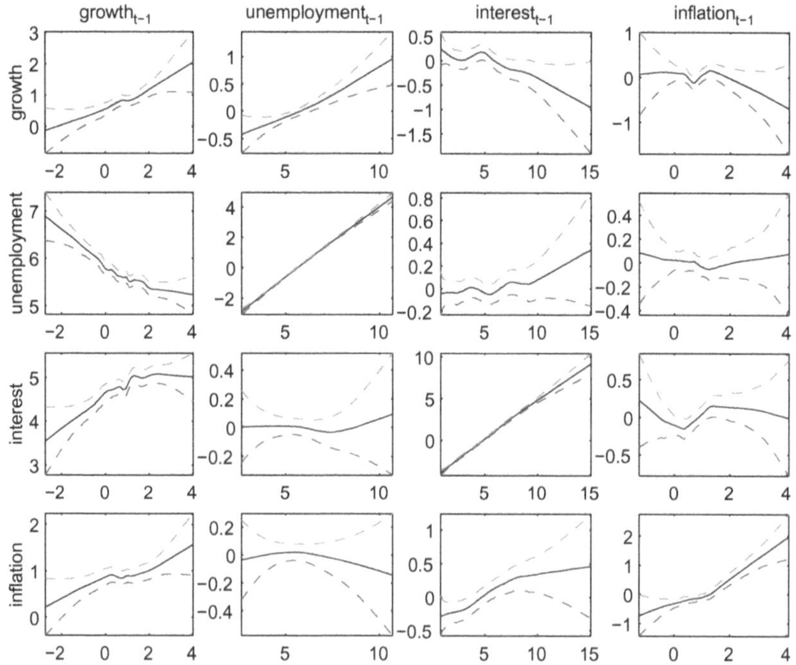

Fig. 5. Full sample estimates from model with three volatility regimes: the rows represent the functions in each equation, columns contain the functions of a given lagged variable across equations.

models and that future research should study such features of the economic relationships more closely.

CONCLUDING REMARKS

This article has examined the specification, estimation, and comparison of NVAR models. Efficient MCMC sampling and model comparison techniques are discussed in the context of a new scheme for identifying the unknown covariate functions, and extensions to heteroskedastic and other settings have been examined. An application of the NPVAR model to U.S. postwar data on GDP growth, unemployment, interest rates, and inflation, has confirmed the presence of distinct volatility regimes in the postwar U.S. macroeconomic series, but has also revealed that important nonlinearities is certain economic relationships may remain undetected by standard regressions. Implementation of these techniques in related settings, such as those considered in the fifth section, is an interesting area for future research.

NOTES

1. In these cases, the modeling involves an additional state variable that can be latent or observed; conditionally on the state, the models are linear, whereas marginalization over the state yields piecewise linear regressions or mixtures of linear regressions.

2. Interested readers are referred to these books for further details on a rich variety of nonparametric modeling approaches such as truncated polynomials, radial basis functions, neural networks, regression trees, wavelets, kernel smoothing, locally weighted polynomials, B-splines, etc., many of which are beyond the scope of this article.

3. It will be sufficient to apply this centering to $r-1$ of the unknown functions in each equation allowing the overall intercept to be absorbed in the remaining function.

4. The sample period excludes the past recession for a number of reasons. Over the last few years, interest rates have approached and stayed very close to their lower bound of zero. This could lead to findings of nonlinearity due to the effects of the lower bound, thereby favoring the methods of the paper over a linear model. Moreover, traditional modeling may be inadequate near the bound, where the distribution of the interest rate process is truncated and exhibits point mass. Appropriate modeling in this case is still an open research problem. Finally, if the "Great Recession" marked a possible structural break, at present there would be insufficient observations estimate the model after the break.

5. A specification including a fourth (year-ago) lag was also considered but it also did not perform competitively with the NPVAR(1) specification.

6. Since the range and level of each function may differ across samples, readers are cautioned to compare those functions over the relevant ranges, keeping in mind that the level of the functions will shift to satisfy the identification constraints.

ACKNOWLEDGMENTS

I am grateful to Tom Fomby, Lutz Killian, Anthony Murphy, two anonymous referees, and my colleagues Dale Poirier, David Brownstone, Fabio Milani, and especially Angela Vossmeyer, for their careful comments on earlier drafts.

REFERENCES

Ahamada, I., & Flachaire, E. (2010). *Non-parametric econometrics*. Oxford: Oxford University Press.

Albert, J., & Chib, S. (1993). Bayesian analysis of binary and polychotomous response data. *Journal of the American Statistical Association, 88*, 669–679.

Andrews, D. F., & Mallows, C. L. (1974). Scale mixtures of normal distributions. *Journal of the Royal Statistical Society – Series B, 36*, 99–102.

Beaudry, P., & Koop, G. (1993). Do recessions permanently change output? *Journal of Monetary Economics, 31*, 149–163.

Belviso, F., & Milani, F. (2006). Structural factor-augmented VARs (SFAVARs) and the effects of monetary policy. *Topics in Macroeconomics, 6*(3). Article 2.

Besag, J., Green, P., Higdon, D., & Mengersen, K. (1995). Bayesian computation and stochastic systems. *Statistical Science, 10*, 3–66.

Canova, F. (1993). Modelling and forecasting exchange rates with a bayesian time-varying coefficient model. *Journal of Economic Dynamics and Control, 17*, 233–261.

Chan, J. C., & Jeliazkov, I. (2009). Efficient simulation and integrated likelihood estimation in state space models. *International Journal of Mathematical Modelling and Numerical Optimisation, 1*, 101–120.

Chauvet, M. (1998). An econometric characterization of business cycle dynamics with factor structure and regime switching. *International Economic Review, 39*, 969–996.

Chib, S. (1995). Marginal likelihood from the gibbs output. *Journal of the American Statistical Association, 90*, 1313–1321.

Chib, S. (1996). Calculating posterior distributions and modal estimates in markov mixture models. *Journal of Econometrics, 75*, 79–97.

Chib, S. (1998). Estimation and comparison of multiple change-point models,. *Journal of Econometrics, 86*, 221–241.

Chib, S., & Greenberg, E. (2007). Analysis of additive instrumental variable models. *Journal of Computational and Graphical Statistics, 16*, 86–114.

Chib, S., Greenberg, E., & Jeliazkov, I. (2009). Estimation of semiparametric models in the presence of endogeneity and sample selection. *Journal of Computational and Graphical Statistics, 18*, 321–348.

Chib, S., & Jeliazkov, I. (2006). Inference in semiparametric dynamic models for binary longitudinal data. *Journal of the American Statistical Association, 101*, 685–700.

Cogley, T., & Sargent, T. J. (2001). Evolving post-world war II U.S. Inflation dynamics. *NBER Macroeconomics Annual, 16*, 331–338.

Dahl, C. M., & Gonzalez-Rivera, G. (2003a). Identifying nonlinear components by random fields in the US GNP growth. Implications for the shape of the business cycle. *Studies in Nonlinear Dynamics & Econometrics, 7*. Article 2.

Dahl, C. M., & Gonzalez-Rivera, G. (2003b). Testing for neglected nonlinearity in regression models based on the theory of random fields. *Journal of Econometrics, 114*, 141–164.

Denison, D. G. T., Mallick, B. K., & Smith, A. F. M. (1998). Automatic bayesian curve fitting. *Journal of the Royal Statistical Society – Series B, 60*, 333–350.

Denison, D. G. T., Holmes, C. C., Mallick, B. K., & Smith, A. F. M. (2002). *Bayesian methods for nonlinear classification and regression*. New York, NY: Wiley.

DiMatteo, I., Genovese, C. R., & Kass, R. E. (2001). Bayesian curve-fitting with free-knot splines. *Biometrika, 88*, 1055–1071.

Dueker, M. (2005). Dynamic forecasts of qualitative variables: A qual var model of U.S. recessions. *Journal of Business & Economic Statistics, 23*, 96–104.

Fahrmeir, L., & Lang, S. (2001). Bayesian inference for generalized additive mixed models based on markov random field priors. *Journal of the Royal Statistical Society – Series C, 50*, 201–220.

Gelfand, A. E. (2000). Discussion to bayesian backfitting. *Statistical Science, 15*, 217–218.

Gelman, A., Carlin, J. B., Stern, H. S., & Rubin, D. B. (2003). *Bayesian data analysis* (2nd ed.). New York, NY: Chapman & Hall.

Greenberg, E. (2008). *Introduction to Bayesian econometrics*. New York, NY: Cambridge University Press.

Hamilton, J. D. (1989). A new approach to the economic analysis of nonstationary time series and the business cycle. *Econometrica, 57*, 357–384.

Hamilton, J. D. (2001). A parametric approach to flexible nonlinear inference. *Econometrica, 69*, 537–573.

Hansen, B. E. (1992). The likelihood ratio test under nonstandard conditions: Testing the markov switching model of GNP. *Journal of Applied Econometrics, 7*, S61–S82.

Härdle, W., & Tsybakov, A. (1997). Local polynomial estimators of the volatility function in nonparametric autoregression. *Journal of Econometrics, 81*, 223–243.

Härdle, W., Tsybakov, A., & Yang, L. (1998). Nonparametric vector autoregression. *Journal of Statistical Planning and Inference, 68*, 221–245.

Hastie, T., & Tibshirani, R. (1990). *Generalized additive models*. New York, NY: Chapman & Hall.

Hastie, T., & Tibshirani, R. (2000). Bayesian backfitting. *Statistical Science, 15*, 196–223.

Holmes, C. C., Denison, D. G. T., & Mallick, B. K. (2002). Accounting for model uncertainty in seemingly unrelated regressions. *Journal of Computational and Graphical Statistics, 11*, 533–551.

Imai, K., & van Dyk, D. (2005). A Bayesian analysis of the multinomial probit model using marginal data augmentation. *Journal of Econometrics, 124*, 311–334.

Jeliazkov, I. (2011). *Specification and inference in nonparametric additive regression*. Working Paper. Department of Economics, University of California, Irvine, CA.

Jeliazkov, I., & Lee, E. H. (2010). MCMC perspectives on simulated likelihood estimation. *Advances in Econometrics: Maximum Simulated Likelihood*, *26*, 3–39.

Kim, C.-J., Morley, J., & Piger, J. (2005). Nonlinearity and the permanent effects of recessions. *Journal of Applied Econometrics*, *20*, 291–309.

Kim, C.-J., & Nelson, C. R. (1999). Has the U.S. Economy become more stable? A Bayesian approach based on a markov-switching model of the business cycle. *Review of Economics and Statistics*, *81*, 608–616.

Koop, G. (2003). *Bayesian econometrics*. New York, NY: Wiley.

Koop, G., & Korobilis, D. (2009). Bayesian multivariate time series methods for empirical macroeconomics. *Foundations and Trends in Econometrics*, *3*, 267–358.

Koop, G., & Poirier, D. J. (2004). Bayesian variants of some classical semiparametric regression techniques. *Journal of Econometrics*, *123*, 259–282.

Koop, G., Poirier, D. J., & Tobias, J. (2005). Bayesian semiparametric inference in multiple equation models. *Journal of Applied Econometrics*, *20*, 723–747.

Kose, M. A., Otrok, C., & Whiteman, C. H. (2003). International business cycles: world, region and country specific factors. *American Economic Review*, *93*, 1216–1239.

Kose, M. A., Otrok, C., & Whiteman, C. H. (2008). Understanding the evolution of world business cycles. *Journal of International Economics*, *75*, 110–130.

Lin, X., & Zhang, D. (1999). Inference in generalized additive mixed models by using smoothing splines. *Journal of the Royal Statistical Society – Series B*, *61*, 381–400.

Lindley, D. V. (1971). *Bayesian statistics: A review*. Philadelphia, PA: SIAM.

Meng, X.-L., & van Dyk, D. (1999). Seeking efficient data augmentation schemes via conditional and marginal augmentation. *Biometrika*, *86*, 301–320.

Pesaran, M. H., & Potter, S. M. (1997). A floor and ceiling model of U.S. output. *Journal of Economic Dynamics and Control*, *21*, 661–695.

Poirier, D. J. (1973). Piecewise regression using cubic spline. *Journal of the American Statistical Association*, *68*, 515–524.

Poirier, D. J. (1998). Revising beliefs in non-identified models. *Econometric Theory*, *14*, 483–509.

Potter, S. M. (1995). A nonlinear approach to US GNP. *Journal of Applied Econometrics*, *10*, 109–125.

Primiceri, G. (2005). Time varying structural vector autoregressions and monetary policy. *Review of Economic Studies*, *72*, 821–852.

Ruppert, D., Wand, M. P., & Carroll, R. J. (2003). *Semiparametric regression*. Cambridge, UK: Cambridge University Press.

Shiller, R. (1984). Smoothness priors and nonlinear regression. *Journal of the American Statistical Association*, *79*, 609–615.

Shively, T. S., Kohn, R., & Wood, S. (1999). Variable selection and function estimation in additive nonparametric regression using a data-based prior. *Journal of the American Statistical Association*, *94*, 777–806.

Silverman, B. (1985). Some aspects of the spline smoothing approach to non-parametric regression curve fitting. *Journal of the Royal Statistical Society – Series B*, *47*, 1–52.

Sims, C. A. (1980). Macroeconomics and reality. *Econometrica*, *48*, 1–48.

Sims, C. A., & Zha, T. (2006). Were there regime switches in U.S. monetary policy? *American Economic Review*, *96*, 54–81.

Smith, M., & Kohn, R. (2000). Nonparametric seemingly unrelated regression. *Journal of Econometrics*, *98*, 257–281.

Stock, J. H., & Watson, M. W. (1996). Evidence on structural instability in macroeconomic time series relations. *Journal of Business and Economic Statistics*, *14*, 11–30.

Stock, J. H. & Watson, M. W. (2003). Has the business cycles changed? Evidence and explanations. In *Monetary policy and uncertainty: Adapting to a changing economy* (pp. 9–56). Kansas City, KS: Federal Reserve Bank of Kansas City.

Tanner, M. A., & Wong, W. H. (1987). The calculation of posterior distributions by data augmentation. *Journal of the American Statistical Association*, *82*, 528–549.

van Dyk, D., & Meng, X.-L. (2001). The art of data augmentation. *Journal of Computational and Graphical Statistics*, *10*, 1–50.

Wahba, G. (1978). Improper priors, spline smoothing and the problem of guarding against model errors in regression. *Journal of the Royal Statistical Society – Series B*, *40*, 364–372.

Wasserman, L. (2006). *All of nonparametric statistics*. New York, NY: Springer.

Whittaker, E. (1923). On a new method of graduation. *Proceedings of the Edinburgh Mathematical Society*, *41*, 63–75.

Wood, S., & Kohn, R. (1998). A Bayesian approach to robust binary nonparametric regression. *Journal of the American Statistical Association*, *93*, 203–213.

Wood, S., Kohn, R., Shively, T., & Jiang, W. (2002). Model selection in spline nonparametric regression. *Journal of the Royal Statistical Society – Series B*, *64*, 119–139.

Yang, L., Härdle, W., & Nielsen, J. (1999). Nonparametric autoregression with multiplicative volatility and additive mean. *Journal of Time Series Analsys*, *20*, 579–604.

TESTING FOR COMMON CYCLES IN NON-STATIONARY VARS WITH VARIED FREQUENCY DATA

Thomas B. Götz, Alain Hecq and Jean-Pierre Urbain

ABSTRACT

This article proposes a new approach to detecting the presence of common cyclical features when several time series are sampled at different frequencies. We generalize the common-frequency approach introduced by Engle and Kozicki (1993) and Vahid and Engle (1993). We start with the mixed-frequency VAR representation investigated in Ghysels (2012) for stationary time series. For non-stationary time series in levels, we show that one has to account for the presence of two sets of long-run relationships. The first set is implied by identities stemming from the fact that the differences of the high-frequency $I(1)$ *regressors are stationary. The second set comes from possible additional long-run relationships between one of the high-frequency series and the low-frequency variables. Our transformed vector error-correction model (VECM) representations extend the results of Ghysels (2012) and are important for determining the correct set of variables to be used in a subsequent common cycle investigation. This fact has implications for the distribution of test statistics and for forecasting. Empirical analyses with*

VAR Models in Macroeconomics – New Developments and Applications: Essays in Honor of Christopher A. Sims
Advances in Econometrics, Volume 32, 361–393

ISSN: 0731-9053/doi:10.1108/S0731-9053(2013)0000031010

quarterly real gross national product (GNP) and monthly industrial production indices for, respectively, the United States and Germany illustrate our new approach. We also conduct a Monte Carlo study which compares our proposed mixed-frequency models with models stemming from classical temporal aggregation methods.

Keywords: Common cycles; mixed frequency VAR; cointegration; MIDAS

JEL classification: C32

INTRODUCTION

Time series are collected and observed at different sampling frequencies. For instance, national account variables such as the gross domestic product (GDP) are often available on a quarterly basis while many other indicators such as the industrial production index, the unemployment rate or price indices are supplied at a monthly frequency. Several variables are only reported annually (e.g., CO_2 emission and health outlook) while most financial time series (stock prices, interest rates, exchange rates, etc.) are recorded almost continuously and are easily obtained at a daily frequency. In this situation it is common in macroeconomics to focus on the lower frequency unless a good proxy for the variable of interest is available at higher frequency. For instance, if one wants to estimate a vector autoregressive (VAR) model with GDP included, one would probably run a VAR model on quarterly time series. The higher frequency variables (monthly, daily, etc.) are either averaged over quarters (for flow variables) or matched at the end of each quarter (for stock variables).

More recently, researchers have developed techniques for modeling, in a unified framework, variables sampled at different frequencies (see notably the mixed data sampling, or MIDAS, approach in Andreou, Ghysels, & Kourtellos, 2010). Most of the mixed-frequency literature which relies on what Ghysels (2012) calls observation-driven models,[1] as introduced in Cox et al. (1981), is concerned with a regression framework where the dependent variable is the low-frequency variable and the high-frequency variables are in the regressors. Only few papers have extended the analysis toward a multivariate setting. Examples are Qian (2013) and Ghysels (2012).

Qian (2013) deals with a mixed-frequency VAR in which some variables are possibly unobserved to account for the fact that some series may be observed at lower frequencies. Ghysels (2012), in contrast, stacks the high- and low-frequency variables into a vector and formulates a VAR in terms of this vector. Interestingly, the model does not involve any latent variables, and thereby no latent shocks, but is formulated exclusively in terms of observable data explaining the term "observation-driven model." Note that due to its observation-driven nature, the model's related impulse response functions are entirely driven by observable shocks. Indeed, the mixed-frequency VAR à la Ghysels (2012) jointly specifies the dynamics of all variables concerned without imposing any a priori restrictions. Furthermore, the ability to apply standard tools of the VAR literature, e.g., impulse response functions, to a mixed-frequency setting gives it a significant edge over univariate approaches such as the aforementioned MIDAS approach.

We follow Ghysels (2012) in this article in specifying an observation-driven rather than parameter-driven model.[2] We deviate from Ghysels (2012) in that we allow for non-stationary and possibly cointegrated variables. An extension of a mixed-frequency VAR model along these lines is important. We show that for non-stationary time series in levels, one has to account for the presence of two sets of long-run relationships. The first set is characterized by identities stemming from the fact that the differences of the high-frequency $I(1)$ regressors are stationary. The second set comes from additional long-run relationships between one of the high-frequency series and the low-frequency variables. Without this latter set of long-run relationships, the VAR in first differences looks very much like in Ghysels (2012). We propose a VECM representation for these models that allows us to extend the common cyclical feature analysis (see Engle & Kozicki, 1993 or Vahid & Engle, 1993) to a mixed-frequency setup. As such, these mixed-frequency VARs present an alternative to temporally aggregating the high-frequency variable in order to obtain a VAR in a common frequency (Marcellino, 1999 or Hecq, 1998).

The rest of the article is organized as follows. We start with the stacked VAR representation proposed in Ghysels (2012) and show how to extend it to possibly cointegrated $I(1)$ variables. In the third section we propose an alternative VAR representation that is useful for testing for the presence of common cyclical features. A suitable likelihood ratio type statistic is proposed in the fourth section. Our new approach is intended for models involving a small number of variables and low observation frequency. The fifth section contains empirical illustrations of our approach based on quarterly real GNP and industrial production for Germany and the United States. The fifth section also contains a small forecasting exercise emphasizing the potential usefulness of our methods compared with classical temporal aggregation methods. The sixth section presents a Monte Carlo study investigating the rejection frequencies of the aforementioned likelihood ratio type test for common features for both, our proposed mixed-frequency and temporal aggregation based models. The seventh section concludes.

A MIXED-FREQUENCY VECM

Let us start from a two variable mixed-frequency system,[3] where y_t, $t=1,...,T$ is the low-frequency variable and $x^{(m)}_{t-i/m}$ the high-frequency variables with m high-frequency observations per low-frequency period t. In a year/month-example $m=12$ and the value of i indicates the specific month under consideration, ranging from January ($x^{(12)}_{t-11/12}$) until December ($x^{(12)}_t$). Obviously, $x^{(m)}_{t-m/m}=x^{(m)}_{t-1}$. Note that as far as lag operators, L, or difference operators, Δ, are concerned, we make a distinction between high- and low-frequency operators. To be more precise, L denotes the low-frequency lag operator, i.e., $Ly_t=y_{t-1}$ or $Lx^{(m)}_{t-i/m}=x^{(m)}_{t-1-i/m}$, whereas $L^{j/m}$ denotes the high-frequency counterpart which is only applicable to the high-frequency variables, i.e., $L^{j/m}x^{(m)}_{t-i/m}=x^{(m)}_{t-i/m-j/m}=x^{(m)}_{t-(i+j)/m}$. Similar rules apply for the difference operators, Δ and $\Delta^{(j/m)}$. Finally, note that $L^{1/m}x^{(m)}_{t-(m-1)/m}=x^{(m)}_{t-1}$ and, by the same logic, $\Delta^{(1/m)}x^{(m)}_{t-(m-1)/m}=x^{(m)}_{t-(m-1)/m}-x^{(m)}_{t-1}$. These notational conventions are similar to the ones used in Clements and Galvão (2008, 2009) or Götz, Hecq, and Urbain (2012a, 2012b). The following table illustrates the notation for a year/month example:

Notation	$t=2011$, $m=12$
$x^{(m)}_{t+1-(m-1)/m}=x^{(12)}_{t+1/12}$	$x^{(12)}_{2012,Jan}$
$x^{(m)}_t$	$x^{(12)}_{2011,Dec}$
$x^{(m)}_{t-1/m}$	$x^{(12)}_{2011,Nov}$
⋮	⋮
$x^{(m)}_{t-(m-1)/m}=x^{(12)}_{t-11/12}$	$x^{(12)}_{2011,Jan}$
$x^{(m)}_{t-m/m}=L^{1/m}x^{(12)}_{t-11/12}=x^{(12)}_{t-1}$	$x^{(12)}_{2010,Dec}$
$x^{(m)}_{t-1-1/m}$	$x^{(12)}_{2010,Nov}$

Considering each high-frequency variable such that $X_t^{(m)} = (x_t^{(m)}, x_{t-1/m}^{(m)}, \ldots, x_{t-(m-1)/m}^{(m)})'$, Ghysels (2012) proposes to difference the $I(1)$ data (e.g., quarterly growth rate of output and the monthly growth rate of the production index) and then construct the VAR(p):

$$Z_t = \Gamma_1 Z_{t-1} + \ldots + \Gamma_p Z_{t-p} + \varepsilon_t \tag{1}$$

where $Z_t = (y_t, X_t^{(m)'})'$ and $\varepsilon_t \sim \text{i.i.d.}\, N(\mathbf{0}, I_{m+1})$.[4] In other words, the high-frequency observations are stacked in a vector, together with the low-frequency variable y_t. Note that compared to Ghysels (2012) we place the low-frequency variable first to facilitate the interpretation of the presence of common cycles we develop in the fourth section.

Eq. (1) is easy to estimate for small systems but relies on the assumption that there is no cointegration. Assuming that series in Z_t are $I(1)$ and that there exists some cointegration between the variables, Eq. (1) in first differences is not correctly specified. This can be seen by computing the VECM representation of Eq. (1) when both the high- and low-frequency variables are $I(1)$ such that

$$\Delta Z_t = \tilde{\Gamma}_1 \Delta Z_{t-1} + \ldots + \tilde{\Gamma}_{p-1} \Delta Z_{t-p+1} + \Pi Z_{t-1} + \varepsilon_t \tag{2}$$

with $\tilde{\Gamma}_i = -\sum_{k=i+1}^{p} \Gamma_k, \quad i = 1, \ldots, p-1$ and $\Pi = -(I - \sum_{j=1}^{p} \Gamma_j) = \alpha\beta'$ with $rank(\Pi) = (r_0 + r_1) < m+1$.

Let us focus on the rank of Π in the presence and absence of cointegration between the low- and high-frequency variables. r_0 denotes a first set of known or prespecified cointegrating vectors stemming from the fact that first differences of the high-frequency $I(1)$ variables are stationary. r_1 will denote the potential additional long-run relationship between the two series. Note that in case of cointegration it is sufficient to consider only one relationship between y and one of the various x's. Indeed, if cointegration between the two variables is present, y_t is cointegrated with each $x_{t-i/m}^{(m)}$, $i = 0, \ldots, m-1$. Hence, it does not matter which of these high-frequency candidates is used to capture the cointegrating relationship as long as one is used (see Götz et al., 2012a for details). For simplicity, we model cointegration employing the end-of-period observation of the high-frequency variable, i.e., $x_t^{(m)}$.

Anticipating the results of the empirical section, when r_0 and the associated cointegrating vectors are known, a test on r_1 will be possible using the approach of Horvath and Watson (1995) for testing for cointegration when some of the cointegrating vectors are prespecified.[5] In particular, we will test the null hypothesis $H_0 : rank(\Pi) = r_0$ against $H_A : rank(\Pi) = r_0 + r_1$, where the cointegrating vectors imposed under the null are precisely the prespecified ones discussed above. Under the alternative, r_1 unknown cointegrating vectors are added to the r_0 prespecified ones. The test itself is then a likelihood ratio statistic. In our case, only one unknown cointegrating vector to capture potential cointegration between the low- and high-frequency variables is added. It is of the form $(1 : -\theta : 0_{1\times(m-1)})$, where the coefficient of y is normalized to one and θ is the cointegrating coefficient.

Our first results can be summarized in the following observation and example.

Observation 1. Suppose that the high-frequency variables, $x^{(m)}_{t-i/m}$, and the low-frequency variable, y_t, are $I(1)$. In the case of no cointegration between the low- and high-frequency variables, $rank(\Pi) = r_0$. $r_0 = m - 1$ in a bivariate analysis. In the case of cointegration between the low- and high-frequency variables, $rank(\Pi) = r_0 + r_1$. $r_1 = 1$ in a bivariate analysis.

Indeed, each high-frequency variable in the subvector X_t of Z_t is trivially cointegrated with the other high-frequency variables. Since there are m such variables in Z_t in the bivariate case, we have $m - 1$ prespecified cointegrating relationships r_0. We model any additional cointegrating relationships between the low- and high-frequency variables by considering one relationship between y_t and $x^{(m)}_t$. This increases the cointegrating rank by one in the bivariate example. These arguments can be extended to a higher dimensional framework.

Example 1. Let us consider a very simple example where the low- and high-frequencies are assumed to be quarterly and monthly, respectively, and where y_t and the elements of $X^{(m)}_t$ are defined as

$$\begin{aligned} y_t &= \theta x^{(3)}_t + \xi_t \text{ with } \xi_t = \rho\xi_{t-1} + \nu_t, \text{ where } |\rho| < 1 \text{ and } \nu_t \sim \text{i.i.d.} N(0,1) \\ x^{(3)}_{t-i/3} &= x^{(3)}_{t-(i+1)/3} + u^{(3)}_{t-i/3} \text{ for } i = 0,1,2 \text{ and where all } u^{(3)}_{t-i/3} \sim \text{i.i.d.} N(0,1), \end{aligned}$$

implying that cointegration between the low- and high-frequency variables is present. Representing this in a VAR framework yields

$$\begin{pmatrix} y_t \\ x^{(3)}_t \\ x^{(3)}_{t-1/3} \\ x^{(3)}_{t-2/3} \end{pmatrix} = \begin{pmatrix} \rho & (1-\rho)\theta & 0 & 0 \\ 0 & 1 & 0 & 0 \\ 0 & 1 & 0 & 0 \\ 0 & 1 & 0 & 0 \end{pmatrix} \begin{pmatrix} y_{t-1} \\ x^{(3)}_{t-1} \\ x^{(3)}_{t-1-1/3} \\ x^{(3)}_{t-1-2/3} \end{pmatrix} + \begin{pmatrix} \theta \sum_{i=0}^{2} u^{(3)}_{t-i/3} + \nu_t \\ \sum_{i=0}^{2} u^{(3)}_{t-i/3} \\ u^{(3)}_{t-1/3} + u^{(3)}_{t-2/3} \\ u^{(3)}_{t-2/3} \end{pmatrix}$$

The corresponding VECM representation looks as follows:

$$\begin{pmatrix} \Delta y_t \\ \Delta x_t^{(3)} \\ \Delta x_{t-1/3}^{(3)} \\ \Delta x_{t-2/3}^{(3)} \end{pmatrix} = \begin{pmatrix} \rho-1 & (1-\rho)\theta & 0 & 0 \\ 0 & 0 & 0 & 0 \\ 0 & 1 & -1 & 0 \\ 0 & 1 & 0 & -1 \end{pmatrix} \begin{pmatrix} y_{t-1} \\ x_{t-1}^{(3)} \\ x_{t-1-1/3}^{(3)} \\ x_{t-1-2/3}^{(3)} \end{pmatrix} + \begin{pmatrix} \theta \sum_{i=0}^{2} u_{t-i/3}^{(3)} + \nu_t \\ \sum_{i=0}^{2} u_{t-i/3}^{(3)} \\ u_{t-1/3}^{(3)} + u_{t-2/3}^{(3)} \\ u_{t-2/3}^{(3)} \end{pmatrix}$$

such that $rank(\Pi) = r_0 + r_1 = 2 + 1 = 3$, with long-run relationships, e.g.,

$$\beta' Z_{t-1} = \begin{pmatrix} \rho-1 & -(\rho-1)\theta & 0 & 0 \\ 0 & 1 & -1 & 0 \\ 0 & 0 & 1 & -1 \end{pmatrix} \begin{pmatrix} y_{t-1} \\ x_{t-1}^{(3)} \\ x_{t-1-1/3}^{(3)} \\ x_{t-1-2/3}^{(3)} \end{pmatrix}$$

If, however, no cointegration existed and both high- and low-frequency variables followed a random walk

$$\begin{aligned} y_t &= y_{t-1} + v_t \\ x_{t-i/3}^{(3)} &= x_{t-(i+1)/3}^{(3)} + u_{t-i/3}^{(3)} \quad \text{for } i = 0, 1, 2, \end{aligned}$$

where v_t and the $u_{t-i/3}^{(3)}$'s are i.i.d. standard normally distributed, we would obtain the VECM representation:

$$\begin{pmatrix} \Delta y_t \\ \Delta x_t^{(3)} \\ \Delta x_{t-1/3}^{(3)} \\ \Delta x_{t-2/3}^{(3)} \end{pmatrix} = \begin{pmatrix} 0 & 0 & 0 & 0 \\ 0 & 0 & 0 & 0 \\ 0 & 1 & -1 & 0 \\ 0 & 1 & 0 & -1 \end{pmatrix} \begin{pmatrix} y_{t-1} \\ x_{t-1}^{(3)} \\ x_{t-1-1/3}^{(3)} \\ x_{t-1-2/3}^{(3)} \end{pmatrix} + \begin{pmatrix} v_t \\ \sum_{i=0}^{2} u_{t-i/3}^{(3)} \\ u_{t-1/3}^{(3)} + u_{t-2/3}^{(3)} \\ u_{t-2/3}^{(3)} \end{pmatrix}$$

such that $rank(\Pi) = r_0 = m - 1 = 2$ and the (prespecified) long-run cointegrating relationships are

$$\beta' Z_{t-1} = \begin{pmatrix} 0 & 1 & -1 & 0 \\ 0 & 0 & 1 & -1 \end{pmatrix} \begin{pmatrix} y_{t-1} \\ x_{t-1}^{(3)} \\ x_{t-1-1/3}^{(3)} \\ x_{t-1-2/3}^{(3)} \end{pmatrix}$$

AN ALTERNATIVE VAR/VECM REPRESENTATION

Starting with the representation (2) an alternative representation of this VAR model may be obtained as illustrated in Hecq and Jacobs (2009). This representation will be used in deriving tests for common cycles. For this prupose we introduce the $(m+1)\times(m+1)$ matrix $C=(M_y:\beta:M_x)'$, with selection matrices $M_y=(1\ \mathbf{0}_{1\times m})'1_{(r=m-1)}$ and $M_x=(\mathbf{0}_{1\times m}1)'$. The indicator function in M_y ensures that we add a corresponding selection row in case y and x are not cointegrated. Premultiplying both sides of Eq. (2) by C gives

$$
\begin{aligned}
C\Delta Z_t &= C\tilde{\Gamma}_1 C^{-1}C\Delta Z_{t-1}+\ldots+C\tilde{\Gamma}_{p-1}C^{-1}C\Delta Z_{t-p+1}+C\Pi C^{-1}CZ_{t-1}+C\varepsilon_t\\
\Leftrightarrow CZ_t &= (I+C\tilde{\Gamma}_1C^{-1}+C\Pi C^{-1})CZ_{t-1}+(C\tilde{\Gamma}_2C^{-1}-C\tilde{\Gamma}_1C^{-1})CZ_{t-2}\\
&\quad+\ldots+(-C\tilde{\Gamma}_{p-1}C^{-1})CZ_{t-p}+C\varepsilon_t
\end{aligned} \tag{3}
$$

Whether or not there is cointegration between y and x, it is possible to rewrite Eq. (3) such that we finally obtain

$$\Delta Z_t^* = \Phi_1\Delta Z_{t-1}^* + \Phi_2\Delta Z_{t-2}^* + \ldots + \Phi_p\tilde{Z}_{t-p} + \tilde{\varepsilon}_t \tag{4}$$

where

$$
\Delta Z_t^* = \begin{pmatrix} \Delta y_t \\ \Delta^{(1/m)}x_t^{(m)} \\ \vdots \\ \Delta^{(1/m)}x_{t-(m-2)/m}^{(m)} \\ \Delta^{(1/m)}x_{t-(m-1)/m}^{(m)} \end{pmatrix},\quad
\tilde{Z}_t = \begin{pmatrix} y_t \\ \Delta^{(1/m)}x_t^{(m)} \\ \vdots \\ \Delta^{(1/m)}x_{t-(m-2)/m}^{(m)} \\ x_{t-(m-1)/m}^{(m)} \end{pmatrix}\ \text{and}\
\tilde{\varepsilon}_t = \begin{pmatrix} \varepsilon_{t,y} \\ \Delta^{(1/m)}\varepsilon_t^{(m)} \\ \vdots \\ \Delta^{(1/m)}\varepsilon_{t-(m-2)/m}^{(m)} \\ \varepsilon_{t-(m-1)/m}^{(m)} \end{pmatrix}
$$

In particular we can write Eq. (4) as

$$\Delta Z_t^* = \Phi_1\Delta Z_{t-1}^* + \Phi_2\Delta Z_{t-2}^* + \ldots + \Phi_p^0\tilde{Z}_{t-p} + \Phi_p^1\tilde{Z}_{t-p} + \tilde{\varepsilon}_t \tag{5}$$

where Φ_p^0 is the matrix with the central part of Φ_p and with zero-coefficients on the first and the last column. In the presence of additional (unknown) r_1

cointegrating vectors $\Phi_p^1 = \Phi_p - \Phi_p^0$ is of rank $r_1 > 0$. In the absence of such cointegrating vectors $\Phi_p = \Phi_p^0$ such that $\Phi_p^1 = \mathbf{0}_{(m+1)\times(m+1)}$.

Remark 1. As far as the error term is concerned, it was necessary to assume that ε_t is i.i.d. and standard normally distributed in order for Eq. (2) not to be misspecified. Premultiplying ε_t by C in Eq. (3), however, causes $\tilde{\varepsilon}_t$ to contain differences of $\varepsilon_t^{(m)}, \ldots, \varepsilon_{t-(m-2)/m}^{(m)}$. Note, however, that these are high-frequency differences, $\Delta^{(1/m)}$, implying that all terms contained in $\tilde{\varepsilon}_t$ correspond to observations within one low-frequency period t. Hence, as the transformed VAR, Eq. (5), is written in terms of the low frequency, premultiplying ε_t by C does not lead to a misspecification of the model.

Example 2. Let us again assume that the high- and low-frequency variables are monthly and quarterly, respectively. Furthermore, suppose first that y and x are not cointegrated and let us consider a VAR(1), $Z_t = \Gamma_1 Z_{t-1} + \varepsilon_t$ (without deterministic components for simplicity), from which we derive a low-frequency VECM representation $\Delta Z_t = \alpha\beta' Z_{t-1} + \varepsilon_t$ as:

$$\Leftrightarrow \begin{pmatrix} \Delta y_t \\ \Delta x_t^{(3)} \\ \Delta x_{t-1/3}^{(3)} \\ \Delta x_{t-2/3}^{(3)} \end{pmatrix} = \begin{pmatrix} \alpha_{11} & \alpha_{12} \\ \alpha_{21} & \alpha_{22} \\ \alpha_{31} & \alpha_{32} \\ \alpha_{41} & \alpha_{42} \end{pmatrix} \begin{pmatrix} 0 & 1 & -1 & 0 \\ 0 & 0 & 1 & -1 \end{pmatrix} \begin{pmatrix} y_{t-1} \\ x_{t-1}^{(3)} \\ x_{t-1-1/3}^{(3)} \\ x_{t-1-2/3}^{(3)} \end{pmatrix} + \begin{pmatrix} \varepsilon_{t,y} \\ \varepsilon_t^{(3)} \\ \varepsilon_{t-1/3}^{(3)} \\ \varepsilon_{t-2/3}^{(3)} \end{pmatrix}$$

Now, we premultiply both sides of the equation by

$$C = \begin{pmatrix} 1 & 0 & 0 & 0 \\ 0 & 1 & -1 & 0 \\ 0 & 0 & 1 & -1 \\ 0 & 0 & 0 & 1 \end{pmatrix}$$

to obtain

$$\begin{aligned} C\Delta Z_t &= C\Pi Z_{t-1} + C\varepsilon_t \\ \Leftrightarrow CZ_t &= CZ_{t-1} + C\Pi C^{-1} CZ_{t-1} + C\varepsilon_t \\ \Leftrightarrow \tilde{Z}_t &= (I + C\Pi C^{-1})\tilde{Z}_{t-1} + \tilde{\varepsilon}_t \end{aligned}$$

Note that

$$\tilde{Z}_t = \begin{pmatrix} y_t \\ \Delta^{(1/m)} x_t^{(3)} \\ \Delta^{(1/m)} x_{t-1/3}^{(3)} \\ x_{t-2/3}^{(3)} \end{pmatrix} = \begin{pmatrix} \Delta y_t \\ \Delta^{(1/m)} x_t^{(3)} \\ \Delta^{(1/m)} x_{t-1/3}^{(3)} \\ \Delta^{(1/m)} x_{t-2/3}^{(3)} \end{pmatrix} + \underbrace{\begin{pmatrix} 1 & 0 & 0 & 0 \\ 0 & 0 & 0 & 0 \\ 0 & 0 & 0 & 0 \\ 0 & 1 & 1 & 1 \end{pmatrix}}_{A} \begin{pmatrix} y_{t-1} \\ \Delta^{(1/m)} x_{t-1}^{(3)} \\ \Delta^{(1/m)} x_{t-1-1/3}^{(3)} \\ x_{t-1-2/3}^{(3)} \end{pmatrix}$$

$$= \Delta Z_t^* + A\tilde{Z}_{t-1}$$

Hence,

$$\tilde{Z}_t = (I + C\Pi C^{-1})\tilde{Z}_{t-1} + \tilde{\varepsilon}_t$$
$$\Leftrightarrow \Delta Z_t^* = \underbrace{(I - A + C\Pi C^{-1})}_{\Phi_1 = \Phi_p^0}\tilde{Z}_{t-1} + \tilde{\varepsilon}_t$$

$$\Leftrightarrow \begin{pmatrix} \Delta y_t \\ \Delta^{(1/m)} x_t^{(3)} \\ \Delta^{(1/m)} x_{t-1/3}^{(3)} \\ \Delta^{(1/m)} x_{t-2/3}^{(3)} \end{pmatrix} = \underbrace{\begin{pmatrix} 0 & \alpha_{11} & \alpha_{12} & 0 \\ 0 & 1 + \alpha_{21} - \alpha_{31} & \alpha_{22} - \alpha_{32} & 0 \\ 0 & \alpha_{31} - \alpha_{41} & 1 + \alpha_{32} - \alpha_{42} & 0 \\ 0 & \alpha_{41} - 1 & \alpha_{42} - 1 & 0 \end{pmatrix}}_{\Phi_1} \begin{pmatrix} y_{t-1} \\ \Delta^{(1/m)} x_{t-1}^{(3)} \\ \Delta^{(1/m)} x_{t-1-1/3}^{(3)} \\ x_{t-1-2/3}^{(3)} \end{pmatrix} + \begin{pmatrix} \varepsilon_{t,y} \\ \Delta^{(1/m)} \varepsilon_t^{(3)} \\ \Delta^{(1/m)} \varepsilon_{t-1/3}^{(3)} \\ \varepsilon_{t-2/3}^{(3)} \end{pmatrix}$$

This shows that in the presence of prespecified cointegrating vectors only, the mixed-frequency stationary VAR specification of Ghysels (2012) does not impose the full set of restrictions on the dynamics. While consistency of the estimators is still achieved, efficiency might be affected when the VAR is not restricted accordingly. In the framework of this article, this changes the set of variables that has to be used in the subsequent common feature analysis.

Example 3. More importantly now, if cointegration between y and x is present, as in

$$\Pi = \alpha\beta' = \begin{pmatrix} \alpha_{11} & \alpha_{12} & \alpha_{13} \\ \alpha_{21} & \alpha_{22} & \alpha_{23} \\ \alpha_{31} & \alpha_{32} & \alpha_{33} \\ \alpha_{41} & \alpha_{42} & \alpha_{43} \end{pmatrix} \begin{pmatrix} 1 & -\theta & 0 & 0 \\ 0 & 1 & -1 & 0 \\ 0 & 0 & 1 & -1 \end{pmatrix}$$

we select $C=(\beta : M_x)'$ with $m=3$. Note that $CZ_t \neq \tilde{Z}_t$. However, $A_\theta CZ_t=\tilde{Z}_t$, where

$$A_\theta = \begin{pmatrix} 1 & \theta & \theta & \theta \\ 0 & 1 & 0 & 0 \\ 0 & 0 & 1 & 0 \\ 0 & 0 & 0 & 1 \end{pmatrix}$$

Hence, we have the following:

$$\begin{aligned} CZ_t &= (I+C\Pi C^{-1})CZ_{t-1}+C\varepsilon_t \\ \Leftrightarrow A_\theta^{-1}\tilde{Z}_t &= (I+C\Pi C^{-1})A_\theta^{-1}\tilde{Z}_{t-1}+A_\theta^{-1}\tilde{\varepsilon}_t \\ \Leftrightarrow \tilde{Z}_t &= A_\theta(I+C\Pi C^{-1})A_\theta^{-1}\tilde{Z}_{t-1}+\tilde{\varepsilon}_t \\ \Leftrightarrow \Delta Z_t^* &= (A_\theta(I+C\Pi C^{-1})A_\theta^{-1}-A)\tilde{Z}_{t-1}+\tilde{\varepsilon}_t \\ &= \underbrace{(I-A+A_\theta C\Pi C^{-1}A_\theta^{-1})}_{\Phi_1}\tilde{Z}_{t-1}+\tilde{\varepsilon}_t \end{aligned}$$

where

$$\Phi_1 = \begin{pmatrix} \alpha_{11} & \alpha_{12}-\alpha_{11} & \alpha_{13}-\theta\alpha_{11} & -\theta\alpha_{11} \\ \alpha_{21}-\alpha_{31} & 1+\alpha_{22}-h-\theta\alpha_{21}+\theta\alpha_{31} & \alpha_{23}-\alpha_{33}-\theta\alpha_{21}+\theta\alpha_{31} & -\theta(\alpha_{21}-\alpha_{31}) \\ \alpha_{31}-\alpha_{41} & \alpha_{32}-\alpha_{42}-\theta\alpha_{31}+\theta\alpha_{41} & 1+\alpha_{33}-\alpha_{43}-\theta\alpha_{31}+\theta\alpha_{41} & -\theta(\alpha_{31}-\alpha_{41}) \\ \alpha_{41} & \alpha_{42}-\theta\alpha_{41}-1 & \alpha_{43}-\theta\alpha_{41}-1 & -\theta\alpha_{41} \end{pmatrix}$$

Note that the first and fourth columns of Φ_1 are linearly dependent, i.e., $\Phi_1 e_1' = -\theta\,\Phi_1 e_4'$, where e_i is a row-vector of size $m+1$ with a value of one at the ith position and zeroes everywhere else. Hence, we can write $\Delta Z_t^* = \Phi_1^0\tilde{Z}_{t-1}+\Phi_1^1\tilde{Z}_{t-1}+\tilde{\varepsilon}_t$, where

$$\Phi_1^0 = \begin{pmatrix} 0 & \alpha_{12}-\alpha_{11} & \alpha_{13}-\theta\alpha_{11} & 0 \\ 0 & 1+\alpha_{22}-\alpha_{32}-\theta\alpha_{21}+\theta\alpha_{31} & \alpha_{23}-\alpha_{33}-\theta\alpha_{21}+\theta\alpha_{31} & 0 \\ 0 & \alpha_{32}-\alpha_{42}-\theta\alpha_{31}+\theta\alpha_{41} & 1+\alpha_{33}-\alpha_{43}-\theta\alpha_{31}+\theta\alpha_{41} & 0 \\ 0 & \alpha_{42}-\theta\alpha_{41}-1 & \alpha_{43}-\theta\alpha_{41}-1 & 0 \end{pmatrix}$$

and, with a particular choice of normalization

$$\Phi_1^1 = \begin{pmatrix} \alpha_{11} & 0 & 0 & -\theta\alpha_{11} \\ \alpha_{21}-\alpha_{31} & 0 & 0 & -\theta(\alpha_{21}-\alpha_{31}) \\ \alpha_{31}-\alpha_{41} & 0 & 0 & -\theta(\alpha_{31}-\alpha_{41}) \\ \alpha_{41} & 0 & 0 & -\theta\alpha_{41} \end{pmatrix} = \begin{pmatrix} \alpha_{11} \\ \alpha_{21}-\alpha_{31} \\ \alpha_{31}-\alpha_{41} \\ \alpha_{41} \end{pmatrix} \begin{pmatrix} 1 & 0 & 0 & -\theta \end{pmatrix}$$

TESTING FOR COMMON CYCLICAL FEATURES IN MIXED-FREQUENCY VARS

The popular common cycle approach (Vahid & Engle, 1993) aims at testing for the presence of common dynamics in VAR/VECM multivariate systems. Allow us to test for the presence of common cyclical features when two or more series are observed at different frequencies.

Let us point out that this article focuses on the particular mixed-frequency case when m is relatively small, say 3 or 12. Indeed, we propose a method based on canonical correlations as well as Full Information Maximum Likelihood (FIML) techniques. It is well known that the latter has shortcomings when the number of variables is large compared to the number of observations T.

Definitions

Define Ω as the set of all outcomes for the stochastic processes under consideration. Let Ω_t be the information set generated by the collection of sigma-fields $F_t = \sigma(Z_s, s \leq t), t \geq 0$, and let Z_t be adapted to that filtration. In other words, Ω_t contains all the information present at moment t, implying that y_t and $x_t^{(m)}$ would be given, but $x_{t+1/m}^{(m)}$ would not.

We define the presence of common cycles in the mixed-frequency VAR (5) as follows:

Definition 1. In a mixed-frequency VAR for the $m+1$ series ΔZ_t^*

$$\Delta Z_t^* = \Phi_1 \Delta Z_{t-1}^* + \Phi_2 \Delta Z_{t-2}^* + \ldots + \Phi_p^0 \tilde{Z}_{t-p} + \Phi_p^1 \tilde{Z}_{t-p} + \tilde{\varepsilon}_t,$$

with $\Delta Z_t^* = (\Delta y_t, \Delta^{(1/m)} x_t^{(m)}, \Delta^{(1/m)} x_{t-1/m}^{(m)}, \ldots, \Delta^{(1/m)} x_{t-(m-1)/m}^{(m)})'$, a Mixed-Frequency Serial Correlation Common Features (MF-SCCF) arises if there exists a matrix δ spanning the common feature space such that

$$\mathbb{E}(\delta' \Delta Z_t^* | \Omega_{t-1}) = 0.$$

This implies that there exists a δ such that $\delta' \Phi_1 = \ldots = \delta' \Phi_{p-1} = \delta' \Phi_p^0 = \delta' \Phi_p^1 = \mathbf{0}$, i.e., δ annihilates the short-run dynamics in the system. In other words, δ lies in the intersection of the left null spaces of all Φ-matrices.

In particular, when δ is a $(m+1)$-dimensional column vector normalized on the first element, we say that two series Δy_t and $\Delta^{(1/m)} x_t^{(m)}$ have a MF-SCCF relationship if there exists a high-frequency-lag polynomial $\gamma(L^{1/m})$ such that

$$\mathbb{E}([\Delta y_t - \gamma(L^{1/m})\Delta^{(1/m)}x_t^{(m)}]|\Omega_{t-1}) = 0$$

Hence, two series Δy_t and $\Delta^{(1/m)}x_t^{(m)}$ having a MF-SCCF imply that

$$\mathbb{E}([\Delta y_t - \gamma_1\Delta^{(1/m)}x_t^{(m)}]|\Omega_{t-1}) = \gamma_2\Delta^{(1/m)}x_{t-1/m}^{(m)} + \ldots + \gamma_m\Delta^{(1/m)}x_{t-(m-1)/m}^{(m)}$$

and there exists a (restricted) polynomial serial correlation common feature relationship of order m, PSCCF(m), as defined in Cubadda and Hecq (2001) between the two series.

Testing Procedure

Now that we have defined MF-SCCF, we can formulate the steps we use to test for mixed-frequency common cycles. Note that as far as testing for cointegration and testing for common cyclical features is concerned we follow a two-step procedure (Vahid & Engle, 1993 or Hecq, Palm, & Urbain, 2006), the validity of which has been established in Paruolo (2006). The steps are as follows:

1. We determine the lag length order in the level (or the log level) of the series for the mixed-frequency VAR in Eq. (1) to obtain p. We rely on usual information criteria, i.e., Akaike information criterion (AIC), Schwarz information criterion (SC) and Hannan-Quinn information criterion (HQ).
2. Given p, we test for the cointegrating rank $r = r_0 + r_1$. In doing so we rely on the Horvath and Watson (1995) approach because the $m - 1 = r_0$ cointegrating vectors are prespecified.
3. In the absence of additional r_1 cointegrating vectors we perform a reduced rank regression in the transformed VAR(p) using a test on the zero eigenvalues computed from the canonical-correlation analysis

$$CanCor\left\{\Delta Z_t^*, \begin{pmatrix} \Delta Z_{t-1}^* \\ \vdots \\ \Delta Z_{t-p+1}^* \\ \tilde{Z}_{t-p}^0 \end{pmatrix}\right\} \tag{6}$$

where $\tilde{Z}_{t-p}^0$ is equivalent to $\tilde{Z}_{t-p}$ except for the first and last elements

$$\tilde{Z}^0_{t-p} = \begin{pmatrix} 0 \\ \Delta^{(1/m)} x^{(m)}_{t-p} \\ \vdots \\ \Delta^{(1/m)} x^{(m)}_{t-p-(m-2)/m} \\ 0 \end{pmatrix}$$

Furthermore, $CanCor\{V_t, W_t\}$ denotes the computation of canonical correlations between the two sets of variables V_t and W_t. The likelihood ratio type test, denoted by ξ_{LR}, considers the null hypothesis that there exist at most s common feature vectors against the alternative that there are more than s, i.e., $H_0 : rank(\delta) \leq s$ versus $H_A : rank(\delta) > s$. The test is obtained as

$$\xi_{LR} = -T \sum_{i=1}^{s} \ln(1 - \hat{\lambda}_i), \qquad s = 1, 2 \tag{7}$$

where $\hat{\lambda}_i$ is the ith smallest squared canonical correlation, i.e., the ith smallest eigenvalue, computed from Eq. (6), i.e., from

$$\hat{\Sigma}^{-1}_{VV} \hat{\Sigma}_{VW} \hat{\Sigma}^{-1}_{WW} \hat{\Sigma}_{WV} \tag{8}$$

or similarly from the symmetric matrix $\hat{\Sigma}^{-1/2}_{VV} \hat{\Sigma}_{VW} \hat{\Sigma}^{-1}_{WW} \hat{\Sigma}_{WV} \hat{\Sigma}^{-1/2}_{VV}$, where $\hat{\Sigma}_{ij}$ are the empirical covariance matrices with i and j being equal to V and/or W. In the mixed-frequency bivariate case and without counting the constant terms, the unrestricted transformed VAR has $(m+1)^2 (p-1) + (m+1-2)(m+1)$ parameters. Note that $s = 1, 2$ because our four-dimensional transformed VAR already contains two known cointegrating relationships such that s is maximum 2.

4. In the presence of additional r_1 cointegrating vectors, eigenvalues and eigenvectors are computed from

$$CanCor\left\{ \Delta Z^*_t, \begin{pmatrix} \Delta Z^*_{t-1} \\ \vdots \\ \Delta Z^*_{t-p+1} \\ \tilde{Z}_{t-p} \end{pmatrix} \right\} \tag{9}$$

where

$$\tilde{Z}_{t-p} = \begin{pmatrix} y_{t-p} \\ \Delta^{(1/m)} x^{(m)}_{t-p} \\ \vdots \\ \Delta^{(1/m)} x^{(m)}_{t-p-(m-2)/m} \\ x^{(m)}_{t-p-(m-1)/m} \end{pmatrix}$$

and we focus on the test for $s=1$. Indeed, due to the additional cointegrating relationship between the low- and high-frequency variables, there is only room for at most one common feature vector in this bivariate example. In the mixed-frequency bivariate case and without counting the constant terms, the unrestricted transformed VAR has $(m+1)^2(p-1)+ (m+1-2+1)(m+1)$ parameters.

5. Alternatively and numerically equivalent is the estimation of the whole system by FIML under the restriction that $rank(\delta)=2$.[6] To be more precise, Vahid and Engle (1993) illustrate that the system can then be written as s so-called pseudo-structural form equations and $n-s=m+1-s$ unconstrained reduced-form equations. In the present context, this representation might be easier for counting the number of restrictions as well as for considering additional restrictions on the common feature vectors. Let us write the pseudo-structural model in the presence of both sets of cointegrating vectors (prespecified and not prespecified) as

$$\begin{pmatrix} I & -\gamma \\ 0 & I \end{pmatrix} \Delta Z_t^* = \begin{pmatrix} 0 \\ \Lambda_1 \end{pmatrix} \Delta Z^*_{t-1} + \ldots + \begin{pmatrix} 0 \\ \Lambda_{p-1} \end{pmatrix} \Delta Z^*_{t-(p-1)} + \begin{pmatrix} 0 \\ \Lambda_p \end{pmatrix} C^* \tilde{Z}_{t-p} + e_t$$

$\Leftrightarrow$

$$\begin{pmatrix} I_s & -\gamma_{s\times(m+1-s)} \\ 0_{(m+1-s)\times s} & I_{(m+1-s)} \end{pmatrix} \begin{pmatrix} \Delta y_t \\ \Delta^{(1/m)} x^{(m)}_t \\ \Delta^{(1/m)} x^{(m)}_{t-1/m} \\ \vdots \\ \Delta^{(1/m)} x^{(m)}_{t-(m-1)/m} \end{pmatrix}$$

$$= \begin{pmatrix} \underbrace{\mathbf{0}}_{s\times(m+1)} \\ \underbrace{\Lambda_1}_{(m+1-s)\times(m+1)} \end{pmatrix} \begin{pmatrix} \Delta y_{t-1} \\ \Delta^{(1/m)} x^{(m)}_{t-1} \\ \Delta^{(1/m)} x^{(m)}_{t-1-1/m} \\ \vdots \\ \Delta^{(1/m)} x^{(m)}_{t-1-(m-1)/m} \end{pmatrix} + \ldots$$

$$+\begin{pmatrix} \underbrace{\mathbf{0}}_{s\times(m+1)} \\ \underbrace{\Lambda_{p-1}}_{(m+1-s)\times(m+1)} \end{pmatrix} \begin{pmatrix} \Delta y_{t-p+1} \\ \Delta^{(1/m)} x^{(m)}_{t-(p-1)} \\ \Delta^{(1/m)} x^{(m)}_{t-(p-1)-1/m} \\ \vdots \\ \Delta^{(1/m)} x^{(m)}_{t-(p-1)-(m-1)/m} \end{pmatrix}$$

$$+\begin{pmatrix} \underbrace{\mathbf{0}}_{s\times(r_0+r_1)} \\ \underbrace{\Lambda_{p}}_{(m+1-s)\times(r_0+r_1)} \end{pmatrix} \underbrace{\begin{pmatrix} 1 & 0 & \cdots & \cdots & 0 & -\hat{\theta} \\ 0 & 1 & 0 & \cdots & \cdots & 0 \\ 0 & 0 & 1 & \cdots & \cdots & 0 \\ \vdots & \vdots & \vdots & \ddots & \vdots & \vdots \\ 0 & \cdots & \cdots & \cdots & 1 & 0 \end{pmatrix}}_{C^*}$$

$$\times \begin{pmatrix} y_{t-p} \\ \Delta^{(1/m)} x^{(m)}_{t-p} \\ \Delta^{(1/m)} x^{(m)}_{t-p-1/m} \\ \vdots \\ \Delta^{(1/m)} x^{(m)}_{t-p-(m-2)/m} \\ x^{(m)}_{t-p-(m-1)/m} \end{pmatrix} + \begin{pmatrix} e_{1t} \\ e_{2t} \\ \\ \vdots \\ e_{m+1t} \end{pmatrix}$$

where $\Lambda_1, \ldots, \Lambda_p$ are the $(m+1-s)$ lower rectangular parts of the corresponding Φ-matrices that are going to be estimated by FIML. Note that the first row of C^* disappears in case $r_1 = 0$. In terms of the terminology introduced by Vahid and Engle (1993), $(I_s - \gamma_{s\times(m+1-s)})\Delta Z_t^*$ are the s pseudo-structural form equations. The system is then completed by adding the $m+1-s$ unconstrained reduced-form equations, i.e., $(0_{(m+1-s)\times s}\ I_{(m+1-s)})\Delta Z_t^*$. This shows that we have to estimate $(m+1-s)\times(m+1)$ parameters in each of the $\Lambda_i, i = 1, \ldots, p-1$. There are, in the presence of one additional cointegrating vector, $(m+1-s)\times m$ parameters in Λ_p. Finally, there are $s\times(m+1-s)$ parameters in γ. Without additional cointegrating vectors, Λ_p has only $(m+1-s)\times(m-1)$ parameters.

6. The null and alternative hypotheses of the common feature test in terms of the pseudo-structural form become

$$H_0 : \delta'\Phi_1 = \ldots = \delta'\Phi_p = 0$$
$$H_A : \delta'\Phi_1 = \ldots = \delta'\Phi_p \neq 0$$

where the rank of δ is implicitly equal to either 1 or 2 as argued before. The test is performed using the following likelihood ratio type test

$$LR_s = 2(l_{H_A} - l_{H_0}) \sim \chi^2(v)$$

where l_{H_A} and l_{H_0} denote the value of the log-likelihood function evaluated under the alternative and the null hypothesis, respectively. The degrees of freedom v are the difference between the number of parameters in the unrestricted mixed-frequency VAR (with cointegration) and the number of coefficients in the pseudo-structural system.

7. Finally, additional restrictions on the vectors of γ can be imposed if necessary for extracting a meaningful relationship between y and x.

The next section applies all these steps in the context of some empirical applications.

APPLICATIONS

To illustrate the methods discussed in the previous sections, we extract two pairs of time series, once for the United States and once for Germany.[7] In both cases y is the quarterly real GNP (ref. GNPC96 and B1GM for the United States and Germany, respectively), seasonally adjusted. For the United States the series is observed from 1948Q1 until 2011Q4, whereas for Germany it is observed from 1991Q1 until 2012Q1. For x we consider, for the same time span, the monthly seasonally adjusted industrial production index (ref. INDPRO and PROD, respectively), or IPI from here on. Since GNP is a quarterly variable and IPI is observed on a monthly basis we have that $m = 3$.

Common Feature Analysis

As a first step, we determine the VAR order in the log-level of the series by relying on information criteria. That implies that we compute the mixed-frequency VAR à la Ghysels (2012) disregarding the possibility that the $I(1)$ series may be cointegrated. After assuming the $I(1)$-ness of both series involved (we indeed do not reject the null of a unit root for all series under consideration), we may test for the cointegrating rank r. As discussed above, due to the presence of r_0 prespecified cointegrating vectors, we follow the approach of Horvath and Watson (1995). In particular, since $m=3$ we test the null hypothesis that merely the two known cointegrating vectors are present against the alternative of one additional cointegrating vector to capture the long-run relationship between GNP and IPI (Götz et al., 2012a). We can formulate the likelihood ratio test statistic as

$$LR_{r_0,r_1} = 2(l_{H_A} - l_{H_0}) = 2(l_{r=r_0+r_1} - l_{r=r_0})$$

where l_{H_0} and l_{H_A} denote the value of the log-likelihood function evaluated under the null and the alternative hypothesis, respectively, i.e., when once only the prespecified cointegrating vectors are imposed and once the prespecified ones and one capturing potential cointegration between y and x. In terms of our empirical applications, this approach involves the log-likelihood of the following two regressions:

$$\text{Under } H_0: \begin{pmatrix} \Delta GNP_t \\ \Delta IPI_t^{(3)} \\ \Delta IPI_{t-1/3}^{(3)} \\ \Delta IPI_{t-2/3}^{(3)} \end{pmatrix} \text{ on } \underbrace{\begin{pmatrix} 0 & 1 & -1 & 0 \\ 0 & 0 & 1 & -1 \end{pmatrix} \begin{pmatrix} GNP_{t-1} \\ IPI_{t-1}^{(3)} \\ IPI_{t-1-1/3}^{(3)} \\ IPI_{t-1-2/3}^{(3)} \end{pmatrix}}_{\begin{pmatrix} \Delta^{(1/m)} IPI_{t-1}^{(3)} \\ \Delta^{(1/m)} IPI_{t-1-1/3}^{(3)} \end{pmatrix}}, \begin{pmatrix} \Delta GNP_{t-1} \\ \Delta IPI_{t-1}^{(3)} \\ \Delta IPI_{t-1-1/3}^{(3)} \\ \Delta IPI_{t-1-2/3}^{(3)} \end{pmatrix}$$

and

$$\text{Under } H_A: \begin{pmatrix} \Delta GNP_t \\ \Delta IPI_t^{(3)} \\ \Delta IPI_{t-1/3}^{(3)} \\ \Delta IPI_{t-2/3}^{(3)} \end{pmatrix} \text{ on } \underbrace{\begin{pmatrix} 1 & -\theta & 0 & 0 \\ 0 & 1 & -1 & 0 \\ 0 & 0 & 1 & -1 \end{pmatrix} \begin{pmatrix} GNP_{t-1} \\ IPI_{t-1}^{(3)} \\ IPI_{t-1-1/3}^{(3)} \\ IPI_{t-1-2/3}^{(3)} \end{pmatrix}}_{\begin{pmatrix} GNP_{t-1} - \theta IPI_{t-1}^{(3)} \\ \Delta^{(1/m)} IPI_{t-1}^{(3)} \\ \Delta^{(1/m)} IPI_{t-1-1/3}^{(3)} \end{pmatrix}}, \begin{pmatrix} \Delta GNP_{t-1} \\ \Delta IPI_{t-1}^{(3)} \\ \Delta IPI_{t-1-1/3}^{(3)} \\ \Delta IPI_{t-1-2/3}^{(3)} \end{pmatrix}$$

According to SC and HQ a lag length of 1 is chosen for both countries under consideration. A VAR(2) is preferred for the United States using AIC.[8] Consequently, we carry out the analysis with both lag lengths for the United States in order to evaluate the robustness of the results.

It turns out that $LR^{US}_{r_0,r_1} = 8.62$ in the VAR(2) and 8.88 for the VAR(1), whereas $LR^{GER}_{r_0,r_1} = 25.41$ in the VAR(1) case such that with a Horvath and Watson (1995) critical value of 18.37 we do not reject the null hypothesis of no cointegration between GNP and IPI for the United States, but do reject it for Germany. The cointegrating vector from the restricted VECM in the Johansen (1991)'s approach corresponding to r_1 is

$$Germany : GNP_t - \underset{(0.048)}{0.751}\, IPI^{(3)}_{t-2/3} \sim I(0)$$

This result will have an impact on how to set up the transformed VAR for both applications which, in turn, has an implication on how to test for common cyclical features thereafter (see Examples 2 and 3).

Let us now look at the presence of additional common cyclical feature restrictions. For the United States, we first look at the VAR(1) in the log-level of the series which corresponds to the following transformed mixed-frequency VAR:

$$\begin{pmatrix} \Delta GNP_t \\ \Delta^{(1/m)} IPI^{(3)}_t \\ \Delta^{(1/m)} IPI^{(3)}_{t-1/3} \\ \Delta^{(1/m)} IPI^{(3)}_{t-2/3} \end{pmatrix} = \begin{pmatrix} 0.005 \\ 0.001 \\ 0.002 \\ 0.001 \end{pmatrix} + \begin{pmatrix} 0 & 0.488 & 0.265 & 0 \\ 0 & 0.247 & 0.023 & 0 \\ 0 & 0.281 & 0.081 & 0 \\ 0 & 0.362 & 0.131 & 0 \end{pmatrix} \begin{pmatrix} GNP_{t-1} \\ \Delta^{(1/m)} IPI^{(3)}_{t-1} \\ \Delta^{(1/m)} IPI^{(3)}_{t-1-1/3} \\ IPI^{(3)}_{t-1-2/3} \end{pmatrix} + e_t$$

with a value of the log-likelihood of 3477.474. Because the rank of the transformed VAR coefficient matrix is 2, we know that, by construction, its orthogonal complement and then the common feature space δ is also of rank 2 in this four-dimensional system. Hence, a normalized common feature matrix that takes the form

$$\delta' = \begin{pmatrix} 1 & 0 & \times & \times \\ 0 & 1 & \times & \times \end{pmatrix}$$

where $\times$ denotes a parameter to be estimated, always exists. Premultiplying the VAR by

$$\begin{pmatrix} \delta' & \\ 0_{(2\times 2)} & I_{(2)} \end{pmatrix}$$

leads to the pseudo-structural form with the same likelihood because no restrictions are added and hence no test is possible.

We may further restrict δ and investigate whether there exists a common cyclical feature relationship between the growth rates of GNP and IPI. Restricting for instance the first possible relationship as in

$$\delta' = \begin{pmatrix} 1 & 0 & 0 & \times \\ 0 & 1 & \times & \times \end{pmatrix}$$

yields a log-likelihood of 3477.067 compared to 3477.474 from the previous mixed-frequency VAR imposing no common feature restrictions. This obviously means a non-rejection of the additional null hypothesis, $H_0 : \delta'\Phi_1 = \ldots, \delta'\Phi_p = 0$ see section Common Feature Analysis. The MF-SCCF relationship is now given by

$$\Delta GNP_t = \underset{(0.0001)}{0.005} + \underset{(0.125)}{1.509}\, \Delta^{(1/3)} IPI^{(3)}_{t-2/3} + w_n$$

i.e., a relationship between the quarterly growth rate of GNP and the monthly difference of IPI for the first month of the corresponding quarter. Still for the United States, but with a VAR(2), as favored by the AIC, we have a log-likelihood of 3480.36 in

$$\begin{pmatrix} \Delta GNP_t \\ \Delta^{(1/3)} IPI^{(3)}_t \\ \Delta^{(1/3)} IPI^{(3)}_{t-1/3} \\ \Delta^{(1/3)} IPI^{(3)}_{t-2/3} \end{pmatrix} = \begin{pmatrix} 0.005 \\ 0.001 \\ 0.001 \\ 0.000 \end{pmatrix} + \begin{pmatrix} 0.043 & 0.478 & 0.255 & 0.004 \\ 0.353 & 0.175 & -0.010 & -0.311 \\ 0.259 & 0.227 & 0.045 & -0.031 \\ 0.113 & 0.330 & 0.100 & 0.021 \end{pmatrix}$$
$$\times \begin{pmatrix} \Delta GNP_{t-1} \\ \Delta^{(1/3)} IPI^{(3)}_{t-1} \\ \Delta^{(1/3)} IPI^{(3)}_{t-1-1/3} \\ \Delta^{(1/3)} IPI^{(3)}_{t-1-2/3} \end{pmatrix} + \begin{pmatrix} 0 & -0.016 & 0.027 & 0 \\ 0 & -0.085 & 0.011 & 0 \\ 0 & -0.096 & -0.046 & 0 \\ 0 & 0.017 & -0.035 & 0 \end{pmatrix} \begin{pmatrix} GNP_{t-2} \\ \Delta^{(1/3)} IPI^{(3)}_{t-2} \\ \Delta^{(1/3)} IPI^{(3)}_{t-2-1/3} \\ IPI^{(3)}_{t-2-2/3} \end{pmatrix} + e_t$$

In this case there does not necessarily exist a matrix that annihilates all of the dynamics, namely the two matrices. For $s = 1$ we have a log-likelihood value of 3479.877, for $s = 2$ we have a log-likelihood value of 3474.863 in the pseudo-structural systems estimated by FIML. Consequently, we do not reject the null of $s=2$ common feature vectors with a likelihood ratio test of $2(3480.36-3474.86)=11 \sim \chi^2(8)$. In addition to $s=2$, we are also able to restrict the first common feature vector as in the $(p=1)$ case. The log likelihood with this additional restriction is 3474.732. The relationship is

$$\Delta GNP_t = \underset{(0.0001)}{0.005} + \underset{(0.128)}{1.488}\, \Delta^{(1/3)} IPI^{(3)}_{t-2/3} + v_n$$

such that very similar results as before emerge, but now with a formal test for MF-SCCF.

For Germany with $r_0 + r_1$ cointegrating vectors we have

$$\begin{pmatrix} \Delta GNP_t \\ \Delta^{(1/3)} IPI^{(3)}_t \\ \Delta^{(1/3)} IPI^{(3)}_{t-1/3} \\ \Delta^{(1/3)} IPI^{(3)}_{t-2/3} \end{pmatrix} = \begin{pmatrix} 0.102 \\ -1.594 \\ -1.008 \\ 0.166 \end{pmatrix} + \begin{pmatrix} 0 & 0.297 & 0.147 & 0 \\ 0 & 0.167 & -0.093 & 0 \\ 0 & 0.143 & 0.067 & 0 \\ 0 & 0.043 & 0.137 & 0 \end{pmatrix} \begin{pmatrix} GNP_{t-1} \\ \Delta^{(1/3)} IPI^{(3)}_{t-1} \\ \Delta^{(1/3)} IPI^{(3)}_{t-1-1/3} \\ IPI^{(3)}_{t-1-2/3} \end{pmatrix}$$
$$+ \begin{pmatrix} -0.010 \\ 0.162 \\ 0.103 \\ -0.017 \end{pmatrix} \left(GNP_{t-1} - 0.751 IPI^{(3)}_{t-1-2/3} \right) + u_t$$

Due to the ranks of these matrices, we know that there always exists a vector that annihilates the dynamics. This is an application of the mixed form reduced rank approach of Hecq et al. (2006). The common feature relationship and the standard errors (in brackets beneath the estimators) obtained by FIML, however, show that there are no cofeature relationships between the GNP and IPI series alone:

$$\Delta GNP_t = \underset{(0.030)}{0.006} + \underset{(16.435)}{2.596}\, \Delta^{(1/3)} IPI^{(3)}_t - \underset{(24.381)}{3.411}\, \Delta^{(1/3)} IPI^{(3)}_{t-1/3}$$
$$+ \underset{(27.25)}{4.912}\, \Delta^{(1/3)} IPI^{(3)}_{t-2/3} + w_n$$

such that the only common cyclical feature relationship involves the growth rate of the GNP which is consequently a white noise process.

Alternative to a common feature analysis in a mixed-frequency setup, the high-frequency variables can be temporally aggregated in order to work with a common (low-)frequency VAR. Note, however, that common features are not invariant to temporal aggregation. Time aggregation of high-frequency variables may not only cause common feature relationships, which are present in mixed frequencies, to disappear in a common frequency framework, but also create additional cofeature vectors (Marcellino, 1999).

Two temporal aggregation schemes are considered here, point-in-time and average sampling. The former takes a specific high-frequency observation as the corresponding low-frequency one and is usually applied to stock variables. The latter simply uses the average of the high-frequency observations as the corresponding t-observation and is usually of interest when flow variables are dealt with (see Marcellino, 1999 or Silvestrini & Veredas, 2008). We only consider the U.S. data for the common feature analysis in these low-frequency VARs. The resulting elasticities (standard errors in brackets) of ΔGNP with respect to $\Delta IPI_t^{(3)}$, $\Delta IPI_{t-1/3}^{(3)}$, $\Delta IPI_{t-2/3}^{(3)}$ and $\Delta\overline{IPI_t^{(3)}}$ are 0.7187 (0.0696), 0.4033 (0.0364), 0.279 (0.016) and 0.4097 (0.0.328), respectively, where $\overline{IPI_t^{(3)}}$ refers to the average of the monthly IPI observations in period t. The corresponding common feature test statistics computed from the pseudo-structural forms of the low-frequency VARs are 10.09, 7.9, 31, and 4.48 for the respective temporal aggregation approaches. At the 5% significance level, the null hypothesis of one common feature is rejected in all cases, implying that there exist two independent, i.e., non-common and non-synchronous, cycles for two variables under consideration. In other words, the commonalities that are present in the mixed-frequency setup are not visible anymore after time aggregation. Finally, note that the potential cofeature elasticities are also very different to the mixed-frequency setup.

Forecasting

In the remainder of this section a small forecasting exercise is presented to emphasize the potential benefits of employing a mixed-frequency approach when dealing with common cycles. Potential gains of incorporating common feature restrictions for forecasting already exist in the literature. Vahid and Issler (2002) show that the costs in terms of forecast accuracy associated with ignoring common feature relationships in VARs may be considerable. Athanasopoulos, de Carvalho Guilln, Issler, and Vahid (2011) also show that models imposing restrictions associated with common cycles outperform unrestricted VAR models for different sample sizes, especially if the data are generated by comparably severe rank restrictions. Anderson and Vahid (2010) argue that VARs "with cointegration and common cycles can be [...] viewed as observable factor models [...that...] have potential for forecasting" (p. 5). In a set of Monte Carlo studies the authors illustrate that over- and underspecification of the associated rank restrictions deteriorate the forecast performance in almost all cases. While the cointegration structure seems to be more crucial in general, misspecification of the common cycle rank restrictions particularly worsens the forecast accuracy in the short run. In their conclusion, Anderson and Vahid (2010) advise to consider reduced rank error-correction models for forecasting if there are compelling reasons to consider common trends or cycles in a particular data set.

We consider 28 out-of-sample one-step-ahead forecasts of the real U.S. GNP growth rate corresponding to the period from 2005Q1 until 2011Q4. We recursively update the estimation period and select the lag length at each moment in time. Furthermore, we test for the number of common feature vectors at each moment in time. Note that we do not use real-time data, but final-vintage data. Indeed, real-time data would require a vintage-specific analysis of which models to employ, which lag length to choose, cointegration properties, and so on (see Götz, Hecq, & Urbain, 2012b for an example of mixed-frequency models in a real-time data set). As already investigated above, the null hypothesis of no cointegration cannot be rejected for the United States such that we assume no cointegration to be present for each estimation period under consideration.

We are interested in the forecast accuracy of three mixed-frequency stationary VAR models. First, the model without restrictions on the short-run parameters, denoted by MF_VAR1_FULL. As pointed out in the second and the third sections, this is the model suggested by Ghysels (2012). Second, the model imposing zero-restrictions corresponding to the absence of cointegration between GNP and IPI in the transformed VAR, denoted by MF_VAR1_ZEROS. Finally, the model imposing, on top of the just mentioned zero restrictions, a set of common feature vectors, denoted by MF_VAR1_CF. In order to investigate whether the development of such mixed-frequency VAR models leads to an improved forecast performance compared to VARs stemming from traditional aggregation methods, we confront our three MF-VARs with two low-frequency VARs, resulting from applying average sampling to IPI on the one hand and point-in-time sampling using the end-of-period observations, $IPI_t^{(3)}$, on the other hand. In both cases a VAR in first differences, denoted by AV_VAR1_DIFF and PIT_VAR1_DIFF, respectively, are considered. RMSFE errors are computed for each model.

As far as the lag length selection is concerned, it turns out that for both, the mixed-frequency and time aggregated models, a VAR(1) model in first differences is chosen by SC or HQ at each moment in time. When testing for common features in the mixed-frequency VARs, the null hypothesis of $s = 2$ common feature vectors could not be rejected for all estimation periods considered. Hence, MF_VAR1_CF imposes two common feature vectors at each step of the forecasting exercise, whereby the first vector is further restricted as done in the previous subsection. Even though no common feature could have been detected for the low-frequency VARs at the 5% significance level at each moment in time, we include the models imposing one common feature vector into the analysis, i.e., AV_VAR1_1CF and PIT_VAR1_1CF. As a justification, although no common feature was detected for average sampled IPI at the 5% significance level, the null hypothesis of one common feature could not be rejected at the 1% significance level.

Fig. 1 shows the forecasts of the real U.S. GNP growth rate from 2005Q1 to 2011Q4 corresponding to each of the seven models just introduced as well as the actual data (solid black line). Note that the black dotted or dashed lines correspond to the mixed-frequency transformed VARs, the gray dotted or dashed ones to the VARs based on averaging the high-frequency variable and, finally, the gray solid lines to the VARs based on taking the last monthly observation of IPI per quarter (point-in-time).

It becomes clear from Fig. 1 that the forecasts based on the mixed-frequency models are, compared to the competing time aggregated models, able to track the actual data rather well, especially the trough associated

Fig. 1. Out-of-sample one-step-ahead static forecasts of real GNP growth in the united states from 2005Q1 to 2011Q4.

Table 1. RMSFEs of the Seven Competing Models.

Model	RMSFE
MF_VAR1_ZEROS	0.005671
MF_VAR1_CF	0.005672
MF_VAR1_FULL	0.005766
PIT_VAR1_DIFF	0.006527
PIT_VAR1_1CF	0.006591
AV_VAR1_1CF	0.007589
AV_VAR1_DIFF	0.00766

Note: The figures represent the RMSFEs of the seven competing models: The model (1) involving no restrictions, (2) imposing zero-restrictions, (3) imposing zero-restrictions and two common features, (4) and (5) based on point-in-time sampling with and without a common feature imposed, respectively, and (6) and (7) based on averaging with and without a common feature imposed.

with the financial crisis in 2008. The forecasts based on VARs stemming from temporally aggregating the high-frequency variables severely understate this drop in the GNP growth rate. This is reflected in the RMSFEs summarized in Table 1.

As is evident in the figures, the forecast accuracy of the mixed-frequency transformed VARs is significantly better than the one of the VARs based on classical temporal aggregation methods. Hence, in the case of predicting the real U.S. GNP growth rate from 2005Q1 to 2011Q1 it seems advantageous to consider mixed-frequency VAR models instead of temporally aggregating the high-frequency variable.

MONTE CARLO STUDY

Motivated by the results of the forecasting exercise in the fifth section, we now investigate more closely the behavior of mixed-frequency VAR models in this section. Given a certain lag length, we are interested in the size and power of the LR type test in Eq. (7) or the one based on Eq. (9), depending on whether there are additional cointegrating vectors above the prespecified ones.

To this end we conduct a Monte Carlo study, in which we simulate a mixed-frequency VAR that matches some of the characteristics that we have obtained using U.S. data, namely a VAR of order $p=2$ in the level of the $I(1)$ series and no additional cointegrating vectors above the known ones. Hence, it is a two-variable system with $m=3$ for the high-frequency series, in which the following parameter values in the transformed VAR representation are chosen:

$$\begin{pmatrix} \Delta y_t \\ \Delta^{(1/3)} x_t^{(3)} \\ \Delta^{(1/3)} x_{t-1/3}^{(3)} \\ \Delta^{(1/3)} x_{t-2/3}^{(3)} \end{pmatrix} = \begin{pmatrix} 0.05 \\ 0.05 \\ 0.05 \\ 0.05 \end{pmatrix} + \begin{pmatrix} 0.2 & 0.5 \\ -0.2 & 0.2 \\ 0.4 & 0.5 \\ 0.5 & 0.25 \end{pmatrix} \begin{pmatrix} 0.5 & 0.1 \\ 0.6 & 0.5 \\ 0.2 & -0.5 \\ 0.1 & 0.25 \end{pmatrix}' \begin{pmatrix} \Delta y_{t-1} \\ \Delta^{(1/3)} x_{t-1}^{(3)} \\ \Delta^{(1/3)} x_{t-1-1/3}^{(3)} \\ \Delta^{(1/3)} x_{t-1-2/3}^{(3)} \end{pmatrix}$$

$$+ \begin{pmatrix} 0 & 0.2 & 0.5 & 0 \\ 0 & -0.2 & 0.2 & 0 \\ 0 & 0.4 & 0.5 & 0 \\ 0 & 0.5 & 0.25 & 0 \end{pmatrix} \begin{pmatrix} y_{t-2} \\ \Delta^{(1/3)} x_{t-2}^{(3)} \\ \Delta^{(1/3)} x_{t-2-1/3}^{(3)} \\ x_{t-2-2/3}^{(3)} \end{pmatrix} + \begin{pmatrix} \varepsilon_{1t} \\ \varepsilon_{2t} \\ \varepsilon_{3t} \\ \varepsilon_{4t} \end{pmatrix}$$

where

$$\varepsilon_t \sim \text{i.i.d. } N(\mathbf{0}, \Sigma) \text{ with } \Sigma = \begin{pmatrix} 1 & 0.5 & 0.5 & 0.5 \\ 0.5 & 1 & 0.9 & 0.9 \\ 0.5 & 0.9 & 1 & 0.9 \\ 0.5 & 0.9 & 0.9 & 1 \end{pmatrix}$$

The normalized cofeature matrix that annihilates the two coefficient matrices of this DGP is fixed to

$$\delta' = \begin{pmatrix} 1 & 0 & -\frac{4}{3} & \frac{2}{3} \\ 0 & 1 & -1 & 1.2 \end{pmatrix}$$

Note that because the DGP is generated as a mixed-frequency transformed VAR, we know the number of common feature vectors to be $s=2$. Consequently, given a lag length p, whether it is the correct one ($p=2$) or an under- or overestimated one, we can analyze size ($s=2$) and power ($s=3$ or $s=4$) of the LR type test in Eq. (7).

As already mentioned before, an alternative to considering a mixed-frequency VAR is temporally aggregating the high-frequency variables in order to work in a common (low-)frequency setup. As a consequence, we mimic the scenario in which practitioners are unaware of the mixed-frequency nature of the underlying data and work with time aggregated IPI data. Assuming they follow the same testing procedure, i.e., use Eq. (7) to test for common feature relationships, it is of interest to analyze in how much the use of time aggregated data instead of high-frequency ones changes the results.

Note that as soon as data are temporally aggregated, the dynamics of the underlying system change completely (see Marcellino, 1999, Silvestrini & Veredas, 2008 or Lütkepohl, 2005). Indeed, an n-dimensional VAR(p) can be written in a final equation system (Zellner & Palm, 1974) such that each component is at most an ARMA$(np, (n-1)p)$ process. In the presence of s SCCF-restrictions the n-dimensional VAR(p) yields at most ARMA $((n-s)p, (n-s)p)$ models (Cubadda, Hecq, & Palm, 2009). In our case, the VAR(2) becomes a VARMA(4,4) after temporal aggregation of IPI (because $n=4$ and $s=2$), which can be written as VAR(∞). Note that for practical reasons, the VARMA(4,4) is usually approximated by a finite order VAR. Consequently, as soon as variables are temporally aggregated, we cannot refer to a certain lag length as correct, under- or over-estimated anymore. Note that we do not derive the final equation system corresponding to the mixed-frequency transformed VAR because it depends not only on the particular parameter values chosen, but also on n, m, r and s, making it a tedious and challenging task that could be tackled in future work.

As already mentioned before, time aggregation also has an impact on the analysis of common cycles (Marcellino, 1999). For example, it may cause common feature relationships, that are present in mixed frequencies, to disappear in a common frequency framework as illustrated before. In our case, temporal aggregation of IPI leads to a two-dimensional low-frequency VAR, for which the corresponding common feature test can be performed and the rejection frequencies for $s=1$ or $s=2$ can be computed. The rejection frequencies for $s=2$ represent power against a bivariate white noise system. The ones for $s=1$, however, cannot clearly be interpreted as size or power, because we do not know whether one synchronous common feature is present after time aggregation (in which case it would measure size) or none (in which case it would correspond to power). Again, the precise impact of temporal aggregation in a mixed-frequency VAR on common cycles analyses is left for further research.

For both, the mixed-frequency and time aggregated models, we consider $T=40$ and 100 corresponding to respectively 10 and 25 years of quarterly data. An additional burning-in period of 100 observations is used to initialize the process and are discarded after generating the data. All figures are based on 10,000 replications.

For the mixed-frequency models, Table 2 reports the rejection frequencies (in percent) associated with the likelihood ratio type test in Eq. (7). Concerning the common feature test statistics, we rely on the estimation of the squared canonical correlation reviewed in the fourth section. We estimate mixed-frequency VARs of orders from one to six but without imposing the zero-restrictions in the last matrix, i.e., in Φ_p for $p=1,\ldots,6$. These models are denoted by MF_VAR. Models imposing zero-restrictions are denoted by MF_VAR_0. As argued before, the size of the test should be interpreted as the rejection frequency for $s=2$, power can be observed for $s=3$ and $s=4$.

Table 2. Rejection Frequencies (in Percent) of the LR Type Test in Eq. (7): Mixed-Frequency Models.

		$T=40$					
	p	1	2	3	4	5	6
MF_VAR	$s=1$	0.2	0.6	1	2.3	5.7	15.4
	$s=2$	1.4	8.2	17.7	25.4	44.3	68.5
	$s=3$	20.4	75.7	79.1	87.3	94.2	98.7
	$s=4$	100	100	100	100	100	100
MF_VAR_0	$s=1$	–	0.4	0.7	1.4	3.8	9.4
	$s=2$	–	6.5	10.3	18.9	34.2	55.9
	$s=3$	18	77.5	76.6	82.8	90.8	97.3
	$s=4$	99.8	100	100	100	100	100
		$T=100$					
MF_VAR	$s=1$	0.2	0.3	0.3	0.4	0.5	0.8
	$s=2$	2	6.6	8.2	9.9	12.4	16.6
	$s=3$	37	99.4	98.5	97.4	96.9	96.9
	$s=4$	100	100	100	100	100	100
MF_VAR_0	$s=1$	–	0.3	0.3	0.4	0.5	0.7
	$s=2$	–	5.7	7.5	9.2	11.1	14.6
	$s=3$	45.6	99.7	99	97.9	97.1	96.8
	$s=4$	100	100	100	100	100	100

Note: The figures represent rejection frequencies associated with the likelihood ratio type test in Eq. (7). The DGP is a VAR(2) in the level of two $I(1)$ series, where no additional cointegrating relationships above the known ones and two common feature vectors are present. Models with VAR orders from 1 to 6 are estimated where zero-restrictions, corresponding to the absence of additional cointegrating relationships, are once imposed (denoted MF_VAR_0) and once not imposed (denoted MF_VAR). The sample size is $T=40$ (top) or $T=100$ (bottom). Size should be interpreted as the rejection frequency for $s=2$; power can be observed for $s=3$ and $s=4$.

Using the correct lag length $p=2$, the size distortion is small, especially if one correctly accounts for the zeros in the last coefficient matrix. Size distortions are amplified when the lag length is overestimated. This is reflected by the figures in the columns corresponding to $p>2$. The test is undersized when p is underestimated (see also Hecq et al., 2006). With respect to size unadjusted power, note that for $T=40$ we reject the presence of three common feature vectors 75% of the cases when the lag length is chosen correctly and up to almost 100% of the cases when it is overestimated. When $p=1$, however, the power deteriorates. The presence of four common feature vectors is rejected in all cases. As soon as the sample size increases, we reject both $s=3$ and $s=4$ in almost all cases (except for $p=1$ and $s=3$).

Table 3 reports the rejection frequencies of the low-frequency counterpart to the common feature test statistic in Eq. (7). The two time aggregation methods already introduced before are considered, i.e., average and point-in-time sampling. Note that as far as the latter is concerned, we only

Table 3. Rejection Frequencies (in Percent) of the LR Type Test in Eq. (7): Low-Frequency Models.

		$T=40$					
	p	1	2	3	4	5	6
Average	$s=1$	3.8	7.1	8.1	10.3	11.1	13.3
	$s=2$	68.4	88.5	87.4	87.4	86.6	86.8
PiT_$x_t^{(3)}$	$s=1$	4.5	9	9.8	14.1	14.4	16.5
	$s=2$	81.3	97.2	96.2	97.6	97.1	96.8
PiT_$x_{t-2/3}^{(3)}$	$s=1$	5	22.5	23.8	25.4	25.6	28.3
	$s=2$	89.2	92.8	95.2	95	94.3	94.8
		$T=100$					
Average	$s=1$	4.4	8.8	10.7	12.2	12.1	12.4
	$s=2$	97.7	100	100	100	100	100
PiT_$x_t^{(3)}$	$s=1$	11.9	21.3	22.1	30.8	28	27.5
	$s=2$	100	100	100	100	100	100
PiT_$x_{t-2/3}^{(3)}$	$s=1$	5.8	55.7	58.5	56.3	54.2	53.3
	$s=2$	100	100	100	100	100	100

Note: The figures represent rejection frequencies associated with the likelihood ratio type test in Eq. (7). The DGP is a VAR(2) in the level of two $I(1)$ series, where no additional cointegrating relationships above the known ones and two common feature vectors are present. Common low-frequency models obtained via three different temporal aggregation methods are considered: (a) average sampling (denoted Average), (b) point-in-time sampling using the end-of-period observation (denoted PiT_$x_t^{(3)}$), and (c) point-in-time sampling using the start-of-period observation (denoted PiT_$x_{t-2/3}^{(3)}$). The sample size is $T=40$ (top) or $T=100$ (bottom). Frequencies for $s=2$ represent power against a bivariate white noise system.

report the results for point-in-time sampling on the end-of-period ($\mathrm{PiT}\_x_t^{(3)}$) or start-of-period ($\mathrm{PiT}\_x_{t-2/3}^{(3)}$) observations to save space (results available upon request). As argued before, note that the frequencies corresponding to $s=2$ represent power against a bivariate white noise system, whereas the ones for $s=1$ cannot clearly be interpreted as size or power. Furthermore, we cannot classify the lag lengths as correct, under- or overestimated anymore, because the dynamics of the underlying system change due to temporal aggregation of the high-frequency variable.

For $T=40$, power against a bivariate white noise system is lower when average sampling is employed (68–87%, depending on p) than when point-in-time sampling is used (81–95%, depending on p and whether the end- or start-of-period observation is concerned). When the sample size increases, this power is equal to 100% in almost all cases. The rejection frequencies corresponding to one common feature relationship, i.e., $s=1$, depend on the temporal aggregation method chosen, p and T. For a given T, the frequencies increase with the lag length chosen for all three time aggregation approaches. Given p, the rejection frequencies also increase with the sample size (except for average sampling and $p=6$). Finally, given p and T, the rejection frequencies are almost always highest when point-in-time sampling on the start-of-period observation is performed, i.e., $\mathrm{PiT}\_x_{t-2/3}^{(3)}$ (the exception being $p=1$ and $T=100$).

CONCLUSION

In this article a new approach to common cycle analyses in multivariate time series processes was proposed when variables are sampled at mixed frequencies. This test allows users to examine commonalities in the business cycle of different countries when the frequency of observations is not the same. It also helps to detect co-movements when the economic indicators are not available at the same sampling frequency as in the applications that were given in this article.

It was shown that, in the presence of non-stationary variables, the user needs to account for two kinds of possible long-run relationships in a VAR; one stemming from the long-run relationship between the low- and high-frequency variables and the others. A transformed VAR representation was introduced, from which the set of variables was derived that was used to construct an LR type test for the presence of common cyclical features. An empirical analysis and a small forecasting exercise with the quarterly GNP and monthly IPI illustrated this new approach. Finally, a Monte Carlo study was presented in order to investigate the rejection frequencies of the previously mentioned LR type common feature test. Using a mixed-frequency VAR as DGP, this analysis was done for our mixed-frequency models and several common low-frequency models obtained by temporal aggregation of the high-frequency variables.

From the empirical section as well as the Monte Carlo experiments it became clear that if the underlying data are of a mixed-frequency nature, relying on mixed-frequency models instead of temporally aggregating the high-frequency variables may have advantageous effects. Note that, given the correct lag length, the common feature test behaved very well. Also, common cycles, whose existence was not necessarily visible when working with a low-frequency model, were detected. Consequently, accounting for them in the model led to gains in terms of forecasting. In other words, if useful restrictions are present in the variables, there is some gain in making use of them and incorporating them adequately in a mixed-frequency VAR model.

Some issues were not treated in this article and are currently under study. A common trend/common cycle decomposition could be easily obtained given that the MF-SCCF can be seen as a special case of the PSCCF modeling. Naturally, the development of a strategy for higher dimensional systems is of importance. Finally, as mentioned in the Monte Carlo study, an extension of Cubadda et al. (2009) and Marcellino (1999) to a mixed-frequency transformed VAR is of interest in order to investigate and generalize the effects of temporal aggregation on the lag length selection and the analysis of common cycles in a mixed-frequency setup. A detailed study of the latter would also allow us to state what difference time aggregation of stock or flow variables makes in our framework. Furthermore, it could help answering the question whether mixed-frequency data help to detect the state of the business cycle earlier than with low-frequency data only.

NOTES

1. In contrast to the parameter-driven models such as state-space models à la Zadrozny (1988). See Seong, Ahn, and Zadrozny (2013) for VECMs in state-space forms.

2. Foroni, Ghysels, and Marcellino (2013) emphasize that state-space models are unrivalled only under ideal conditions including correct model specification and no

parameter estimation errors. These ideal conditions are difficult to satisfy in practice. For example, correct specification is difficult to achieve due to missing observations for the regressand. Moreover, specification errors in one equation will spread to the entire system.

3. An extension to higher dimensional multivariate systems is straightforward.

4. It is possible to include deterministic components such as a constant, trends, or seasonal dummies here as well. For illustrative purposes we omit the deterministic regressors in the theoretical sections of the article. In the application and Monte Carlo study, however, a vector of intercepts is included.

5. Horvath and Watson (1995) show that accounting for the fact that some cointegrating vectors are prespecified leads to improved power results in terms of testing for potential further, unknown, cointegrating relationships.

6. This can be easily accomplished in Eviews or in OxMetrics, for instance.

7. From respectively http://alfred.stlouisfed.org/ and http://ec.europa.eu/eurostat/

8. Note that 0 was specified as lower and 8 as upper bound on the lag order.

ACKNOWLEDGMENTS

We thank Lutz Kilian, Tom Fomby, and the participants of the 12th Annual Advances in Econometrics Conference in Dallas, the 8th Netherlands Econometric Study Group Meeting in Amsterdam 2013 and the 67th European Meeting of the Econometric Society in Gothenburg 2013 for useful suggestions and comments on earlier versions of the article. The usual disclaimer applies.

REFERENCES

Anderson, H. M., & Vahid, F. (2010). *Vars, cointegration and common cycle restrictions*. Monash Econometrics and Business Statistics Working Papers 14/10, Monash University, Department of Econometrics and Business Statistics.

Andreou, E., Ghysels, E., & Kourtellos, A. (2010). Regression models with mixed sampling frequencies. *Journal of Econometrics*, *158*(2), 246–261.

Athanasopoulos, G., de Carvalho Guilln, O. T., Issler, J. V., & Vahid, F. (2011). Model selection, estimation and forecasting in var models with short-run and long-run restrictions. *Journal of Econometrics*, *164*(1), 116–129.

Clements, M., & Galvão, A. B. (2008). Macroeconomic forecasting with mixed-frequency data. *Journal of Business & Economic Statistics*, *26*, 546–554.

Clements, M., & Galvão, A. B. (2009). Forecasting U.S. output growth using leading indicators: An appraisal using Midas models. *Journal of Applied Econometrics*, *24*(7), 1187–1206.

Cox, D. R., Gudmundsson, G., Lindgren, G., Bondesson, L., Harsaae, E., Laake, P., ... Lauritzen, S. L. (1981). Statistical analysis of time series: Some recent developments [with discussion and reply]. *Scandinavian Journal of Statistics*, *8*(2), 93–115.

Cubadda, G., & Hecq, A. (2001). On non-contemporaneous short-run co-movements. *Economics Letters*, *73*(3), 389–397.

Cubadda, G., Hecq, A., & Palm, F. C. (2009). Studying co-movements in large multivariate data prior to multivariate modelling. *Journal of Econometrics*, *148*(1), 25–35.

Engle, R. F., & Kozicki, S. (1993). Testing for common features. *Journal of Business & Economic Statistics*, *11*(4), 369–380.

Foroni, C., Ghysels, E., & Marcellino, M. (2013). Mixed-frequency vector autoregressive models. In T. B. Fomby, L. Kilian, & A. Murphy (Eds.), *Van Models in Macroeconomics – New developments and applications: Essays in honor of Christopher A. Sims*. Bingley, UK: Emerald Group.

Ghysels, E. (2012). *Macroeconomics and the reality of mixed frequency data*. Working Paper, University of North Carolina.

Götz, T. B., Hecq, A., & Urbain, J.-P. (2012a). *Forecasting mixed frequency time series with ecm-midas models*. Research Memoranda 012, Maastricht: METEOR, Maastricht Research School of Economics of Technology and Organization.

Götz, T. B., Hecq, A., & Urbain, J.-P. (2012b). *Real-time forecast density combinations (forecasting U.S. GDP growth using mixed-frequency data)*. Research Memoranda 021, Maastricht: METEOR, Maastricht Research School of Economics of Technology and Organization.

Hecq, A. (1998). Does seasonal adjustment induce common cycles? *Economics Letters*, *59*(3), 289–297.

Hecq, A., & Jacobs, J. P. A. M. (2009). *On the var-vecm representation of real time data*. Discussion Paper.

Hecq, A., Palm, F. C., & Urbain, J.-P. (2006). Common cyclical features analysis in var models with cointegration. *Journal of Econometrics*, *132*(1), 117–141.

Horvath, M. T., & Watson, M. W. (1995). Testing for cointegration when some of the cointegrating vectors are prespecified. *Econometric Theory*, *11*(05), 984–1014.

Johansen, S. (1991). Estimation and hypothesis testing of cointegration vectors in gaussian vector autoregressive models. *Econometrica*, *59*(6), 1551–1580.

Lütkepohl, H. (2005). *New introduction to multiple time series analysis*. Berlin Heidelberg: Springer.

Marcellino, M. (1999). Some consequences of temporal aggregation in empirical analysis. *Journal of Business & Economic Statistics*, *17*(1), 129–136.

Paruolo, P. (2006). Common trends and cycles in *I*(2) var systems. *Journal of Econometrics*, *132*(1), 143–168.

Qian, H. (2013). *Vector autoregression with mixed frequency data*. MPRA Paper 47856, Munich Personal RePEc Archive.

Seong, B., Ahn, S. K., & Zadrozny, P. A. (2013). Estimation of vector error correction models with mixed-frequency data. *Journal of Time Series Analysis*, *34*(2), 194–205.

Silvestrini, A., & Veredas, D. (2008). Temporal aggregation of univariate and multivariate time series models: A survey. *Journal of Economic Surveys*, *22*(3), 458–497.

Vahid, F., & Engle, R. F. (1993). Common trends and common cycles. *Journal of Applied Econometrics*, *8*(4), 341–360.

Vahid, F., & Issler, J. V. (2002). The importance of common cyclical features in var analysis: A Monte Carlo study. *Journal of Econometrics*, *109*(2), 341–363.

Zadrozny, P. (1988). Gaussian likelihood of continuous-time Armax models when data are stocks and flows at different frequencies. *Econometric Theory*, *4*(1), 108–124.

Zellner, A., & Palm, F. (1974). Time series analysis and simultaneous equation econometric models. *Journal of Econometrics*, *2*(1), 17–54.

MULTIVARIATE DYNAMIC PROBIT MODELS: AN APPLICATION TO FINANCIAL CRISES MUTATION

Bertrand Candelon, Elena-Ivona Dumitrescu, Christophe Hurlin and Franz C. Palm

ABSTRACT

In this article we propose a multivariate dynamic probit model. Our model can be viewed as a nonlinear VAR model for the latent variables associated with correlated binary time-series data. To estimate it, we implement an exact maximum likelihood approach, hence providing a solution to the problem generally encountered in the formulation of multivariate probit models. Our framework allows us to study the predictive relationships among the binary processes under analysis. Finally, an empirical study of three financial crises is conducted.

Keywords: Nonlinear VAR; multivariate dynamic probit models; exact maximum likelihood; impulse-response function; financial crises

JEL classifications: C35; F37

VAR Models in Macroeconomics – New Developments and Applications: Essays in Honor of Christopher A. Sims
Advances in Econometrics, Volume 32, 395–427

ISSN: 0731-9053/doi:10.1108/S0731-9053(2013)0000031011

INTRODUCTION

Since the pioneering work of Sims (1980), vector autoregressive (VAR) models have been the main tool used to analyze a set of macroeconomic time-series. This has paved the way for numerous papers proposing modifications of the standard linear VAR model. One research direction focuses on cointegrated time-series (e.g., Engle & Granger, 1987; Johansen, 1988, 1991), leading to Vector Error Correction Models, while another line of research concentrates on adapting VAR specifications to the case of nonlinear time-series (Gallant, Rossi, & Tauchen, 1993; Killian & Vigfusson, 2011; Koop, Pesaran, & Potter, 1996, among others).

Apart from models for continuous variables, discrete-choice dependent variables are commonly used, especially for forecasting. There are two main approaches. The standard approach used to forecast the probability of occurrence of such events consists of simple logit/probit models. These models have been used, for example, by Berg and Pattillo (1999), Estrella and Trubin (2006), and Fuertes and Kalotychou (2007) to forecast recessions as well as financial (currency, banking, sovereign debt, stock market,...) crises. An alternative, more recent approach, consists of a dynamic univariate framework estimated by an exact maximum likelihood (EML) estimation method (see Candelon, Dumitrescu, & Hurlin, 2010 or Nyberg, 2010, 2011; Kauppi & Saikonnen, 2008, inter alia). Forecasts from these models have been shown to outperform forecasts from static models, that is, models that rely only on the dynamics of exogenous variables. In a similar vein, Dueker (2005) estimates by simulation methods a "mixed" Qual-VAR model including as dependent variables the latent variable underlying a binary business cycle indicator along with standard continuous indicators.

Modeling the dynamics of qualitative variables may be as important as that of continuous variables (for the improvement of Early Warning Systems, for example). Nevertheless, no multivariate extension of these dynamic binary models exists as far as we know. This omission is surprising because one would conjecture that such models could improve the accuracy of forecasts of discrete events in light of the correlation between different binary indicators in macroeconomics. To our knowledge, all existing multivariate binary models are static. Indeed, although numerous macroeconomic studies involve discrete-choice dependent variables, this particular type of nonlinearity has received little attention in the literature on time-series models. The main difficulty lies in formulating a multivariate binary model and in evaluating the likelihood function. In this context, Carey, Zeger, and Diggle (1993) and Glonek and Mccullagh (1995) proposed generalizations of the binary logistic model to multivariate outcomes by selecting a particular parametrization for the correlations, while Ashford and Sowden (1970) and Amemiya (1972) focused on generalizing the binary probit model. Some attempts have subsequently been made to solve the computational difficulty of evaluating multivariate normal integrals. Chib and Greenberg (1998) developed a simulation-based Bayesian and non-Bayesian approach and Song and Lee (2005) relied on the Expectation-Maximization algorithm to evaluate the likelihood function for a multivariate probit model. More recently, Huguenin, Pelgrin, and Holly (2009) have shown that a multivariate probit model cannot be accurately estimated using simulation methods, as generally done in the literature. Its estimation instead requires the derivation of the EML function.

The objective of this article is to extend the estimation methodology proposed by Huguenin et al. (2009) for non-dynamic multivariate probit models to the case of dynamic multivariate probit models. We introduce a multivariate dynamic probit model which is estimated by an EML estimation approach to produce dynamic forecasts of binary time-series correlated variables. Our model takes the form of a nonlinear VAR for the latent variables associated with the binary indicators under analysis. It is easy to implement and provides a solution to the problem encountered in the estimation of multivariate probit models. For this, in a first step we extend the decomposition of higher-order integrals into integrals of lower order proposed by Huguenin et al. (2009) to the case of our multivariate dynamic model. In the second step, we evaluate the lower-order finite-range multiple integrals by using quadrature-rules over bounded intervals.

Our framework allows us to introduce dynamics in several ways. First, dynamics may be included in the form of a set of lagged binary variables. Notice the existence of threshold effects in this case, as the event under analysis occurs only if the latent variable goes beyond a certain threshold. Second, they can be introduced via the past latent variables associated with the binary indicators. This comes down to an autoregressive (AR) model, where the lagged latent variable summarizes all the past information of the system. Finally, both types of dynamics can be simultaneously considered. We thus generalize the univariate dynamic probit model developed by Kauppi and Saikkonen (2008) to a multivariate setting. We derive its exact likelihood and propose consistent and efficient estimates.

In an empirical application on three types of financial crises, namely currency, banking, and sovereign debt crises, we investigate the potential spillover from one crisis to another within a number of emerging countries. It appears that in the bivariate case mutations of a banking crisis into a currency crisis (and vice versa) have been quite common, confirming other results in the financial crises literature (e.g., Glick & Hutchinson, 1999). More importantly, for the two countries (Ecuador and South Africa) which suffered from all three types of crises, the trivariate model turns out to be more parsimonious than existing models.

The rest of the article is organized as follows. The second section presents the multivariate dynamic probit model. In the third section we describe the EML method as well as some numerical procedures to estimate the multivariate dynamic probit model, while in the fourth section the multivariate dynamic probit model is estimated for 17 emerging countries in its bivariate and trivariate form.

A MULTIVARIATE DYNAMIC PROBIT MODEL

In this section we describe the multivariate dynamic probit model and highlight its similarities with traditional VAR models. Let us denote by $y_{m,t}$, $m=\{1,2,\ldots,M\}$, M binary variables taking the value one if the event under analysis occurs at time t and zero otherwise. Let $y^*_{m,t}$ be the normal latent continuous variable associated with $y_{m,t}$, and define $F_{t-1}=\sigma[(y'_s,x'_v)'|\ s<t, v\leq t]$ as the information set available at time t.

The corresponding multivariate probit model takes the form

$$y^*_{m,t}=\pi_{m,t}+\varepsilon_{m,t},\quad \forall m=\{1,2,\ldots,M\}$$
$$y_{m,t}=\begin{cases}1, & \text{if } y^*_{m,t}>0\\ 0, & \text{otherwise}\end{cases} \tag{1}$$

where $\pi_{m,t}$ denotes the index, and the innovations $\varepsilon_t=\{\varepsilon_{1,t},\ldots,\varepsilon_{m,t}\}$ verify

$$\varepsilon_t|F_{t-1}\sim \text{IIN}(0,\Omega) \tag{2}$$

such that $V(\varepsilon)=I_T\otimes\Omega$, where I_T is the identity matrix of order T. Ω stands for the covariance matrix of ε_t, with $\Omega=(\sigma_{m,m'})$ and $\sigma_{m,m'}=\rho_{m,m'}\sigma_m\sigma_{m'}$, where $\rho_{m,m'}$ represents the correlation coefficient between the mth and m'th binary processes and σ_m and $\sigma_{m'}$ are the associated standard deviations. The m events under analysis are related at time t through the associated innovations, but there is no dependence in time among these innovations (V is block-diagonal). In this model, the probability of occurrence of the mth event is equal to

$$p_{m,t} = \Pr(-\varepsilon_{m,t} \leq \pi_{m,t} | F_{t-1}) = \Phi(\pi_{m,t}), \quad \forall m = \{1, 2, \ldots, M\} \tag{3}$$

The objective of this article is to propose a dynamic multivariate modeling of these M qualitative variables. We specify the dynamics of each stochastic process through its associated index $\pi_{m,t}$. Traditionally, the index only depends on exogenous explanatory variables $x_t = \{x_{1,t}, \ldots, x_{K,t}\}$, where K is the number of exogenous variables in the model. But in a dynamic model, it can also depend on the past information on the dependent variable. Formally, for a given event m, the dynamics can be introduced in two ways: either through the lagged binary variables $y_{m,t}$, or through the lagged latent variable $y^*_{m,t}$. The first equation of our multivariate dynamic probit model in Eq. (1) is hence given by

$$y^*_{m,t} = \alpha_m + \beta_m x_{t-v} + \sum_{m'=1}^{M} \Delta_{m,m'} y_{m',t-s} + \sum_{m'=1}^{M} \Gamma_{m,m'} y^*_{m',t-s} + \varepsilon_{m,t} \tag{4}$$

where $\alpha_m \in \mathbb{R}$ is the intercept, $\beta_m = \{\beta_{1,m}, \ldots, \beta_{K,m}\}$, $\beta_m \in \mathbb{R}^K$ is the vector of parameters associated with the explanatory variables x[1] and $\Delta_{m,m'} \in \mathbb{R}$ and $\Gamma_{m,m'} \in \mathbb{R}$ are the parameters of the predetermined variables giving the dynamics of the mth equation of the model. $s, v > 0$ are the lags associated with the predetermined right-hand-side variables. For $s, v \geq h$ it is possible to obtain directly the h-step-ahead forecast made at time $t-h$. An iterative approach could also be envisaged. However, the computation of the forecasts becomes more burdensome since all the possible paths leading to the occurrence of the M events in h-periods' time must be taken into account (for more details on iterated forecasts in a univariate discrete-choice analysis see Kauppi & Saikonnen, 2008).

Denote by $\theta_m = (\alpha_m; \beta_m; \Delta_{m,m'}; \Gamma_{m,m'})'$ the vector of parameters for equation/event m in Eq. (4), with $\theta = (\theta_1'; \theta_2'; \ldots; \theta_M')'$, that will be used in the estimation of the model (see section "Exact Maximum Likelihood Estimation").

It is clear that our model has the usual VAR-X structure, as it assumes a linear relation between the latent variables $y^*_{m,t}$ and their past. Its dynamics is then enriched by the nonlinear relation between the latent variables $y^*_{m,t}$ and the observed binary ones $y_{m,t}$, which in turn depend on $y^*_{m,t-s}$ (see the second equation in Eq. (1)). In other words, our dynamic probit model differs from a standard VAR in two ways. First, it introduces both a linear and a nonlinear dynamics. Indeed, the dynamics of the mth process/event can be modeled by considering that the latent variable $y^*_{m,t}$ depends either on its lagged value via the $\Gamma_{m,m}$ coefficient, or on the past regime (0/1) through $\Delta_{m,m}$.

Second, the analysis of the interdependence, that is, predictive relationship[2] between the M qualitative variables is more complicated, as it passes through several channels.

1. Unobserved common factors can be taken into account through the contemporaneous dependence of the innovation terms ($\mathbb{E}(\varepsilon_m \varepsilon_{m'}) = \sigma_{mm'} \neq 0$ for $m \neq m'$).
2. For an event m, the unobservable latent variable $y^*_{m,t}$ depends on past values of other processes $y^*_{m',t-s}$ (where $m \neq m'$), themselves unobservable.
3. The latent variable may depend on past realizations of the other events, that is, $y_{m',t-s} = 1$. Formally, $y^*_{m,t}$ depends on past values of the observable variable $y_{m',t-s}$, $s > 0$, where $m \neq m'$.
4. It is possible to combine the two previous cases, assuming that $y^*_{m,t}$ depends on both the latent variable $y^*_{m',t-s}$, and past values of the observable variable, $y_{m',t-s}$, for other binary processes ($m \neq m'$).

From this perspective, our model in Eq. (4) is a multivariate extension of the univariate dynamic probit model recently proposed by Kauppi and Saikkonen (2008).

Another novelty introduced by this nonlinear VAR framework relies on the fact that our new specification enables us to compute not only marginal but also joint and conditional probabilities. The traditional marginal probabilities are associated with each binary response, $\Pr(y_m = 1 | y^*_m) = \Phi(y^*_m)$, and rely on univariate discrete-choice models. In contrast, joint and conditional probabilities, that is, $\Pr(y_1 = 1, y_2 = 1, \ldots, y_M = 1 | y^*) = \Phi_M(y^*)$, and $\Pr(y_m = 1 | y^*_{m'}) = \Phi_M(y^*)/\Phi_{M-1}(y^*_{m'})$, for $m, m' \in \{1, 2, \ldots, M\}$, where Φ and Φ_M represent the univariate and M-variate normal cumulative distribution functions respectively can also be obtained here.

First-order dynamics, that is, $s = 1$, are most common in empirical applications with binary-dependent indicators that stand for regime-switches, including the empirical example we present in section "Empirical Application." In vector notation, the process in Eq. (4) hence becomes

$$y^*_t = \alpha + Bx_{t-v} + \Delta y_{t-1} + \Gamma y^*_{t-1} + \varepsilon_t \tag{5}$$

where $\alpha = (\alpha_1, \ldots, \alpha_m)'$, B is a $M \times K$ matrix, and Δ and Γ are $M \times M$ matrices of parameters. Note that the matrices Δ and Γ summarize useful information about the dynamics of the binary processes, in particular about their persistence and predictability.

One the one hand, as previously discussed, the diagonal terms of Γ specify the persistence of each process. These parameters correspond to a first-order AR representation of each latent variable. An increase in the latent variable during a certain period is always transmitted to the next period, hence always increasing the probability of realization of the event (observing a value of 1). The closer these parameters are to 1, the more persistent the processes are. Notice that the diagonal elements of this matrix typically are constrained to be strictly smaller than 1 to exclude the case where the latent variable $y^*_{m,t}$ follows a random walk, which is not an interesting case in this context.

At the same time, the diagonal terms of Δ also deliver information about persistence. Indeed, they indicate to what extent the probability of the occurrence of an event (defined by the *m*th binary indicator) depends on the regime prevailing the period before. In this situation we observe the existence of threshold effects, as the regime defined by a value of one for the binary process lasts more than one spell only if the latent variable soars sufficiently to exceed the threshold which initiated this regime in the previous period.

Altogether, we can distinguish between a linear persistence of the phenomenon, captured through the diagonal terms of Γ, and a nonlinear, threshold-based one, apprehended by the diagonal terms of Δ.

On the other hand, predictability is taken into account in the off-diagonal elements of the two matrices Γ and Δ. This predictability between the *M* binary stochastic processes play a key role in the analysis of numerous economic events (e.g., financial crises, economic cycles). As in the analysis of the persistence of a binary process, both a linear and a nonlinear, threshold-effect transmission can be identified. A significant off-diagonal Γ element means that no sooner the latent variable for the *m* binary indicator soars, then the one for another process, say m', rises. By contrast, a Δ term reveals the presence of causality only if the corresponding latent variable is high enough to impact the occurrence of another binary event.

EXACT MAXIMUM LIKELIHOOD ESTIMATION

The EML estimator for the multivariate dynamic probit model cannot be obtained as a simple extension of the univariate model. For this reason, the maximum likelihood method is generally considered. Huguenin et al. (2009) prove that this estimation method causes bias in the estimation of the correlation coefficients as well as in their standard deviations. Therefore, they advocate the use of EML estimation. This section deals with this objective.

The Maximum Likelihood

Following Greene (2002), full information maximum likelihood (FIML) estimates are obtained by maximizing the log likelihood $\log L(Y|Z;\theta,\Omega)$, where θ is the vector of identified parameters and Ω is the covariance matrix.[3] Under the usual regularity conditions[4] (Lesaffre & Kauffmann, 1992), the FIML is given by the joint density of observed outcomes

$$L(y|z,\theta;\Omega)=\prod_{t=1}^{T} L_t(y_t|z_{t-1},\theta;\Omega) \tag{6}$$

where $y_t=(y_{1,t},\ldots,y_{M,t})'$ and $y=[y_1,\ldots,y_T]$. The individual likelihood $L_t(.)$ is given in Lemma 1.

Lemma 1. The likelihood of observation t is the cumulative density function, evaluated at the vector w_t of an M-variate standardized normal vector with a covariance matrix $Q_t\,\Omega\,Q_t$

$$L_t(y_t|z_{t-1},\theta;\Omega)=\Pr(y_1=y_{1,t},\ldots,y_M=y_{M,t})=\Phi_{M,\varepsilon_t}(w_t;Q_t\Omega Q_t) \tag{7}$$

where Q_t is a diagonal matrix whose main diagonal elements are $q_{m,t}=2y_{m,t}-1$ and thus depends on the realization or not of the events ($q_{m,t}=1$ if $y_{m,t}=1$ and $q_{m,t}=-1$ if $y_{m,t}=0$, $\forall\ m=\{1,2,\ldots,M\}$). Besides, the elements of the vector $w_t=[w_{1,t},\ldots,w_{M,t}]$ are given by $w_{m,t}=q_{m,t}\pi_{m,t}$, where $\pi_{m,t}$ is the index associated with the mth binary variable.

For a complete proof of Lemma 1, see Appendix A. Thus, the FIML estimates are obtained by maximizing the log likelihood

$$\log \mathrm{L}(y|z,\theta;\Omega)=\sum_{t}^{T}\log\Phi_{M,\varepsilon}(w_t;Q_t\Omega Q_t) \tag{8}$$

with respect to θ and Ω.

The main problem with FIML is that it requires the evaluation of higher-order multivariate normal integrals. Existing methods for this purpose are not sufficient to allow accurate and efficient evaluation for more than two variables (see Greene, 2002, p. 714). Indeed, Greene (2002) argues that the existing quadrature methods to approximate trivariate or higher-order integrals are far from being exact. To tackle this problem in the case of a probit model, Huguenin et al. (2009) decompose the triple integral into simple and double integrals, allowing an EML estimation by means of computing double integrals. They prove that the EML increases the numerical accuracy of both the slope and covariance parameters estimates, which outperform the maximum simulated likelihood method of McFadden (1989) which is generally used for the estimation of multivariate probit models. Here we extend the decomposition proposed by Huguenin et al. (2009) to the case of our multivariate dynamic model so as to obtain a direct approximation of the trivariate normal cumulative distribution function.

The EML log likelihood function is given by

$$\log \mathrm{L}(y|z,\theta;\Omega) = \sum_{t=1}^{T} \log\left[\prod_{m=1}^{M} \Phi(w_{m,t}) + G\right] \tag{9}$$

where $\Phi(w_t)$ is the univariate normal cumulative distribution function of w_t. Indeed, the log likelihood function depends on the product of the marginal distributions (w_t) and the correction term G which captures the dependence between the m events analyzed.

The maximum likelihood estimators $\{\hat{\theta};\hat{\Omega}\}_{\mathrm{EML}}$ are the values of θ and Ω which maximize Eq. (9)

$$\{\hat{\theta};\hat{\Omega}\}_{\mathrm{EML}} = \mathrm{Arg}\max_{\theta;\Omega} \sum_{m=1}^{M} \log \mathrm{L}(.) \tag{10}$$

with log L(.) given in Eq. (9).

Under the regularity conditions of Lesaffre and Kaufman (1992), the EML estimator of a multivariate probit model exists and is unique. Besides, the estimates $\{\hat{\theta};\hat{\Omega}\}_{\mathrm{EML}}$ are asymptotically normally distributed and are consistent and efficient estimators of the slope and covariance parameters. It is worth noting that in a correctly specified model for which the error terms are independent across the m equations the EML function corresponds to $\sum_{t=1}^{T}\prod_{m=1}^{M}\Phi(w_{m,t})$, since the probability correction term G in Eq. (9) tends toward zero.

The Empirical Procedure

Most of the empirical applications involving correlated time-series binary data (e.g., financial crises, economic cycles, etc.) tackle only two or three such events at a time. Accordingly, henceforth we restrict our attention to the bivariate and trivariate form of the model. This presentation simplifies the exposition of the EML estimation method and corresponds to the empirical analysis performed in section "Empirical Application." Further details are provided in Appendices A–C.

For the bivariate model, we get

$$\Phi_2(w_t; Q_t\Omega Q_t)=\Phi(w_{1,t})\Phi(w_{2,t})\frac{1}{2\pi}\int_0^{\rho_{12}} \exp\left(-\frac{1}{2}\frac{w_{1,t}^2+w_{2,t}^2-2w_{1,t}w_{2,t}}{1-\lambda_{12}^2}\right)\frac{d\lambda_{12}}{\sqrt{1-\lambda_{12}^2}} \tag{11}$$

and for the trivariate model

$$\begin{aligned}\Phi_3(w_t; Q_t\Omega Q_t) &= \prod_{m=1}^{3}\Phi(w_{m,t})+G\\ &=\Phi(w_{1,t})\Phi(w_{2,t})\Phi(w_{3,t})+\Phi(w_{3,t})\int_0^{\rho_{12}}\phi_2(w_{1,t}, w_{2,t}; \lambda_{12})\ d\lambda_{12}\\ &\quad+\Phi(w_{2,t})\int_0^{\rho_{13}}\phi_2(w_{1,t}, w_{3,t}; \lambda_{13})\ d\lambda_{13}\\ &\quad+\Phi(w_{1,t})\int_0^{\rho_{23}}\phi_2(w_{2,t}, w_{3,t}; \lambda_{23})\ d\lambda_{23}\\ &\quad+\int_0^{\rho_{12}}\int_0^{\rho_{13}}\frac{\partial\phi_3(w_t; \lambda_{12}, \lambda_{13}, 0)}{\partial w_{1,t}}\ d\lambda_{12}d\lambda_{13}\\ &\quad+\int_0^{\rho_{12}}\int_0^{\rho_{23}}\frac{\partial\phi_3(\dot{w}_t; \lambda_{12}, 0, \lambda_{23})}{\partial w_{2,t}}\ d\lambda_{12}d\lambda_{23}\\ &\quad+\int_0^{\rho_{13}}\int_0^{\rho_{23}}\frac{\partial\phi_3(\ddot{w}_t; 0, \lambda_{13}, \lambda_{23})}{\partial w_{3,t}}\ d\lambda_{13}d\lambda_{23}\\ &\quad+\int_0^{\rho_{12}}\int_0^{\rho_{13}}\int_0^{\rho_{23}}\frac{\partial^3\phi_3(\dddot{w}_t; \lambda_{12}, \lambda_{13}, \lambda_{23})}{\partial w_{1,t}\partial w_{2,t}\partial w_{3,t}}\ d\lambda_{12}d\lambda_{13}d\lambda_{23}\end{aligned} \tag{12}$$

where ρ are the non-diagonal elements of the $Q_t\Omega Q_t$ matrix and λ are the non-diagonal elements of a theoretical 2×2 matrix and respectively a 3×3

matrix in which one of the correlation coefficients is null. Moreover, $\dot{w}_t$ is a vector of indices obtained by changing the order of the elements to $(w_{2,t}, w_{3,t}, w_{1,t})$. Similarly $\ddot{w}_t$ corresponds to a vector of indices of the form $(w_{3,t}, w_{1,t}, w_{2,t})$. Finally, $\dddot{w}_t$ corresponds to w_t, $\dot{w}_t$, or $\ddot{w}_t$ respectively, depending on the way the last integral is decomposed. The computation of the last term is not trivial. However, this integral can be decomposed in a nonunique way as follows:

$$
\begin{aligned}
&\int_0^{\rho_{12}}\int_0^{\rho_{12}}\int_0^{\bar{\rho}_{23}} \frac{\partial^3 \phi_3(\dddot{w}_t;\lambda_{12},\lambda_{13},\lambda_{23})}{\partial w_{1,t}\partial w_{2,t}\partial w_{3,t}}\, d\lambda_{12}d\lambda_{13}d\lambda_{23} \\
&\quad = \int_0^{\rho_{13}}\int_0^{\rho_{23}} \frac{\partial \phi_3(\ddot{w}_t;\lambda_{12},\lambda_{13},\lambda_{23})}{\partial w_{3,t}}\, d\lambda_{13}d\lambda_{23} - \int_0^{\rho_{13}}\int_0^{\rho_{23}} \frac{\partial \phi_3(\ddot{w}_t;0,\lambda_{13},\lambda_{23})}{\partial w_{3,t}}\, d\lambda_{13}d\lambda_{23} \\
&\quad = \int_0^{\rho_{12}}\int_0^{\rho_{23}} \frac{\partial \phi_3(\dot{w}_t;\lambda_{12},\lambda_{13},\lambda_{23})}{\partial w_{2,t}}\, d\lambda_{12}d\bar{\rho}_{23} - \int_0^{\rho_{12}}\int_0^{\rho_{23}} \frac{\partial \phi_3(\dot{w}_t;\lambda_{12},0,\lambda_{23})}{\partial w_{2,t}}\, d\lambda_{12}d\lambda_{23} \\
&\quad = \int_0^{\lambda_{12}}\int_0^{\lambda_{13}} \frac{\partial \phi_3(w_t;\lambda_{12},\lambda_{13},\lambda_{23})}{\partial w_{1,t}}\, d\lambda_{12}d\lambda_{13} - \int_0^{\rho_{12}}\int_0^{\rho_{13}} \frac{\partial \phi_3(w_t;\lambda_{12},\lambda_{13},0)}{\partial w_{1,t}}\, d\lambda_{12}d\lambda_{13}
\end{aligned}
\tag{13}
$$

These finite-range multiple integrals are numerically evaluated by using a Gauss-Legendre quadrature rule[5] over bounded intervals. In such a context, two possibilities can be considered: whether the likelihood function is directly maximized, or the first-order conditions[6] are derived so as to obtain an exact score vector. As stressed by Huguenin et al. (2009), the two methods may not lead to the same results if the objective function is not sufficiently smooth. We also tackle the autocorrelation problem induced by some binary time-series variables by considering a Gallant (1987) correction for the covariance matrix of the parameters.

EMPIRICAL APPLICATION

This section aims at implementing the multivariate dynamic probit methodology presented above to a system composed by three types of financial crises, that ia, currency, banking, and sovereign debt crises. As historical events have proven, most of the time crises do not remain restricted to a single market, but tend to spillover into another one. We evaluate the probability of mutation of one type of crisis into another one. After a short data description, we estimate bivariate models that include only currency and banking crises, as previously done in a panel framework by Glick and Hutchison (1999). This constitutes a benchmark for the second part where the sovereign debt crises are included in the system.

The Database

Monthly macroeconomic indicators expressed in US dollars covering the period from January 1985 to June 2010 have been extracted for 17 emerging countries[7] from the IMF-IFS database as well as from the national banks of the countries under analysis via Datastream.[8]

The three types of crises are identified by relying on popular measures, generally considered in the literature.[9] As in Lestano and Jacobs (2004) and Candelon, Dumitrescu, and Hurlin (2012), we use a modified version of the pressure index proposed by Kaminsky, Lizondo, and Reinhart (1998) to date currency crises. Besides, the money market pressure index proposed by Hagen and Ho (2004) is considered for the monthly identification of banking crises, while the nonparametric method based on sovereign debt spread introduced by Pescatori and Sy (2007) is used to detect debt-servicing difficulties. To this aim, the government bond returns are obtained via the JPMorgan EMDB database.

We have selected the main leading indicators used in the literature for the three types of crises that we analyze (see Candelon et al., 2012; Glick & Hutchison, 1999; Hagen & Ho, 2004; Jacobs, Kuper, & Lestano, 2003; Pescatori & Sy, 2007), namely, the one-year growth rate of international reserves, the growth rate of M2 to reserves ratio, one-year growth of domestic credit over GDP ratio, one-year growth of domestic credit, one-year growth of GDP, government deficit, debt service ratio, and external debt ratio.

Remarks

1. As in Kumar, Moorthy, & Perraudin (2003), we dampen the magnitude of every variable using the formula: $f(x_t) = \text{sign}(x_t)\log(1 + |x_t|)$, so as to reduce the impact of extreme values.[10]

2. It should also be noted that the entire sample is used for the identification of currency and banking crises, while the identification of debt crises is realized by using data from December 1997 (see Table 1) since the CDS spread series used for the identification of sovereign debt crises are not available before 1997 in the JPMorgan EMDB database. Consequently our empirical analysis will consist of two parts, the first one analyzing the case of twin crises (currency and banking) for which the entire database can be used, while the second part focuses on the interactions between the three types of crises and is thus based on data from 1997 onwards. The data sample actually used for each of the 17 countries and the two types of analyses is available in Table 1.
3. We only retain the countries for which the percentage of crisis periods is higher than 5% (see Table 2).[11]
4. As mentioned in section "A Multivariate Dynamic Probit Model," there are three dynamic multivariate specifications that can be used. As shown by Candelon et al. (2010), the dynamic model including the lagged binary variable seems to be the best choice according to model selection using the Schwarz information criterion. Relying on the same univariate results, we fix the lag-number v of the continuous variables x to one. Since we cannot expect one type of crisis to affect the probability of

Table 1. Database.

Country	Bivariate Model	Trivariate Model
Argentina	February 1988–May 2010	December 1997–May 2010
Brazil	September 1990–May 2010	December 1997–May 2010
Chile	January 1989–May 2009	May 1999–May 2010
Colombia	February 1986–August 2009	December 1997–August 2009
Ecuador	January 1994–November 2007	December 1997–November 2007
Egypt	February 1986–June 2009	July 2001–June 2009
El Salvador	January 1991–November 2008	April 2002–November 2008
Indonesia	January 1989–August 2009	May 2004–August 2009
Lebanon	January 1989-April 2010	April 1998–April 2010
Malaysia	January 1988–March 2010	December 1997–March 2010
Mexico	January 1988–May 2010	December 1997–May 2010
Peru	January 1990–May 2010	December 1997–May 2010
Philippines	January 1995–February 2008	December 1997–February 2008
South Africa	January 1988–August 2009	December 1997–August 2009
Turkey	January 1988–May 2010	December 1997–May 2010
Venezuela	February 1986–November 2009	December 1997–November 2009

Note: Data sample.

Table 2. Percentage of Crisis Periods.

	Bivariate Model		Trivariate Model		
	Currency crisis	Banking crisis	Currency crisis	Banking crisis	Debt crisis
Argentina	**5.13**	**8.90**	4.00	**6.67**	**10.0**
Brazil	3.77	**7.19**	0.00	3.33	2.67
Chile	**6.07**	**10.0**	**5.79**	**5.79**	3.31
Colombia	4.95	**9.90**	**9.22**	**12.8**	0.00
Ecuador	**5.73**	**9.93**	**6.67**	**10.8**	**6.67**
Egypt	**6.76**	**9.96**	4.17	**7.30**	**7.30**
El Salvador	3.65	**9.85**	0.00	0.00	2.50
Indonesia	5.30	**9.90**	0.00	**14.0**	**6.25**
Lebanon	**9.62**	**9.96**	1.38	**8.97**	2.76
Malaysia	3.10	**10.0**	4.05	**6.08**	4.73
Mexico	**6.50**	**9.93**	0.00	**9.33**	0.00
Panama	0.00	**9.89**	0.00	**6.38**	0.00
Peru	4.45	**8.22**	0.00	**10.7**	0.00
Philippines	4.90	**9.80**	**5.69**	**6.50**	3.25
South Africa	**6.71**	**9.89**	**7.09**	**7.80**	4.26
Turkey	4.80	**8.56**	4.00	**6.67**	0.00
Venezuela	**7.33**	**10.1**	4.17	**7.64**	2.78

Note: The entries represent the proportion of crises period over the whole sample. It is indicated in bold as it exceeds 5%.

another type of crisis immediately, in the empirical application we allow for response lags of 3, 6, and respectively 12 months for the bivariate models and of 3 or 6 months for the trivariate models.[12] Therefore, for each type of crisis we construct a lagged variable $y_{m,t-l}$ which takes the value of one if there was crisis in the past l periods or at time t, and the value of 0 otherwise:

$$y_{m,t-l} = \begin{cases} 1, & \text{if } \sum_{j=0}^{l} y_{m,t-j} > 0 \\ 0, & \text{otherwise} \end{cases} \tag{14}$$

5. The significance of the parameters of each model is tested by using simple t-statistics based on robust estimates of standard-errors (which rely on a Gallant kernel, as in Kauppi & Saikkonen, 2008). A special attention is given to the interpretation of cross-effects which stand for the transmission channels of the shocks/crisis. Besides, the joint null of zero contemporaneous correlations between crises is tested using a log likelihood ratio test for the trivariate models.

Bivariate Analysis

Along the lines of Kaminsky et al. (1998) it is possible to find a large number of explanatory variables that may signal the occurrence of a crisis. Nevertheless, Candelon et al. (2010) showed that a univariate dynamic probit model presents the advantage of yielding plausible results while being fairly parsimoniously parametrized. Indeed, a large part of the information is integrated either in the past state variable or in the lagged latent variable and thus, only a few explanatory variables turn out to be significant. Therefore, we consider the first lag of the four explanatory variables which are significant in Candelon et al. (2010), that is, one-year growth of international reserves, one-year growth of M2 to reserves for currency crises as well as one-year growth of domestic credit over GDP and one-year growth of domestic credit for banking crises, resulting in four different specifications including one explanatory variable for each type of crisis. Three different lags (3, 6, and 12 months) are considered for the lagged binary variable $y_{m,t-l}$. The dynamic probit model is estimated country-by-country using the EML.[13] This model is indeed a simplification as contagion (or spillovers) from one country to another are not taken into account. A panel version of the model would lead to several problems. First, as shown by van de Berg, Candelon, and Urbain (2008) heterogeneity due to country specificities would have to be accounted for. Second, the estimation of a fixed effect panel would be biased without a correction on the score vector.[14] Third, in a country-by-country analysis contagion has to be ignored. For all these reasons, we consider this extension to be beyond the scope of this article and leave it for future research.

Each model is estimated via maximum likelihood, the bivariate normal cumulative distribution function being approximated using the Gauss-Legendre quadrature, as proposed by Huguenin et al. (2009). However, the quadrature specified in Matlab by default, that is, the adaptive Simpson quadrature, has been considered as a benchmark.

Information criterion BIC is used to identify the best model for each country; the specification with the lagged binary variable turns out to be preferred. Lag lengths are determined similarly. It is worth stressing that the results are generally robust to the choice of explanatory variables and even to the choice of lags.

A summary of the results for the selected models is given in Table 3.

First of all, it seems that most of the models exhibit dynamics, whatever the lag used to construct the "past crisis" variable is. This result confirms the findings of Candelon et al. (2010) and Bussière (2007), showing that

Table 3. Bivariate Analysis.

Country		3 Months		6 Months		12 Months	
		Δ	Ω	Δ	Ω	Δ	Ω
Argentina	Currency banking	$\begin{bmatrix} + & + \\ . & + \end{bmatrix}$	$\begin{bmatrix} 1 & + \\ + & 1 \end{bmatrix}$	$\begin{bmatrix} . & + \\ . & + \end{bmatrix}$	$\begin{bmatrix} 1 & + \\ + & 1 \end{bmatrix}$	$\begin{bmatrix} . & . \\ . & + \end{bmatrix}$	$\begin{bmatrix} 1 & . \\ . & 1 \end{bmatrix}$
Chile	Currency banking	$\begin{bmatrix} + & . \\ . & . \end{bmatrix}$	$\begin{bmatrix} 1 & + \\ + & 1 \end{bmatrix}$	$\begin{bmatrix} . & . \\ . & . \end{bmatrix}$	$\begin{bmatrix} 1 & + \\ + & 1 \end{bmatrix}$	$\begin{bmatrix} . & - \\ . & + \end{bmatrix}$	$\begin{bmatrix} 1 & + \\ + & 1 \end{bmatrix}$
Ecuador	Currency banking	$\begin{bmatrix} . & . \\ . & + \end{bmatrix}$	$\begin{bmatrix} 1 & . \\ . & 1 \end{bmatrix}$	$\begin{bmatrix} . & . \\ . & + \end{bmatrix}$	$\begin{bmatrix} 1 & . \\ . & 1 \end{bmatrix}$	$\begin{bmatrix} . & . \\ . & + \end{bmatrix}$	$\begin{bmatrix} 1 & . \\ . & 1 \end{bmatrix}$
Egypt	Currency banking	$\begin{bmatrix} + & . \\ - & + \end{bmatrix}$	$\begin{bmatrix} 1 & . \\ . & 1 \end{bmatrix}$	$\begin{bmatrix} + & . \\ - & + \end{bmatrix}$	$\begin{bmatrix} 1 & - \\ - & 1 \end{bmatrix}$	$\begin{bmatrix} + & . \\ . & + \end{bmatrix}$	$\begin{bmatrix} 1 & - \\ - & 1 \end{bmatrix}$
Lebanon	Currency banking	$\begin{bmatrix} + & . \\ - & + \end{bmatrix}$	$\begin{bmatrix} 1 & . \\ . & 1 \end{bmatrix}$	$\begin{bmatrix} . & . \\ . & + \end{bmatrix}$	$\begin{bmatrix} 1 & + \\ + & 1 \end{bmatrix}$	$\begin{bmatrix} + & . \\ . & + \end{bmatrix}$	$\begin{bmatrix} 1 & + \\ + & 1 \end{bmatrix}$
Mexico	Currency banking	$\begin{bmatrix} + & . \\ . & + \end{bmatrix}$	$\begin{bmatrix} 1 & . \\ . & 1 \end{bmatrix}$	$\begin{bmatrix} + & . \\ . & . \end{bmatrix}$	$\begin{bmatrix} 1 & . \\ . & 1 \end{bmatrix}$	$\begin{bmatrix} + & . \\ . & . \end{bmatrix}$	$\begin{bmatrix} 1 & + \\ + & 1 \end{bmatrix}$
South Africa	Currency banking	$\begin{bmatrix} + & . \\ . & + \end{bmatrix}$	$\begin{bmatrix} 1 & . \\ . & 1 \end{bmatrix}$	$\begin{bmatrix} + & . \\ . & + \end{bmatrix}$	$\begin{bmatrix} 1 & . \\ . & 1 \end{bmatrix}$	$\begin{bmatrix} . & . \\ . & + \end{bmatrix}$	$\begin{bmatrix} 1 & . \\ . & 1 \end{bmatrix}$
Venezuela	Currency banking	$\begin{bmatrix} + & . \\ . & + \end{bmatrix}$	$\begin{bmatrix} 1 & + \\ + & 1 \end{bmatrix}$	$\begin{bmatrix} + & . \\ . & + \end{bmatrix}$	$\begin{bmatrix} 1 & + \\ + & 1 \end{bmatrix}$	$\begin{bmatrix} . & . \\ . & + \end{bmatrix}$	$\begin{bmatrix} 1 & + \\ + & 1 \end{bmatrix}$

Note: Three different lags of the dependent variable are used, namely 3, 6, and 12 months. "Δ" stands for the parameters of the lagged crisis variables, while Ω represents the covariance matrix. A "+"/"−" sign means that the coefficient is significant and positive/negative, while a "." indicates its nonsignificance. For example, in the case of Argentina, 3 months, all the parameters are positive and significative except for the impact of a currency crisis on the probability of occurrence of banking crises. Similarly, the correlation coefficient between currency and banking crises is significative.

crises exhibit a regime dependence: if the country is proven to be more vulnerable than investors had initially thought, investors will start withdrawing their investments, thus increasing the probability of a new crisis. More precisely, most of the countries are found to have experienced banking and currency crises, with a significant AR coefficient, that is, the crisis variable depends on its own past, for example, Argentina, Egypt, Lebanon, Mexico, South Africa, Venezuela. Besides, only for a small number of cases, only one of the two types of crises is best reproduced by a dynamic model (currency crises in Chile (3 and 12 months), Mexico (6 and 12 months); banking crises in Argentina (6 and 12 months), Ecuador, Lebanon (6 months), South Africa (12 months), and Venezuela (12 months)). Actually, in Chile a past currency crisis has only a short-term positive impact on the emergence of another currency crisis, whereas a banking crisis has just a long-term effect on the probability of occurrence of another banking crisis. Mexico, however, seems to be more prone to recurring currency crises than banking crises as the former type of crisis has a long-term impact on the probability of experiencing a new crisis, whereas the latter has a positive effect only in the short run. On the contrary, for Argentina, South Africa, and Venezuela the impact of past banking crises on currency crises is longer (up to one year) as opposed to that of past currency crises on banking ones (up to 3 and 6 months, respectively).

Second, for the majority of the countries (Argentina, Chile, Lebanon, Mexico, and Venezuela), currency and banking crises are interconnected. This link between crises can take two forms. On the one hand, a certain type of crisis increases (or diminishes) the probability of occurrence of the other type of crisis. This strong link between banking to currency crisis was emphasized by Glick and Hutchinson (1999) within a panel framework. Nevertheless, there is no reason for the transmission of shocks to be symmetric. Indeed, our country per country analysis reveals that for some countries like Argentina (3 and 6 months) a banking crisis in the past months increased the probability of a currency crisis at time t. At the same time, a banking crisis in Chile in the last 12 months reduced the probability of experiencing a currency crisis. Conversely, a currency crisis in Egypt and in Lebanon (3 months) diminished the probability of a banking crisis.

On the other hand, crisis can be contemporaneously positively correlated. This feature seems to be very stable across models (independent of the lag used). The only exceptions are Egypt and Lebanon, for which there is no instantaneous correlation in the model with 3-month lagged binary variables and Mexico, for which such a correlation appears only for the 12-month lag. To sum up, but for Egypt, all countries are characterized by a positive instantaneous correlation between currency and banking crises variables, corroborating the previous findings of Glick and Hutchinson (1999).

Third, the macroeconomic variables are rarely significant.[15] These results corroborate our previous findings (see Candelon et al., 2010): the lag variable captures most of the information summarized by the exogenous variables. Furthermore, when the coefficients of the exogenous variable are significant, they have the expected sign (an increase in the growth of international reserves diminishes the probability of a crisis in the next periods, while a surprise in the rest of indicators soars the probability of a crisis).

To summarize, these results confirm the presence of interaction between the banking and currency crises. The twin crisis phenomenon is thus confirmed empirically. Our findings are also robust to the quadrature choice and the lags considered when constructing the dynamic binary variables.

Trivariate Analysis

But is it really enough to look at two crises only? This subsection extends the previous analysis to the trivariate case by modeling simultaneously the occurrence of currency, banking, and debt crises. However, only two countries experienced these three events during a sufficiently long period. Ecuador presents for our sample an ex post probability larger than 5% for each type of crisis. Such a result is not surprising if one remembers that Ecuador faced a strong financial turmoil in the late 1990, affecting first the banking sector,[16] then the Sucre,[17] and the government budget. Jacone (2004) showed that institutional weaknesses, rigidities in public finances, and high financial dollarization have amplified this crisis. South Africa constitutes a borderline case as the sovereign debt crisis probability is slightly below 5%.

Each of the models is estimated for these countries using both the methodology proposed by Huguenin et al. (2009) based on the Gauss-Legendre quadrature and the direct approximation of a triple integral based on the adaptive Simpson quadrature that Matlab uses by default. Similar results are obtained for the two methods.[18] However, the latter implies a significant gain in time without any loss in accuracy proving that recently developed quadrature methods are good approximations of the normal cumulative distribution function. Besides, 6 and 12 month-lags of the crisis variable are considered.

In the case of Ecuador, the results corroborate our bivariate findings: the banking crises are persistent, while currency crises are not. Nevertheless, it is clear that the bivariate model is misspecified, since it cannot capture the

Table 4. Trivariate Analysis.

Country		3 Months		6 Months	
		Δ	Ω	Δ	Ω
Ecuador	Currency banking sovereign	$\begin{bmatrix} . & . & + \\ . & + & . \\ . & . & + \end{bmatrix}$	$\begin{bmatrix} 1 & . & . \\ . & 1 & . \\ . & . & 1 \end{bmatrix}$	$\begin{bmatrix} . & + & + \\ . & + & . \\ . & . & + \end{bmatrix}$	$\begin{bmatrix} 1 & . & . \\ . & 1 & . \\ . & . & 1 \end{bmatrix}$
South Africa	Currency banking sovereign	$\begin{bmatrix} + & . & . \\ . & . & . \\ . & . & + \end{bmatrix}$	$\begin{bmatrix} 1 & . & + \\ . & 1 & . \\ + & . & 1 \end{bmatrix}$	$\begin{bmatrix} + & . & . \\ . & . & . \\ . & . & + \end{bmatrix}$	$\begin{bmatrix} 1 & . & + \\ . & 1 & . \\ + & . & 1 \end{bmatrix}$

Note: Two different lags of the dependent variable are used, namely 3 and 6 months. "Δ" stands for the parameters of the lagged crisis variables, while Ω represents the variance–covariance matrix. A "+"/"−" sign means that the coefficient is significant and positive/negative, while a "." indicates its nonsignificance. For example, in the case of Ecuador, 3 months, sovereign debt crises have a positive and significative impact on the probability of occurrence of currency crises.

impact of a banking crisis on the occurrence of a currency crisis when using the 6-month lagged binary variables to account for the dynamics of these phenomena (see Table 4).

Moreover, the trivariate model turns out to be more parsimonious in terms of parameters to be estimated since the latent variable of past debt crisis has a positive effect on the probability of both a currency and a debt crises occurring. We also observe that the contemporaneous correlation matrix is diagonal, ruling out common shocks. Crises in Ecuador turn out to be exclusively driven by transmission channels, as in the late 1990, when the banking distress was diffused to the currency and the government budget.

In the case of South Africa, both currency and debt crises are dynamic. There is no evidence of causality between the different types of crises, but significant contemporaneous correlation. It highlights the fact that contrary to Ecuador, South African crises did not mutate but they originated from a common shock. It is worth noting that in the sensitivity analyses performed the results are found to be robust to the choice of macroeconomic variables and the use of different lags for the past crisis variables.

Further Results

To grasp better the properties of the models estimated and selected, a conditional probability analysis is provided. For sake of space, we only report the results obtained for Ecuador.[19]

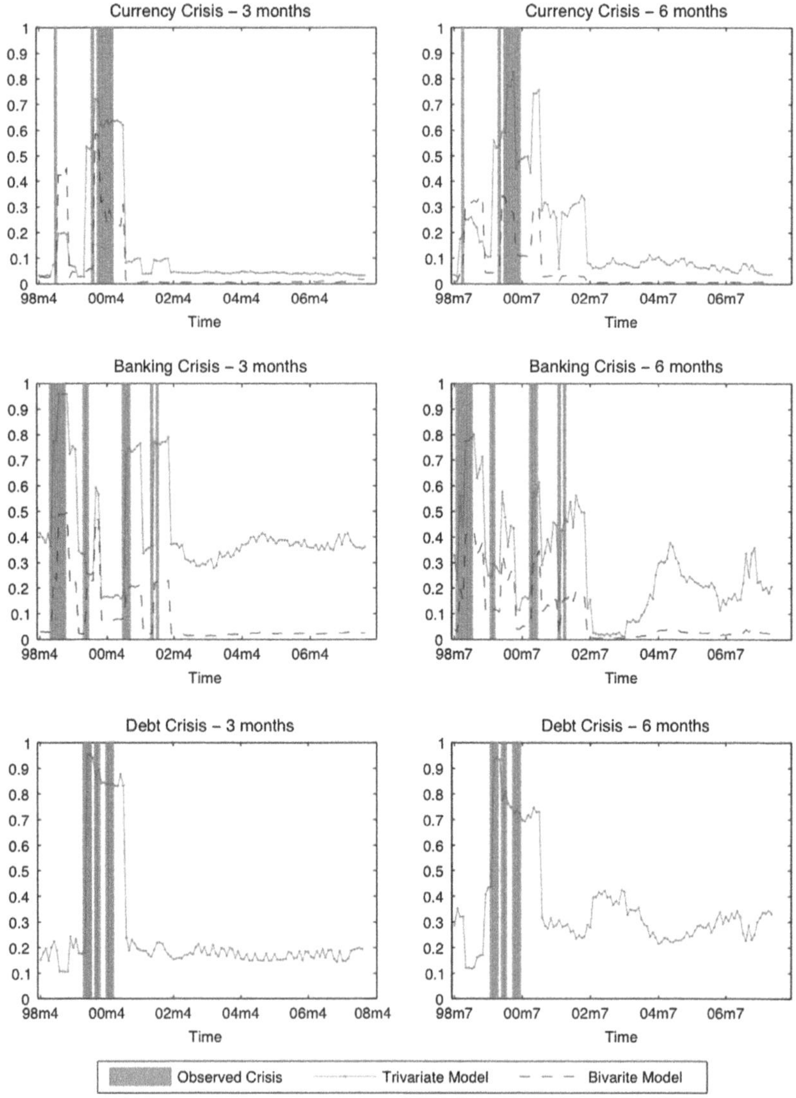

Fig. 1. Conditional crisis probabilities – Ecuador. *Note*: Probabilities at time *t* are calculated including observed information prior 3 or 6 months.

First, Fig. 1 reports the conditional probabilities for each type of crisis obtained from both the bi- and trivariate models considering a forecast horizon of 3 and 6 months. To allow a fair comparison, both models are estimated from the same sample, that is, from 1997 onwards. It goes without saying that the bivariate model does not provide any conditional probabilities for sovereign debt crisis.

It turns out that in this application the trivariate model yields better results than the bivariate one whatever the forecast horizon is, that is, the conditional probabilities issued from the trivariate model are higher than those obtained from the bivariate model during observed crisis periods, while they appear to be similar for calm periods. An evaluation strategy of these series of probabilities along the lines of Candelon et al. (2012) (including the percentage of correctly identified crisis and calm periods and evaluation criteria) supports this inference.[20] Besides, the conditional probabilities obtained from the trivariate model do not immediately collapse after the occurrence of the crisis, which is the case for the bivariate model. It stresses hence the vulnerability of the economy after the exit from a turmoil in particular if it affects the foreign exchange market.

Overall, the conditional probability analysis stresses the superiority of the trivariate model to scrutinize the diffusion mechanisms that occurred in Ecuador after the banking crisis in 1998. Strong interactions between the three types of crises are clearly present in particular between banking and other crises. From a more general perspective, we show that a crisis model should take into account the whole sequence of crises to be accurate.

CONCLUSION

This article proposes a multivariate dynamic probit model to produce dynamic forecasts of binary time-series correlated variables. It is easy to implement and relies on an EML estimation approach, hence providing a solution to the problem generally encountered in the estimation of multivariate probit models. For this, higher-order integrals are decomposed into lower-order finite-range multiple integrals, that are subsequently evaluated using quadrature-rules over bounded intervals. Our framework allows us to apprehend dynamics and predictability in several ways, namely through the lagged binary indicators or the lagged latent variables associated with the qualitative variables.

To illustrate this methodology we consider an application to three types of financial crises for a sample of emerging countries. We investigate the potential mutations of one crisis into another within each country and find that in the bivariate case that one crisis makes the other more likely in expectation. More importantly, for the two countries, Ecuador and South Africa, which suffered from all the three types of crises, the trivariate model turns out to be the best performing in term of conditional probabilities and comprehension of the reasons why a specific crisis mutates to another one: this can be due to either common shocks (as in South Africa) or to a strong predictive relationship (as in Ecuador).

NOTES

1. As generally assumed in discrete-choice forecasting models, the continuous variables x have predictive content v-steps-ahead (see Kauppi & Saikkonen, 2008 for a discussion on this point).

2. In a conditional model this predictability could also be interpreted as Granger causality.

3. Note that to identify the slope and covariance parameters, we impose that the diagonal elements of Ω to be standardized, that is, equal to one.

4. If the parameters θ are estimated while the correlation coefficients are assumed constant, the log-likelihood function is concave. In this case the MLE exists and it is unique. Nevertheless, when θ and ρ are jointly estimated (as in our model), the likelihood function is not (strictly) log-concave as a function of ρ. Thus, the MLE exists only if the log likelihood is not identically $-\infty$ and $E(z^T z|\rho)$ is upper semi-continuous finite and not identically 0. Furthermore, if no $\theta \neq 0$ fulfills the first-order conditions for a maximum, the MLE of (θ, ρ) for the multivariate probit model exists and for each covariance matrix not on the boundary of the definition interval, the MLE is unique.

5. Details about this quadrature are available in Appendix B.

6. The score vector of the trivariate probit model is presented in Appendix C.

7. Argentina, Brazil, Chile, Colombia, Ecuador, Egypt, El Salvador, Indonesia, Lebanon, Malaysia, Mexico, Panama, Peru, Philippines, South Africa, Turkey, and Venezuela.

8. We choose not to include any European country, as (i) only few of them have suffered from the three types of crises and (ii) if this is the case it corresponds to a single episode: the recent turmoil.

9. For a more detailed description of the three dating methods see the Candelon, Dumitrescu, Hurlin, and Palm (2011).

10. Missing values of the series are replaced by cubic spline interpolation.

11. Argentina, Chile, Ecuador, Egypt, Indonesia, Lebanon, Mexico, South Africa, and Venezuela are included in the bivariate analysis, whereas a trivariate model is specified for Ecuador and South Africa. Since the threshold has been arbitrarily set to 5%, we have also checked the borderline countries, like Colombia or

Turkey in the bivariate analysis and Egypt in the trivariate analysis respectively, and similar results have been obtained.

12. A 12-month lag is not used in the case of trivariate models since it would significantly reduce the already small number of observations we have at our disposal.
13. Initial conditions are introduced as given by the univariate static probit.
14. See Candelon et al. (2010) for a discussion about this point.
15. These results are available upon request.
16. Sixteen out of the 40 banks existing in 1997 faced liquidity problems.
17. The Ecuadorian currency was replaced by the U.S. dollar on March 13, 2000.
18. The results for Ecuador when considering a 6-month lag have been obtained with Matlab's quadrature since the model based on the Gauss-Legendre quadrature did not converge.
19. For South Africa, crisis mutation is exclusively driven by the contemporaneous correlation matrix as indicated in Table 4. Otherwise we can see that currency and sovereign debt crises are more persistent than banking ones. All figures are available from the authors upon request.
20. These results are available upon request.

ACKNOWLEDGMENTS

The authors thank Lutz Kilian, Anthony Murphy, and Thomas Fomby, the editors of Advances in Econometrics for comments and discussions. We also benefited from the reaction of audiences at 12th Annual Advances in Econometrics Conference Vector Autoregressive Models: New Developments and Applications in Dallas, the 6th Method in International Finance Network Congress in Sydney, the 65th European Meeting of the Econometric Society in Oslo as well as at internal seminars at the European University Institute, the research department of the IMF, the Bundesbank-European Central Bank-Frankfurt University, the University Catholique de Louvain-CORE, the National Bank of Tunisia, the National Bank of Serbia. The usual disclaimers apply.

REFERENCES

Amemiya, T. (1972). Bivariate probit analysis: Minimum chi-square methods. *Journal of the American Statistical Association*, *69*, 940–944.

Ashford, J. R., & Sowden, R. R. (1970). Multivariate probit analysis. *Biometrics*, *26*, 535–46.

Berg, A., & Pattillo, C. (1999). Predicting currency crises: The indicators approach and an alternative. *Journal of International Money and Finance*, *18*, 561–586.

Bussière, M. (2007). *Balance of payment crises in emerging markets. How early were the "Early" warning signals?* ECB Working Paper No. 713. European Central Bank, Frankfurt.

Candelon, B., Dumitrescu, E. I., & Hurlin, C. (2010). *Currency crises early warning systems: Why they should be dynamic?* METEOR Research memorandum RM/10/047.

Candelon, B., Dumitrescu, E. I., & Hurlin, C. (2012). How to evaluate an early warning system? Towards a unified statistical framework for assessing financial crises forecasting methods. *IMF Economic Review*, *60(1)*, 75–113.

Candelon, B., Dumitrescu, E. I., Hurlin, C., & Palm, F. (2011). *Modeling financial crisis mutation*. Document de Recherche Laboratoire d'Economie d'Orléans n. 2011–17.

Carey, V., Zeger, S. L., & Diggle, P. (1993). Modelling multivariate binary data with alternating logistic regressions. *Biometrika*, *80*, 517–526.

Chib, S., & Greenberg, E. (1998). Analysis of multivariate probit models. *Biometrika*, *85*(2), 37–361.

Dueker, M. (2005). Dynamic forecasts of qualitative variables: A Qual VAR model of U.S. recessions. *Journal of Business and Economic Statistics*, *23*, 96–104.

Engle, R. F., & Granger, C. W. J. (1987). Co-integration and error correction: Representation, estimation, and testing. *Econometrica*, *55*(2), 251–276.

Estrella, A., & Trubin, M. R. (2006). The yield curve as a leading indicator: Some practical issues. *Current Issues in Economics and Finance*, *12*(5), 1–7.

Fuertes, A.-M., & Kalotychou, E. (2007). Optimal design of early warning systems for sovereign debt crises. *International Journal of Forecasting*, *23*(1), 85–100.

Gallant, A. R. (1987). *Nonlinear statistical models*. New York, NY: Wiley.

Gallant, A. R., Rossi, P. E., & Tauchen, G. (1993). Nonlinear dynamic structures. *Econometrica*, *61*, 871–908.

Glick, R., & Hutchison, M. (1999). *Banking and currency crises: How common are the twins?* Working Papers 012000, Hong Kong Institute for Monetary Research.

Glonek, G. F. V., & Mccullagh, P. (1995). Multivariate logistic models. *Journal of the Royal Statistical Society: Series B*, *57*, 533–546.

Greene, W. H. (2002). *Econometric analysis* (5th ed.). Upper Saddle River, NJ: Prentice Hall.

Hagen, T. K., & Ho, J. (2004). *Money market pressure and the determinants of banking crises*. CEPR 4651.

Huguenin, J., Pelgrin, F., & Holly, A. (2009). *Estimation of multivariate probit models by exact maximum likelihood*. Working Paper No. 09-02. Institute of Health Economics and Management, University of Lausanne.

Jacobs, J., Kuper, G., & Lestano. (2003). *Indicators of financial crises do work! An early warning system for six Asian countries*. University of Groningen, CCSO Centre for Economic Research.

Jacone, L. I. (2004). *The late 1990s financial crisis in Ecuador: Institutional weaknesses, fiscal rigidities, and financial dollarization at work*. I.M.F. Working Paper No. 04/12. International Monetary Fund working paper Washington D.C.

Johansen, S. (1988). Statistical analysis of cointegration vectors. *Journal of Economic Dynamics and Control*, *12*(2–3), 231–254.

Johansen, S. (1991). Estimation and hypothesis testing of cointegration vectors in Gaussian vector autoregressive models. *Econometrica*, *59*(6), 1551–1580.

Kaminsky, J., Lizondo, J., & Reinhart, C. (1998). Leading indicators of currency crisis. *IMF Staff Papers*, *45*(1), 1–48.

Kauppi, H., & Saikkonen, P. (2008). Predicting U.S. recessions with dynamic binary response models. *The Review of Economics and Statistics*, *90*(4), 777–791.

Killian, L., & Vigfusson, R. J. (2011). Are the responses of the U.S. economy asymmetric in energy price increases and decreases? *Quantitative Economics*, *2*(3), 419–453.

Koop, G., Pesaran, M. H., & Potter, S. M. (1996). Impulse response analysis in nonlinear multivariate models. *Journal of Econometrics*, *74*(1), 119–147.

Kumar, M., Moorthy, U., & Perraudin, W. (2003). Predicting emerging market currency Crashes. *Journal of Empirical Finance*, *10*, 427–454.

Lesaffre, E., & Kauffmann, H. (1992). Existence and uniqueness of the maximum likelihood estimator for a multivariate probit model. *Journal of the American Statistical Association*, *87*(419), 805–811.

Lestano, & Jacobs, J. (2004). *A comparison of currency crisis dating methods: East Asia 1970–2002*. CCSO Working Papers 200412, CCSO Centre for Economic Research. Database, IMF Working Paper No. 08/224.

McFadden, D. (1989). A method of simulated moments for estimation of discrete response models without numerical integration? *Econometrica*, *57*, 995–1026.

Nyberg, H. (2010). Dynamic probit models and financial variables in recession forecasting. *Journal of Forecasting*, *29*, 215–230.

Nyberg, H. (2011). Forecasting the direction of the US stock market with dynamic binary probit models. *International Journal of Forecasting*, *27*(2), 561–578.

Pescatori, A., & Sy, A. N. R. (2007). Are debt crises adequately defined? *IMF Staff Papers*, *54*(2), 306–337.

Sims, C. (1980). Macroeconomics and reality. *Econometrica*, *48*, 1–48.

Song, X. Y., & Lee, S. Y. (2005). A multivariate probit latent variable model for analyzing dichotomous responses. *Statistica Sinica*, *15*, 645–664.

van de Berg, J. B., Candelon, B., & Urbain, J. P. (2008). A cautious note on the use of panel models to predict financial crises. *Economics Letters*, *101*(1), 80–83.

APPENDIX A: PROOF OF LEMMA 1

By definition, the likelihood of observation t is given by

$$\begin{aligned} L_t(yt|z_{t-1},\theta;\Omega) &= \Pr((-q_{1,t}y_{1,t}^* \leq 0), \ldots, (-q_{M,t}y_{M,t}^* \leq 0)) \\ &= \Pr(-q_{1,t}\varepsilon_{1,t} \leq q_{1,t}\pi_{1,t}, \ldots, -q_{M,t}\pi_{M,t} \leq q_{M,t}\pi_{M,t}) \\ &= \Phi_{M,-Q_t\varepsilon_t}(w_t|0_M;\Omega) \\ &= \int_{-\infty}^{w_{M,t}} \cdots \int_{-\infty}^{w_{1,t}} \phi_{M,-Q_t\varepsilon_t}(Q_t\varepsilon_t,\Omega) \prod_{m=1}^{M} d\varepsilon_{m,t} \end{aligned}$$

Since each $q_{m,t}$ takes only the values $\{-1, 1\}$, it is straightforward to show that $Q_t = Q_t^{-1}$ and $|Q_t\Omega Q_t| = |\Omega|$. Moreover, the density of an M-variate standardized normal vector $-Q_t\varepsilon_t$ with covariance matrix Ω may be rewritten as the density of an M-variate standardized normal vector ε_t with variance–covariance matrix $Q_t\Omega Q_t$

$$\begin{aligned} \phi_{M,-Q_t\varepsilon_t}(Q_t\varepsilon_t;\Omega) &= |2\pi\Omega|^{\frac{-1}{2}} \exp\left\{\frac{-1}{2}(-Q_t\varepsilon_t)'\Omega^{-1}(-Q_t\varepsilon_t)\right\} \\ &= |2\pi(Q_t\Omega Q_t)|^{\frac{-1}{2}} \exp\left\{\frac{-1}{2}\varepsilon_t'(Q_t\Omega Q_t)^{-1}\varepsilon_t\right\} \\ &= \phi_{M,\varepsilon_t}(\varepsilon_t;Q_t\Omega Q_t) \end{aligned}$$

Therefore, the likelihood of observation t is given by

$$\begin{aligned} L_t(yt|Z_{t-1},\theta;\Omega) &= \int_{-\infty}^{q_{M,t}\pi_{M,t}} \cdots \int_{-\infty}^{q_{1,t}\pi_{1,t}} \phi_{M,\varepsilon_t}(\varepsilon_t;Q_t\Omega Q_t) \prod_{m=1}^{M} d\varepsilon_{m,t} \\ &= \Phi_{M,\varepsilon_t}(Q_t\pi_t;Q_t\Omega Q_t) \end{aligned}$$

APPENDIX B: THE GAUSS-LEGENDRE QUADRATURE RULE

The goal of the Gauss-Legendre quadrature rule is to provide an approximation of the following integral:

$$\int_a^b f(x)\ dx \tag{B.1}$$

In a first step, the bounds of the integral must be changed from $[a, b]$ to $[-1,1]$ before applying the Gaussian quadrature rule

$$\int_a^b f(x)\ dx = \frac{b-a}{2}\int_{-1}^{1} f(z)\ dz \tag{B.2}$$

where $z_i = \frac{b-a}{2}abs_i + \frac{b+a}{2}$ and the nodes abs_i, $i \in \{1, 2, \ldots, p\}$ are zeros of the Legendre polynomial $P_p(abs)$.

Definition 1. Then, the standard p-point Gauss-Legendre quadrature rule over a bounded arbitrary interval [a,b] is given by the following approximation:

$$\int_a^b f(x)\ dx \approx \frac{b-a}{2}\sum_{i=1}^{p} v_i f(z_i) + R_p \tag{B.3}$$

where v_i are the corresponding weights, $v_i = \frac{2}{(1-abs_i^2)\left(\frac{\partial P_p(abs)}{\partial abs}|_{abs_i}\right)^2}$, $\sum_{i=1}^{p} v_i = 2$, and R_p is the error term, $R_p = Q_p f^{(2p)}(\xi) = \frac{(b-a)^{2p+1}(p!)^4}{(2p+1)(2p!)^3} f^{2p}(\xi)$, with $\xi \in (a, b)$.

APPENDIX C: THE EML SCORE VECTOR FOR A TRIVARIATE DYNAMIC PROBIT MODEL

For ease of notation, let us denote by $\rho_{i,j}$, $i,j = 1, 2, 3$, $i \neq j$ the correlation coefficients associated to the Ω matrix. The likelihood of observation t may be written as:

$$\begin{aligned} P_t &= \Phi_3(q_1\pi_{1,t}, q_2\pi_{2,t}, q_3\pi_{3,t}, q_1q_2\rho_{12}, q_1q_3\rho_{13}, q_2q_3\rho_{23}) \\ &= \Phi(q_1\pi_{1,t})\Phi(q_2\pi_{2,t})\Phi(q_3\pi_{3,t}) + q_1q_2\Phi(q_3\pi_{3,t})\Psi_2(\pi_{1,t}, \pi_{2,t}, \rho_{12}) \\ &\quad + q_1q_3\Phi(q_2\pi_{2,t})\Psi_2(\pi_{1,t}, \pi_{3,t}, \rho_{13}) + q_2q_3\Phi(q_1\pi_{1,t})\Psi_2(\pi_{2,t}, \pi_{3,t}, \rho_{23}) \\ &\quad + q_1q_2q_3\Psi_3(\pi_{3,t}, \pi_{1,t}, \pi_{2,t}, \rho_{13}, \rho_{23}, 0) + q_1q_2q_3\Psi_3(\pi_{2,t}, \pi_{3,t}, \pi_{1,t}, \rho_{23}, \rho_{12}, 0) \\ &\quad + q_1q_2q_3\Psi_3(\pi_{1,t}, \pi_{2,t}, \pi_{3,t}, \rho_{12}, \rho_{13}, \rho_{23}) \end{aligned} \tag{C.1}$$

where

$$\Psi_2(\pi_{1,t}, \pi_{2,t}, \rho_{12}) = \int_0^{\rho_{12}} \phi_2(\pi_{1,t}, \pi_{2,t}, \lambda_{12})\ d\lambda_{12}$$

$$\Psi_2(\pi_{1,t}, \pi_{3,t}, \rho_{13}) = \int_0^{\rho_{13}} \phi_2(\pi_{1,t}, \pi_{3,t}, \lambda_{13})\ d\lambda_{13}$$

$$\Psi_2(\pi_{2,t}, \pi_{3,t}, \rho_{23}) = \int_0^{\rho_{23}} \phi_2(\pi_{2,t}, \pi_{3,t}, \lambda_{23})\ d\lambda_{23},$$

and

$$\Psi_3(\pi_{3,t}, \pi_{1,t}, \pi_{2,t}, \rho_{13}, \rho_{23}, 0) = \int_0^{\rho_{13}} \int_0^{\rho_{23}} \frac{-\pi_{3,t} + \lambda_{13}\pi_{1,t} + \lambda_{23}\pi_{2,t}}{1 - \lambda_{13}^2 - \lambda_{23}^2} \phi_3(\pi_{3,t}, \pi_{1,t}, \pi_{2,t}, \lambda_{13}, \lambda_{23}, 0)\ d\lambda_{13} d\lambda_{23}$$

$$\Psi_3(\pi_{2,t}, \pi_{3,t}, \pi_{1,t}, \rho_{23}, \rho_{12}, 0) = \int_0^{\rho_{23}} \int_0^{\rho_{12}} \frac{-\pi_{2,t} + \lambda_{23}\pi_{3,t} + \lambda_{12}\pi_{1,t}}{1 - \lambda_{23}^2 - \lambda_{12}^2} \phi_3(\pi_{2,t}, \pi_{3,t}, \pi_{1,t}, \lambda_{23}, \lambda_{12}, 0)\ d\lambda_{23} d\lambda_{12}$$

$$\Psi_3(\pi_{1,t}, \pi_{2,t}, \pi_{3,t}, \rho_{12}, \rho_{13}, \rho_{23}) = \int_0^{\rho_{12}} \int_0^{\rho_{13}} \frac{-(1-\rho_{23}^2)\pi_{1,t} + (\lambda_{12} - \lambda_{13}\rho_{23})\pi_{2,t} + (\lambda_{13} - \lambda_{12}\rho_{23})\pi_{3,t}}{1 - \lambda_{12}^2 - \lambda_{13}^2 - \rho_{23}^2 + 2\lambda_{12}\lambda_{13}\rho_{23}} \times \phi_3(\pi_{1,t}, \pi_{2,t}, \pi_{3,t}, \lambda_{12}, \lambda_{13}, \rho_{23})\ d\lambda_{12} d\lambda_{23}$$

Therefore, the first-order partial derivatives can be obtained as follows:

$$\begin{aligned} \frac{\partial}{\partial \pi_1} P_t = {} & q_1 \phi(\pi_{1,t}) \Phi(q_2\pi_{2,t}) \Phi(q_3\pi_{3,t}) + q_1 q_2 \Phi(q_3\pi_{3,t}) \frac{\partial}{\partial \pi_1} \Psi_2(\pi_{1,t}, \pi_{2,t}, \rho_{12}) \\ & + q_1 q_3 \Phi(q_2\pi_{2,t}) \frac{\partial}{\partial \pi_1} \Psi_2(\pi_{1,t}, \pi_{3,t}, \rho_{13}) + q_1 q_2 q_3 \phi(\pi_{1,t}) \Psi_2(\pi_{2,t}, \pi_{3,t}, \rho_{23}) \\ & + q_1 q_2 q_3 \frac{\partial}{\partial \pi_1} \Psi_3(\pi_{3,t}, \pi_{1,t}, \pi_{2,t}, \rho_{13}, \rho_{23}, 0) \\ & + q_1 q_2 q_3 \frac{\partial}{\partial \pi_1} \Psi_3(\pi_{2,t}, \pi_{3,t}, \pi_{1,t}, \rho_{23}, \rho_{12}, 0) \\ & + q_1 q_2 q_3 \frac{\partial}{\partial \pi_1} \Psi_3(\pi_{1,t}, \pi_{2,t}, \pi_{3,t}, \rho_{12}, \rho_{13}, \rho_{23}) \end{aligned} \tag{C.2}$$

$$\frac{\partial}{\partial \pi_2} P_t = q_2 \phi(\pi_{2,t}) \Phi(q_1 \pi_{1,t}) \Phi(q_3 \pi_{3,t}) + q_1 q_2 \Phi(q_3 \pi_{3,t}) \frac{\partial}{\partial \pi_2} \Psi_2(\pi_{1,t}, \pi_{2,t}, \rho_{12})$$

$$+ q_1 q_2 q_3 \phi(\pi_{2,t}) \Psi_2(\pi_{1,t}, \pi_{3,t}, \rho_{13}) + q_2 q_3 \Phi(q_1 \pi_{1,t}) \frac{\partial}{\partial \pi_2} \Psi_2(\pi_{2,t}, \pi_{3,t}, \rho_{23})$$

$$+ q_1 q_2 q_3 \frac{\partial}{\partial \pi_2} \Psi_3(\pi_{3,t}, \pi_{1,t}, \pi_{2,t}, \rho_{13}, \rho_{23}, 0)$$

$$+ q_1 q_2 q_3 \frac{\partial}{\partial \pi_2} \Psi_3(\pi_{2,t}, \pi_{3,t}, \pi_{1,t}, \rho_{23}, \rho_{12}, 0)$$

$$+ q_1 q_2 q_3 \frac{\partial}{\partial \pi_2} \Psi_3(\pi_{1,t}, \pi_{2,t}, \pi_{3,t}, \rho_{12}, \rho_{13}, \rho_{23}) \qquad \text{(C.3)}$$

$$\frac{\partial}{\partial \pi_3} P_t = q_1 \phi(\pi_{3,t}) \Phi(q_1 \pi_{1,t}) \Phi(q_2 \pi_{2,t}) + q_1 q_2 q_3 \phi(\pi_{3,t}) \Psi_2(\pi_{1,t}, \pi_{2,t}, \rho_{12})$$

$$+ q_1 q_3 \Phi(q_2 \pi_{2,t}) \frac{\partial}{\partial \pi_3} \Psi_2(\pi_{1,t}, \pi_{3,t}, \rho_{13})$$

$$+ q_2 q_3 \Phi(q_1 \pi_{1,t}) \frac{\partial}{\partial \pi_3} \Psi_2(\pi_{2,t}, \pi_{3,t}, \rho_{23})$$

$$+ q_1 q_2 q_3 \frac{\partial}{\partial \pi_3} \Psi_3(\pi_{3,t}, \pi_{1,t}, \pi_{2,t}, \rho_{13}, \rho_{23}, 0)$$

$$+ q_1 q_2 q_3 \frac{\partial}{\partial \pi_3} \Psi_3(\pi_{2,t}, \pi_{3,t}, \pi_{1,t}, \rho_{23}, \rho_{12}, 0)$$

$$+ q_1 q_2 q_3 \frac{\partial}{\partial \pi_3} \Psi_3(\pi_{1,t}, \pi_{2,t}, \pi_{3,t}, \rho_{12}, \rho_{13}, \rho_{23}) \qquad \text{(C.4)}$$

$$\frac{\partial}{\partial \rho_{12}} P_t = q_1 q_2 \Phi(q_3 \pi_{3,t}) \frac{\partial}{\partial \rho_{12}} \Psi_2(\pi_{1,t}, \pi_{2,t}, \rho_{12})$$

$$+ q_1 q_2 q_3 \frac{\partial}{\partial \rho_{12}} \Psi_3(\pi_{2,t}, \pi_{3,t}, \pi_{1,t}, \rho_{23}, \rho_{12}, 0)$$

$$+ q_1 q_2 q_3 \frac{\partial}{\partial \rho_{12}} \Psi_3(\pi_{1,t}, \pi_{2,t}, \pi_{3,t}, \rho_{12}, \rho_{13}, \rho_{23}) \qquad \text{(C.5)}$$

$$\begin{aligned}\frac{\partial}{\partial\rho_{13}}P_t &= q_1q_3\Phi(q_2\pi_{2,t})\frac{\partial}{\partial\rho_{13}}\Psi_2(\pi_{1,t},\pi_{3,t},\rho_{13})\\ &+ q_1q_2q_3\frac{\partial}{\partial\rho_{13}}\Psi_3(\pi_{3,t},\pi_{1,t},\pi_{2,t},\rho_{13},\rho_{23},0)\\ &+ q_1q_2q_3\frac{\partial}{\partial\rho_{13}}\Psi_3(\pi_{1,t},\pi_{2,t},\pi_{3,t},\rho_{12},\rho_{13},\rho_{23})\end{aligned} \tag{C.6}$$

$$\begin{aligned}\frac{\partial}{\partial\rho_{23}}P_t &= q_2q_3\Phi(q_1\pi_{1,t})\frac{\partial}{\partial\rho_{23}}\Psi_2(\pi_{2,t},\pi_{3,t},\rho_{23})\\ &+ q_1q_2q_3\frac{\partial}{\partial\rho_{23}}\Psi_3(\pi_{3,t},\pi_{1,t},\pi_{2,t},\rho_{13},\rho_{23},0)\\ &+ q_1q_2q_3\frac{\partial}{\partial\rho_{23}}\Psi_3(\pi_{2,t},\pi_{3,t},\pi_{1,t},\rho_{23},\rho_{12},0)\\ &+ q_1q_2q_3\frac{\partial}{\partial\rho_{23}}\Psi_3(\pi_{1,t},\pi_{2,t},\pi_{3,t},\rho_{12},\rho_{13},\rho_{23})\end{aligned} \tag{C.7}$$

where

$$\begin{aligned}\frac{\partial}{\partial\pi_1}\Psi_3(\pi_{3,t},\pi_{1,t},\pi_{2,t},\rho_{13},\rho_{23},0) &= \int_0^{\rho_{23}}\int_0^{\rho_{13}}\frac{\partial}{\partial\lambda_{13}}\phi_3(\pi_{3,t},\pi_{1,t},\pi_{2,t},\lambda_{13},\lambda_{23},0)\ d\lambda_{13}\lambda_{23}\\ &= \int_0^{\rho_{23}}\phi_3(\pi_{3,t},\pi_{1,t},\pi_{2,t},\rho_{13},\lambda_{23},0)d\lambda_{23},\end{aligned}$$

$$\begin{aligned}\frac{\partial}{\partial\pi_2}\Psi_3(\pi_{3,t},\pi_{1,t},\pi_{2,t},\rho_{13},\rho_{23},0) &= \int_0^{\rho_{13}}\int_0^{\rho_{23}}\frac{\partial}{\partial\lambda_{23}}\phi_3(\pi_{3,t},\pi_{1,t},\pi_{2,t},\lambda_{13},\lambda_{23},0)\ d\lambda_{23}\lambda_{13}\\ &= \int_0^{\rho_{13}}\phi_3(\pi_{3,t},\pi_{1,t},\pi_{2,t},\lambda_{13},\rho_{23},0)\ d\lambda_{13}\end{aligned}$$

$$\begin{aligned}&\frac{\partial}{\partial\pi_3}\Psi_3(\pi_{3,t},\pi_{1,t},\pi_{2,t},\rho_{13},\rho_{23},0)\\ &= \int_0^{\rho_{13}}\int_0^{\rho_{23}}[(\pi_{3,t}-\lambda_{13}\pi_{1,t}-\lambda_{23}\pi_{2,t})^2-(1-\lambda_{13}^2-\lambda_{23}^2)]\\ &\quad\times\frac{1}{(1-\lambda_{13}^2-\lambda_{23}^2)^2}\phi_3(\pi_{3,t},\pi_{1,t},\pi_{2,t},\lambda_{13},\lambda_{23},0)\ \ d\lambda_{13}d\lambda_{23}\end{aligned}$$

$$\frac{\partial}{\partial\rho_{13}}\Psi_3(\pi_{3,t},\pi_{1,t},\pi_{2,t},\rho_{13},\rho_{23},0)$$
$$= \int_0^{\rho_{23}} \frac{-\pi_{3,t}+\rho_{13}\pi_{3,t}+\lambda_{23}\pi_{2,t}}{1-\rho_{13}^2-\lambda_{23}^2}\phi_3(\pi_3,\pi_1,\pi_2,\rho_{13},\lambda_{23},0)\ d\lambda_{23}$$

$$\frac{\partial}{\partial\rho_{23}}\Psi_3(\pi_{3,t},\pi_{1,t},\pi_{2,t},\rho_{13},\rho_{23},0)$$
$$= \int_0^{\rho_{13}} \frac{-\pi_{3,t}+\lambda_{13}\pi_{3,t}+\rho_{23}\pi_{2,t}}{1-\lambda_{13}^2-\rho_{23}^2}\phi_3(\pi_3,\pi_1,\pi_2,\lambda_{13},\rho_{23},0)\ d\lambda_{13}$$

$$\frac{\partial}{\partial\pi_1}\Psi_3(\pi_{2,t},\pi_{3,t},\pi_{1,t},\rho_{23},\rho_{12},0)$$
$$= \int_0^{\rho_{23}}\int_0^{\rho_{12}} \frac{\partial}{\partial\lambda_{12}}\phi_3(\pi_{2,t},\pi_{3,t},\pi_{1,t},\lambda_{23},\lambda_{12},0)\ d\lambda_{12}\lambda_{23}$$
$$= \int_0^{\rho_{23}} \phi_3(\pi_{2,t},\pi_{3,t},\pi_{1,t,},\lambda_{23},\rho_{12},0)\ d\lambda_{23}$$

$$\frac{\partial}{\partial\pi_2}\Psi_3(\pi_{2,t},\pi_{3,t,},\pi_{1,t},\rho_{23},\rho_{12},0)$$
$$= \int_0^{\rho_{23}}\int_0^{\rho_{12}} [(\pi_{2,t}-\lambda_{23}\pi_{3,t}-\lambda_{12}\pi_{1,t})^2-(1-\lambda_{23}^2-\lambda_{12}^2)]$$
$$\times\frac{1}{(1-\lambda_{23}^2-\lambda_{12}^2)^2}\phi_3(\pi_{2,t},\pi_{3,t},\pi_{1,t},\lambda_{23},\lambda_{12},0)d\lambda_{23}\ \ d\lambda_{12}$$

$$\frac{\partial}{\partial\pi_3}\Psi_3(\pi_{2,t},\pi_{3,t},\pi_{1,t},\rho_{23},\rho_{12},0) = \int_0^{\rho_{12}}\int_0^{\rho_{23}} \frac{\partial}{\partial\lambda_{23}}\phi_3(\pi_{2,t},\pi_{3,t,},\pi_{1,t},\lambda_{23},\lambda_{12},0)\ d\lambda_{23}\lambda_{12}$$
$$= \int_0^{\rho_{12}} \phi_3(\pi_{2,t},\pi_{3,t},\pi_{1,t},\rho_{23},\lambda_{12},0)\ d\lambda_{12}$$

$$\frac{\partial}{\partial\rho_{12}}\Psi_3(\pi_{2,t},\pi_{3,t},\pi_{1,t},\rho_{23},\rho_{12},0)$$
$$= \int_0^{\rho_{23}} \frac{-\pi_{2,t}+\lambda_{23}\pi_{3,t}+\rho_{12}\pi_{1,t}}{1-\lambda_{23}^2-\rho_{12}^2}\phi_3(\pi_{2,t},\pi_{3,t},\pi_{1,t},\lambda_{23},\rho_{12},0)\ d\lambda_{23}$$

$$\frac{\partial}{\partial\rho_{23}}\Psi_3(\pi_{2,t,},\pi_{3,t},\pi_{1,t},\rho_{23},\rho_{12},0)$$
$$=\int_0^{\rho_{12}}\frac{-\pi_{2,t}+\rho_{23}\pi_{3,t}+\lambda_{12}\pi_{1,t}}{1-\rho_{23}^2-\lambda_{12}^2}\phi_3(\pi_{2,t},\pi_{3,t},\pi_{1,t},\rho_{23},\lambda_{12},0)\ d\lambda_{12}$$

$$\frac{\partial}{\partial\pi_1}\Psi_3(\pi_{1,t},\pi_{2,t},\pi_{3,t},\rho_{12},\rho_{13},\rho_{23})$$
$$=\int_0^{\rho_{12}}\int_0^{\rho_{13}}[(1-\rho_{23}^2)\pi_{1,\mathrm{t}}-(\lambda_{12}-\lambda_{13}\rho_{23})\pi_{2,\mathrm{t}}-(\lambda_{13}-\lambda_{12}\lambda_{23})\pi_{3,\mathrm{t}}]^2$$
$$-(1-\rho_{23}^2)(1-\lambda_{12}^2-\lambda_{13}^3-\rho_{23}^2+2\lambda_{12}\lambda_{13}\rho_{23})\}$$
$$\times\frac{1}{(1-\lambda_{12}^2-\lambda_{13}^2-\rho_{23}^2+2\lambda_{12}\lambda_{13}\rho_{23})^2}\phi_3(\pi_{1,t},\pi_{2,t},\pi_{3,t},\lambda_{12},\lambda_{13},\rho_{23})\ d\lambda_{12}d\lambda_{13}$$

$$\frac{\partial}{\partial\pi_2}\Psi_3(\pi_{1,t},\pi_{2,t},\pi_{3,t},\rho_{12},\rho_{13},\rho_{23})$$
$$=\int_0^{\rho_{13}}\int_0^{\rho_{12}}\frac{\partial}{\partial\lambda_{12}}\phi_3(\pi_{1,t},\pi_{2,t},\pi_{3,t,},\lambda_{12},\lambda_{13},\rho_{23})\ d\lambda_{12}\lambda_{13}$$
$$=\int_0^{\rho_{13}}\phi_3(\pi_{1,t},\pi_{2,t},\pi_{3,t},\rho_{12},\lambda_{13},\rho_{23})\ d\lambda_{13}$$

$$\frac{\partial}{\partial\pi_3}\Psi_3(\pi_{1,t},\pi_{2,t},\pi_{3,t},\rho_{12},\rho_{13},\rho_{23})$$
$$=\int_0^{\rho_{12}}\int_0^{\rho_{13}}\frac{\partial}{\partial\lambda_{13}}\phi_3(\pi_{1,t},\pi_{2,t},\pi_{3,t},\lambda_{12},\lambda_{13},\rho_{23})\ d\lambda_{13}\lambda_{12}$$
$$=\int_0^{\rho_{12}}\phi_3(\pi_{1,t},\pi_{2,t},\pi_{3,t},\lambda_{12},\rho_{13},\rho_{23})\ d\lambda_{12}$$

$$\frac{\partial}{\partial\rho_{12}}\Psi_3(\pi_{1,t},\pi_{2,t},\pi_{3,t},\rho_{12},\rho_{13},\rho_{23})$$
$$=\int_0^{\rho_{13}}\frac{(1-\rho_{23}^2)\pi_{1,t}+(\rho_{12}-\lambda_{13}\rho_{23})\pi_{2,t}+(\lambda_{13}-\rho_{12}\rho_{23})\pi_{3,t}}{1-\rho_{12}^2-\lambda_{13}^2-\rho_{23}^2+2\rho_{12}\lambda_{13}\rho_{23}}$$
$$\times\phi_3(\pi_{1,t},\pi_{2,t},\pi_{3,t},\rho_{12},\lambda_{13},\rho_{23}d\lambda_{13})$$

$$\frac{\partial}{\partial\rho_{13}}\Psi_3(\pi_{1,t},\pi_{2,t},\pi_{3,t},\rho_{12},\rho_{13},\rho_{23})$$
$$=\int_0^{\rho_{12}}\frac{(1-\rho_{23}^2)\pi_{1,t}+(\lambda_{12}-\rho_{13}\rho_{23})\pi_{2,t}+(\rho_{13}-\lambda_{12}\rho_{23})\pi_{3,t}}{1-\lambda_{12}^2-\rho_{13}^2-\rho_{23}^2+2\lambda_{12}\rho_{13}\rho_{23}}$$
$$\times\phi_3(\pi_{1,t},\pi_{2,t},\pi_{3,t},\lambda_{12},\rho_{13},\rho_{23}d\lambda_{12})$$

$$\frac{\partial}{\partial\rho_{23}}\Psi_3(\pi_{1,t},\pi_{2,t},\pi_{3,t},\rho_{12},\rho_{13},\rho_{23})$$
$$=\int_0^{\rho_{12}}\int_0^{\rho_{13}}\frac{\partial^2}{\partial\pi_{2,t}\partial\lambda_{13}}\phi_3(\pi_{1,t},\pi_{2,t},\pi_{3,t},\lambda_{12},\lambda_{13},\rho_{23})\ d\lambda_{12}d\lambda_{13}$$
$$=\int_0^{\rho_{12}}\frac{-(1-\rho_{13}^2)\pi_{2,t}+(\lambda_{12}-\rho_{13}\rho_{23})\pi_{1,t}+(\rho_{23}-\lambda_{12}\rho_{13})\pi_{3,t}}{1-\lambda_{12}^2-\rho_{13}^2-\rho_{23}^2+2\lambda_{12}\rho_{13}\rho_{23}}$$
$$\times\phi_3(\pi_{1,t},\pi_{2,t},\pi_{3,t},\lambda_{12},\rho_{13},\rho_{23})d\lambda_{12}$$
$$=\int_0^{\rho_{12}}\int_0^{\rho_{13}}\frac{\partial^2}{\partial\pi_{3,t}\partial\lambda_{12}}\phi_3(\pi_{1,t},\pi_{2,t},\pi_{3,t},\lambda_{12},\lambda_{13},\rho_{23})\ d\lambda_{12}d\lambda_{13}$$
$$=\int_0^{\rho_{13}}\frac{-(1-\rho_{12}^2)\pi_{3,t}+(\lambda_{13}-\rho_{12}\rho_{23})\pi_{1,t}+(\rho_{23}-\rho_{12}\lambda_{13})\pi_{2,t}}{1-\rho_{12}^2-\lambda_{13}^2-\rho_{23}^2+2\rho_{12}\lambda_{13}\rho_{23}}$$
$$\times\phi_3(\pi_{1,t},\pi_{2,t},\pi_{3,t},\rho_{12},\lambda_{13},\rho_{23})\ d\lambda_{13}$$